W9-DFE-263

ELEMENTARY GEOMETRY
FROM AN
ADVANCED STANDPOINT

Second Edition

EDWIN E. MOISE
Harvard University

ELEMENTARY GEOMETRY FROM AN ADVANCED STANDPOINT

Second Edition

ADDISON-WESLEY PUBLISHING COMPANY
Reading, Massachusetts · Menlo Park, California
London · Amsterdam · Don Mills, Ontario · Sydney

 Printed in the United States of America. Published simultaneously in Canada. Library of Congress Catalog Card No. 73-2347.

ISBN 0-201-04793-4
STUVWXYZ-AL-89

Preface to the Second Edition

The preface to the first edition will serve equally well as an explanation of the intent of the second, since the revisions have consisted merely of a large number of minor improvements. Some of these are as follows.

1. Many problems have been added.
2. The chapter on rigid motion has been expanded, so as to include a discussion of isometries and dilations of a plane. Here it was desirable to use coordinate systems, and so the latter are now discussed just before rigid motion, rather than just after them, as in the first edition.
3. Various minor errors and obscurities have been eliminated. I must confess that the first edition was less accurate than the reader had a right to expect, but I hope that this is not true of the new edition.

Thanks are due to a very large number of colleagues who have written to me in the past few years, pointing out errors and suggesting revisions. These letters were very helpful in the preparation of the second edition, and I am accordingly grateful.

New York, N.Y. E. E. M.
September 1973

Preface to the First Edition

The title of this book is the best brief description of its content and purposes that the author was able to think of. These purposes are rather different from those of most books on "higher geometry" or the foundations of geometry. The difference that is involved here is somewhat like the difference between the two types of advanced calculus courses now commonly taught. Some courses in advanced calculus teach material which has not appeared in the preceding courses at all. There are others which might more accurately be called courses in elementary calculus from an advanced standpoint: their purpose is to clean up behind the introductory courses, furnishing valid definitions and valid proofs for concepts and theorems which were already known, at least in some sense and in some form. One of the purposes of the present book is to reexamine elementary geometry in the same spirit.

If we grant that elementary geometry deserves to be thoroughly understood, then it is plain that such a job needs to be done; and no such job is done in any college course now widely taught. The usual senior-level courses in higher geometry proceed on the very doubtful assumption that the foundations are well understood. And courses in the foundations (when they are taught at all) are usually based on such delicate postulate sets, and move so slowly, that they cover little of the substance of the theory. The upshot of this is that mathematics students commonly leave college with an understanding of elementary geometry which is not much better than the understanding that they acquired in high school.

The purpose of this book is to elucidate, as thoroughly as possible, both this elementary material and its surrounding folklore. My own experience, in teaching the course to good classes, indicates that it is not safe to presuppose an exact knowledge of anything. Moreover, the style and the language of traditional geometry courses are rather incongruous with the style and the language of the rest of mathematics today. This means that ideas which are, essentially, well understood may need to be reformulated before we proceed. (See, for example, Chapter 6, on Congruences Between Triangles.) In some cases, the reasons for reformulation are more compelling. For example, the theory of geometric inequalities is used in the chapter on hyperbolic geometry; and this would hardly be reasonable if the student had not seen these theorems proved without the use of the Euclidean parallel postulate.

For these reasons, the book begins at the beginning. Some of the chapters are quite easy, for a strong class, and can simply be assigned as outside reading. Others, such as Chapters 20 and 24, are more difficult. These differences are due to the nature of the material: it is not always possible to get from one place to another by walking along a path of constant slope.

This unevenness in level of difficulty makes the book rather flexible. A well-prepared class may go rapidly through Chapters 1–7, and devote most of its time to the sort of material presented in Chapters 8, 10, 14, 19, and 20. A poorly prepared class may go carefully through Chapters 1–7, omit such chapters as 14 and 20, and still not get to Chapter 25.

The book is virtually self-contained. The necessary fragments of algebra and the theory of numbers are presented in Chapters 29 and 30, at the end. At many points, ideas from algebra and analysis are needed in the discussion of the geometry. These ideas are explained in full, on the ground that it is easier to skip explanations of things that are known than to find convenient and readable references. The only exception to this is in Chapter 25, where it seemed safe to assume that epsilon-delta limits are understood.

In some chapters, especially Chapters 20 and 25, we give full expositions of topics which are commonly dismissed rather briefly and almost casually. The process by which the real numbers are introduced into an Archimedean geometry, for purposes of measurement, is highly important and far from trivial. The same is true of the consistency proof for the hyperbolic postulates. Here (as elsewhere) the purpose of this book is to explain, with all possible lucidity and thoroughness, ideas which are widely alluded to but not so widely understood.

Cambridge, Mass. E. E. M.
October 1962

Contents

Chapter 1
The Algebra of the Real Numbers

1.1 INTRODUCTION

In this book, we shall be concerned mainly with geometry. We shall begin at the beginning, and base our work on carefully stated postulates. Thus, in a way, we shall be using the scheme which has been used in the study of geometry ever since Euclid wrote the *Elements*.

But our postulates will be different from Euclid's. The main reason for this is that in modern mathematics, geometry does not stand alone as it did at the time when the *Elements* were written. In modern mathematics, the real number system plays a central role, and geometry is far easier to deal with and to understand if real numbers are allowed to play their natural part. For this reason, the first step in our program is to put the real number system on the same solid basis that we intend to provide for geometry, by first stating our assumptions clearly and then building on them.

1.2 ADDITION AND MULTIPLICATION OF REAL NUMBERS

We shall think of the real numbers as being arranged as points on a line, like this:

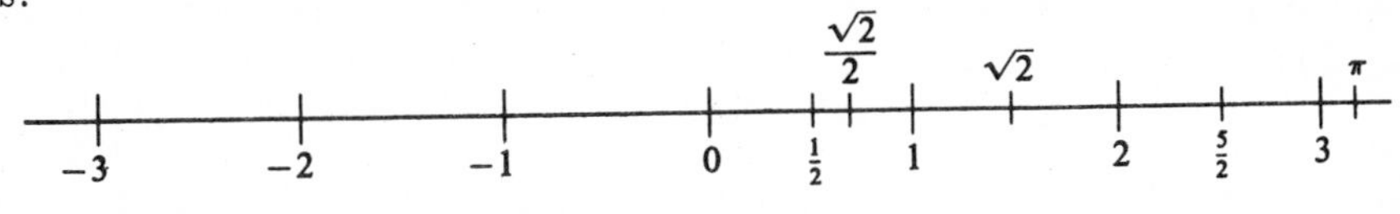

FIGURE 1.1

The real numbers include, at least, all of the following:

(1) the positive integers 1, 2, 3, . . . , and so on;

(2) the integer 0;

(3) the negative integers $-1, -2, -3, \ldots$, and so on;

(4) the numbers expressible as fractions with integers as numerators and integers different from 0 as denominators. For example, $\frac{1}{2}$, $-\frac{3}{5}$ and 1,000,000/1 are numbers of this type.

Note that this fourth kind of number includes the first three, because every integer n is equal to $n/1$. What we have so far, then, are the numbers of the form p/q, where p and q are integers and q is not 0. These are called the *rational* numbers. This term is not meant to suggest that any other kind of number must be crazy. It merely refers to the fact that a rational number is the *ratio* of two integers. As you know, there are many real numbers that are not of this type. For example, $\sqrt{2}$ is not the ratio of any two integers. Such numbers are called *irrational.*

We proceed to state the basic properties of the real number system in the form of postulates. We have given a set **R**, whose elements are called *real numbers* (or simply *numbers*, if the context makes it clear what is meant). We have given two operations, addition and multiplication, denoted by $+$ and $\cdot$. Thus the algebraic structure that we are dealing with is a triplet

$$[\mathbf{R}, +, \cdot].$$

The properties of the system are as follows:

A-1. **R** is closed under addition. That is, if a and b belong to **R**, then $a + b$ also belongs to **R**.

A-2. Addition in **R** is associative. That is, if a, b, and c belong to **R**, then

$$a + (b + c) = (a + b) + c.$$

A-3. There is exactly one element of **R**, denoted by 0, such that

$$a + 0 = 0 + a = a$$

for every a in **R**.

A-4. For every a in **R** there is exactly one number $-a$ in **R**, called the *negative* of a, such that

$$a + (-a) = (-a) + a = 0.$$

A-5. Addition in **R** is commutative. That is, if a and b belong to **R**, then

$$a + b = b + a.$$

These postulates are numbered A-1 through A-5 because they are the postulates that deal with addition. We now move on to multiplication.

M-1. **R** is closed under multiplication. That is, if a and b belong to **R**, then ab belongs to **R**.

(Here and hereafter, we denote the product $a \cdot b$ simply as ab. This is a matter of convenience, and we shall not do it consistently. For example, when we write 26, we mean twenty-six, not twelve.)

M-2. Multiplication in **R** is associative. That is, if a, b, and c belong to **R**, then

$$a(bc) = (ab)c.$$

M-3. There is exactly one element of **R**, denoted by 1, such that

$$a1 = 1a = a$$

for every a in **R**.

M-4. For every a in **R**, other than 0, there is exactly one number a^{-1}, called the *reciprocal* of a, such that

$$aa^{-1} = a^{-1}a = 1.$$

M-5. Multiplication in **R** is commutative. That is, if a and b belong to **R**, then

$$ab = ba.$$

M-6. 1 is different from 0.

This postulate may look peculiar, but it is necessary. Under the preceding postulates, we have no guarantee that **R** contains any number at all except 0.

The postulates, so far, have dealt with addition and multiplication separately. These two operations are connected by the following postulate.

AM-1. *The Distributive Law.* If a, b, and c belong to **R**, then

$$a(b + c) = ab + ac.$$

In addition to these postulates, you may feel the need for the following two statements:

E-1. If

$$a = b \quad \text{and} \quad c = d,$$

then

$$a + c = b + d.$$

E-2. If

$$a = b \quad \text{and} \quad c = d,$$

then

$$ac = bd.$$

Here it should be understood that a, b, c and d belong to **R**. But these statements are not really postulates for the real number system. They merely serve to remind us of what addition and multiplication are all about. The first one says that the sum of two numbers depends only on the numbers, and does not depend on the letters that we use to denote the numbers; similarly for the second "law" E-2.

Throughout this book, the symbol "$=$" will always mean "is the same as." As usual, the symbol "$\neq$" means "is different from."

Subtraction is defined by means of the negatives given by A-4. That is,

$$a - b = a + (-b),$$

by definition. Similarly, division is defined by means of the reciprocals given by

M-4. Thus, if $b \neq 0$, then

$$\frac{a}{b} = a \div b = ab^{-1},$$

by definition.

From the above postulates, all of the usual laws governing addition and multiplication can be derived. We start the process as follows.

Theorem 1. $a0 = 0$ for every a.

Proof. By A-3, we have

$$1 = 1 + 0.$$

Therefore

$$a \cdot 1 = a(1 + 0).$$

Hence

$$\begin{aligned} a &= a1 + a0, \\ a &= a + a0, \\ (-a) + a &= (-a) + (a + a0), \\ 0 &= [(-a) + a] + a0, \\ 0 &= 0 + a0, \end{aligned}$$

and

$$0 = a0,$$

which was to be proved. (You should be able to give the reasons for each step by citing the appropriate postulates.)

Theorem 2. If $ab = 0$, then either $a = 0$ or $b = 0$.

Proof. Given $ab = 0$. We need to show that if $a \neq 0$, then $b = 0$. If $a \neq 0$, then a has a reciprocal a^{-1}. Therefore

$$\begin{aligned} a^{-1}(ab) &= a^{-1}0 \\ &= 0. \end{aligned}$$

But

$$\begin{aligned} a^{-1}(ab) &= (a^{-1}a)b \\ &= 1b = b. \end{aligned}$$

Thus $b = 0$, which was to be proved.

This, of course, is the theorem that you use when you solve equations by factoring. If

$$(x - 1)(x - 2) = 0,$$

then either $x = 1$ or $x = 2$, because the product of the numbers $x - 1$ and $x - 2$ cannot be 0 unless one of the factors $x - 1$ and $x - 2$ is 0. You need this principle to be sure that nobody is going to find an extra root by investigating the equation by some other method.

Theorem 3. 0 has no reciprocal. That is, there is no number x such that $0x = 1$.

Proof. We know that $0x = 0$ for every x. If we had $0x = 1$ for some x, it would follow that $0 = 1$. This is impossible, because M-6 tells us that $0 \neq 1$.

This theorem gives the reason why division by 0 is impossible. If division by 0 meant anything, it would mean multiplication by the "reciprocal of 0." Since there is no such reciprocal, there is no such operation as division by 0.

Theorem 4. *The Cancellation Law of Addition.* If $a + b = a + c$, then $b = c$.

Proof. If

$$a + b = a + c,$$

then

$$(-a) + (a + b) = (-a) + (a + c),$$
$$[(-a) + a] + b = [(-a) + a] + c,$$
$$0 + b = 0 + c,$$

and

$$b = c.$$

Theorem 5. *The Cancellation Law of Multiplication.* If $ab = ac$, and $a \neq 0$, then $b = c$.

Proof. If $ab = ac$, and $a \neq 0$, then a has a reciprocal a^{-1}. Therefore

$$a^{-1}(ab) = a^{-1}(ac),$$
$$(a^{-1}a)b = (a^{-1}a)c,$$
$$1b = 1c,$$

and

$$b = c.$$

Theorem 6. $-(-a) = a$, for every a.

Proof. By definition of the negative, the number $-(-a)$ is the number x such that

$$(-a) + x = x + (-a) = 0.$$

The number a has this property, because

$$(-a) + a = a + (-a) = 0.$$

But A-4 tells us that every number has *exactly one* negative. Therefore a is the negative of $-a$, which was to be proved.

Theorem 7. $(-a)b = -(ab)$, for every a and b.

Proof. What we need to show is that

$$(-a)b + ab = ab + (-a)b = 0,$$

because this is what it means to say that $(-a)b$ is the negative of ab. By the commutative law, it will be sufficient to show that

$$(-a)b + ab = 0.$$

By the distributive law,

$$(-a)b + ab = [(-a) + a]b.$$

Since $(-a) + a = 0$, and $0b = 0$, we have

$$(-a)b + ab = 0,$$

which was to be proved.

This theorem gives the "rule of signs" under which $(-2) \cdot 4 = -8$, $(-7) \cdot 4 = -28$.

Theorem 8. $(-a)(-b) = ab$ for every a and b.

Proof.

$$\begin{aligned}(-a)(-b) &= -[a(-b)] \\ &= -[(-b)a] \\ &= -[-(ba)] \\ &= ba \\ &= ab.\end{aligned}$$

(What is the reason for each step?)

This, of course, is the second "rule of signs," which tells us that $(-3)(-4) = 12$.

Theorem 9. The reciprocal of the product is the product of the reciprocals. That is,

$$(ab)^{-1} = a^{-1}b^{-1}$$

for every $a \neq 0$, $b \neq 0$.

Proof. We need to show that

$$(ab)(a^{-1}b^{-1}) = 1.$$

Now

$$\begin{aligned}(ab)(a^{-1}b^{-1}) &= a[b(a^{-1}b^{-1})] \\ &= a[b(b^{-1}a^{-1})] \\ &= a[(bb^{-1})a^{-1}] \\ &= a[1a^{-1}] \\ &= aa^{-1} \\ &= 1.\end{aligned}$$

Theorem 10. The negative of the sum is the sum of the negatives. That is,

$$-(a + b) = (-a) + (-b).$$

Proof. We need to show that

$$(a + b) + [(-a) + (-b)] = 0.$$

To avoid an excessive accumulation of parentheses, let us agree, in this proof only, to denote $-x$ by x'. We then have

$$\begin{aligned}(a + b) + [(-a) + (-b)] &= (a + b) + (a' + b') \\ &= a + [b + (a' + b')] \\ &= a + [b + (b' + a')] \\ &= a + [(b + b') + a'] \\ &= a + [0 + a'] \\ &= a + a' \\ &= 0.\end{aligned}$$

Note that this proof is precisely analogous to the proof of the preceding theorem.

Obviously, we could go on proving theorems like this indefinitely. In fact, if you stop to think, you will realize that nearly every time you have performed an algebraic calculation, you have in effect proved a theorem of this sort. For example, when you factor $x^2 - a^2$, and get $(x - a)(x + a)$, you are claiming that the following theorem holds.

Theorem 11. For every x, a, we have

$$(x - a)(x + a) = x^2 - a^2.$$

Proof?

An equation which holds for all real numbers is called an *algebraic identity.* Stated in this language, the two associative laws say that the equations

$$\begin{aligned}a + (b + c) &= (a + b) + c, \\ a(bc) &= (ab)c,\end{aligned}$$

are algebraic identities; the distributive law says that the equation

$$a(b + c) = ab + ac$$

is an algebraic identity, and so on.

In the following exercises, you may use the two associative laws without comment. In fact, since it doesn't matter how the terms or factors are grouped, we don't need to indicate a grouping at all; we can write $a + b + c$ to denote $a + (b + c)$ or $(a + b) + c$, and similarly for multiplication. We can do the same for n-fold products of the form $a_1a_2 \ldots a_n$, although the justification for this is more complicated than you might think. (See Section 1.10.)

As usual, a^2 means aa, a^3 means aaa, and so on. Similarly, 2 means $1 + 1$, 3 means $2 + 1 = 1 + 1 + 1$, and so on.

Problem Set 1.2

Show that the following equations are algebraic identities. All these statements are to be regarded as theorems, and should be proved on the basis of the postulates and the theorems that have previously been proved. Give a reason for each step in the proofs.

1. $b(-a) = -(ab)$
2. $(-a)(-b) = ba$
3. $a(b + c) = ca + ba$
4. $a(b - c) = ab - ac$
5. $-0 = 0$
6. $a - 0 = a$
7. $a^3b = ba^3$ (Try to get a very short proof.)
8. $a + a = 2a$
9. $(-a) + (-a) = (-2)a$
10. $a^2(b^2 + c^2) = a^2b^2 + a^2c^2$
11. $a^2(b^2 - c^2) = -a^2c^2 + a^2b^2$
12. $(a + b)(c + d) = ac + bc + ad + bd$
13. $(a + b)^2 = a^2 + 2ab + b^2$

The following are discussion questions.

14. Suppose that subtraction is regarded as an operation. Does this operation obey the associative law? That is, is the equation $(a - b) - c = a - (b - c)$ an algebraic identity? If not, under what condition does the equation hold? (It is not necessary to answer this question on the basis of the postulates; you are free to use all the algebra that you know.)

15. Suppose that division is regarded as an operation. Does this operation obey the associative law? That is, is the equation, $(a/b)/c = a/(b/c)$, an algebraic identity? If not, under what conditions does the equation hold?

16. Does subtraction obey the commutative law? How about division?

The answers to the preceding three questions indicate why we do not regard subtraction and division as basic operations when we are formulating the basic properties of the real numbers.

17. Postulate M-6 (which says that $1 \neq 0$) may seem superfluous. Is it? Can it be proved, on the basis of the other postulates, that there is any number at all other than 0?

18. Suppose that the only elements of **R** were 0 and 1 with addition and multiplication defined by these tables. Which of the postulates would hold true?

+	0	1
0	0	1
1	1	0

·	0	1
0	0	0
1	0	1

19. Suppose that we replace A-3 by the following:

A-3′. There is at least one element 0 of **R** such that $0 + a = a + 0 = a$ for every a.

Show that A-3 can be proved as a theorem, on the basis of A-3′ and the other postulates. To do this, you need to show that if $0 + a = a + 0 = a$ and $0' + a = a + 0' = a$, for every a, then $0 = 0'$.

20. Similarly, suppose that we replace A-4 by the following:

A-4′. For each a there is at least one number $-a$ such that $a + (-a) = (-a) + a = 0$.

Show that A-4 can be proved as a theorem. To do this, you need to show that if $a + x = x + a = 0$, then $x = -a$.

1.3 FIELDS

An algebraic structure satisfying the postulates of the preceding section is called a *field*. Since we have explained that **R** denotes the set of real numbers, it may be worth while to state the definition of a field over again from the beginning, allowing the possibility that the field may not be the real number system.

Given a set **F**, of objects called *numbers*, with two operations $+$ and $\cdot$, called addition and multiplication. The structure

$$[\mathbf{F}, +, \cdot]$$

is called a *field* if the following conditions hold.

A-1. **F** is closed under addition.

A-2. Addition in **F** is associative.

A-3. **F** contains exactly one number 0 such that

$$a + 0 = 0 + a = a$$

for each a in **F**.

A-4. Every a in **F** has exactly one negative $-a$ in F such that

$$a + (-a) = (-a) + a = 0.$$

A-5. Addition in **F** is commutative.

M-1. **F** is closed under multiplication.

M-2. Multiplication in **F** is associative.

M-3. **F** contains exactly one element 1, such that $a1 = 1a = a$ for every a in **F**.

M-4. Every $a \neq 0$ in **F** has exactly one reciprocal a^{-1} such that

$$aa^{-1} = a^{-1}a = 1.$$

M-5. Multiplication in **F** is commutative.

M-6. $1 \neq 0$.

AM-1. For every a, b, c in **F**, we have

$$a(b + c) = ab + ac.$$

All the theorems of the preceding section were proved merely on the basis of the above postulates. It follows that all these theorems hold true not merely in the real number system but in any field. For example, they all hold true in the algebraic system described in Problem 18 of Problem Set 1.2.

Problem Set 1.3

The purpose of this problem set is merely to clarify the meaning of the postulates for a field. In answering the questions below, you may make free use of all the algebra which in fact you know. In the following section of the text, we shall return to our "official" mathematics, based on postulates.

1. Let **F** be the set of all numbers of the form $p/2^q$, where p and q are integers and $q \geqq 0$. These numbers are called *dyadic rationals.* Do the dyadic rationals form a field, under the usual definition of $+$ and $\cdot$? Which, if any, of the field postulates fail to hold?

2. Let **F** be the set of all complex numbers with absolute value $= 1$. Does **F** form a field, under the usual definitions of $+$ and $\cdot$? Which, if any, of the field postulates fail to hold? (You may assume, of course, that the set of *all* complex numbers forms a field; it does.)

3. Same question, for the set of all positive real numbers.

4. Assume that $\sqrt{2}$ is irrational. Show that if a and b are rational, and $a + b\sqrt{2} = 0$, then $a = b = 0$.

5. Show that if a, b, c, and d are rational, and

$$a + b\sqrt{2} = c + d\sqrt{2},$$

then $a = c$ and $b = d$.

6. Let **F** be the set of all real numbers of the form $a + b\sqrt{2}$, where a and b are rational. Is $[\mathbf{F}, +, \cdot]$ a field?

7. An algebraic structure $[\mathbf{F}, +, \cdot]$ is called a *commutative ring with unity* if it satisfies all of the field postulates except possibly for M-4. Obviously every field is a commutative ring with unity, but not every commutative ring with unity is a field. Exactly one of the algebraic structures described in the preceding problems forms a commutative ring with unity but does not form a field. Which one is it?

8. In the algebra of the real numbers, the following theorem holds.

Theorem. If $a_1b_2 - a_2b_1 \neq 0$, then the system of equations

$$a_1x + b_1y = c_1, \qquad a_2x + b_2y = c_2,$$

is satisfied by exactly one pair of numbers (x, y).

Does this theorem hold true in any commutative ring with unity? Does it hold true in any field?

9. Consider a coordinate plane, with points identified by pairs (x, y) of numbers. We define the "sum" of two points (u, v) and (x, y) to be the point $(u + x, v + y)$.

Does this system satisfy A-1 through A-5? Is it possible to define the "product" of two points in such a way as to get a field? If so, how?

1.4 THE ORDERING OF THE REAL NUMBERS

We remember that the real numbers can be thought of, informally, as being arranged on a line, like this:

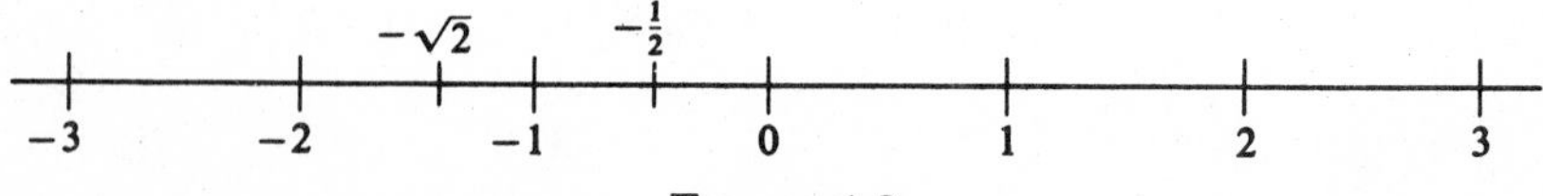

Figure 1.2

When we write $a < b$, this means, graphically speaking, that a lies to the left of b on the number scale. Thus $-2 < 1$, and $-1{,}000{,}000 < \frac{1}{10}$.

The laws governing the relation $<$ are as follows.

O-1. *Trichotomy.* Every pair of real numbers a, b satisfies one, and only one, of the conditions $a < b$, $a = b$, $b < a$.

O-2. *Transitivity.* If $a < b$ and $b < c$, then $a < c$.

The expression $a < b$ is pronounced "a is less than b." When we write $b > a$, this means (by definition) that $a < b$. When we write

$$a \leqq b,$$

this means that either $a < b$ or $a = b$. The relation $<$ is connected up with addition and multiplication by the following conditions.

MO-1. If $a > 0$ and $b > 0$, then $ab > 0$.

AO-1. If $a < b$, then $a + c < b + c$ for every c.

From these four conditions (together with our other postulates), all of the laws governing inequalities can be derived. Let us take some examples.

Theorem 1. Any two inequalities can be added. That is, if

$$a < b$$

and

$$c < d,$$

then

$$a + c < b + d.$$

Proof. By AO-1,

$$a + c < b + c;$$

by AO-1,

$$b + c < b + d.$$

(From now on, we are going to use the commutative law and similar principles without comment.) By O-2, this means that

$$a + c < b + d,$$

which was to be proved.

Theorem 2. $a < b$ if and only if $b - a > 0$.

Proof. If $a < b$, then $a - a < b - a$, by AO-1. Therefore $b - a > 0$. Conversely, if $b - a > 0$, then $b - a + a > a$, and $b > a$.

Theorem 3. An inequality is *preserved* if we multiply both sides by the same positive number. That is, if

$$a < b,$$

and

$$c > 0,$$

then

$$ac < bc.$$

Proof. Since $a < b$, we have
$$b - a > 0.$$
Therefore
$$c(b - a) > 0,$$
by MO-1. Hence
$$bc - ac > 0;$$
and by Theorem 2 this means that $ac < bc$, which was to be proved.

Theorem 4. If $a > 0$, $-a < 0$.

Proof. If $a > 0$, then $a - a > 0 - a$, by AO-1. Thus $0 > -a$, and $-a < 0$, which was to be proved.

Theorem 5. If $a < 0$, then $-a > 0$.

Proof. If $a < 0$, then $0 - a > 0$, by Theorem 2. Therefore $-a > 0$, which was to be proved.

Theorem 6. $1 > 0$.

(It may seem strange to prove a statement like this. But we need to use the fact, and if we didn't prove it, we would have to state it as a postulate.)

Proof. Suppose that the theorem is false. Since we know that $1 \neq 0$, it follows by O-1 that $1 < 0$. By Theorem 5 it follows that $-1 > 0$. By MO-1, $(-1)^2 > 0$. Since $(-1)^2 = 1$, we have $1 > 0$, which contradicts the assumption $0 < 1$.

Theorem 7. An inequality is *reversed* if we multiply both sides by the same negative number. That is, if $a < b$, and $c < 0$, then $ac > bc$.

Proof. If $a < b$, then
$$b - a > 0,$$
by Theorem 2. If
$$c < 0,$$
then
$$-c > 0,$$
by Theorem 5. Therefore
$$-bc + ac > 0,$$
by MO-1. Therefore
$$ac > bc,$$
by Theorem 2.

Suppose that we have an inequality involving an unknown number x, such as
$$2x - 5 < 7x + 3.$$
Every number x either satisfies the inequality or doesn't. For example, $x = 1$ satisfies the inequality, because $-3 < 10$; but $x = -2$ does not, because $-9 > -11$. An expression of this sort, involving a letter for which we can substitute anything we want, is called an *open sentence.* When we substitute 1 for x, we get the statement $-3 < 10$, which is true. When we substitute -2 for x, we get the statement $-9 < -11$, which is false. The set of all numbers

which give true statements when substituted for x is called the *solution set* of the open sentence. Here are a few examples.

Open sentence	*Solution set*
$x + 2 = 7$	$\{5\}$
$x + 0 = x$	$\mathbf{R}$
$x^2 - 4 = 0$	$\{2, -2\}$
$x + 2 = 2 + x$	$\mathbf{R}$

In the column on the right, $\{5\}$ denotes the set whose only element is 5, and $\{2, -2\}$ denotes the set whose elements are 2 and -2. The same notation is used whenever we want to describe a finite set by giving a complete list of its elements. For example,

$$\{1, 3, 5, 7, 9\}$$

is the set of all positive odd integers less than 10. The braces are used to describe *sets*, rather than *sequences*, and so the order in which the elements are listed makes no difference. For example,

$$\{1, 3, 5, 7, 9\} = \{7, 3, 9, 1, 5\};$$

the sets described are exactly the same.

It sometimes happens, of course, that an open sentence never becomes a true statement, no matter what you substitute for x. For example, the equation $(x + 1)^2 = x^2 + 2 \cdot x$ has no roots. In this case, the solution set is the empty set, that is, the set that has no elements. The empty set is denoted by $\emptyset$, to avoid confusion with the number 0. Here are more examples:

Open sentence	*Solution set*
$x^2 + 2x + 2 = (x + 1)^2$	$\emptyset$
$x^2 = 0$	$\{0\}$
$x < x$	$\emptyset$
$2x = x$	$\{0\}$

There is a short notation for the solution set of an open sentence. When we write

$$\{x \mid x^2 = 0\},$$

this means the set of all real numbers x such that $x^2 = 0$. Thus

$$\{x \mid x^2 = 0\} = \{0\},$$
$$\{x \mid x^2 - 5x + 6 = 0\} = \{2, 3\},$$

and so on.

To *solve* an equation or an inequality means to find the solution set of the corresponding open sentence. For inequalities, the "answer" usually takes the form of a second open sentence which is simpler and easier to interpret than the first.

The simplification process might look like this: If

$$(1) \qquad 2x - 5 < 7x + 3,$$

then by AO-1 we have

$$(2) \qquad -5x - 5 < 3,$$

and

$$(3) \qquad -5x < 8.$$

By Theorem 7, we have

$$(4) \qquad x > -\tfrac{8}{5}.$$

(We have multiplied by the *negative number* $-\frac{1}{5}$.) Therefore *every number* x *which satisfies* (1) *also satisfies* (4). Conversely if (4) holds, then so does (3); if (3) holds, then so does (2), and if (2) holds, then so does (1). The inequalities (1) and (4) are called *equivalent;* by this we mean that every number which satisfies one of them also satisfies the other. Expression (4) is called the *solution* of (1). The process that we have gone through can be written in an abbreviated form as follows:

$$\begin{aligned} 2x - 5 &< 7x + 3, \\ \Leftrightarrow -5x - 5 &< 3, \\ \Leftrightarrow -5x &< 8, \\ \Leftrightarrow x &> -\tfrac{8}{5}. \end{aligned}$$

The double-headed arrow on the left should be pronounced "is equivalent to"; when we write

$$2x - 5 < 7x + 3 \Leftrightarrow x > -\tfrac{8}{5},$$

we mean that the open sentences connected by the symbol $\Leftrightarrow$ have exactly the same solution set. The advantage of the abbreviation is that it makes it easy to write at each stage exactly what we have on our minds. (When we write long strings of formulas, it is not always easy to tell or to remember what the logical connection between them is supposed to be.) The result of our work on the problem above can be written as follows:

$$\{x | 2x - 5 < 7x - 3\} = \{x | x > -\tfrac{8}{5}\}.$$

We use a single-headed arrow to indicate that one condition implies another. For example, when we write

$$x > 2 \Rightarrow x^2 > 4,$$

we mean that *if* $x > 2$, *then* $x^2 > 4$. This is true, because if $x > 2$, then $x^2 > 2x$, by Theorem 3. Also by Theorem 3, if $x > 2$, then $2x > 4$. By O-2, $x^2 > 4$, which was to be proved.

Note that it is *not* true that

$$x > 2 \Leftrightarrow x^2 > 4,$$

because any number less than -2 satisfies the second inequality but not the first.

(As one step in the proof above, we showed that $x > 2 \Rightarrow x^2 > 2x$. Is it true that $x^2 > 2x \Rightarrow x > 2$? Why or why not?)

The *absolute value* of a number x is denoted by $|x|$. It is defined by the following two conditions.

(1) If $x \geqq 0$, then $|x| = x$.

(2) If $x < 0$, then $|x| = -x$.

For example, $|2| = 2$, because $2 \geqq 0$, and $|-2| = -(-2) = 2$, because $-2 < 0$. In other words, the absolute value of a positive number is the same positive number, and the absolute value of a negative number x is the corresponding positive number, which is $-x$.

Theorem 8. For every x, $|x| \geqq 0$.

Proof. (1) If $x \geqq 0$, then $|x| \geqq 0$, because in this case $|x| = x$.
(2) If $x < 0$, then $-x > 0$. Therefore $|x| > 0$, because $|x| = -x$.

Theorem 9. $|-x| = |x|$ for every x.

Proof. (1) If $x \geqq 0$, then $-x \leqq 0$. Thus

$$|x| = x,$$

and

$$|-x| = -(-x) = x.$$

Therefore, in this case, $|-x| = |x|$.

(2) If $x < 0$, then $-x > 0$. Therefore

$$|x| = -x,$$

and

$$|-x| = -x.$$

Hence, in this case also, $|-x| = |x|$.

Theorem 10. $|x| \geqq x$ for every x.

Proof. If $x \geqq 0$, this is true because $x \geqq x$. If $x < 0$, then $x < |x|$, because $|x| \geqq 0$.

Theorem 11. $|xy| = |x| \cdot |y|$ for every x and y.

Proof. When x is replaced by $-x$, both sides of the equation are unchanged; therefore we can assume that $x \geqq 0$. For the same reason, we can assume that $y \geqq 0$. If $x \geqq 0$ and $y \geqq 0$, the equation takes the form $xy = xy$.

When we write a "double inequality"

$$a < b < c,$$

we mean that both of the inequalities $a < b$ and $b < c$ hold true.

Theorem 12. Let a be > 0. Then

$$|x| < a$$

if and only if

$$-a < x < a.$$

Graphically speaking, this theorem says that the numbers that satisfy the inequality $|x| < a$ are the numbers between a and $-a$ like this:

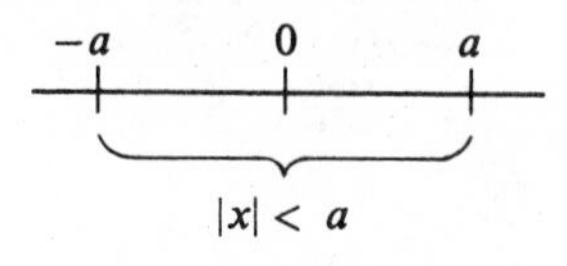

FIGURE 1.3

Proof. (1) If $x \geqq 0$, then $|x| < a$ means that $x < a$. Therefore $|x| < a$ is true when $0 \leqq x < a$.

(2) If $x < 0$, then $|x| < a$ means that $-x < a$, or $-a < x$. Hence $|x| < a$ is true when $-a < x < 0$.

Therefore $|x| < a$ holds whenever $-a < x < a$. It is easy to check, conversely, that if $|x| < a$, then $-a < x < a$. (There are two cases to be considered, as in Conditions (1) and (2) above.) Therefore

$$|x| < a \Leftrightarrow -a < x < a,$$

which was to be proved.

Theorem 13. For every a, b,

$$|a + b| \leqq |a| + |b|.$$

Proof. CASE 1. Suppose that $a + b \geqq 0$. In this case,

$$|a + b| = a + b.$$

By Theorem 10,

$$a \leqq |a|, \qquad \text{and} \qquad b \leqq |b|.$$

Therefore

$$a + b \leqq |a| + |b|,$$

and since $a + b = |a + b|$ in Case 1, the theorem follows.

CASE 2. Suppose that $a + b < 0$. Then $(-a) + (-b) > 0$. By our result for Case 1, we have

$$|(-a) + (-b)| \leqq |-a| + |-b|.$$

But by Theorem 9, we know that

$$|-a - b| = |a + b|, \qquad |-a| = |a|, \qquad |-b| = |b|.$$

Substituting, we get

$$|a + b| \leqq |a| + |b|,$$

which was to be proved.

Problem Set 1.4

1. Show that if $a > 0$, then $a^{-1} > 0$.

2. Show that if $a < 0$, then $a^{-1} < 0$.

3. Given $x > 0$ and $y > 0$, show that $x^3 = y^3 \Rightarrow x = y$. Does this hold for every x and y?

4. Solve the following inequalities. The answers should be in one of the forms

$$\text{——} \Leftrightarrow \text{——} \qquad \text{or} \qquad \{x | \cdots\} = \{x | \text{——}\}.$$

(a) $5 - 3 \cdot x > 17 + x$
(b) $5 \cdot x - 3 < 17 \cdot x + 1$
(c) $x + 5 > 6 - x$
(d) $|x| < 1$
(e) $|x - 3| < 2$
(f) $|x - 5| < 5$

5. Is it true that $|x^2| = |x|^2$ for every x? Why or why not?

6. Is it true that $|x^3| = |x|^3$ for every x? Why or why not?

7. Show that $x^2 - 2x + 1 \geqq 0$ for every x.

8. For what numbers x (if any) does each of the following conditions hold?

(a) $|x^2 - 5 \cdot x + 6| = |x - 3| \cdot |x - 2|$
(b) $|x^2 - 5 \cdot x + 6| = x^2 - 5 \cdot x + 6$
(c) $|x - 5| = |2 \cdot x - 3|$
(d) $|x + 1| = |1 - x|$
(e) $\sqrt{x^2 + 1} = x$
(f) $\sqrt{x^2 - 1} = x$
(g) $|2 \cdot x - 1| + |x + 3| \geqq |3 \cdot x + 2|$
(h) $|7 \cdot x + 3| + |3 - x| \geqq 6|x + 1|$

9. Indicate graphically, on a number scale, the places where the following conditions hold.

(a) $|x| < 2$
(b) $|x - 2| < \frac{1}{2}$
(c) $|2 \cdot x - 3| < \frac{1}{2}$
(d) $|x - 1| < 2$ and (also) $|x - 2| < 1$
(e) $|3 - 2 \cdot x| < \frac{1}{2}$
(f) $|x - 2| < \frac{1}{4}$ and $x > 2$

10. Show that if $b \neq 0$, then

$$\left|\frac{1}{b}\right| = \frac{1}{|b|}.$$

11. Show that if $b \neq 0$, then

$$\left|\frac{a}{b}\right| = \frac{|a|}{|b|}.$$

12. Show that for every a and b, $|a - b| \geqq |a| - |b|$.

13. Show that for every a and b, $|a + b| \geqq |a| - |b|$.

14. For what numbers a is the fraction $a/|a|$ defined? What is this fraction equal to, for various values of a?

1.5 ORDER RELATIONS AND ORDERED FIELDS

So far, we have described the properties of the real number system relative to addition, multiplication, and order. A system which satisfies all the postulates that we have stated so far is called an *ordered field*. We repeat the definition of an ordered field in its general form, as follows.

Given a set **F**. Let $*$ be a relation defined on **F**, satisfying the following two conditions.

O-1. Every pair a, b of elements of **F** satisfies one and only one of the conditions $a * b$, $a = b$, and $b * a$.

O-2. If $a * b$ and $b * c$, then $a * c$.

Then $*$ is called an *order relation*.

Order relations are usually denoted by the symbol $<$. But we need a more general notation, such as $*$, because we may want to talk about two different relations defined on the same set.

Suppose now that we have a field

$$[\mathbf{F}, +, \cdot].$$

Suppose that we also have given an order relation $<$, defined on **F**, satisfying the following two conditions.

MO-1. If $a > 0$ and $b > 0$, then $ab > 0$.

AO-1. If $a < b$, then $a + c < b + c$ for every c. Then the structure

$$[\mathbf{F}, +, \cdot, <]$$

is called an *ordered field*.

Thus what we have said, so far, about the real number system is that it forms an ordered field.

It should be emphasized that an ordered field is not merely a field which is somehow arranged in an order. To know that we have an ordered field, we need to know that the order relation $<$ is related to multiplication and addition by Conditions MO-1 and AO-1.

In the preceding section, all the theorems were proved on the basis of Conditions O-1, O-2, MO-1, and AO-1. Therefore these theorems hold true in any ordered field. You are free to use them in solving the following problems.

Problem Set 1.5

1. In Problem 6 of Problem Set 1.3 you showed that the real numbers of the form $a + b\sqrt{2}$, where a and b are rational, form a field. Is this an ordered field, under the usual order relation? Why or why not?

2. Consider the field **F** described in Problem 18 of Problem Set 1.2. Here **F** = {0, 1}, and addition and multiplication are described by the following tables.

+	0	1
0	0	1
1	1	0

·	0	1
0	0	0
1	0	1

Is it possible to define an order relation in such a way as to make this system an ordered field?

3. Show that an order relation can be defined for the set of points (x, y) of a coordinate plane. (The verification of O-2 requires a discussion of four cases.)

4. Show that an order relation can be defined for the complex numbers.

5. Show that it is *not* possible to define an order relation for the complex numbers in such a way as to get an ordered field. [*Hint:* Suppose that such an order has been defined. Show that each of the conditions $i > 0$, and $i < 0$ leads to a contradiction of one of the postulates or one of the theorems for ordered fields.]

1.6 THE POSITIVE INTEGERS AND THE INDUCTION PRINCIPLE

We know that $1 > 0$. We get the rest of the positive integers by starting with 1, and then adding 1 as often as we like. Thus the first few positive integers are

$$\begin{aligned} &1, \\ &2 = 1 + 1, \\ &3 = 2 + 1 = 1 + 1 + 1, \\ &4 = 3 + 1 = 1 + 1 + 1 + 1, \end{aligned}$$

and so on. We let **N** be the set of all positive integers. (Here **N** stands for *natural;* the positive integers are often referred to as the *natural numbers.*)

The above common-sense remarks, about the way we get positive integers by adding 1 to other positive integers, suggest the pattern of an exact definition of the set **N**. The set **N** is defined by the following three conditions.

(1) 1 belongs to **N**.

(2) **N** is closed under the operation of adding 1. That is, if n belongs to **N**, then so also does $n + 1$.

(3) Of all sets of numbers satisfying (1) and (2), **N** is the smallest. That is, **N** is the intersection, or common part, of all sets of numbers satisfying (1) and (2).

From Condition (3) we get the following result immediately.

Theorem 1. *The Induction Principle.* Let S be a set of numbers. If (1) S contains 1, and (2) S is closed under the operation of adding 1, then (3) S contains all of the positive integers.

The reason is simple. Since **N** is the smallest set that satisfies (1) and (2), it follows that every other such set contains **N**.

Let us see how the induction principle, in the form in which we have stated it, can be put to work.

Theorem A. For every positive integer n, the sum of the squares of the first n positive integers is $(n/6)(n+1)(2n+1)$. That is, for every n we have

$$1^2 + 2^2 + \cdots + n^2 = \frac{n}{6}(n+1)(2n+1).$$

The induction proof is as follows. Let S be the set of all positive integers n for which it is true that

$$1^2 + 2^2 + \cdots + n^2 = \frac{n}{6}(n+1)(2n+1).$$

Thus, if we say that 1 belongs to S, this means that

$$1^2 = \tfrac{1}{6}(1+1)(2 \cdot 1 + 1).$$

To say that 2 belongs to S means that

$$1^2 + 2^2 = \tfrac{2}{6}(2+1)(2 \cdot 2 + 1),$$

and so on.

We shall show that (1) S contains 1, and (2) S is closed under the operation of adding 1.

(1) S contains 1, because the equation

$$1^2 = \tfrac{1}{6}(1+1)(2 \cdot 1 + 1)$$

is true.

(2) To prove (2), we must show that *if* a given integer n belongs to S, *then* so also does $n+1$. Thus we must show that *if*

$$\text{(a)} \quad 1^2 + 2^2 + \cdots + n^2 = \frac{n}{6}(n+1)(2n+1),$$

then

$$\text{(b)} \quad 1^2 + 2^2 + \cdots + n^2 + (n+1)^2 = \frac{n+1}{6}(n+2)(2n+3).$$

(The first equation tells us that n belongs to S, and the second tells us that $n+1$ belongs to S.)

Given that (a) holds, it follows that

$$\begin{aligned} \text{(c)} \quad 1^2 + 2^2 + \cdots + n^2 + (n+1)^2 &= \frac{n}{6}(n+1)(2n+1) + (n+1)^2 \\ &= \frac{n+1}{6}(2n^2 + n + 6n + 6) \\ &= \frac{n+1}{6}(2n^2 + 7n + 6) \\ &= \frac{n+1}{6}(n+2)(2n+3). \end{aligned}$$

Therefore (b) holds.

This is the way a proof based on Theorem 1 always works. Always, you are proving that a certain open sentence gives true statements for every positive integer n. Always, you start by letting S be the set of all positive integers in the solution set. You then show that your set S satisfies (1) and (2) of Theorem 1. Then you conclude from Theorem 1 that your solution set S contains all of the positive integers.

The following alternative form of the induction principle may be more familiar.

Theorem 2. Let

$$P_1, P_2, \ldots$$

be a sequence of propositions (one proposition P_n for each positive integer n). If

(1) P_1 is true, and

(2) for each n, P_n implies P_{n+1},

then

(3) all of the propositions $P_1, P_2, \ldots$ are true.

For example, we might consider the case where P_n says that

$$1^2 + 2^2 + \cdots + n^2 = \frac{n}{6}(n+1)(2n+1).$$

Thus the first few propositions in the sequence would be the following:

$$\begin{aligned} P_1:&\quad 1^2 = \tfrac{1}{6}(1+1)(2\cdot 1+1), \\ P_2:&\quad 1^2+2^2 = \tfrac{2}{6}(2+1)(2\cdot 2+1), \\ P_3:&\quad 1^2+2^2+3^2 = \tfrac{3}{6}(3+1)(2\cdot 3+1), \end{aligned}$$

and so on.

Theorem 2 is a consequence of Theorem 1. To prove this, we begin the same way that we always begin when applying Theorem 1. We let S be the set of all positive integers n for which P_n is true. Statement (1) now tells us that S contains 1. Statement (2) tells us that S is closed under the operation of adding 1. By Theorem 1, S contains all positive integers. Therefore all the propositions P_1, $P_2, \ldots$ are true, which was to be proved.

There is a third form of the induction principle, known as the *well-ordering principle*, which is useful for some purposes. It asserts that every non-empty set of positive integers has a least element. To prove it, we need some preliminary results.

Theorem 3. Every positive integer is either $= 1$ or $= k + 1$ for some positive integer k.

Proof. Let S be the set of all positive integers satisfying the conditions of the theorem. Then 1 belongs to S. If n belongs to S, and $n = 1$, then $n + 1$ belongs to S, with $k = 1$. If n belongs to S, and $n \neq 1$, then $n = k + 1$ for some positive integer k, and $n + 1 = (k + 1) + 1$, so that $n + 1$ belongs to S. By the induction principle, the theorem follows.

Theorem 4. 1 is the least positive integer. That is, if $n \neq 1$, then $n > 1$.

Proof. If $n \neq 1$, then $n = k + 1$, for some positive integer k. Therefore $n - 1 = k > 0$, and $n > 1$, by Theorem 2 of Section 1.4.

Theorem 5. For each positive integer n, $n + 1$ is the smallest positive integer that is greater than n.

Proof. Let S be the set of all positive integers for which this holds.

(1) 1 belongs to S. Proof: Suppose that there is a positive integer p such that

$$1 < p < 1 + 1.$$

Since $p > 1$, it follows that $p = k + 1$ for some positive integer k. Thus $p - 1 = k > 0$. Therefore

$$0 < p - 1 < 1,$$

and

$$0 < k < 1,$$

which contradicts Theorem 4.

(2) If n belongs to S, then $n + 1$ belongs to S. The proof is like that of (1). Suppose that $n + 1$ does not belong to S. Then there is a positive integer p such that

$$n + 1 < p < (n + 1) + 1,$$

and $p = k + 1$ for some positive integer k. Therefore

$$n < p - 1 = k < n + 1,$$

and n does not belong to S.

By the induction principle, the theorem follows.

Theorem 6. *The Well-Ordering Principle.* Every nonempty set of positive integers has a smallest element.

Proof. Let K be a nonempty subset of $\mathbf{N}$. If K contains 1, then K has a least element, namely 1, and so there is nothing to prove.

Suppose then that K does not contain 1. Let S be the set of all positive integers n for which it is true that K contains *none* of the integers $1, 2, \ldots, n$. (For example, if K were the set $\{10, 20, 30, \ldots\}$, S would be the set $\{1, 2, 3, \ldots, 9\}$. Fifteen would not belong to S, because $10 < 15$, and 10 belongs to K.)

We know that (1) S contains 1, because K does not contain 1. If it were true that (2) S is closed under the operation of adding 1, then S would contain all of the positive integers, and K would be empty. Therefore S must *not* be closed under the operation of adding 1. Hence there is an integer n such that n belongs to S and $n + 1$ does not. This means that K contains none of the numbers $1, 2, \ldots,$

n, but does contain $n+1$. It follows that $n+1$ is the smallest element of K.
element of K.

(In the example given above, the smallest element of K is obviously 10. You should check to see that 10 is the number that we get when we apply the general proof to this particular set K.)

The choice between Theorems 1 and 2 is merely a matter of taste. But there are times when the well-ordering principle is easier to use than either of them. (See, for example, the chapter on the theory of numbers.)

PROBLEM SET 1.6

1. Show by induction that for every $n > 0$,

$$1 + 2 + \cdots + n = \frac{n}{2}(n+1).$$

2. Show that the sum of the first n odd positive integers is n^2. That is,

$$1 + 3 + 5 + \cdots + (2n - 1) = n^2.$$

3. Show that for every $n > 0$,

$$1^3 + 2^3 + \cdots + n^3 = \left(\frac{n}{2}(n+1)\right)^2.$$

4. Assume that for every positive real number x there is a positive integer $n > x$. (This is true; it is a consequence of the *Archimedean* property of the real numbers, discussed in Section 1.8.) Show that for every positive real number x there is a nonnegative integer n such that $n \leqq x < n + 1$.

5. Now show that for every real number x there is an integer n such that $n \leqq x < n + 1$.

*6. The game known as the Towers of Hanoi is played as follows. We have three spindles A, B, and C, of the sort used as targets in quoits. On spindle A we have a stack of n disks, diminishing in size from bottom to top. These are numbered $1, 2, \ldots, n$, in the order from top to bottom. A legal move in the game consists in taking the topmost disk from one spindle and placing it at the top of the stack on another spindle, provided that we do not, at any stage, place a disk above a disk which is smaller. (Thus, at the outset, there are exactly two legal moves: disk 1 can be moved either to spindle B or to spindle C.) The object of the game is to move all the disks to spindle B. Show that for every positive integer n, the game can be completed.

*7. Let p_n be the number of moves required to complete the game with n disks. Show that for each n,

$$p_{n+1} = 2p_n + 1.$$

8. Given that $p_1 = 1$, and that $p_{n+1} = 2p_n + 1$ for each n, show that

$$p_n = 2^n - 1.$$

(Since $2^{10} = 1024$, this means that the game with 20 disks requires more than a million moves.)

1.7 THE INTEGERS AND THE RATIONAL NUMBERS

If we add to the set **N** the number 0 and then all of the negatives of the numbers in **N**, we get all of the integers. The set of integers is denoted by **Z**. Thus

$$\mathbf{Z} = \{\ldots, -3, -2, -1, 0, 1, 2, 3, \ldots\}.$$

If a number x can be expressed in the form of p/q, where p and q are integers and $q \neq 0$, then x is called a *rational* number. The set of all rational numbers is denoted by **Q**. (Here **Q** stands for *quotient;* the rational numbers are those which are quotients of integers.)

We would now like to prove the well-known fact that the rational numbers form a field. Under the scheme that we have been using in this chapter, the proof involves an unexpected difficulty. Following a procedure which is the reverse of the usual one, we have defined the positive integers in terms of the real numbers; and at the present stage we do not officially know that sums and products of integers are always integers. This can be proved, but we postpone the proof until the end of this chapter; in the meantime we regard the closure of the integers as a postulate.

CL. *The Closure Postulate.* The integers are closed under addition and multiplication.

The following theorem is now easy.

Theorem 1. The rational numbers form an ordered field.

Proof. We shall verify the field postulates one at a time.

A-1. *Closure Under Addition.*

$$\frac{p}{q} + \frac{r}{s} = \frac{p \cdot s}{q \cdot s} + \frac{r}{s} = \frac{p \cdot s}{q \cdot s} + \frac{q \cdot r}{q \cdot s} = \frac{1}{q \cdot s}(p \cdot s + q \cdot r) = \frac{p \cdot s + q \cdot r}{q \cdot s},$$

which is rational.

A-2. Since addition is associative for real numbers in general, it follows that addition is associative for rational numbers in particular.

(This is an instance of a general principle. If a postulate says that a certain equation is an algebraic identity, then this postulate automatically holds in any subsystem of the given system.)

A-3. Zero is rational, because $0 = 0/1$.

A-4. Given a rational number p/q, we have $-(p/q) = (-p)/q$, which is rational.

A-5. Since addition is commutative for all real numbers, it is commutative for all rational numbers.

M-1. $p/q \cdot r/s = pr/qs$, which is rational.

M-2. See the verifications of A-2 and A-5.

M-3. 1 is rational, because $1 = 1/1$.

M-4. If $p/q \neq 0$, then $p \neq 0$. Therefore $(p/q)^{-1} = q/p$, which is rational.

M-5. See the verifications of A-2, A-5, and M-2.

AM-1. The distributive law holds for rational numbers, because it holds for all real numbers.

Thus **Q** forms a field. And the order relation $<$, given for all real numbers, applies in particular to the rational numbers, and the order postulates automatically hold.

Finally we remark, by way of preparation for some of the problems below, that if a number is rational, $=p/q$, then it can be expressed as a fraction in lowest terms. That is, p and q can be chosen in such a way that no positive integer other than 1 is a factor of both of them. Thus, for example, if $x = p/q$, then x can be expressed as a fraction r/s, where r and s are not both even, are not both divisible by 3, and so on. Here we are really appealing to a theorem in the theory of numbers, to be proved in Appendix B at the end of the book.

Problem Set 1.7

1. A positive integer n is *even* if $n = 2k$, where k is an integer; n is *odd* if $n = 2j + 1$, where j is an integer. Show that every positive integer is either even or odd. [*Hint:* Let S be the set of all positive integers which are either even or odd. What you need to show is that $S = \mathbf{N}$. Verify that S satisfies conditions (1) and (2) of Theorem 1 of Section 1.6.]

2. Show that if n is odd, then n^2 is odd.

3. Show that if n^2 is even, then n is even.

4. Show that if $\sqrt{2} = p/q$, then p is even.

5. Show that if $\sqrt{2} = p/q$, then q is also even.

6. Show that $\sqrt{2}$ is not $= p/q$ for any integers p and q.

7. Show that every positive integer n has one of the three forms $3k$, $3j + 1$, or $3m + 2$.

8. Show that if $n = 3j + 1$, then n^2 has the same form.

9. Show that if $n = 3m + 2$, then n^2 has the form $3k + 1$.

10. Show that if n^2 is divisible by 3, then so also is n.

11. Show that $\sqrt{3}$ is irrational.

12. Now try to use the same pattern of proof to "prove" that $\sqrt{4}$ is irrational. The "proof" must break down at some point, because the theorem is ridiculous. Where does the proof break down?

13. Show that if a is rational and x is irrational, then $a + x$ is irrational.

14. Show that if a and b are rational, with $b \neq 0$, and x is irrational, then $a + bx$ is irrational.

The results of the preceding two problems show that irrational numbers are not scarce: given one of them, we can find lots of others.

*15. Let n and p be positive integers. Show that n can always be expressed in the form

$$n = pq + r,$$

where $0 \leqq r < p$. (Two of the preceding exercises assert that this is true for $p = 2$ and $p = 3$.)

1.8 THE ARCHIMEDEAN POSTULATE; EUCLIDEAN COMPLETENESS

It may appear that the postulates for an ordered field are an adequate description of the real number system. But this is far from true; our postulates, so far, allow some strange possibilities indeed. (See Chapter 28.) We shall not discuss these, but merely state postulates which rule them out.

Throughout this section, **F** is an ordered field.

The easiest way to see the meaning of the following postulate is to think of it geometrically. Suppose we have given two linear segments, like this:

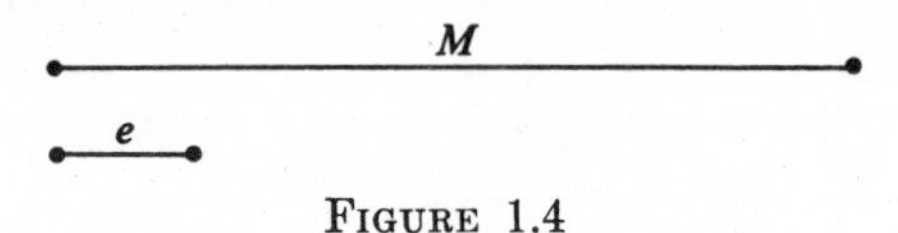

FIGURE 1.4

The case of interest is the one in which the first segment is "very long" and the second is "very short." It is reasonable to suppose that if you take enough copies of the second segment, and lay them end to end, you get a segment longer than the first one. And this should be true no matter how long the first may be, and no matter how short the second may be. If the lengths of the segments are the real numbers M and e, as indicated in the figure, and n copies of the second segment are enough, then we have $ne > M$.

The algebraic form of this statement follows.

A. *The Archimedean Postulate.* Let M and e be any two positive numbers. Then there is a positive integer n such that

$$ne > M.$$

An ordered field which satisfies this condition is called *Archimedean.* Henceforth we shall assume that the real number system forms an Archimedean ordered field.

Note that if a certain integer n gives us $ne > M$, then any larger integer has the same property. Therefore the postulate might equally well have gone on to say that we have $ne > M$ for every integer n greater than or equal to a certain n_0.

Even our latest postulate, however, is still not enough for the purposes either of algebra or of geometry. The easiest way to see this is to observe that the field **Q** of rational numbers satisfies all our postulates so far, and in **Q** the number 2 has no square root. We need to know that our number field is *complete* in such a sense as to permit the ordinary processes of algebra. For a long time to come, it will be sufficient for us to know that every positive number has a square root.

If $a > 0$, then x *is a square root of* a if $x^2 = a$. Obviously, if x is a square root of a, then so also is $-x$. Therefore, if a number has one square root, it must have two. On the other hand, no number a has two different positive square roots x_1, and x_2; if this were so, we would have

$$x_1^2 = a = x_2^2, \qquad x_1^2 - x_2^2 = 0, \qquad (x_1 - x_2)(x_1 + x_2) = 0.$$

Here $x_1 - x_2 \neq 0$, because $x_1 \neq x_2$, and $x_1 + x_2 > 0$, because $x_1 > 0$ and $x_2 > 0$. Therefore the product $(x_1 - x_2)(x_1 + x_2)$ cannot be $= 0$.

An ordered field is called *Euclidean* if it satisfies the following condition.

C-1. *The Euclidean Completeness Postulate.* Every positive number has a positive square root.

We call this the Euclidean postulate because of the part that it will play in geometry. Eventually, this postulate will ensure that circles will intersect lines, and intersect each other, in the ways that we would expect.

It follows, as shown above, that every $a > 0$ has *exactly* one positive square root. This is denoted by $\sqrt{a}$. The other square root of a is $-\sqrt{a}$. We agree that $\sqrt{0} = 0$.

This terminology may be confusing. Consider the following statements:

(1) x is a square root of a.

(2) $x = \sqrt{a}$.

The second of these statements is not merely a shorthand transcription of the first. Statement (1) means merely that $x^2 = a$. Statement (2) means not only that $x^2 = a$ but also that $x \geqq 0$. Strictly speaking, it is never correct to speak of "*the* square root of a," except when $a = 0$, because every $a \neq 0$ has either two square roots or none at all (in the real number system). One way to avoid this confusion is to pronounce the symbol $\sqrt{a}$ as "root a," thus warning people that you are pronouncing a formula.

Much later, we shall need another completeness postulate, to guarantee, for example, the existence of π. We shall postpone this discussion until we need it.

The following trivial looking observations turn out to be surprisingly useful.

Theorem 1. For every real number a there is an integer $n > a$ and an integer $m < a$.

Proof. In getting n, we can assume $a > 0$. In the Archimedean postulate, take $M = a$, $e = 1$. This gives an n such that $n \cdot 1 > a$, as desired. To get $m < a$, we merely take $n > -a$, and let $m = -n$.

Theorem 2. Between any two real numbers, there is at least one rational number.

(Obviously there are more.)

Proof. Given $x < y$. If there is a rational number r, with $x + n < r < y + n$, then there is a rational number $r' = r - n$ between x and y. We may therefore suppose that

$$1 < x < y.$$

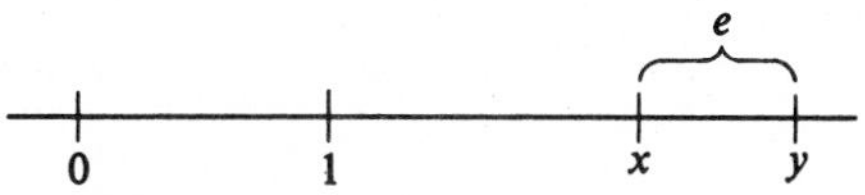

FIGURE 1.5

Let $e = y - x$. By the Archimedean postulate, we have

$$pe > 1$$

for some integer p. Thus

$$\frac{1}{p} < e.$$

The rational numbers with denominator p now divide the whole number line into segments of length $1/p$, like this:

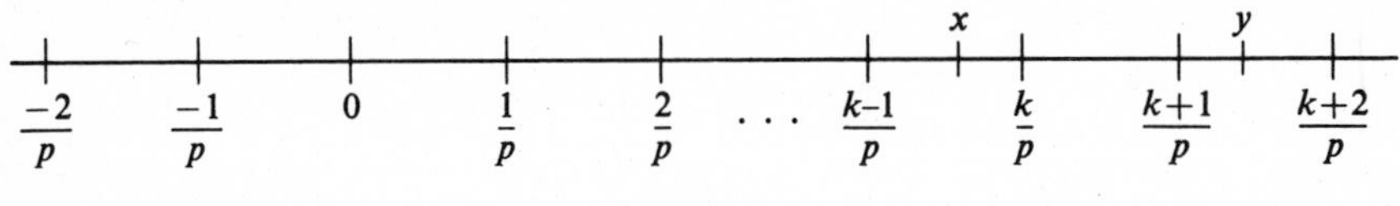

FIGURE 1.6

If k/p is the first one of them that lies to the right of x, then k/p ought to be between x and y, because

$$\frac{1}{p} < e = y - x.$$

To be more precise, let

$$K = \left\{ n \,\middle|\, \frac{n}{p} > x \right\}.$$

By the well-ordering principle, K has a least element k. Thus

$$\frac{k}{p} > x,$$

but

$$\frac{k-1}{p} \leqq x.$$

Therefore

$$\begin{aligned} \frac{k}{p} &\leqq x + \frac{1}{p} < x + e \\ &\leqq x + (y - x) \\ &= y. \end{aligned}$$

Therefore

$$x < \frac{k}{p} < y,$$

which was to be proved.

In using the Archimedean postulate to prove this trivial looking theorem, we are not making any sort of joke; the postulate is needed. There are ordered fields in which the postulate fails. (See Chapter 28.) In such fields, Theorems 1 and 2 do not hold true either; in them, some numbers x and y are greater than every integer and hence greater than every rational number. For many of the purposes of geometry we need the Archimedean postulate to rule out such phenomena.

Problem Set 1.8

1. Show that if $0 < x < y$, then $x^2 < y^2$. Does this conclusion follow if we know only that $x < y$? Why or why not?

2. Show that if $x, y > 0$, and $x^2 < y^2$, then $x < y$.

3. Show that if $0 < a < b$, then $\sqrt{a} < \sqrt{b}$.

4. Show that there is such a number as $\sqrt{1 + \sqrt{2}}$.

5. Same as Problem 4, for $\sqrt{2 - \sqrt{2}}$.

6. Same as Problem 4, for $\sqrt{(3 - \sqrt{2})/(7 - \sqrt{13})}$.

7. Show that $\sqrt{\sqrt{2}}$ cannot be expressed in the form $a + b\sqrt{2}$, where a and b are rational. [*Hint:* You need a theorem from Problem Set 1.3.]

8. For each x, let C_x be the set of all rational numbers less than x. Show that if $C_x = C_y$, then $x = y$.

9. Show that if p^3 is even, then p is even.

10. Show that $\sqrt[3]{2}$ is irrational.

11. Show that $\sqrt[3]{2}$ cannot be expressed in the form $a + b\sqrt{2}$, where a and b are rational.

12. Show that for every $\epsilon > 0$ there is a positive integer n_0 such that

$$n > n_0 \Rightarrow \frac{1}{n} < \epsilon.$$

1.9 THE LANGUAGE AND NOTATION OF SETS

So far, we have been using the language of sets rather sparingly and with a minimum of special notation. There is a standard shorthand, however, which is worth learning partly because it is widely used, and partly because it enables us to be both brief and exact in notebooks and on blackboards.

Throughout this section, capital letters denote sets. If a is an element of A, then we write

$$a \in A.$$

The symbol $\in$ is pronounced "belongs to." When we write $a \notin A$, this means that a does not belong to A. If every element of A is also an element of B, then A is called a *subset* of B, and we write

$$A \subset B,$$
$$B \supset A.$$

Note that here we are allowing the possibility that $A = B$; that is, every set is a subset of itself.

The *intersection* of A and B is the set of all objects that are elements of A and also elements of B. The intersection is denoted by $A \cap B$. (This is pronounced "A cap B," because the symbol $\cap$ looks vaguely like a cap.) Thus

$$A \cap B = \{x | x \in A \quad \text{and} \quad x \in B\}.$$

A word of caution is in order about the use of the word *intersection*. When we speak of the *intersection* of A and B, and write $A \cap B$, this allows the possibility that $A \cap B$ is the empty set $\emptyset$. But when we say that two sets A and B *intersect*, we always mean that A and B have at least one element in common. This distinction in usage, between the noun and the verb, is not logical, but it is convenient; besides, it is nearly universal, and there is not much to be done about it.

The *union* of A and B is the set of all objects that are elements either of A or of B, or of both. The union is denoted by $A \cup B$. (This is pronounced "A cup B," because the symbol looks vaguely like a cup.) Thus

$$A \cup B = \{x | x \in A \quad \text{or} \quad x \in B\}.$$

(Here, and everywhere else in mathematics, when we say "either . . . or . . . ," we allow the possibility that both of the stated conditions hold. If we really mean ". . . but not both," we have to say so.)

The *difference* between two sets A and B is the set of all objects that belong to A but not to B. The difference is denoted by $A - B$. (This is pronounced "A minus B.") Thus

$$A - B = \{x | x \in A \quad \text{and} \quad x \notin B\}.$$

Some books have been written making very free use of this symbolism, but this book is not one of them. Most of the time, we shall use words. We shall, of course, make constant use of the *concepts* represented by the symbols $\in$, $\notin$, $\subset$, $\supset$, $\cap$, $\cup$, and $-$. The following problem set is designed merely to give you some practice in writing and interpreting the symbols. These problems should be worked out on the basis of your "common sense" knowledge of how sets behave; in this book, we shall make no attempt to treat sets formally by means of postulates.

Finally, we mention two common and useful blackboard abbreviations:

(1) $\exists$ means "there exists."

(2) $\ni$ means "such that."

For example, the Euclidean completeness postulate C-1 might be stated as follows:

If $a \in R$ and $a > 0$, then $\exists\, x \ni x > 0$ and $x^2 = a$.

The symbol $\nexists$ means "there does not exist."

Problem Set 1.9

Which of the following statements hold true for all sets A, B, C, . . . ?

1. $A \subset A \cup B$
2. $A \supset A \cap B$
3. $A \subset A \cap B$
4. $A \cap (B - A) = \emptyset$
5. If $A \subset B$, then $x \in A \Rightarrow x \in B$
6. If $A \subset B$ and $B \subset C$, then $A \subset C$
7. Either $A \subset B$ or $\exists\, a \ni a \in A$ and $a \notin B$
8. $A - B \subset A$
9. If $A = A \cap B$, then $A \subset B$
10. If $A \subset B$, then $A = A \cap B$
11. If $A = A \cup B$, then $B \subset A$
12. $(A - B) \cap (A - C) = A - (B \cup C)$
13. $(A - B) \cap (A \cup B) = A \cap B$

1.10 n-FOLD SUMS AND PRODUCTS; THE GENERALIZED ASSOCIATIVE LAW

There is a certain trouble with the associative laws of addition and multiplication. As they stand, they are not adequate to justify the things that we talk about, and the things we do, when we do algebra. At the end of Section 1.2, we remarked that it was quite all right to write triple products abc, because $(ab)c$ is always the same number as $a(bc)$; and similarly for addition. In practice, however, as soor as you get past Chapter 1 of anybody's book, you are writing n-fold sums

$$a_1 + a_2 + \cdots + a_n,$$

and n-fold products

$$a_1a_2 \cdots a_n,$$

for $n > 3$. We insert and delete parentheses in these sums and products, at will. All this is fine, but it has not been connected up, so far, with the operations that are supposed to be given for *pairs* of numbers (a, b) and with the associative laws for triplets (a, b, c). It would be a pity if mathematics appeared to be split down the middle, with the postulates and definitions on one side, and the mathematical content on the other. Let us therefore bridge the gap between our postulates and the things that we intend to do.

The key to our problem is the induction idea. What is given in a field is the twofold product ab, for every a and b in the field. Given a_1, a_2, a_3, we define the threefold product by the formula

$$a_1a_2a_3 = (a_1a_2)a_3.$$

Similarly, we define

$$a_1a_2a_3a_4 = (a_1a_2a_3)a_4,$$

where the parenthesis on the right is defined by the preceding equation. In general,

$$a_1a_2 \ldots a_na_{n+1} = (a_1a_2 \ldots a_n)a_{n+1}.$$

That is, to form an $(n + 1)$-fold product, we first form the n-fold product (of the first n factors) and then multiply the result by the last factor.

This is our official definition of the n-fold product. But there is another scheme that we might have used. We might have defined the triple product as

$$a_1a_2a_3 = a_1(a_2a_3).$$

We could then have said, in general, that

$$a_1a_2 \ldots a_na_{n+1} = a_1(a_2a_3 \ldots a_na_{n+1});$$

that is, to form an $(n + 1)$-fold product, we could first form the product of the *last* n factors and then multiply the result by the first factor. In fact, the first thing that we need to prove is that the choice between these two schemes makes no difference. In the following theorem, we change our notation slightly, to avoid getting peculiar looking statements for $n = 1$ and $n = 2$.

Theorem 1. For every positive integer n,

$$aba_1a_2 \ldots a_n = a(ba_1a_2 \ldots a_n).$$

Proof. Let S be the set of all positive integers for which this formula holds. We shall show, by induction, that S contains all of the positive integers. Thus we need to show two things:

(1) S contains 1.

(2) S is closed under the operation of adding 1.

Proof of (1). S contains 1 if

$$aba_1 = a(ba_1).$$

Now $aba_1 = (ab)a_1$, by definition. And $(ab)a_1 = a(ba_1)$, by the ordinary associative law for triplets. Therefore S contains 1.

Proof of (2). Here we need to show that if S contains n, then S also contains $n + 1$. This means that *if*

(i) $aba_1a_2 \ldots a_n = a(ba_1a_2 \ldots a_n)$,

then

(ii) $aba_1a_2 \ldots a_na_{n+1} = a(ba_1a_2 \ldots a_na_{n+1})$.

This is shown as follows. We have

$$aba_1a_2 \ldots a_na_{n+1} = (aba_1a_2 \ldots a_n)a_{n+1},$$

by definition. By (i), the expression on the right is

$$= [a(ba_1a_2 \ldots a_n)]a_{n+1}.$$

By the associative law, this is

$$= a[(ba_1a_2 \ldots a_n)a_{n+1}].$$

By definition of the $(n + 2)$-fold product, the expression on the right is

$$= a(ba_1a_2 \ldots a_na_{n+1}).$$

Therefore (ii) holds. With the aid of this theorem, we shall prove the following.

Theorem 2. *The General Associative Law.* In any n-fold product, the insertion of one pair of parentheses leaves the value of the product unchanged.

Proof. Let S be the set of all positive integers n for which it is true that parentheses can be inserted in any n-fold product without changing its value. To prove our theorem, we need to show that

(1) S contains 1, and

(2) S is closed under the operation of adding 1.

Proof of (1). Obviously $a_1 = (a_1)$ for every a. Therefore S contains 1.

Proof of (2). Given an $(n + 1)$-fold product

$$a_1a_2 \ldots a_na_{n+1}.$$

Suppose that we insert a pair of parentheses. There are three cases to be considered.

(i) The opening parenthesis comes somewhere after a_1, as

$$a_1a_2 \dots a_i(a_{i+1} \dots a_k) \dots a_{n+1}.$$

(Here we allow the possibility that $k = n + 1$.)

(ii) The closing parenthesis comes somewhere before a_{n+1}.

(iii) The parentheses enclose the entire product.

Obviously, in Case (iii) there is nothing to prove. In Case (i),

$$a_1a_2 \dots a_i(a_{i+1} \dots a_k) \dots a_{n+1} = a_1[a_2 \dots a_i(a_{i+1} \dots a_k) \dots a_{n+1}],$$

by Theorem 1. This is

$$= a_1[a_2 \dots a_{n+1}],$$

because S contains n, and in turn, this becomes

$$= a_1a_2 \dots a_{n+1},$$

by Theorem 1. Thus, in Case (i), if S contains n, it follows that S contains $n + 1$.

In Case (ii), we have

$$a_1a_2 \dots a_i(a_{i+1} \dots a_k) \dots a_{n+1},$$

where $k < n + 1$, but $i + 1$ may be $= 1$. By definition of an $(n + 1)$-fold product, this becomes

$$= [a_1a_2 \dots (a_{i+1} \dots a_k) \dots a_n]a_{n+1};$$

which in turn becomes

$$= [a_1a_2 \dots a_n]a_{n+1},$$

because S contains n; this is

$$= a_1a_2 \dots a_na_{n+1},$$

by definition of an $(n + 1)$-fold product. This completes our induction proof.

The things that we ordinarily do with n-fold products can be justified by repeated applications of this theorem. For example,

$$ab\,\frac{1}{b}\,c\,\frac{1}{c}\,d = ad.$$

Proof. The expression on the left is equal to

$$a\left(b\,\frac{1}{b}\right)\left(c\,\frac{1}{c}\right)d,$$

by two applications of Theorem 2. This is

$$= a11d$$
$$= (a1)(1d)$$
$$= ad.$$

We define n-fold sums in exactly the same way and conclude by the same proof that n-fold sums satisfy the general associative law. That is, insertion of one pair of parentheses in an n-fold sum,

$$a_1 + a_2 + \cdots + a_n,$$

leaves the value of the sum unchanged. Finally we observe that we always have

$$a(b_1 + b_2 + \cdots + b_n) = ab_1 + ab_2 + \cdots + ab_n.$$

The proof is by an easy induction. For $n = 1$, we have $ab_1 = ab_1$. Given that

$$a(b_1 + b_2 + \cdots + b_n) = ab_1 + ab_2 + \cdots + ab_n,$$

it follows that

$$\begin{aligned} a(b_1 + b_2 + \cdots + b_n + b_{n+1}) &= a[(b_1 + b_2 + \cdots + b_n) + b_{n+1}] \\ &= a(b_1 + b_2 + \cdots + b_n) + ab_{n+1} \\ &= (ab_1 + ab_2 + \cdots + ab_n) + ab_{n+1} \\ &= ab_1 + ab_2 + \cdots + ab_n + ab_{n+1}. \end{aligned}$$

1.11 THE CLOSURE OF THE INTEGERS UNDER ADDITION AND MULTIPLICATION

We found, in Section 1.7, that to prove that the set **Q** of rational numbers forms a field, we needed to know that sums and products of integers are always integers. Under our definition of integers, this requires proof; the proof is merely a series of exercises in the use of induction.

We recall that the set **N** of positive integers was defined by the following three conditions:

(1) **N** contains 1,

(2) **N** is closed under the operation of adding 1, and

(3) of all sets of numbers satisfying (1) and (2), **N** is the smallest.

To get the set **Z** of integers, we added to **N** the number 0, and also the negatives of all of the positive integers.

Theorem 1. If a and n are positive integers, then so also is $a + n$.

Proof. Let a be fixed. Let S be the set of all positive integers n for which $a + n \in \mathbf{N}$. Then (1) $1 \in S$, because **N** is closed under the operation of adding 1, and (2) if $n \in S$, then $n + 1 \in S$. For if $a + n \in \mathbf{N}$, we have

$$a + (n + 1) = (a + n) + 1,$$

which belongs to **N**.

Theorem 2. If a and n are positive integers, then so also is an.

Proof. Let a be fixed, and let

$$S = \{n | an \in N\}.$$

Then (1) $1 \in S$, because $a1 = a$. (2) If $n \in S$, then $n + 1 \in S$. For

$$a(n + 1) = an + a1 = an + a,$$

which is the sum of two positive integers.

Theorem 3. If x and y are integers, then so also is xy.

Proof. If $x, y > 0$, this follows from Theorem 2. If $x > 0$ and $y < 0$, then

$$xy = -[x(-y)],$$

which is the negative of the positive integer $x(-y)$. The case $x < 0, y > 0$ is the same. If $x, y < 0$, then xy is the positive integer $(-x)(-y)$. Finally, if $x = 0$ or $y = 0$, we have $xy = 0$, which is an integer. This takes care of closure under multiplication. Rather oddly, addition is more troublesome.

Theorem 4. If $n \in \mathbf{Z}$, then $n - 1 \in \mathbf{Z}$.

Proof. (1) If $n > 1$, then $n = k + 1$ for some k in $\mathbf{N}$. (Theorem 6 of Section 1.6.) Therefore $n - 1 = k \in \mathbf{Z}$.

(2) If $n = 1$, then $n - 1 = 0 \in \mathbf{Z}$.

(3) If $n = 0$, then $n - 1 = -1 \in \mathbf{Z}$.

(4) If $n < 0$, then $n = -k$, where $k > 0$. Therefore $n - 1 = (-k) + (-1) = -(k + 1)$, which is the negative of a positive integer.

Theorem 5. If $a \in \mathbf{Z}$ and $n \in \mathbf{N}$, then $a - n \in \mathbf{Z}$.

Proof. Let $S = \{n | a - n \in \mathbf{Z}\}$. Then (1) $1 \in S$, by Theorem 4, and (2) if $n \in S$, then $n + 1 \in S$. For if $a - n \in \mathbf{Z}$, then

$$a - (n + 1) = (a - n) - 1,$$

which belongs to $\mathbf{Z}$, by Theorem 4.

Theorem 6. If $x, y \in \mathbf{Z}$, then $x + y \in \mathbf{Z}$.

Proof. CASE 1. If either $x = 0$ or $y = 0$, this holds trivially.

CASE 2. If $x, y > 0$, then $x + y \in \mathbf{N}$, and so belongs to $\mathbf{Z}$.

CASE 3. If $x, y < 0$, then $x + y = -[(-x) + (-y)]$, which is the negative of the positive integer $(-x) + (-y)$.

CASE 4. If $x < 0 < y$, let $n = -x$. Then $n > 0$, and $x + y = y - n$. We know by Theorem 6 that $y - n \in \mathbf{Z}$.

These verifications are tedious, but they are needed. We need to know about the integers, the real numbers, and the relation between them. One way to do this is first to set up the integers, and then to build the real numbers from them. (For such a treatment see, for example, Landau's *Foundations of Analysis*.) In this chapter we have first stated postulates for the real numbers and then "moved from the top downward" to get the integers. The latter scheme is by far the quicker and easier, but no scheme can reduce our technical difficulties to zero.

Chapter 2
Incidence Geometry in Planes and Space

You will recall that when we started to discuss the real numbers from a postulational point of view, we began with three things: a set **R** (whose elements were called *numbers*) and two laws of combination (called *addition* and *multiplication*, and denoted by $+$ and $\cdot$). Thus, in Section 1.2, the structure that we were working with was a triplet $[\mathbf{R}, +, \cdot]$, where **R** was a set and $+$ and $\cdot$ were operations defined in **R**. A little later, we assumed that we had an order relation $<$, defined in **R** and subject to certain conditions. Thus, at the end of Chapter 1, the structure that we were working with was a quadruplet $[\mathbf{R}, +, \cdot, <]$; and all of our postulates were stated in terms of these four objects.

We shall follow the same scheme in our postulational treatment of the geometry of planes and space. In the scheme that we shall be using, space will be regarded as a set S; the *points* of space will be the elements of this set. We will also have given a collection of subsets of S, called *lines*, and another collection of subsets of S, called *planes*. Thus the structure that we start with is a triplet

$$[S, \mathcal{L}, \mathcal{P}],$$

where the elements of S, $\mathcal{L}$, and $\mathcal{P}$ are called *points*, *lines* and *planes*, respectively. Later, we shall add to this structure, just as we added to our algebraic structure in the latter part of Chapter 1. For the present, however, our postulates are going to be stated in terms of the sets S, $\mathcal{L}$, and $\mathcal{P}$.

The above presentation is equivalent to one in which we say that the terms *point*, *line*, and *plane* are taken as undefined.

In our formal mathematics, we are going to use postulates; and the only things that we shall claim to know about points, lines, and planes will be the things stated in the postulates. Informally, however, it may be a good idea to remind ourselves of the sort of things that lines and planes will turn out to be. A *line* is going to stretch out infinitely far in both directions, like this:

FIGURE 2.1

Here the arrowheads are supposed to indicate that the line doesn't stop where the picture of it stops. We shall have another term, *segment*, for a figure which looks like this:

FIGURE 2.2

If the end points are P and Q, then this figure will be called the *segment from P to Q*. For a "line" to stretch out infinitely far in only one direction is not enough:

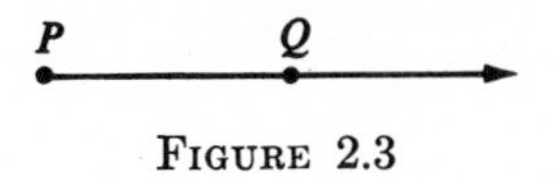

FIGURE 2.3

A figure like this will be called a *ray*. Similarly, a *plane* stretches out infinitely far in every direction. Thus the floor of your room would not form a plane, even if it were perfectly flat. It would form a *part* of a plane.

Logically speaking, we are getting ahead of ourselves when we draw these pictures. The postulates of this section are nowhere nearly enough to guarantee that lines look like our pictures, as you will see in the next set of problems.

Our first postulate is merely a reminder.

I-0. All lines and planes are sets of points.

If a line L is a subset of a plane E, then we shall say that *L lies in E*. (The same term is used in general, to mean that one set is a subset of another.) If a point P belongs to a line L, then we may say that *P lies on L* or that *L passes through P*. Similarly, if P belongs to a plane E, then we may say that *P lies in E* or that *E passes through P*. (Here we are merely defining the familiar language of geometry in terms of the set-theoretic apparatus that is used in our postulates.) By a *figure* we mean a set of points.

Points lying on one line are called *collinear*, and points lying in one plane are *coplanar*.

I-1. Given any two different points, there is exactly one line containing them.

If the points are P and Q, then the line containing them is denoted by $\overleftrightarrow{PQ}$. The arrowheads are meant to remind us of the usual representation of lines in figures.

I-2. Given any three different noncollinear points, there is exactly one plane containing them.

If the three points are P, Q, and R, then the plane containing them is denoted by $\overleftrightarrow{PQR}$.

I-3. If two points lie in a plane, then the line containing them lies in the plane.

I-4. If two planes intersect, then their intersection is a line.

If you check back carefully, you will see that postulates I-0 through I-4 are satisfied by the "geometry" in which there is exactly one point P in S, and this point P is both a line and a plane. To rule out such cases, we state another postulate immediately.

I-5. Every line contains at least two points. S contains at least three noncollinear points. Every plane contains at least three noncollinear points. And S contains at least four noncoplanar points.

(Throughout this book, if we say that "P and Q are points," we allow the possibility that $P = Q$. But if we speak of "two points," we mean that there are really *two* of them; that is, the points must be different; similarly for planes, and so forth. Sometimes we may speak of "two different points," as in I-1, but this is merely for emphasis.)

Theorem 1. Two different lines intersect in at most one point.

Proof. Let L_1 and L_2 be two lines, and suppose that their intersection contains two points P and Q. This is impossible by Postulate I-1, because I-1 says that there is exactly one line, and hence *only* one line, containing P and Q.

Theorem 2. If a line intersects a plane not containing it, then the intersection is a single point.

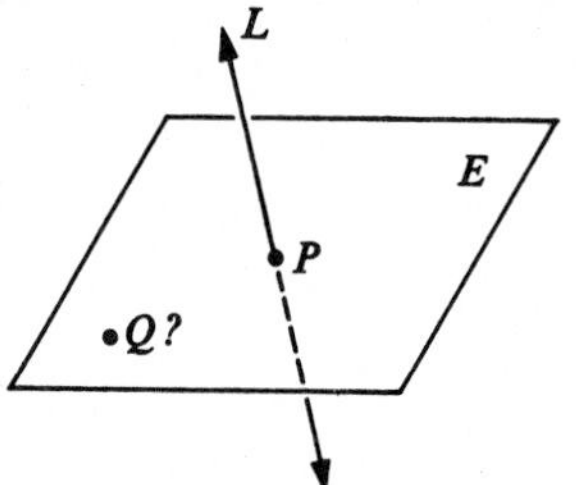

FIGURE 2.4

Proof. Let L be a line intersecting a plane E, but not lying in E. We have given that $L \cap E$ contains at least one point P; and we need to prove that $L \cap E$ contains no other point Q.

Suppose that there is a second point Q in $L \cap E$. Then $L = \overleftrightarrow{PQ}$, by Theorem 1. By I-3, $\overleftrightarrow{PQ}$ lies in E. Therefore L lies in E, which contradicts the hypothesis for L.

Theorem 3. Given a line and a point not on the line, there is exactly one plane containing both of them.

RESTATEMENT. Let L be a line, and let P be a point not on L. Then there is one and only one plane containing $L \cup P$.

(Here we introduce a device which will be convenient later. Whenever we can, we shall state theorems in ordinary English, with little or no notation. This way, the theorems are easier to read and to remember. The restatement furnishes us with the notation that will be used in the proof, and in some cases it may remove some vagueness or ambiguity.)

Proof. (1) By I-5, *L contains at least two points Q and R.*

(2) *P, Q, and R are not collinear.* The reason is that by I-1, L is the only line that contains Q and R; and L does not contain P. Therefore no line contains P, Q, and R.

(3) By (2) and I-2, there is a plane $E = \overleftrightarrow{PQR}$, containing P, Q, and R. By I-3, E also contains L.

Thus there is at least one plane containing $L \cup P$. If there were two such planes, then both of them would contain P, Q, and R. This is impossible, by I-2, because P, Q, and R are noncollinear.

Theorem 4. If two lines intersect, then their union lies in exactly one plane.

Let L and L' be two intersecting lines. The following statements are the main steps in the proof; you should be able to supply the reasons for each of these statements.

(1) $L \cap L'$ is a point P.

(2) L' contains a point $Q \neq P$.

(3) There is a plane E, containing L and Q.

(4) E contains $L \cup L'$.

(5) No other plane contains $L \cup L'$.

Theorems of the kind that we have just been proving are called *incidence* theorems; such a theorem deals with the question whether two sets intersect (and if so, how?) or the question whether one set lies in another. Incidence theorems are used constantly, but the incidence postulates on which they are based do not go very far in describing space geometry, as Problem 1 below will indicate.

Problem Set 2.1

1. Consider the system $[S, \mathcal{L}, \mathcal{P}]$, where S contains exactly four points A, B, C, and D, the lines are the sets with exactly two points, and the planes are the sets with exactly three points. This "space" is illustrated by the following figure:

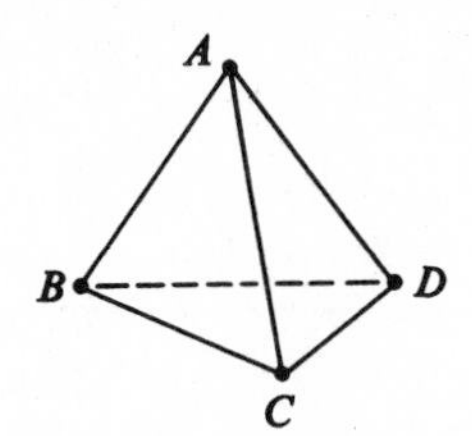

Figure 2.5

Here it should be remembered that A, B, C, and D are the only points that count. Verify that all the incidence postulates hold in this system.

2. Let $P_1, P_2, \ldots, P_5$ be five points, no three of which are collinear. How many lines contain two of these five points?

3. If no four of the five points are coplanar, how many planes contain three of the five points?

4. Given $P_1, P_2, \ldots, P_n$, all different such that no three of them are collinear and no four of them are coplanar. How many lines contain two of them? How many planes contain three of them?

5. Show that under our incidence postulates, S cannot be a line.

6. Show that there is at least one plane.

7. Show that there are at least two planes.

Chapter 3
Distance and Congruence

3.1 THE IDEA OF A FUNCTION

The word *function* is most commonly used in connection with calculus and its various elaborations, but the idea occurs, often without the word, in nearly all mathematics. In fact, the first two chapters of this book have been full of functions, as we shall now see.

(1) In a field $\mathbf{F}$, to every element a there corresponds a unique negative, $-a$. Here we have a function

$$\mathbf{F} \to \mathbf{F}$$

under which

$$a \mapsto -a$$

for every a. (We use $\to$ between sets, and $\mapsto$ between elements of the sets.)

(2) In an ordered field $\mathbf{F}$, to every element x there corresponds a unique number $|x|$, called the absolute value of x. The rule of correspondence is that if $x \geqq 0$, then the number corresponding to x is x itself, and if $x < 0$, then the number corresponding to x is $-x$. Thus we have a function,

$$\mathbf{F} \to \mathbf{F},$$

under which

$$x \mapsto |x|$$

for every x.

(3) Suppose that $\mathbf{F}$ is a Euclidean ordered field. Let $\mathbf{F}^+$ be the set of all elements of $\mathbf{F}$ that are $\geqq 0$. To each element a of $\mathbf{F}^+$ there corresponds a unique element $\sqrt{a}$ of $\mathbf{F}^+$. (Recall the Euclidean completeness postulate and the definition of $\sqrt{a}$.) Here we have a function,

$$\mathbf{F}^+ \to \mathbf{F}^+,$$

under which

$$a \mapsto \sqrt{a}$$

for every a in $\mathbf{F}^+$.

(4) The operation of addition in a field $\mathbf{F}$ can be considered as a function, once we have the idea of the product of two sets. For any pair of sets A, B, the *product* $A \times B$ is the set of all ordered pairs (a, b), where $a \in A$ and $b \in B$. We allow the possibility that $A = B$. Thus, when we identify a point P of a coordinate plane by giving a pair of coordinates (x, y), we are associating with P an element of the product $\mathbf{R} \times \mathbf{R}$ of the real numbers with themselves.

Consider now the operation of addition in a field $\mathbf{F}$. Under this operation, to every pair (a, b) of numbers in $\mathbf{F}$ there corresponds a number $a + b$, called their *sum*. This can be regarded as a function,

$$\mathbf{F} \times \mathbf{F} \to \mathbf{F},$$

where

$$(a, b) \mapsto a + b$$

for every (a, b) in $\mathbf{F} \times \mathbf{F}$.

Obviously multiplication can be regarded in the same way.

Note that in these situations there are always three objects involved: first, a set A of objects *to which* things are going to correspond; second, a set B which contains the objects that correspond to elements of A; and third, the correspondence itself, which associates with every element of A a unique element of B. The set A is called the *domain of definition*, or simply the *domain*. The set B is called the *range*. The correspondence itself is called the *function*. In the examples that we have been discussing, these are as follows.

TABLE 3.1

Domain	Range	Law
$\mathbf{F}$	$\mathbf{F}$	$a \mapsto -a$
$\mathbf{F}$	$\mathbf{F}$	$a \mapsto \lvert a \rvert$
$\mathbf{F}^+$	$\mathbf{F}^+$	$a \mapsto \sqrt{a}$
$\mathbf{F} \times \mathbf{F}$	$\mathbf{F}$	$(a, b) \mapsto a + b$

In the third column, we have described the function by giving the law of correspondence.

A less simple example would be the function which assigns to every positive real number its common logarithm. Here the domain A is the set of all positive real numbers, the range B is the set of all real numbers, and the law of correspondence is $x \mapsto \log_{10} x$. Here the expression $\log_{10} x$ is an example of functional notation. If the function itself is denoted by f, then $f(x)$ denotes the object corresponding to x. For example, if f is the absolute value function, then

$$f(1) = 1, \qquad f(-1) = 1, \qquad f(-5) = 5,$$

and so on. Similarly, if g is the "positive square root function," then

$$g(4) = 2, \qquad g(16) = 4, \qquad g(8) = 2g(2) = 2 \cdot \sqrt{2},$$

and so on. We can also use functional notation for addition, if we want to (which we usually don't). If s is the "sum function," then

$$s(a, b) = a + b,$$

so that

$$s(2, 3) = 5, \qquad s(5, 4) = 9,$$

and so on. Similarly, if p is the "product function," then

$$p(a, b) = ab,$$

so that

$$p(5, 4) = 20 \qquad \text{and} \qquad p(7, 5) = 35.$$

To sum up, a function f is defined if we describe three things: (1) a set A, called the *domain*, (2) a set B called the *range*, and (3) a law of correspondence under which to every element a of A there corresponds a unique element b of B. If $a \in A$, then $f(a)$ denotes the corresponding element of B. We indicate the function f, the domain A, and the range B by writing

$$f : A \to B,$$

and we say that f is a *function of A into B*.

We define composition of functions in the way which is familiar from calculus. Thus, given

$$f : A \to B$$

and

$$g : B \to C,$$

we can write

$$c = g(f(a));$$

this means that c corresponds to $f(a)$ under the function g. For example, if we are using the functions s and p to describe sums and products, then

$$s(p(a, b),\ p(a, c))$$

means $ab + ac$, and

$$p(a, s(b, c))$$

means

$$a(b + c).$$

Finally, we define two special types of function which have special importance. If every b in B is $= f(a)$ for at least one a in A, then we say that f is a function of A *onto* B. If every b in B is $= f(a)$ for *exactly* one a in A, then we say that f is a *one-to-one correspondence between A and B*, and we write

$$f : A \leftrightarrow B.$$

For example, the function $F : \mathbf{R} \to \mathbf{R}$, $x \mapsto x^3$ is a one-to-one correspondence. The function $g : \mathbf{R} \to \mathbf{R}$, $x \mapsto x^2$ is not a one-to-one correspondence because, in the range, every positive number appears twice, and no negative number appears at all. The function under which $x \mapsto -x$ is a one-to-one correspondence. (Proof? You need to check that each number y is $= -x$ for exactly one number x.) Similarly, the function $x \to 1/x$ is a one-to-one correspondence; here

$$A = B = \{x \mid x \neq 0\}.$$

If f is a one-to-one correspondence, then there is a function,

$$f^{-1}: B \leftrightarrow A,$$

called the *inverse* of f, which reverses the action of f. That is, $f^{-1}(b) = a$ if $f(a) = b$. The symbol f^{-1} is pronounced "f-inverse." When we say that a function has an inverse, this is merely another way of saying that the function is a one-to-one correspondence.

Given a function

$$f: A \to B.$$

The *image* of A is the set of all elements of B that appear as values of the function. Thus the image is

$$\{b | a \in A \text{ and } b = f(a)\}.$$

In other words, the image is the smallest set that might have been used as the range of the function. For example, if the function

$$f: \mathbf{R} \to \mathbf{R}$$

is defined by the condition $f(x) = x^2$ for every x, then the range is the set $\mathbf{R}$ of all real numbers, and the image is the set of all nonnegative real numbers.

The question may arise why we define functions in such a way as to permit the range to be a bigger set than the image. We might have stated the definition in such a way that every function would be *onto*. But such a definition would be unmanageable. For example, suppose that we define a function, in calculus, by the equation

$$f(x) = x^4 - 7x^3 + 3x^2 - 17x + 3.$$

This is a function of $\mathbf{R}$ *into* $\mathbf{R}$. To find out what the image is, we would have to find out where this function assumes its minimum; this is a problem in calculus, leading to a difficult problem in algebra. If we required that the image be known for the function to be properly defined, we could not state our calculus problem without first solving it; and this would be an awkward proceeding.

Problem Set 3.1

1. Using the functional notations $s(a, b)$ and $p(a, b)$ for sums and products, rewrite the associative, commutative, and distributive laws for an ordered field. Now rewrite the Postulates MO-1 and AO-1 which related the field structure to the order relation in an ordered field.

2. We recall that $\mathbf{Z}$ is the set of all integers. For each i, j in $\mathbf{Z}$, let $f(i, j)$ be the larger of the two integers i and j. Do these remarks define a function? If so, what are the domain and the image?

3. Let $f: \mathbf{R} \to \mathbf{R}$ be defined by the condition $f(x) = x^2$. Does f have an inverse? Why or why not?

4. Let $\mathbf{R}^+$ be the set of all nonnegative real numbers. Let $g: \mathbf{R}^+ \to \mathbf{R}^+$ be defined by the condition $g(x) = x^2$. Does g have an inverse? Why or why not?

5. The same question, for $f: \mathbf{R} \to \mathbf{R}$, $f(x) = \sin x$.
6. Let A be the closed interval $[-\pi/2, \pi/2]$. That is,

$$A = \left\{x \mid x \in \mathbf{R} \quad \text{and} \quad -\frac{\pi}{2} \leqq x \leqq \frac{\pi}{2}\right\}.$$

Let

$$B = [-1, 1] = \{x \mid x \in \mathbf{R} \quad \text{and} \quad -1 \leqq x \leqq 1\}.$$

Let

$$g: A \to B$$

be the function defined by the condition

$$g(x) = \sin x.$$

Does g have an inverse? Why or why not?

3.2 THE SET-THEORETIC INTERPRETATION OF FUNCTIONS AND RELATIONS

In the preceding section, we have explained, with numerous examples, what people are talking about when they talk about functions. And in fact the idea of a function, in the form in which we have explained it, is adequate for nearly all the uses that you will need to make of it for a long time to come.

You will find, however, if you re-read the last section carefully, that at no point have we given a straightforward definition of a function; we have explained the conditions under which a function is *defined*, but we have not said what kind of object a function *is*. This we shall now do. But first we shall give some preliminary discussion, to indicate the idea behind the definition.

Consider first the case where the domain is a finite set, say,

$$A = \{0, 1, 2, 3, 4, 5\}.$$

For any function f with A as domain, we can write down a complete table, giving the values of the function f.

a	$f(a)$
0	0
1	1
2	4
3	3
4	4
5	1

There was a system used in making up the table; and you may be able to figure out what this system was. But even if you can't figure it out, the function is defined by the table. To define a function, you have to explain what its value is, for each element of the domain A, but you don't necessarily have to give this explanation tersely. In fact, if you remember the kind of functions that were important in calculus, you will recall that it took quite a while to explain what they were. The expression $\sin x$ was not a formula for the sine function, but merely a *name* for the sine function; and the explanation of what $\sin x$ really meant was given in words, at considerable length. For a function like the sine, you could write down only a partial table of values, because the domain A was the set of all real numbers, which is infinite. Under the definition, however, a correspondence was defined under which to every x in R there corresponded a unique real number y which was $\sin x$.

From a finite table, such as the one that we have written above, it is easy to read off a set of ordered pairs which describe the function. Each line of the table gives us an ordered pair (a, b), in which a is in A and b is the corresponding element of B. From our table for f, we get the pairs

$$(0, 0), \quad (1, 1), \quad (2, 4), \quad (3, 3), \quad (4, 4), \quad (5, 1).$$

If the set A is infinite, then so is the table.

A partial table for the sine might look like this.

x	$\sin x$
0	0
π	0
$\pi/2$	1
$\pi/3$	$\sqrt{3}/2$
$\pi/4$	$\sqrt{2}/2$
$-\pi$	0
$-\pi/2$	-1

From this table we could read off a partial collection of ordered pairs

$$\left\{\begin{matrix} (0, 0), & (\pi, 0), & (\pi/2, 1), & (\pi/3, \sqrt{3}/2), \\ (\pi/4, \sqrt{2}/2), & (-\pi, 0), & (-\pi/2, -1) & \end{matrix}\right\}.$$

If we formed the set of *all* ordered pairs of the type $(x, \sin x)$, then this infinite collection would describe the sine function completely. Similarly, *every* function can be described by a collection of ordered pairs. If the domain of the function is finite, then so is the collection, and if A is infinite, then so is the collection. If the collection describes a function, then every a in A must appear as the first term of exactly one pair in the collection, because the function assigns a *unique* value to a.

Thus, given a function $f\colon A \to B$, we have a collection of ordered pairs (a, b), where (1) $a \in A$, (2) $b \in B$, and (3) every element of A appears exactly once as the first term of an ordered pair in the collection. And conversely, given a collection of ordered pairs (a, b), satisfying (1), (2), and (3), we always have a function $f: A \to B$. These observations are the basis of the following definition. In this definition, we merely are saying that the function *is* the sort of collection of ordered pairs that we have been discussing.

DEFINITION. Let A and B be sets. *A function with domain A and range B* is a collection f of ordered pairs (a, b), such that

(1) for each (a, b) in f, $a \in A$;

(2) each a in A is the first term of exactly one pair (a, b) in f; and

(3) for each (a, b) in f, $b \in B$.

When we write $b = f(a)$, we mean that (a, b) belongs to the collection f. From here on, we proceed to handle functions exactly as before.

A somewhat similar device enables us to give an explicit definition of the idea of a *relation defined on a set A*. We have been using this idea informally, writing $a < b$ to mean that a has the relation $<$ to b, and, more generally, $a * b$ is written to mean that a has the relation $*$ to b. Now, given a relation $*$, defined on the set A, we can form the collection

$$\{(a, b) | a * b\}.$$

Conversely, given any collection of ordered pairs of elements of A, we can define a relation $*$, by saying that $a * b$ if the pair (a, b) belongs to the collection. In the following definition, we are saying that the relation *is* the collection. Recall, of course, that $A \times A$ is the set of all ordered pairs of elements of A.

DEFINITION. A *relation defined on a set A* is a subset of $A \times A$.

For example, let

$$A = \{1, 2, 3\},$$

and let

$$* = \{(1, 2), (1, 3), (2, 3)\}.$$

Then $*$ is a relation. (It is, in fact, the usual relation $<$.)

It is not necessary, of course, to denote relations by peculiar symbols. For example, if $A = \{1, 2, 3\}$, as before, we may let

$$G = \{(2, 1), (3, 1), (3, 2)\}.$$

Thus $2G1$, $3G1$ and $3G2$, because $(2, 1)$, $(3, 1)$, and $(3, 2)$ belong to G. (In fact, G is the relation $>$.)

PROBLEM SET 3.2

1. Let $A = \{1, 2, 3, 4\}$. Let

$$G = \{(4, 2), (4, 1), (4, 3), (2, 1), (2, 3), (1, 3)\}.$$

Is G a relation? Is G an order relation?

2. Let A be as before, and let G be the set of all ordered pairs (a, b) such that a and b belong to A and $a \neq b$. Is G a relation? Is G an order relation?

3. Is the following collection a function? If so, what are its domain and image?

$$\{(0, 0), (1, 1), (2, 4), (3, 2), (4, 2), (5, 4), (6, 1)\}$$

Can you see a systematic way in which this collection might have been constructed?

4. Is the following collection a function?

$$\{(0, 1), (1, 0), (0, 0)\}$$

5. Let f be the set of all ordered pairs (x, y) such that x and y belong to R and $y = x^2$. Is this a function?

6. The same question, for the set of all ordered pairs (x, y) such that x and y belong to R and $x = y^2$.

7. Consider a rectangular coordinate system in the plane, in the usual sense of analytic geometry. Every point has a pair of coordinates (x, y). For the purposes of this question, let us regard points as indistinguishable from the ordered pairs (x, y) that describe them. Thus every figure, that is, every set of points, becomes a collection of ordered pairs of real numbers. Under what conditions, if any, do the following figures represent functions?

(a) a triangle
(b) a single point
(c) a line
(d) a circle
(e) a semicircle, including the end points
(f) an ellipse

What, in general, is the geometric condition that a figure in the coordinate plane must satisfy, to be a function?

3.3 THE DISTANCE FUNCTION

So far, the structure dealt with in our geometry has been the triplet

$$[S, \mathcal{L}, \mathcal{P}].$$

We shall now add to the structure by introducing the idea of distance. To each pair of points there will correspond a real number called the distance between them. Thus we want a distance function d, subject to the following postulates.

D-0. d is a function

$$d: S \times S \to \mathbf{R}.$$

D-1. For every P, Q, $d(P, Q) \geqq 0$.

D-2. $d(P, Q) = 0$ if and only if $P = Q$.

D-3. $d(P, Q) = d(Q, P)$ for every P and Q in S.

Here we have numbered our first postulate D-0 because it is never going to be cited in proofs; it merely explains what sort of object d is. Of course $d(P, Q)$ will be called the distance between P and Q, and, for the sake of brevity, we shall write $d(P, Q)$ simply as PQ. (We shall be using distances so often that we ought to reserve for them the simplest notation available.)

Any reasonable notion of distance ought to satisfy D-1 through D-3. We might have required also that

$$PQ + QR \geqq PR,$$

which would say, approximately, that "a straight line is the shortest distance between two points." But as it happens, we don't need to make this statement a postulate, because it can be proved on the basis of other geometric postulates, to be stated later.

Henceforth, until further notice, the distance function d is going to be part of our structure. Thus the structure, at the present stage, is

$$[S, \mathcal{L}, \mathcal{P}, d].$$

The distance function is connected up with the rest of the geometry by the ruler postulate D-4, which we shall state presently.

We ordinarily think of the real numbers as being arranged on a line, like this:

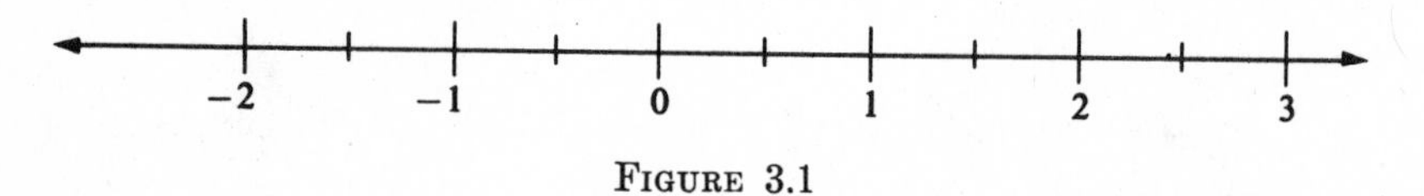

FIGURE 3.1

If the "lines" in our geometry, that is, the elements of $\mathcal{L}$, really "behave like lines," then we ought to be able to apply the same process in reverse and label the points of any line L with numbers in the way that we label the points of the x-axis in analytic geometry:

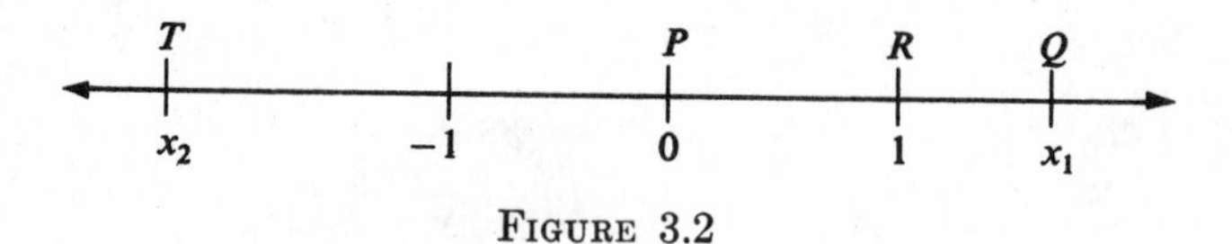

FIGURE 3.2

If this is done in the usual way, then we have a one-to-one correspondence,

$$f: L \leftrightarrow \mathbf{R},$$

between the points of L and the real numbers. This correspondence will turn out to be a *coordinate system*, in a sense which we shall soon define. Meanwhile, therefore, if $x = f(P)$, we shall refer to x as the *coordinate* of P. In the figure, the coordinates of P, Q, R, and T are 0, x_1, 1, and x_2. If the coordinates are related to distance in the usual way, then

$$PQ = |x_1| \quad \text{and} \quad PT = |x_2|.$$

In fact, no matter where Q and T may lie on the line, we will always have

$$QT = |x_2 - x_1|.$$

(You can check this for the cases $x_2 < x_1 < 0$, $x_2 < 0 < x_1$, and $0 < x_2 < x_1$. There is no harm in assuming that $x_2 < x_1$, because when x_1 and x_2 are interchanged, both sides of our equation are unchanged.)

Obviously nothing can be proved by this discussion, because the postulates that we have stated so far do not describe any connection at all between the distance function and lines. All that we have been trying to do is to indicate why the following definition, and the following postulate, are reasonable.

DEFINITION. Let

$$f: L \leftrightarrow \mathbf{R}$$

be a one-to-one correspondence between a line L and the real numbers. If for all points P, Q of L, we have

$$PQ = |f(P) - f(Q)|,$$

then f is a *coordinate system* for L. For each point P of L, the number $x = f(P)$ is called the *coordinate* of P.

D-4. *The Ruler Postulate.* Every line has a coordinate system.

The postulate D-4 is called the ruler postulate because, in effect, it furnishes us with an infinite ruler which can be laid down on any line and used to measure distances along the line. This kind of ruler is not available in classical Euclidean geometry. When we speak of "ruler-and-compass constructions" in classical geometry, the first of these abstract drawing instruments is not really a ruler, because it has no marks on it. It is, properly speaking, merely a straight-edge. You can use it to draw the line containing two different points, but you can't use it to measure distances with numbers, or even to tell whether two distances PQ, RT are the same.

As it stands, D-4 says merely that every line has at least one coordinate system. It is easy to show, however, that there are lots of others.

Theorem 1. If f is a coordinate system for L, and

$$g(P) = -f(P)$$

for each point P of L, then g is a coordinate system for L.

Proof. It is plain that the condition $g(P) = -f(P)$ defines a function $L \to \mathbf{R}$. And this function is one to one, because if $x = g(P)$, it follows that $-x = f(P)$, and $P = f^{-1}(-x)$, so that P is uniquely determined by x.

It remains to check the distance formula. Given that

$$x = g(P), \qquad y = g(Q),$$

we want to prove that

$$PQ = |x - y|.$$

We know that

$$-x = f(P), \qquad -y = f(Q).$$

Since f is a coordinate system, it follows that

$$PQ = |(-x) - (-y)|.$$

Therefore

$$\begin{aligned} PQ &= |y - x| \\ &= |x - y|. \end{aligned}$$

which was to be proved.

Theorem 1 amounts to a statement that if we reverse the direction of the coordinate system, then we get another coordinate system. We can also shift the coordinates to left or right.

Theorem 2. Let f be a coordinate system for the line L. Let a be any real number, and for each $P \in L$, let

$$g(P) = f(P) + a.$$

Then $g\colon L \to \mathbf{R}$ is a coordinate system for L.

The proof is very similar to that of the preceding theorem. Combining the two, we get the following theorem.

Theorem 3. *The Ruler Placement Theorem.* Let L be a line, and let P and Q be any two points of L. Then L has a coordinate system in which the coordinate of P is 0 and the coordinate of Q is positive.

Proof. Let f be any coordinate system for L. Let $a = f(P)$; and for each point T of L, let $g(T) = f(T) - a$.

Then g is a coordinate system for L, and $g(P) = 0$. If $g(Q) > 0$, then g is the system that we were looking for. If $g(Q) < 0$, let $h(T) = -g(T)$ for every $T \in L$. Then h satisfies the conditions of the theorem.

PROBLEM SET 3.3

1. Show that D-1, D-2, and D-3 are consequences of the ruler postulate.

3.4 BETWEENNESS

One of the simplest ideas in geometry is that of betweenness for points on a line. In fact, Euclid seems to have regarded it as too simple to analyze at all, and he uses it, without comment, in proofs, but doesn't mention it at all in his postulates.

Roughly speaking, B is between A and C on the line L if the points are situated like this:

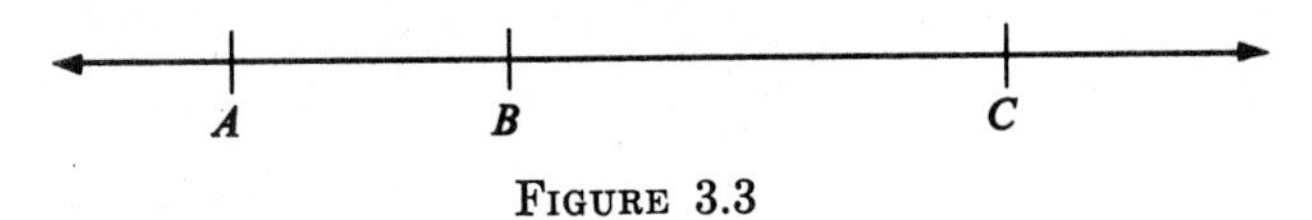

FIGURE 3.3

or like this:

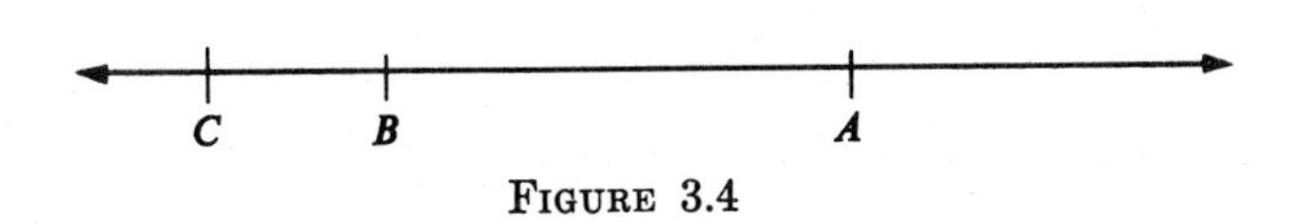

FIGURE 3.4

(Logically speaking, of course, the second figure is superfluous, because on a line, there is no way to tell left from right or up from down.) What we need, to handle betweenness mathematically, is an exact definition which conveys our common-sense idea of what betweenness ought to mean. One such definition is as follows.

DEFINITION. Let A, B, and C be three collinear points. If

$$AB + BC = AC,$$

then B *is between* A *and* C. In this case we write A-B-C.

As we shall see, this definition is workable. It enables us to prove that betweenness has the properties that it ought to have.

Theorem B-1. If A-B-C, then C-B-A.

This is a triviality. If $AB + BC = AC$, then $CB + BA = CA$.

The rest of the basic theorems on betweenness are going to depend essentially on the ruler postulate.

Betweenness for real numbers is defined in the expected way; y is between x and z if either $x < y < z$ or $z < y < x$. In this case we write x-y-z. (Confusion with subtraction is unlikely to occur, because "x minus y minus z" would be ambiguous anyway.)

Lemma 1. Given a line L with a coordinate system f and three points A, B, C with coordinates x, y, z, respectively. If x-y-z, then A-B-C.

Proof of lemma. (1) If $x < y < z$, then

$$AB = |y - x| = y - x,$$

because $y - x > 0$. For the same reasons,

$$BC = |z - y| = z - y$$

and

$$AC = |z - x| = z - x.$$

Therefore

$$\begin{aligned} AB + BC &= (y - x) + (z - y) \\ &= z - x \\ &= |z - x| \\ &= AC \end{aligned}$$

so that A-B-C.

(2) If $z < y < x$, it follows by a precisely similar argument that C-B-A, which means that A-B-C, as before.

Theorem B-2. Of any three points on a line exactly one is between the other two.

Proof. (1) Let f be a coordinate system for the line; and let x, y, z be the coordinates of the points A, B, C. One of the numbers x, y, z is between the other two. By Lemma 1, this means that the corresponding point A, B, or C is between the other two points.

(2) We now need to prove that if A-B-C, then neither of the conditions B-A-C, A-C-B holds. If B-A-C, we have

$$BA + AC = BC.$$

But we have given that

$$AB + BC = AC.$$

By addition, we get

$$BA + AC + AB + BC = BC + AC,$$

or

$$2AB = 0.$$

Therefore $AB = 0$. This is impossible, because $A \neq B$.

The proof that both A-B-C and A-C-B cannot hold is precisely analogous.

Consider now four points A, B, C, D of a line L. In the list below, we indicate the four possible triplets that can be formed from these; opposite each triplet we have listed the three possible betweenness relations.

$$\begin{array}{llll} A, B, C: & \overline{A\text{-}B\text{-}C}, & A\text{-}C\text{-}B, & B\text{-}A\text{-}C, \\ A, B, D: & \overline{A\text{-}B\text{-}D}, & A\text{-}D\text{-}B, & B\text{-}A\text{-}D, \\ A, C, D: & \overline{A\text{-}C\text{-}D}, & A\text{-}D\text{-}C, & C\text{-}A\text{-}D, \\ B, C, D: & \overline{B\text{-}C\text{-}D}, & B\text{-}D\text{-}C, & C\text{-}B\text{-}D. \end{array}$$

When we write

$$A\text{-}B\text{-}C\text{-}D,$$

we mean *that all the overscored betweenness relations A-B-C, A-B-D, A-C-D and B-C-D hold, but none of the other eight relations hold.* (Thus A-B-C-D is an efficient shorthand.) The scheme is easy to remember; the relations that hold are the ones that you get by leaving out one of the letters in the expression A-B-C-D.

Theorem B-3. Any four points of a line can be named in an order A, B, C, D, in such a way that A-B-C-D.

Proof. Let f be a coordinate system for the line that contains our four points P, Q, R, S. The coordinates of our points are four numbers; and these appear in some order

$$w < x < y < z.$$

Here w, x, y, and z are $f(P)$, $f(Q)$, $f(R)$, $f(S)$, but not necessarily respectively. Let

$$A = f^{-1}(w), \qquad B = f^{-1}(x), \qquad C = f^{-1}(y), \qquad D = f^{-1}(z)$$

From the double inequalities $w < x < y$, $w < x < z$, $w < y < z$, $x < y < z$, we get (by Lemma 1) the betweenness relations A-B-C, A-B-D, A-C-D, B-C-D. Thus, for each three of our four points, we have a betweenness relation; and Theorem B-2 tells us that every three points stand in only one betweenness relation. Therefore our list is complete, and A-B-C-D, which was to be proved.

Theorem B-4. If A and B are any two points, then (1) there is a point C such that A-B-C, and (2) there is a point D such that A-D-B.

Proof. Take a coordinate system f for the line AB that contains A and B.

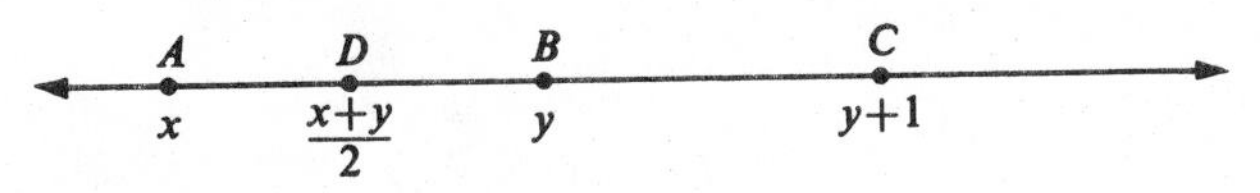

FIGURE 3.5

There is no loss of generality in supposing that $x < y$. (See Theorem 1, Section 3.3.) As in the figure, let

$$C = f^{-1}(y + 1).$$

Then A-B-C, because $x < y < y + 1$.

As in the figure again, let

$$D = f^{-1}\left(\frac{x + y}{2}\right).$$

Since $x < y$; we have

$$2x < x + y < 2y.$$

(Why?) Therefore

$$x < \frac{x + y}{2} < y,$$

so that A-D-B.

In the next few chapters, we shall want to handle betweenness by referring only to the *theorems* of this section, without going back to the definition. (The reasons for this will be explained much later.) It turns out that the theorems above are adequate, if we include the following trivial one.

Theorem B-5. If A-B-C, then A, B, and C are three different points of the same line.

This held, of course, under our original definition of the relation A-B-C. For convenience of reference, we list the basic properties of betweenness.

B-1. If A-B-C, then C-B-A.

B-2. Of any three points on a line, exactly one is between the other two.

B-3. Any four points of a line can be named in an order A, B, C, D, in such a way that A-B-C-D.

B-4. If A and B are any two points, then (1) there is a point C such that A-C-B, and (2) there is a point D such that A-B-D.

B-5. If A-B-C, then A, B, and C are three different points of the same line.

Problem Set 3.4

1. Show that if A-B-C and B-C-D, then A-B-D and A-C-D.
2. Show that if A-B-C and A-D-C, then A-B-D-C, A-D-B-C, or $B = D$.
3. Given four spherical beads of different colors. In how many different ways is it possible to arrange them in a trough, in order from left to right? (This is a problem in *order*.)
4. Given four beads as in Problem 3. In how many essentially different ways is it possible to arrange them on a rigid symmetrical rod? (This is a problem in *betweenness*.)
5. Given four beads as in the preceding problems. In how many essentially different ways is it possible to arrange them on a string so as to make a four-bead necklace? (The string is so thin that the knot can slip through the holes in the beads. This is a problem in "betweenness on a circle," and the answer indicates that the idea of "betweenness on a circle" is more peculiar than one might have supposed.)
6. Prove the following converse of Lemma 1.

Lemma 2. Given a line L with a coordinate system f, and three points A, B, and C with coordinates x, y, and z, respectively. If A-B-C, then x-y-z.

7. In this section, we defined a betweenness relation for real numbers, by saying that x-y-z if either $x < y < z$ or $z < y < x$. Show that, for this betweenness relation, Conditions B-1 through B-4 hold true.

3.5 SEGMENTS, RAYS, ANGLES, AND TRIANGLES

If A and B are two points, then the *segment between A and B* is the set whose points are A and B, together with all points between A and B. By B-5, the segment lies on the line $\overleftrightarrow{AB}$, and we have the figure below. As indicated, the segment is denoted by $\overline{AB}$.

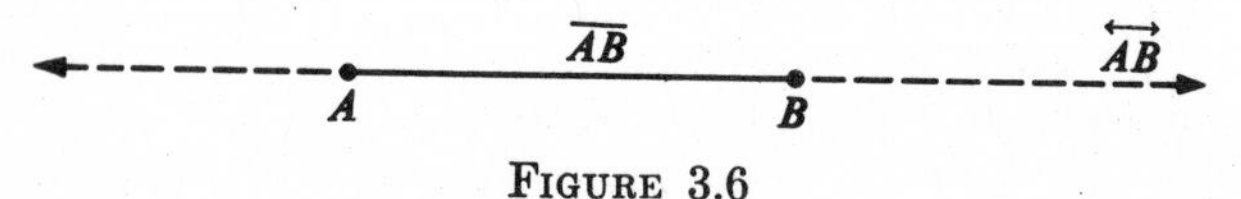

Figure 3.6

If A and B are two points, then the *ray from A through B* is the figure that looks like this:

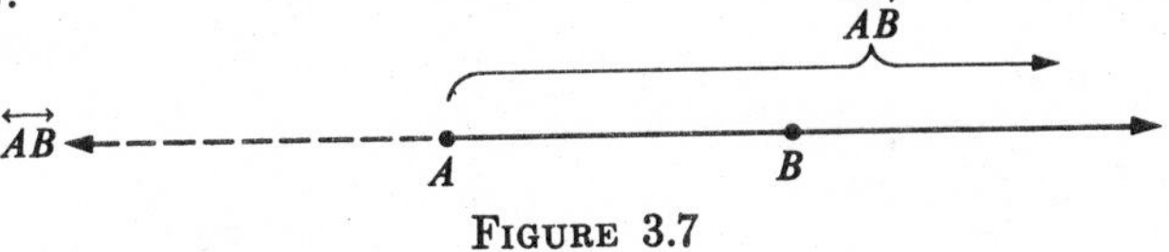

FIGURE 3.7

As indicated, the ray is denoted by $\overrightarrow{AB}$.

More precisely, the ray $\overrightarrow{AB}$ is the set of all points C of the line $\overleftrightarrow{AB}$ such that A is *not* between C and B. The point A is called the *end point* of the ray $\overrightarrow{AB}$.

If this definition looks peculiar, you should check it against the figure to make sure that it agrees with our rough notion of what points of the line ought to be on the ray. It is fairly easy to see that $\overrightarrow{AB}$ is the union of (1) the segment $\overline{AB}$, and (2) the set of all points C such that A-B-C. If this latter description seems more natural to you, you may regard it as the definition of a ray.

Roughly speaking, an *angle* is a figure that looks like this:

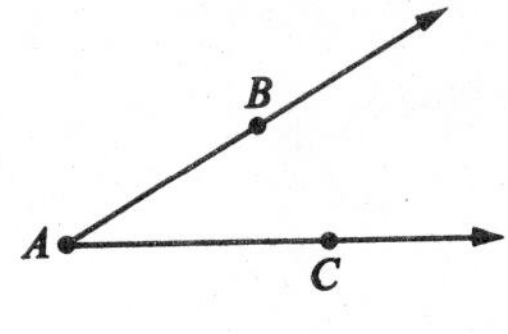

FIGURE 3.8

More precisely, an *angle* is a figure which is the union of two rays which have the same end point, but do not lie on the same line. If the angle is the union of $\overrightarrow{AB}$ and $\overrightarrow{AC}$, then these rays are called the *sides* of the angle; the point A is called the *vertex*; and the angle itself is denoted by the symbol

$$\angle BAC.$$

Notice that we always have $\angle BAC = \angle CAB$.

Finally, if A, B, and C are three noncollinear points, then the set

$$\overline{AB} \cup \overline{BC} \cup \overline{AC}$$

is called a *triangle*.

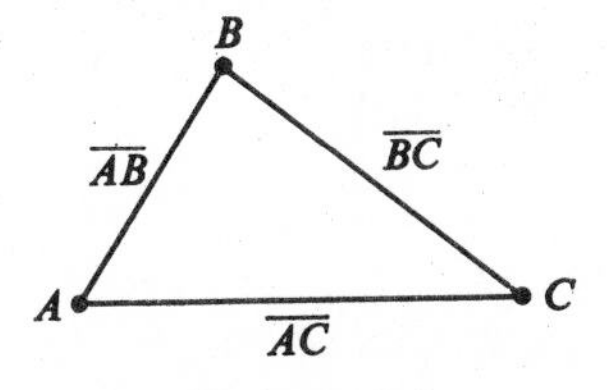

FIGURE 3.9

The three segments $\overline{AB}$, $\overline{BC}$, and $\overline{AC}$ are called its *sides*; and the points A, B, and C are called its *vertices*. (The English plural *vertexes* is used by some authors.)

The triangle itself is denoted by the symbol

$$\triangle ABC.$$

The *angles* of $\triangle ABC$ are $\angle BAC$, $\angle ACB$, and $\angle ABC$. Note that $\triangle ABC$ contains none of these three angles, because the sides of an angle are rays and the sides of a triangle are segments. If we drew in all the angles, the figure would look like this:

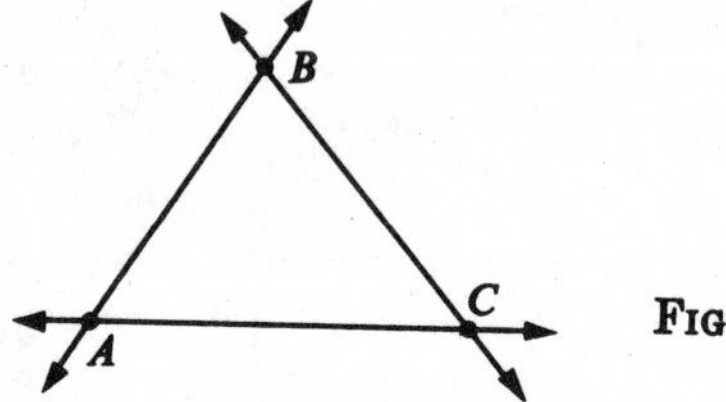

FIGURE 3.10

The following theorems look easy, but some of them are not.

Theorem 1. If A and B are any two points, then $\overline{AB} = \overline{BA}$.

Theorem 2. If C is a point of $\overrightarrow{AB}$, other than A, then $\overrightarrow{AB} = \overrightarrow{AC}$.

Theorem 3. If B_1 and C_1 are points of $\overrightarrow{AB}$ and $\overrightarrow{AC}$, other than A, then $\angle BAC = \angle B_1AC_1$.

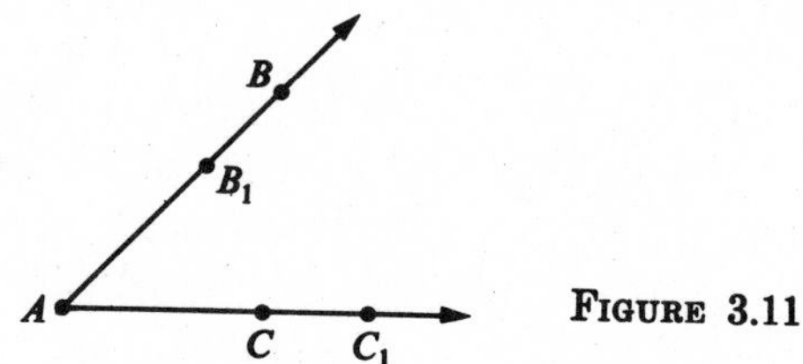

FIGURE 3.11

Theorem 4. If $\overline{AB} = \overline{CD}$, then the points A, B are the same as the points C, D, in some order. (That is, the end points of a segment are uniquely determined by the segment.)

Theorem 5. If $\triangle ABC = \triangle DEF$, then the points A, B, and C are the same as the points D, E, and F, in some order. (That is, the vertices of a triangle are uniquely determined by the triangle.)

If you review the definitions of $\overline{AB}$, $\overrightarrow{AB}$, $\angle BAC$, and $\triangle ABC$, you will see that all of these definitions are based on the idea of betweenness. The proofs of Theorems 1 through 5 must, therefore, be based on Theorems B-1 through B-5.

A word of caution: Here and hereafter, the symbol $=$ is going to be used in one and only one sense; it means "is exactly the same as." Thus, when we write $\overline{AB} = \overline{BA}$, we mean that the sets $\overline{AB}$ and $\overline{BA}$ have exactly the same elements.

Finally, a few remarks may be in order about the way in which we have defined the idea of an angle. Under our definition, an angle is simply a set which is the union of two noncollinear rays with the same end-point. Angles, in this sense, are quite adequate for the purposes of Euclidean geometry.

Much later, in analytic geometry and in trigonometry, we shall need to talk about *directed* angles in which the initial side can be distinguished from the terminal side, like this:

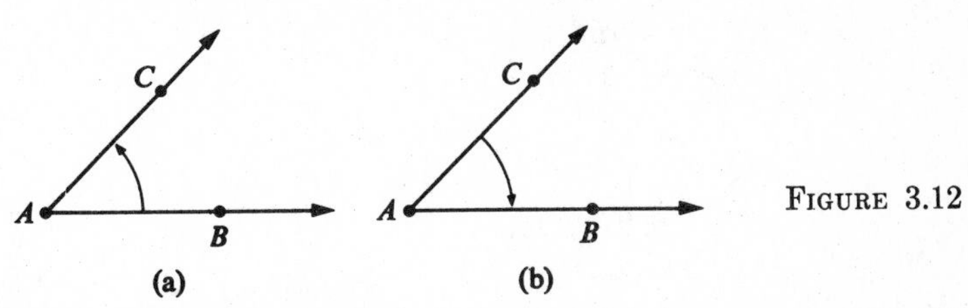

FIGURE 3.12

An angle, in this sense, is not a set of points, but rather an *ordered* pair $(\overrightarrow{AB}, \overrightarrow{AC})$ of rays; thus $(\overrightarrow{AB}, \overrightarrow{AC})$ is different from $(\overrightarrow{AC}, \overrightarrow{AB})$. For directed angles, we allow the possibility that the sides are collinear, and also the possibility that the sides are the same. We have not used this more complicated idea of an angle, because at the present stage we have no use for it. For example, the angles of a triangle never consist of two collinear rays, and there is no natural way to assign directions to them.

For the purposes of this book, there are good reasons for ruling out "zero angles" and "straight angles." In the first place, these terms are superfluous: a zero angle is simply a ray, and a straight angle is a line. In the second place, angles, rays, and lines are different figures, in important ways; and if we used the same word "angle" to apply to all three, then we would continually be involved in discussions of special cases. (Contrary to popular impression, Euclid did not use "straight angles.")

PROBLEM SET 3.5

1. Prove Theorem 1.

2. Show that, given a ray $\overrightarrow{AB}$, there is a coordinate system f on the line $\overleftrightarrow{AB}$ such that

$$\overrightarrow{AB} = \{P \mid f(P) \geqq 0\}.$$

3. Prove Theorem 2.

4. Prove Theorem 3.

*5. Prove the following. Let A and B be two points, and let D, E, and F be three noncollinear points. If $\overleftrightarrow{AB}$ contains only one of the points D, E, or F, then each of the lines $\overleftrightarrow{DE}$, $\overleftrightarrow{DF}$, $\overleftrightarrow{EF}$ intersects $\overleftrightarrow{AB}$ in at most one point.

*6. Prove the following. If $\triangle ABC = \triangle DEF$, then each of the lines $\overleftrightarrow{AB}$, $\overleftrightarrow{BC}$, $\overleftrightarrow{AC}$ contains two of the points D, E, and F.

*7. Show that for any $\triangle ABC$, we have $\overleftrightarrow{AB} \cap \triangle ABC = \overline{AB}$.

That is, the only points of $\overleftrightarrow{AB}$ that lie on the triangle are the points of the side $\overline{AB}$.

*8. Prove the following. If $\triangle ABC = \triangle DEF$, then each side of $\triangle ABC$ contains two of the points D, E, and F.

9. Show that A is not between any two points of $\triangle ABC$.

10. Prove Theorem 5.

3.6 CONGRUENCE OF SEGMENTS

The intuitive idea of congruence, for any two figures at all, is always the same. Two figures F and G are *congruent* if one can be moved so as to coincide with the other. Thus two equilateral triangles of the same size are always congruent; two circles of the same radius are always congruent; two squares of the same size are always congruent, and so on.

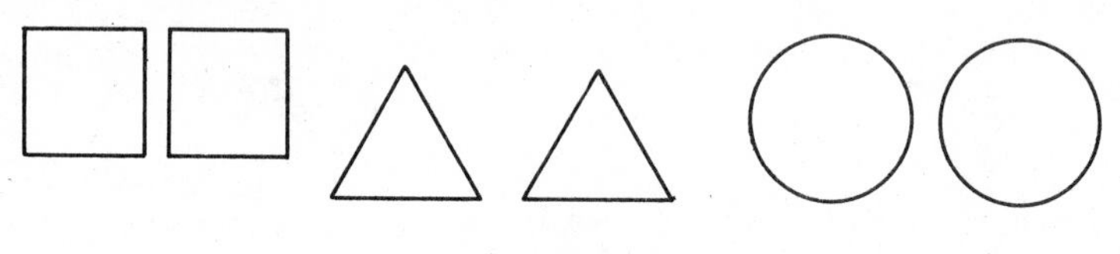

FIGURE 3.13

In the same way, two segments of the same length are always congruent.

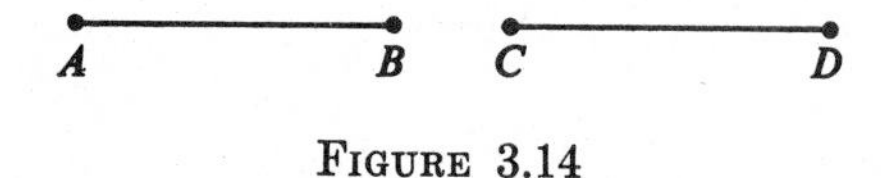

FIGURE 3.14

Here, by the *length* of a segment, we mean the distance between its end points.

Our problem, in our mathematical study of congruence, is to formulate the idea in sufficiently exact form to be able to prove things about it. In the present section, we shall do this for the case in which the figures are segments. Later we shall do the same for the case in which the figures are angles; and still later, we shall discuss triangles. Finally, in the chapter on rigid motion, we shall discuss congruence in a form sufficiently general to apply to *any* two sets of points.

We start with our official definition.

DEFINITION. Let $\overline{AB}$ and $\overline{CD}$ be segments. If $AB = CD$, then the segments are called *congruent*, and we write $\overline{AB} \cong \overline{CD}$.

On the basis of this definition, it is easy to prove the familiar and fairly trivial facts about congruence of segments.

A relation $\sim$, defined on a set A, is called an *equivalence* relation if the following conditions hold.

(1) *Reflexity.* $a \sim a$, for every a.

(2) *Symmetry.* If $a \sim b$, then $b \sim a$.

(3) *Transitivity.* If $a \sim b$ and $b \sim c$, then $a \sim c$.

Theorem C-1. For segments, congruence is an equivalence relation.

That is, every segment is congruent to itself; if $\overline{AB} \cong \overline{CD}$, then $\overline{CD} \cong \overline{AB}$; if $\overline{AB} \cong \overline{CD}$ and $\overline{CD} \cong \overline{EF}$, then $\overline{AB} \cong \overline{EF}$.

Proof?

Theorem C-2. *The Segment-Construction Theorem.* Given a segment $\overline{AB}$ and a ray $\overrightarrow{CD}$. There is exactly one point E of $\overrightarrow{CD}$ such that $\overline{AB} \cong \overline{CE}$.

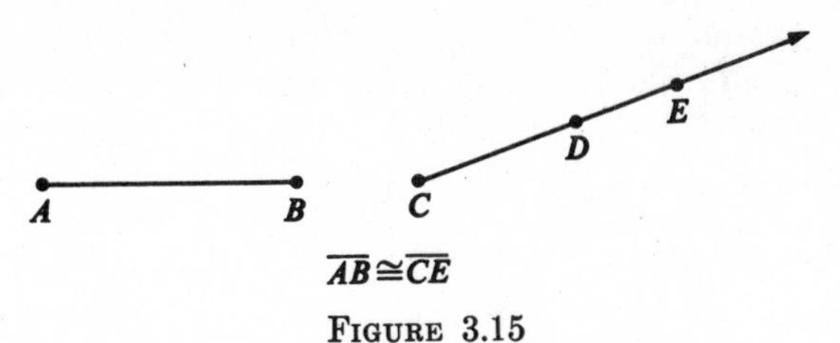

FIGURE 3.15

That is, starting at the end point of a ray, you can measure off a segment of any desired length, and the resulting segment is unique.

Proof. By the ruler placement theorem, set up a coordinate system f for the line $\overleftrightarrow{CD}$, in such a way that $f(C) = 0$ and $f(D) > 0$.

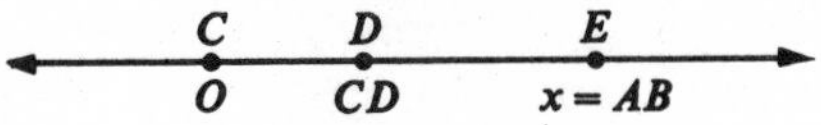

FIGURE 3.16

In the figure, we have indicated that the number CD is the coordinate of the point D, and this is correct, because $f(D) > 0$. If E is a point of $\overrightarrow{CD}$, then $\overline{CE} \cong \overline{AB}$ if and only if $f(E) = AB$ as in the figure. Thus $\overline{CE} \cong \overline{AB}$ if and only if $E = f^{-1}(AB)$. There is exactly one such point $f^{-1}(AB)$, and therefore there is exactly one such point E.

The following theorem says, in effect, that if congruent segments are laid end to end, the resulting segments are congruent.

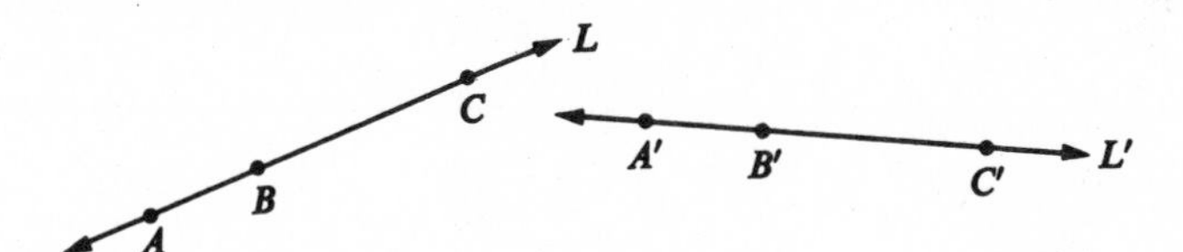

FIGURE 3.17

Theorem C-3. *The Segment-Addition Theorem.* If

(1) A-B-C,

(2) A'-B'-C',

(3) $\overline{AB} \cong \overline{A'B'}$,

and

(4) $\overline{BC} \cong \overline{B'C'}$,

then

(5) $\overline{AC} \cong \overline{A'C'}$.

We also have a converse.

Theorem C-4. *The Segment-Subtraction Theorem.* If (1) A-B-C, (2) A'-B'-C', (3) $\overline{AB} \cong \overline{A'B'}$, and (4) $\overline{AC} \cong \overline{A'C'}$, then (5) $\overline{BC} \cong \overline{B'C'}$.

These theorems can most conveniently be proved by means of the definition of betweenness. You should work out the proofs in full.

Note that we have called Theorem C-4 *a* converse of Theorem C-3, rather than *the* converse of Theorem C-3. The reason is that most theorems have more than one converse (each of which, of course, may or may not be true). For statements of the form $P \Rightarrow Q$, where P and Q are propositions, the situation is simple. The converse of the implication $P \Rightarrow Q$ is the implication $Q \Rightarrow P$. A theorem may, however, be stated in the following form: "If (a), (b), and (c), then (d), (e), and (f)." This says that

$$[(a) \text{ and } (b) \text{ and } (c)] \Rightarrow [(d) \text{ and } (e) \text{ and } (f)].$$

In this case, any statement that you get by interchanging part of the hypothesis and part of the conclusion is called *a* converse. Thus, to get Theorem C-4 from Theorem C-3, we moved (5) into the hypothesis, and moved (4) into the conclusion. Theorem C-3 has three more converses. You should state them and find out which of them are true.

If A-B-C, and $\overline{AB} \cong \overline{BC}$, then B is a *midpoint* of $\overline{AC}$. The following theorem justifies us in referring to B as *the* midpoint.

Theorem C-5. Every segment has exactly one midpoint.

Proof. Given $\overline{AC}$. By the ruler placement theorem, take a coordinate system f, for the line $\overleftrightarrow{AC}$, such that $f(A) = 0$ and $f(C) > 0$.

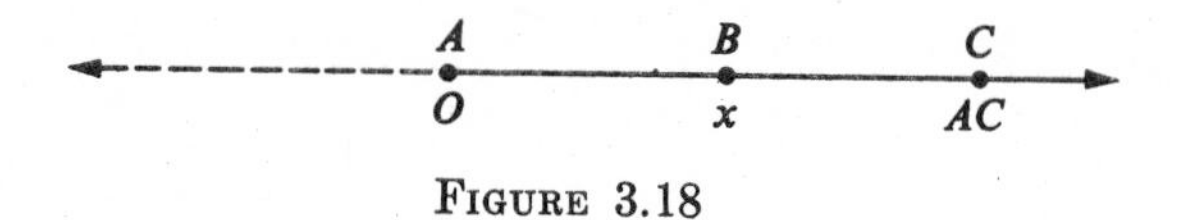

FIGURE 3.18

If B is between A and C, then

$$AB = |x - 0| = x$$

and

$$BC = |AC - x| = AC - x.$$

Thus, for the case where A-B-C, the condition $\overline{AB} \cong \overline{BC}$ is equivalent to the condition

$$x = AC - x,$$

or

$$2x = AC,$$

or

$$x = \frac{AC}{2}.$$

There is exactly one such number x, and therefore there is exactly one such point B.

Chapter 4
Separation in Planes and Space

4.1 CONVEXITY AND SEPARATION

A set A is called *convex* if for every two points P, Q of A, the entire segment $\overline{PQ}$ lies in A. For example, the three figures below are convex:

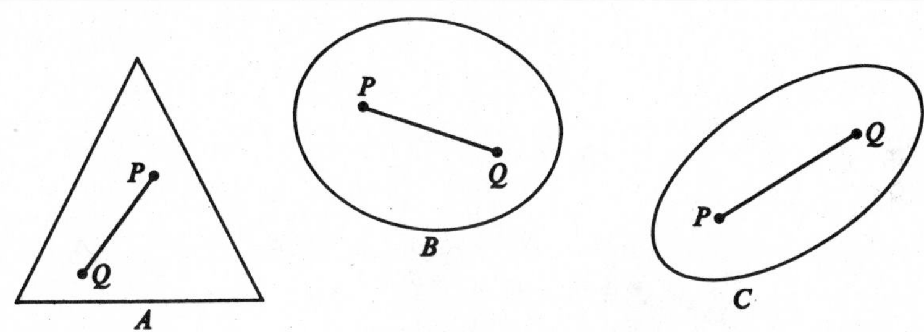

FIGURE 4.1

Here each of the sets A, B, and C is a region in the plane; for example, A is the union of a triangle and the set of all points that lie inside the triangle. We have illustrated the convexity of the sets A, B, and C by drawing in some of the segments $\overline{PQ}$.

On the other hand, none of the sets D, E, and F below are convex:

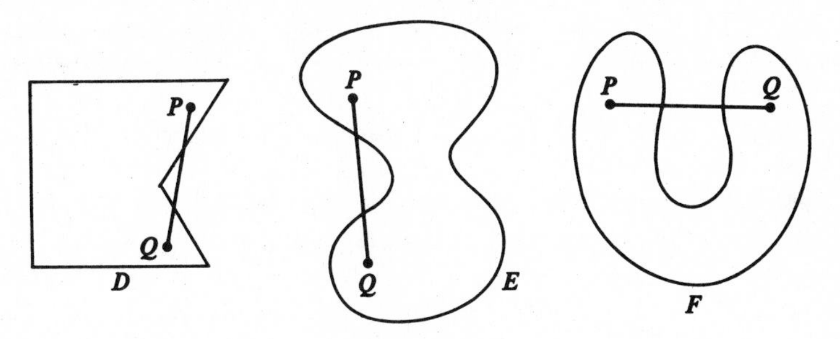

FIGURE 4.2

To show that a set, say D, is *not* convex, you have to show that there are two points P and Q, both belonging to D, such that $\overline{PQ}$ does not lie in D. This is what we have indicated, for each of our last three figures.

A convex set may be "thin and small." For example, every segment $\overline{PQ}$ is a convex set. In fact, a set with only one point is convex. (Since such a set does not contain *any* two points, it follows that *every* two points of it have any property we feel like mentioning.)

A convex set may also be large. For example, the whole space S is a convex set; and all lines and planes are convex. (Proof?) Given a line L in a plane E, the parts of E that lie on the two sides of L are both convex.

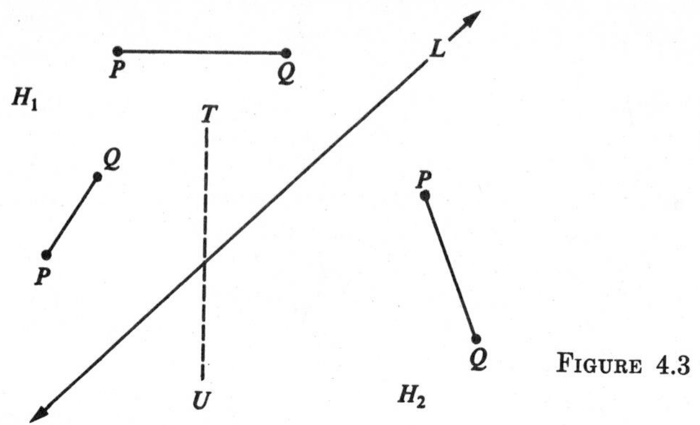

FIGURE 4.3

In Fig. 4.3, H_1 is the part of the plane lying above and to the left of the line L, and H_2 is the part of the plane that lies below and to the right of L. The sets H_1 and H_2 are called *half planes*. As before, we have illustrated their convexity by showing a few sample segments $\overline{PQ}$. We notice, of course, that if T belongs to H_1 and U belongs to H_2, then the segment $\overline{TU}$ always intersects the line. The situation described in this discussion is fundamental in plane geometry. It is covered by the following postulate.

PS-1. *The Plane-Separation Postulate.* Given a line and a plane containing it, the set of all points of the plane that do not lie on the line is the union of two disjoint sets such that (1) each of the sets is convex, and (2) if P belongs to one of the sets and Q belongs to the other, then the segment $\overline{PQ}$ intersects the line.

We can now begin to state definitions based on our postulates. If E and L are a plane and a line, as in the postulate, and H_1 and H_2 are the two sets given by the postulate, then each of the sets H_1 and H_2 is called a *half plane*, and L is called the *edge* of each of them.

Obviously there is no natural way to decide which of the half planes should be mentioned first, but except for this question of order, the two half planes are uniquely determined by E and L. To see this, we observe that if $P \in H_1$, then the points of H_1 are the point P and the points Q such that $\overline{PQ} \cap L = \emptyset$. Similarly, $H_2 = \{Q | Q \in E - L \text{ and } \overline{PQ} \cap L \neq \emptyset\}$.

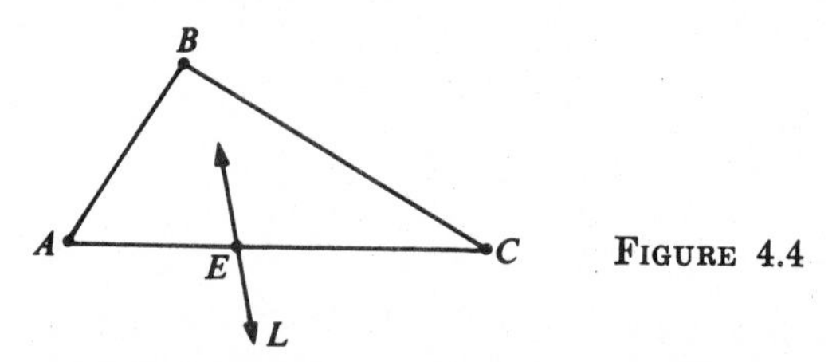

FIGURE 4.4

Theorem 1. *The Postulate of Pasch.* Given a triangle $\triangle ABC$, and a line L in the same plane. If L contains a point E, between A and C, then L intersects either $\overline{AB}$ or $\overline{BC}$.

(In the work of Pasch this statement was used as a postulate, in place of PS-1 above.)

Proof. Suppose not. Then (1) A and B are on the same side of L, and (2) B and C are on the same side of L. Therefore (3) A and C are on the same side of L. This is impossible, because A-E-C.

PROBLEM SET 4.1

Prove the following theorems. In all these theorems, it should be understood that E, L, H_1, and H_2 are a plane, a line, and the two half planes given by PS-1. The proofs are of a sort that may not be at all familiar to you. Obviously, PS-1 uses the idea of a segment, and segments were defined in terms of betweenness. Therefore the chances are that you will have to appeal to a fair number of postulates (and theorems) other than PS-1. Use of the ruler postulate or the ruler placement theorem is not allowed; the proofs should use, instead, the theorems B-1 through B-5 which were based on them.

1. **Theorem 2.** The sets H_1 and H_2 are not both empty.

2. **Theorem 3.** Neither of the sets H_1 and H_2 is empty.

3. **Theorem 4.** H_1 contains at least two points.

4. **Theorem 5.** H_1 contains at least three noncollinear points.

5. **Theorem 6.** E is uniquely determined by H_1. That is, every half plane lies in only one plane.

6. **Theorem 7.** L is uniquely determined by H_1. That is, every half plane has only one edge.

7. **Theorem 8.** If A and B are convex, then so also is $A \cap B$.

8. **Theorem 9.** If G is any collection of convex sets g_i, then the intersection of all of the sets g_i in the collection is convex.

The *convex hull* of a set A is the intersection of all convex sets that contain A.

9. **Theorem 10.** If A is any set of points, then the convex hull of A is convex.

10. **Theorem 11.** $H_1 \cup L$ is convex.

11. **Theorem 12.** Every ray is convex.

12. Let A be a set of points. Let B be the union of all segments of the form $\overline{PQ}$, where P and Q belong to A. Does it follow that B is convex? Why or why not?

13. **Theorem 13.** Given a triangle $\triangle ABC$, and a line L in the same plane. If L contains no vertex of the triangle, then L cannot intersect all of the three sides.

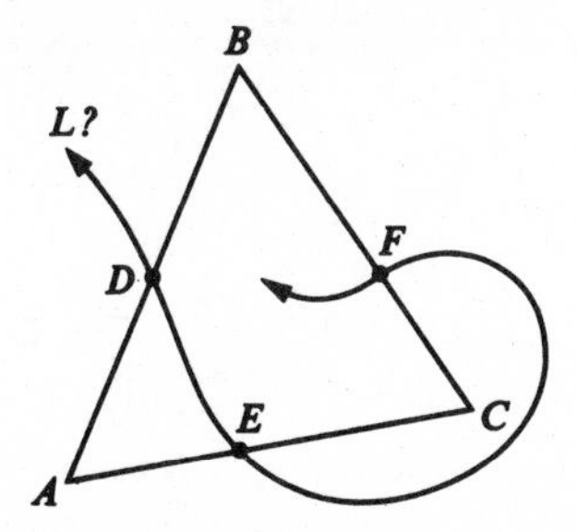

FIGURE 4.5

14. Show that if the Postulate of Pasch is taken as a postulate, then Theorem 13 can be proved as a theorem.

15. Show that if the Postulate of Pasch is used as a postulate, then PS-1 can be proved as a theorem.

4.2 INCIDENCE THEOREMS BASED ON THE PLANE-SEPARATION POSTULATE

If
$$E - L = H_1 \cup H_2,$$
as in the plane-separation postulate, then we say that the sets H_1 and H_2 are *half planes of* L or *sides of* L. Notice that every line has two sides in every plane that contains it, but if P and Q are on the same side of a line L, this automatically means that L, P, and Q are coplanar. On the other hand, to say that P and Q are on *different* sides of L, in space, may mean merely that no one plane contains L, P, and Q. If $E - L = H_1 \cup H_2$, as in the plane-separation postulate, then H_1 and H_2 are called *opposite* sides of L; and if P belongs to H_1 and Q belongs to H_2, we say that P and Q are on *opposite* sides of L.

The following two theorems are easy.

Theorem 1. If P and Q are on opposite sides of the line L, and Q and T are on opposite sides of L, then P and T are on the same side of L.

Theorem 2. If P and Q are on opposite sides of the line L, and Q and T are on the same side of L, then P and T are on opposite sides of L.

We use a similar terminology for the "sides of a point" on a line. That is, if A-B-C, then the rays $\overrightarrow{BA}$ and $\overrightarrow{BC}$ are called *opposite rays*.

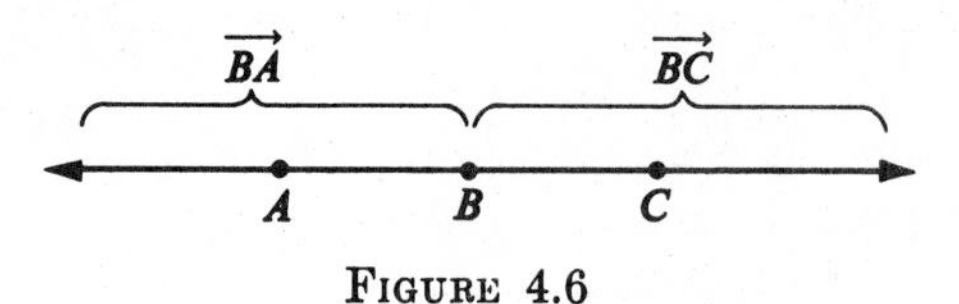

FIGURE 4.6

Theorem 3. Given a line, and a ray which has its end point on the line but does not lie on the line. Then all points of the ray, except for the end point, are on the same side of the line.

Proof. Let L be the line, and let $\overrightarrow{AB}$ be the ray, with $A \in L$.

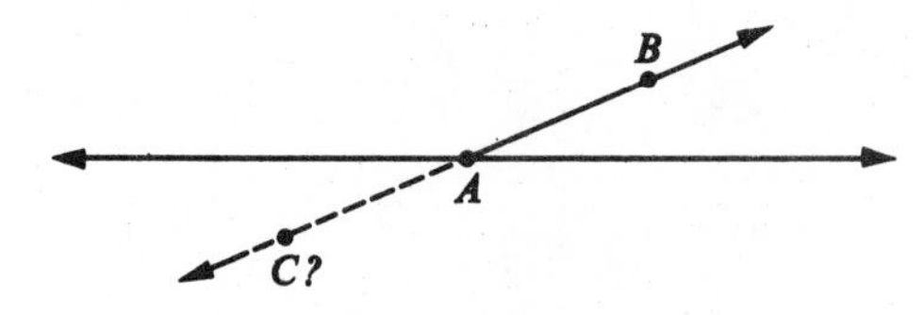

FIGURE 4.7

Suppose that $\overrightarrow{AB}$ contains a point C such that B and C are on opposite sides of L (in the plane that contains L and $\overleftrightarrow{AB}$). Then $\overline{BC}$ intersects L in some point, and this point must be A, because $\overline{BC}$ lies in $\overleftrightarrow{AB}$, and $\overleftrightarrow{AB}$ intersects L only in A. Therefore C-A-B. But this is impossible. By definition, the ray $\overrightarrow{AB}$ is the set of all points C of the line AB for which it is *not* true that C-A-B. Therefore all points of the ray, other than A, are on the same side of L, namely, the side that contains B.

Similarly for segments:

Theorem 4. Let L be a line, let A be a point of L, and let B be a point not on L. Then all points of $\overline{AB} - A$ lie on the same side of L.

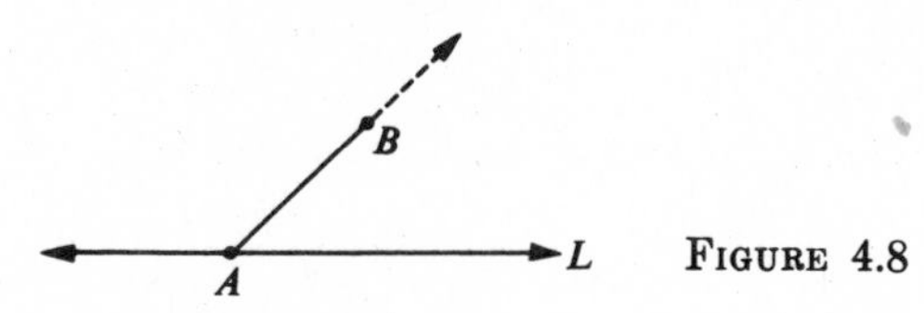

FIGURE 4.8

This is true because $\overline{AB} - A$ lies in $\overrightarrow{AB} - A$.

Given $\angle BAC$.

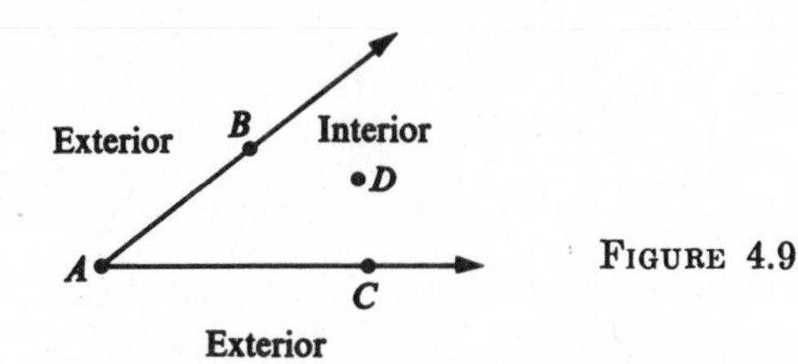

FIGURE 4.9

Roughly speaking, the interior of the angle is the set of all points that lie inside it, and the exterior is the set of all points that lie outside it. We can make this idea precise in the following way.

The *interior* of $\angle BAC$ is the intersection of the side of $\overleftrightarrow{AC}$ that contains B, and the side of $\overleftrightarrow{AB}$ that contains C. Thus a point D lies in the interior (1) if D and B are on the same side of $\overleftrightarrow{AC}$, and (2) if D and C are on the same side of $\overleftrightarrow{AB}$.

For this definition to be valid, it has to depend only on the *angle* that we started with, and not on the points B and C that we happened to choose to describe the angle. Thus, in the figure below, it would be sad if our definition gave us two different interiors for $\angle B'AC'$ and $\angle BAC$:

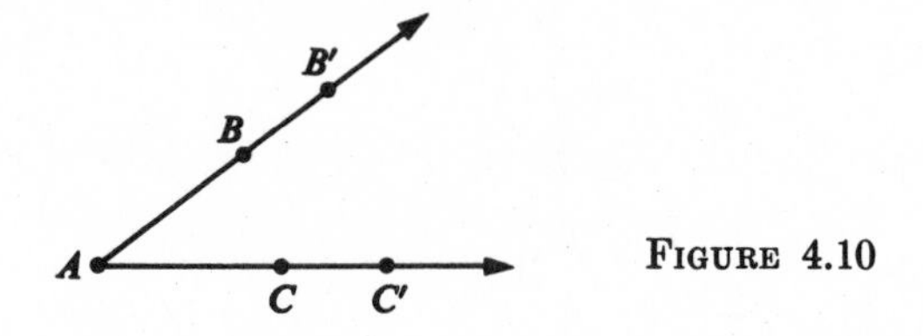

FIGURE 4.10

Theorem 3 shows, however, that our definition depends only on the angle, because B and B' are on the same side of $\overleftrightarrow{AC}$, and C and C' are on the same side of $\overleftrightarrow{AB}$.

Given an angle $\angle ABC$, there is exactly one plane E that contains it. The *exterior* of the angle is the set of all points of E that lie neither on the angle nor in its interior.

Theorem 5. Every side of a triangle lies, except for its end points, in the interior of the opposite angle.

Here we are using the ordinary terminology; that is, in $\triangle ABC$, the angle $\angle A = \angle BAC$ is *opposite* the side $\overline{BC}$.

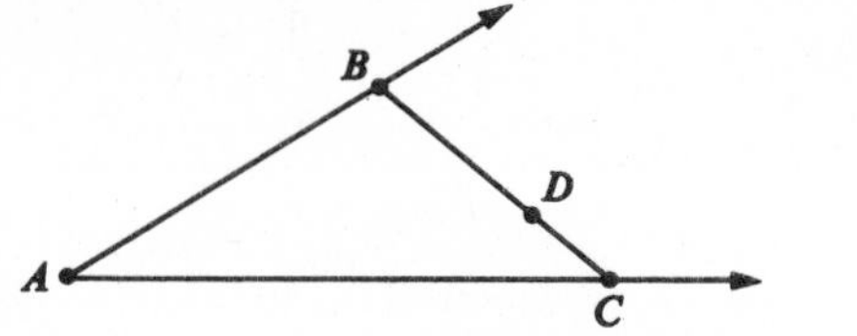

FIGURE 4.11

Proof. (1) First we apply Theorem 4 to the line $\overleftrightarrow{AC}$ and the segment $\overline{BC}$. By Theorem 4, $\overline{BC} - C$ lies on the side of $\overleftrightarrow{AC}$ that contains B.

(2) Next we apply Theorem 4 to the line $\overleftrightarrow{AB}$ and the segment $\overline{BC}$. By Theorem 4, $\overline{BC} - B$ lies on the side of $\overleftrightarrow{AB}$ that contains C.

(3) By (1) and (2), $\overline{BC} - \{B, C\}$ lies in the interior of $\angle BAC$.

Theorem 6. If F is in the interior of $\angle BAC$, then $\overrightarrow{AF} - F$ lies in the interior of $\angle BAC$.

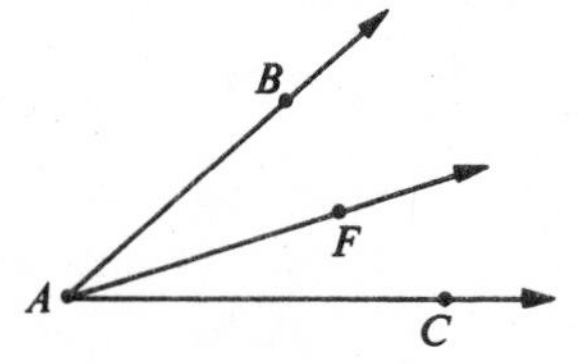

FIGURE 4.12

Proof. (1) By definition of the interior of an angle, F and B are on the same side of $\overleftrightarrow{AC}$. By Theorem 3, $\overrightarrow{AF} - F$ lies on the side of $\overleftrightarrow{AC}$ that contains F. Therefore $\overrightarrow{AF} - F$ lies on the side of $\overleftrightarrow{AC}$ that contains B.

(2) By definition of the interior of an angle, F and C are on the same side of $\overleftrightarrow{AB}$. By Theorem 3, $\overrightarrow{AF} - F$ lies on the side of $\overleftrightarrow{AB}$ that contains F. Therefore $\overrightarrow{AF} - F$ lies on the side of $\overleftrightarrow{AB}$ that contains C.

By (1) and (2) it follows that $\overrightarrow{AF} - F$ lies in the interior of $\angle BAC$.

Theorem 7. Let $\triangle ABC$ be a triangle, and let F, D, and G be points, such that B-F-C, A-C-D, and A-F-G. Then G is in the interior of $\angle BCD$.

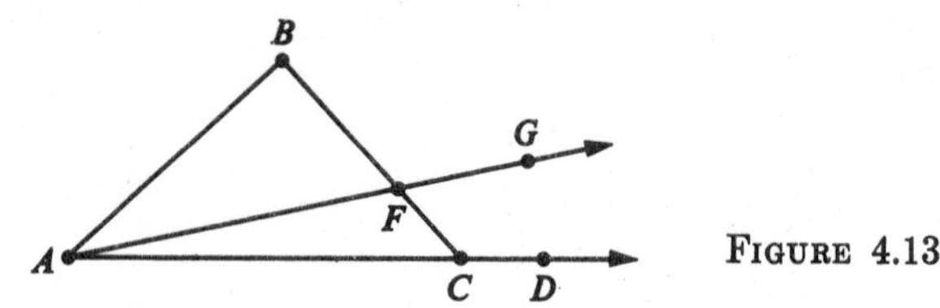

FIGURE 4.13

Proof. (1) Since A-F-G, G lies on $\overleftrightarrow{AF}$, and A is not between G and F. Therefore G lies on $\overrightarrow{AF}$. Since $G \neq A$, G lies on $\overrightarrow{AF} - A$.

(2) By Theorem 5, F is in the interior of $\angle BAC$. It follows by Theorem 6 that $\overrightarrow{AF} - A$ lies in the interior of $\angle BAC$. Therefore G and B lie on the same side of $\overleftrightarrow{AC}$ (= $\overleftrightarrow{CD}$.)

(3) A and G are on opposite sides of $\overleftrightarrow{BC}$, and A and D are on opposite sides of $\overleftrightarrow{BC}$. Therefore G and D are on the same side of $\overleftrightarrow{BC}$.

By (2) and (3), G is in the interior of $\angle BCD$.

Throughout this section, we have been using figures to help us keep track of what is going on. (Here *us*, of course, includes the author. All authors use figures, whether or not they show these to the reader.) You should watch closely, however, to be sure that the figures are playing merely their legitimate part as memoranda. It is customary, in elementary texts, for the reader to be assured that "the proofs do not depend on the figure," but these promises are almost never kept. (Whether such promises *ought* to be kept, in an elementary course, is another question, and the answer should probably be "No.") In a mathematically thorough treatment, however, the hypothesis and conclusion ought to be stated in such a way that no figure is actually necessary to make them plain; and in the same way, the proofs ought to rest on the postulates and the previous theorems. This point is especially relevant in the present context, because in most informal treatments of geometry it is customary to convey betweenness relations and separation properties *only* by figures, without ever mentioning them in words at all.

You may be able to remember a situation, in elementary geometry, where Theorem 7 is needed.

The *interior* of $\triangle ABC$ is defined as the intersection of the following three sets:

(1) The side of $\overleftrightarrow{AB}$ that contains C.

(2) The side of $\overleftrightarrow{AC}$ that contains B.

(3) The side of $\overleftrightarrow{BC}$ that contains A.

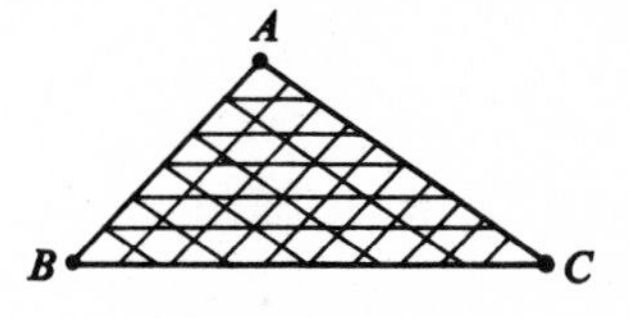

FIGURE 4.14

Theorem 8. The interior of a triangle is always a convex set.

Proof?

Theorem 9. The interior of a triangle is the intersection of the interiors of its angles.

Proof?

4.3 INCIDENCE THEOREMS CONTINUED

In the figure below, D is supposed to be in the interior of $\angle BAC$.

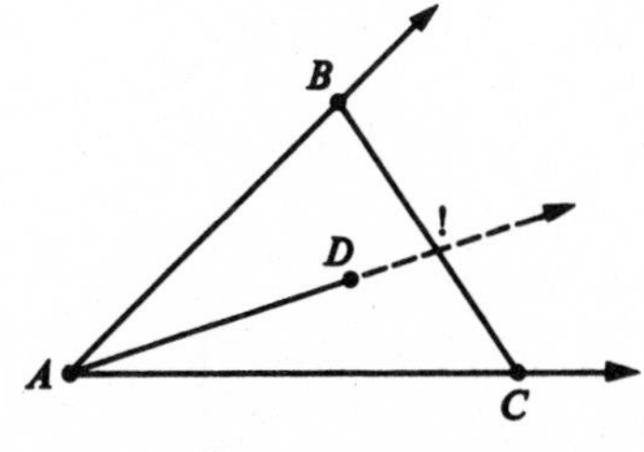

FIGURE 4.15

It is intuitively clear that $\overrightarrow{AD}$ must intersect $\overline{BC}$, as the figure suggests. But it is not obvious that this can be proved on the basis of the postulates that we have stated so far, and in fact the proof is hard. We shall need some preliminary results.

Theorem 1. Let L be a line, let A and F be two (different) points of L, and let B and G be points on opposite sides of L. Then $\overline{FB}$ does not intersect $\overrightarrow{AG}$.

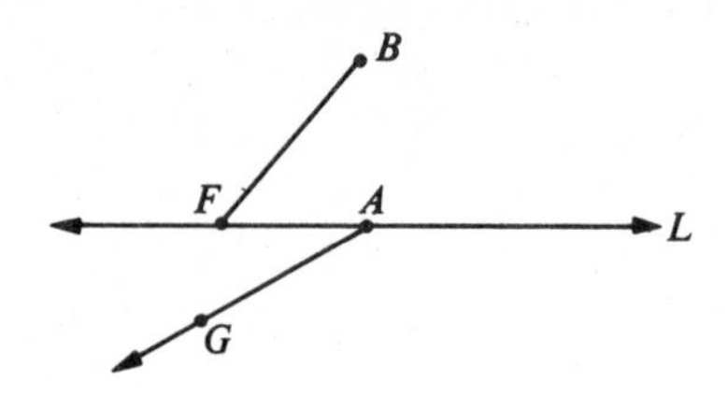

FIGURE 4.16

Proof. (1) By Theorem 3 of Section 4.2, $\overrightarrow{AG} - A$ lies on the side of L that contains G.

(2) By Theorem 4 of Section 4.2, $\overline{FB} - F$ lies on the side of L that contains B.

(3) By (1) and (2) it follows that $\overrightarrow{AG} - A$ does not intersect $\overline{FB} - F$. Therefore $\overline{FB}$ and $\overrightarrow{AG}$ cannot intersect, except possibly at F or A. But this is not possible: A does not lie on $\overline{FB}$, and F does not lie on $\overrightarrow{AG}$. The theorem follows.

The following theorem is a stronger form of the Postulate of Pasch.

Theorem 2. In $\triangle FBC$, let A be a point between F and C, and let D be a point such that D and B are on the same side of $\overleftrightarrow{FC}$. Then $\overrightarrow{AD}$ intersects either $\overline{FB}$ or $\overline{BC}$.

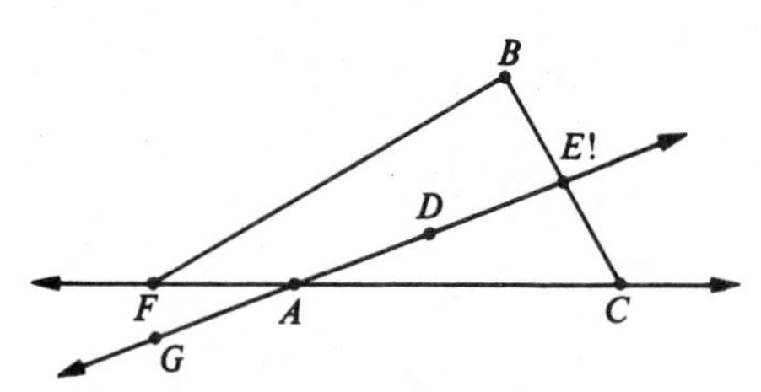

FIGURE 4.17

Proof. (1) Let G be a point such that G-A-D. Then G and D are on opposite sides of $\overleftrightarrow{FC}$, and so G and B are on opposite sides of $\overleftrightarrow{FC}$. Evidently

$$\overleftrightarrow{AD} = \overrightarrow{AD} \cup \overrightarrow{AG}.$$

(2) We apply Theorem 1 to the line $\overleftrightarrow{FC}$, the segment $\overline{FB}$, and the ray $\overrightarrow{AG}$. It follows that $\overrightarrow{AG}$ does not intersect $\overline{FB}$.

(3) In exactly the same way, we conclude that $\overrightarrow{AG}$ does not intersect $\overline{BC}$.

(4) We know by the Postulate of Pasch that the *line* $\overleftrightarrow{AD}$ intersects either $\overline{FB}$ or $\overline{BC}$. Since $\overrightarrow{AG}$ intersects neither of these segments, it follows that $\overrightarrow{AD}$ intersects one of them, which was to be proved.

Theorem 3. *The Crossbar Theorem.* If D is in the interior of $\angle BAC$, then $\overrightarrow{AD}$ intersects $\overline{BC}$, in a point between B and C.

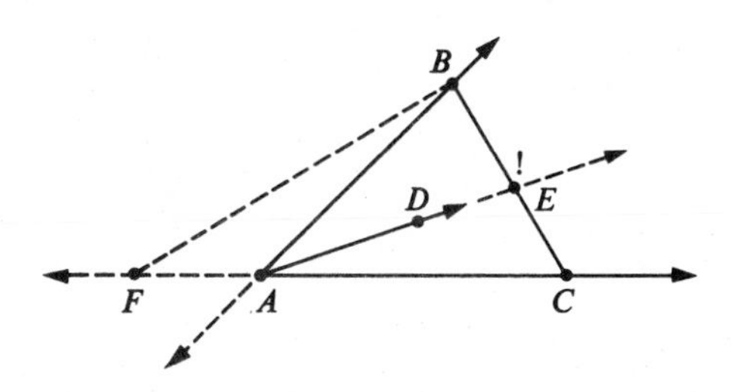

FIGURE 4.18

Proof. (1) Let F be a point such that F-A-C. Then $\overleftrightarrow{FC} = \overleftrightarrow{AC}$, and F and C are on opposite sides of $\overleftrightarrow{AB}$.

(2) Since D is in the interior of $\angle BAC$, it follows that B and D are on the same side of $\overleftrightarrow{AC}$ $(= \overleftrightarrow{FC}.)$ It follows by Theorem 2 that $\overrightarrow{AD}$ intersects either $\overline{FB}$ or $\overline{BC}$.

(3) Since F and C are on opposite sides of $\overleftrightarrow{AB}$, and C and D are on the same side of $\overleftrightarrow{AB}$, it follows that F and D are on opposite sides of $\overleftrightarrow{AB}$.

(4) We now apply Theorem 1 to the line $\overleftrightarrow{AB}$, the segment $\overline{FB}$, and the ray $\overrightarrow{AD}$. By Theorem 1, $\overrightarrow{AD}$ does not intersect $\overline{FB}$.

(5) From (2) and (4) it follows that $\overrightarrow{AD}$ intersects $\overline{BC}$, in a point E which is different from B. If $E = C$, then A, D, and C are collinear, which is false. Therefore B-E-C, which was to be proved.

You may remember that one of the first theorems in plane geometry that most people learn is the one dealing with the base angles of an isosceles triangle. These are always "equal," that is, *congruent* in the sense in which we shall define the latter term later in this book. That is, if $\overline{AB} \cong \overline{AC}$, then $\angle B \cong \angle C$.

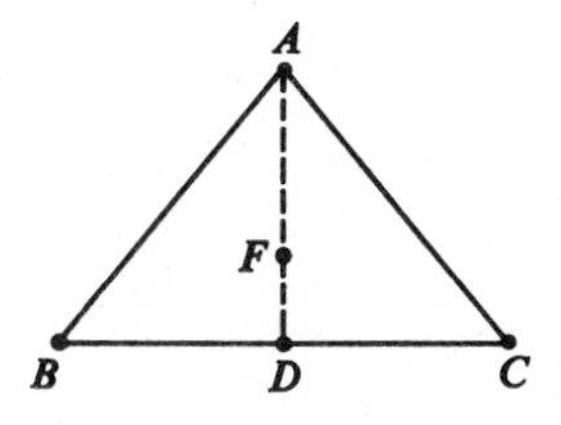

FIGURE 4.19

Although a good proof of the theorem was known in antiquity, it has become customary in later centuries to prove it in needlessly complicated ways; and probably the worst of these rambling detours is the proof that starts by telling you (1) to bisect $\angle BAC$, (2) to "let" D be the point where the bisecting ray $\overrightarrow{AF}$ intersects the base, and (3) to show that $\triangle ADB$ and $\triangle ADC$ are congruent.

Of course, a light-hearted use of the word *let* is no substitute for a proof that $\overrightarrow{AF}$ intersects $\overline{BC}$. This method of proof thus depends essentially on the crossbar theorem. We shall see, however, that for the simple theorem that we have been discussing, the crossbar theorem is not needed.

Problem Set 4.3

1. **Theorem 4.** Given a triangle, and a line lying in the same plane. If the line intersects the interior of the triangle, then it intersects at least one of the sides.

4.4 CONVEX QUADRILATERALS

Given four points A, B, C, and D, such that they all lie in the same plane, but no three are collinear. If the segments $\overline{AB}$, $\overline{BC}$, $\overline{CD}$, and $\overline{DA}$ intersect only at their end points, then their union is called a *quadrilateral*, and is denoted by $\square ABCD$. This notation is not meant to suggest that every quadrilateral is a square, any more than the analogous notation $\triangle ABC$ is meant to suggest that every triangle is equilateral.

The *angles* A, B, C, D of $\square ABCD$ are $\angle DAB$, $\angle ABC$, and so on.

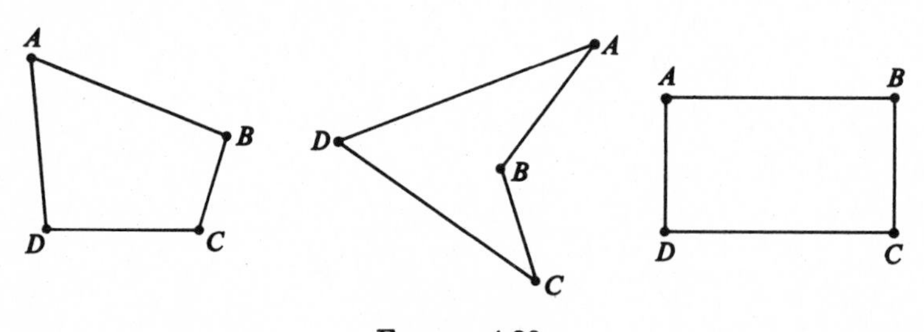

FIGURE 4.20

The *sides* of $\square ABCD$ are $\overline{AB}$, $\overline{BC}$, and so on. Two sides which have a common end point are called *adjacent*; two sides which are not adjacent are *opposite*. Two angles of a quadrilateral are *adjacent* if their intersection contains a side; and two angles which are not adjacent are *opposite*.

The diagonals of $\square ABCD$ are the segments $\overline{AC}$ and $\overline{BD}$.

A quadrilateral is called *convex* if each of its sides lies in one of the half planes determined by the opposite side. Note that if A and B are on the same side of $\overleftrightarrow{CD}$, then all points of $\overline{AB}$ are on the same side of $\overleftrightarrow{CD}$. (The converse is trivial.)

Thus $\square ABCD$ is a convex quadrilateral if and only if all four of the following conditions hold.

(1) A and B are on the same side of $\overleftrightarrow{CD}$.

(2) B and C are on the same side of $\overleftrightarrow{DA}$.

(3) C and D are on the same side of $\overleftrightarrow{AB}$.

(4) D and A are on the same side of $\overleftrightarrow{BC}$.

Note that the use of the word *convex* in geometry is inconsistent; no quadrilateral can possibly form a *convex set* in the sense in which the latter term was defined in Section 4.1. This usage is universal, however.

The following theorem is widely known but seldom proved.

Theorem 1. The diagonals of a convex quadrilateral always intersect each other.

Proof. Let $\square ABCD$ be a convex quadrilateral. We need to show that $\overline{AC}$ intersects $\overline{BD}$.

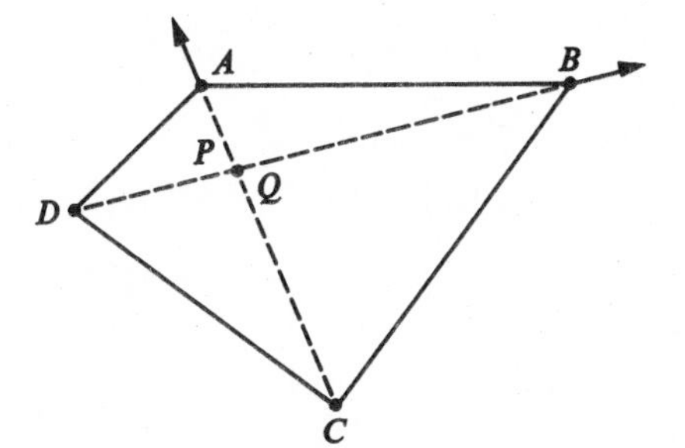

FIGURE 4.21

By conditions (1) and (2) above, we conclude that B is in the interior of $\angle ADC$. Therefore, by the crossbar theorem, $\overrightarrow{DB}$ intersects $\overline{AC}$ at a point P.

Similarly, by conditions (1) and (4) we conclude that A is in the interior of $\angle BCD$. Therefore, by the crossbar theorem, $\overrightarrow{CA}$ intersects $\overline{BD}$ at a point Q.

Since each of these rays and segments lies in the corresponding line, it follows that $\overleftrightarrow{DB}$ intersects $\overleftrightarrow{AC}$ at P and also at Q. Therefore $P = Q$. Since P lies on $\overline{AC}$ and Q lies on $\overline{BD}$, it follows that $\overline{AC}$ and $\overline{BD}$ have a point in common, which was to be proved.

If you review this chapter and observe how much of it is needed for the proof of the above theorem, it will be obvious to you why the proof is commonly omitted in elementary treatments.

PROBLEM SET 4.4

1. Prove the converse of Theorem 1. That is, show that if the diagonals of a quadrilateral intersect each other, then the quadrilateral is convex.

2. Show that if each vertex of a quadrilateral lies in the interior of the opposite angle, then the quadrilateral is convex. (In $\square ABCD$, if the vertex is A, then the opposite angle is $\angle BCD$. Similarly for the other vertices.)

3. Show that for any quadrilateral (convex or not), the lines containing the diagonals always intersect.

4. Show that every quadrilateral (convex or not) has a side $\overline{XY}$ such that the other two vertices lie on the same side of the line $\overleftrightarrow{XY}$.

5. Prove the converse of the theorem stated in Problem 2.

4.5 SEPARATION OF SPACE BY PLANES

The behavior of planes in space, with regard to separation properties, is very closely analogous to the behavior of lines in a plane. We therefore merely state the basic postulate, definitions, and theorems, and leave the verifications to the reader.

> **SS-1.** *The Space-Separation Postulate.* Given a plane in space. The set of all points that do not lie in the plane is the union of two sets H_1, H_2 such that (1) each of the sets is convex, and (2) if P belongs to one of the sets and Q belongs to the other, then the segment $\overline{PQ}$ intersects the plane.

The two sets H_1, H_2 described in SS-1 are called *half spaces*, or *sides* of the plane E, and E is called the *face* of each of them. As in the case of half planes, there is no natural way to decide which of them should be mentioned first; but except for order, the sets H_1 and H_2 are uniquely determined by E. The reason is that if $P \in H_1$, then

$$H_1 = \{P \cup Q | P = Q \quad \text{or} \quad \overline{PQ} \cap E = \emptyset\},$$

and

$$H_2 = \{Q | \overline{PQ} \cap E \neq \emptyset\}.$$

The following theorems are merely the appropriate revisions of some of the theorems in Section 4.1.

Theorem 1. The sets H_1 and H_2 are not both empty.

Theorem 2. Neither of the sets H_1, H_2 is empty.

Theorem 3. Each of the sets H_1 and H_2 contains at least four noncoplanar points.

Theorem 4. E is uniquely determined by H_1. That is, every half space has only one face.

A *dihedral angle* is a figure that looks like this:

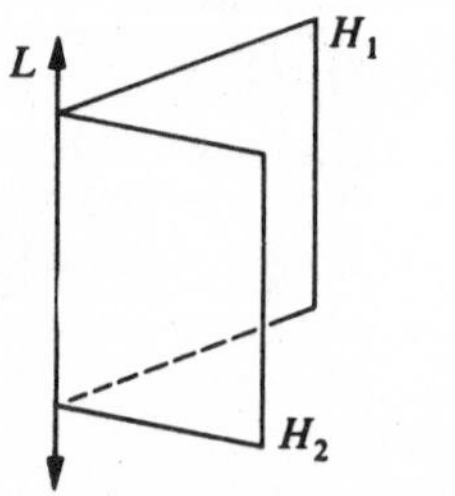

FIGURE 4.22

More precisely, if two half planes H_1 and H_2 have the same edge L, but do not lie in the same plane, then the set $H_1 \cup H_2 \cup L$ is called a *dihedral angle*. The line L is called the *edge* of the dihedral angle, and the sets $H_1 \cup L$ and $H_2 \cup L$ are called its *sides*. (Note that just as the sides of an angle contain their common end point, so the sides of a dihedral angle contain their common edge.)

The following theorem is analogous to Theorem 3, Section 4.2.

Theorem 5. Let H be a half plane with edge L, and let E be a plane which contains L but not H. Then all points of H are on the same side of E.

Given a dihedral angle

$$D = H_1 \cup H_2 \cup L.$$

Let E_1 and E_2 be the planes that contain H_1 and H_2, respectively. Then the *interior* of D is the intersection of (1) the side of E_1 that contains H_2 and (2) the side of E_2 that contains H_1.

Theorem 6. The interior of a dihedral angle is always a convex set.

Theorem 7. If P and Q are in different sides of a dihedral angle, then every point between P and Q is in the interior of the dihedral angle.

This section, of course, is hardly more than an introduction to the following problems.

Problem Set 4.5

1. Prove Theorems 1 through 7.

Chapter 5
Angular Measure

5.1 DEGREE MEASURE FOR ANGLES

You will recall that when we started doing geometry, we began with the structure

$$[S, \mathcal{L}, \mathcal{P}].$$

Later we included in the structure the distance function

$$d: S \times S \to \mathbf{R}.$$

This gave us the structure

$$[S, \mathcal{L}, \mathcal{P}, d].$$

In terms of distances between points, we defined betweenness, and also congruence for segments.

We now complete the structure by introducing measure for angles. This will turn out to be the familiar degree measure. The situation here is closely analogous to that for distance. We could equally well use radians, or any constant multiple of degrees or of radians; but since all of these measures for angles behave in essentially the same way, we may as well simplify the discussion by choosing one of them, once for all, and using it consistently thereafter. (In analysis, radian measure is obligatory, and degree measure is out of the question, but in elementary geometry one unit of measure is as good as another.)

Angular measure is going to be a function m, defined for angles, with real numbers as values of the function. Let $\mathcal{A}$ be the set of all angles. We shall study the structure

$$[S, \mathcal{L}, \mathcal{P}, d, m],$$

where

$$m: \mathcal{A} \to \mathbf{R}$$

is a function of the angles into the real numbers. In the usual functional notation, we would write

$$m(\angle ABC)$$

to denote the measure of $\angle ABC$, but since no confusion with multiplication could possibly occur, we omit the parentheses and write merely

$$m\angle ABC.$$

Since we shall be talking about only one measure function for angles, we can write merely

$$m\angle ABC = 90, \qquad m\angle DBC = 45.$$

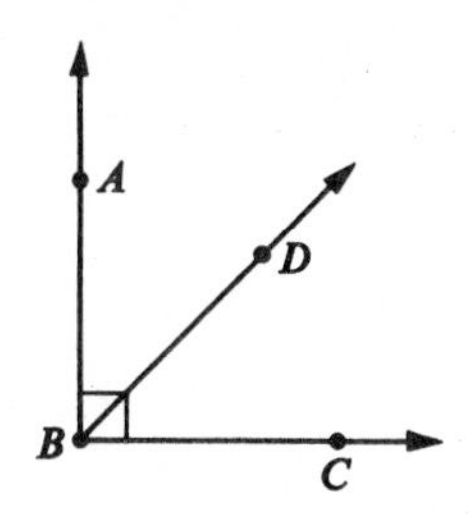

FIGURE 5.1

We do not write $m\angle ABC = 90°$, because the values of the function m are simply real numbers; they stand alone and don't need to carry little flags to indicate where they came from. On the other hand, in labeling figures, it is convenient to use the degree sign merely to indicate that certain letters or numbers are meant to be the degree measures of angles.

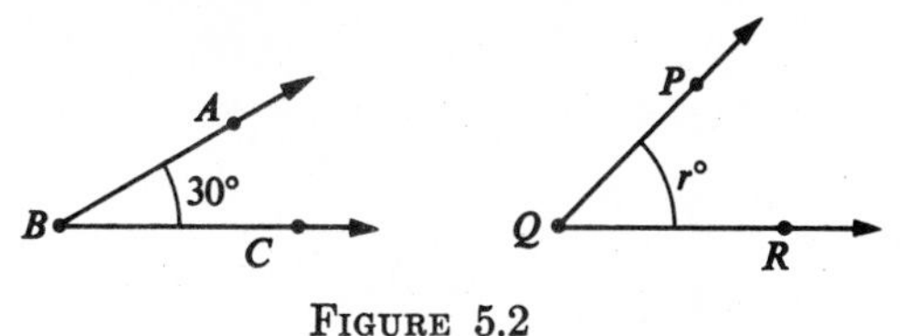

FIGURE 5.2

The figures above tell us that

$$m\angle ABC = 30,$$

and

$$m\angle PQR = r.$$

The postulates governing the function m are merely abstract descriptions of the familiar properties of protractors. With a protractor placed with its edge on the edge of the half plane H, as in the figure below, we can read off the measures of a large number of angles.

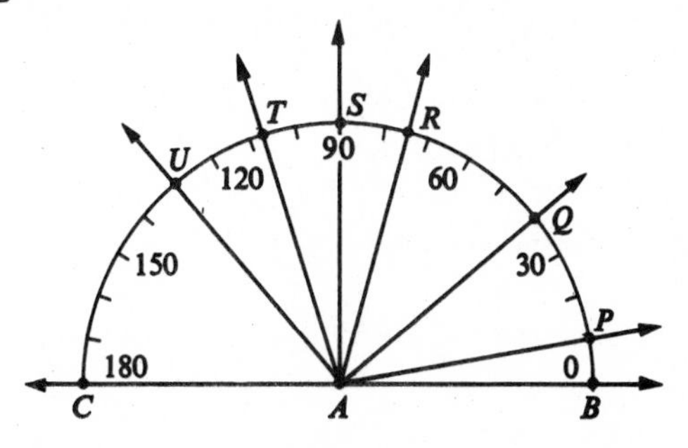

FIGURE 5.3

For example,

$$m\angle PAB = 10,$$
$$m\angle QAB = 40,$$
$$m\angle RAB = 75.$$

By subtraction, we also get

$$m\angle QAP = 40 - 10 = 30,$$
$$m\angle SAR = 90 - 75 = 15,$$
$$m\angle CAU = 180 - 130 = 50,$$

and so on. These and other uses of a protractor are reflected in the following postulates.

M-1. m is a function $\mathcal{A} \to \mathbf{R}$, where $\mathcal{A}$ is the set of all angles, and $\mathbf{R}$ is the set of all real numbers.

M-2. For every angle $\angle A$, $m\angle A$ is between 0 and 180.

M-3. *The Angle-Construction Postulate.* Let $\overrightarrow{AB}$ be a ray on the edge of the half plane H. For every number r between 0 and 180, there is exactly one ray $\overrightarrow{AP}$, with P in H, such that $m\angle PAB = r$.

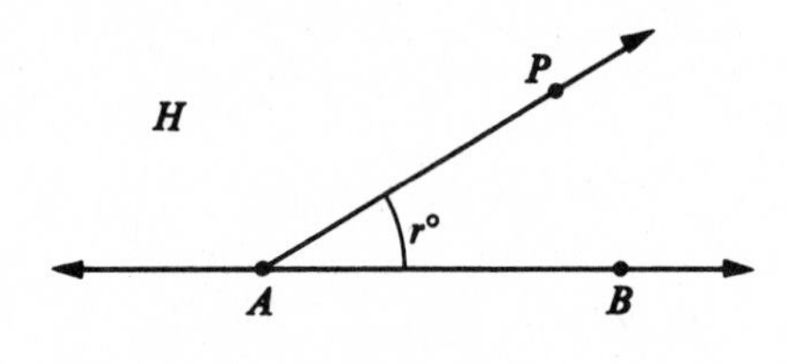

FIGURE 5.4

M-4. *The Angle-Addition Postulate.* If D is in the interior of $\angle BAC$, then

$$m\angle BAC = m\angle BAD + m\angle DAC.$$

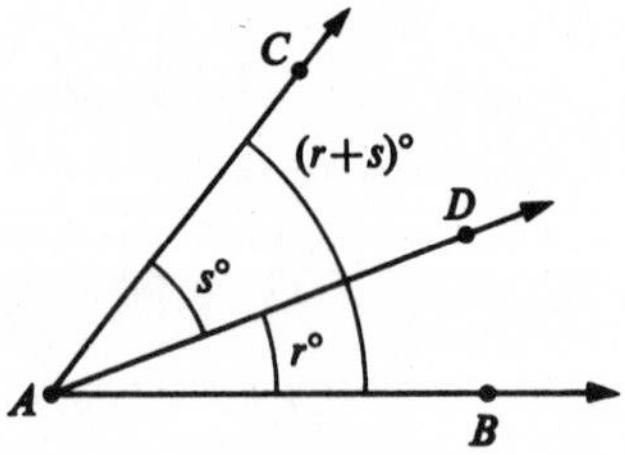

FIGURE 5.5

(This, of course, is the property of m that we use when we compute the measure of an angle by subtraction.)

Two angles form a *linear pair* if they look like this:

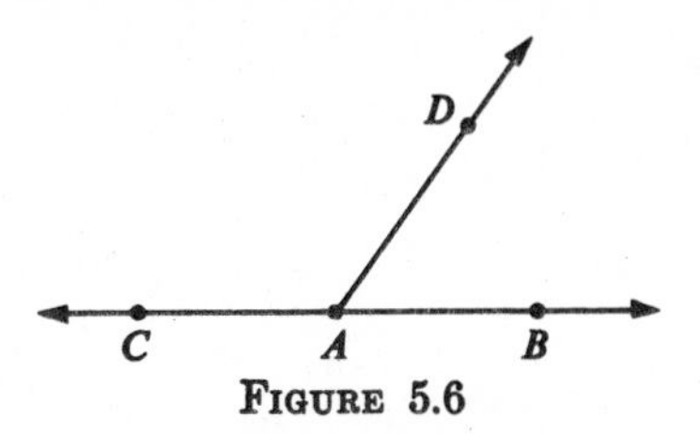

FIGURE 5.6

That is, if $\overrightarrow{AB}$ and $\overrightarrow{AC}$ are opposite rays, and $\overrightarrow{AD}$ is any third ray, then $\angle DAB$ and $\angle DAC$ form a linear pair. If $m\angle ABC + m\angle DEF = 180$, then the two angles are called *supplementary*. Notice that this definition says nothing at all about where the angles are; it deals only with their measures.

M-5. *The Supplement Postulate.* If two angles form a linear pair, then they are supplementary.

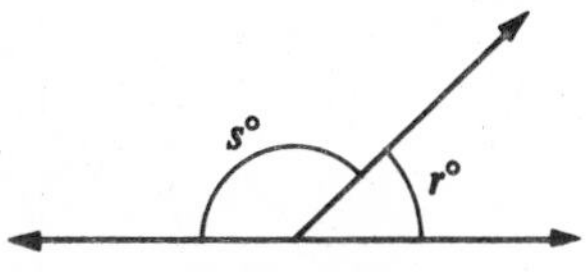

$s°$ $r°$

$r+s=180$

FIGURE 5.7

Just as we defined congruence for segments in terms of distance, so we define congruence for angles in terms of measure. That is, if

$$m\angle ABC = m\angle DEF,$$

then the angles are *congruent*, and we write

$$\angle ABC \cong \angle DEF.$$

If the angles in a linear pair are congruent, then each of them is called a *right* angle. If $\angle ABC$ is a right angle, and $m\angle ABC = r$, then, of course, we have $r = 90$; the reason is that since the angles in a linear pair are always supplementary, we must have $r + r = 180$. The converse is also true (and easy). Thus an angle is a right angle if and only if its measure is 90.

The following theorems are closely analogous to our first few theorems on congruence for segments. To see the analogy, we observe that a sort of "betweenness" relation can be defined for rays with the same end point; we might say that $\overrightarrow{AD}$ is "between" $\overrightarrow{AB}$ and $\overrightarrow{AC}$ if $\overrightarrow{AD} - A$ lies in the interior of $\angle BAC$.

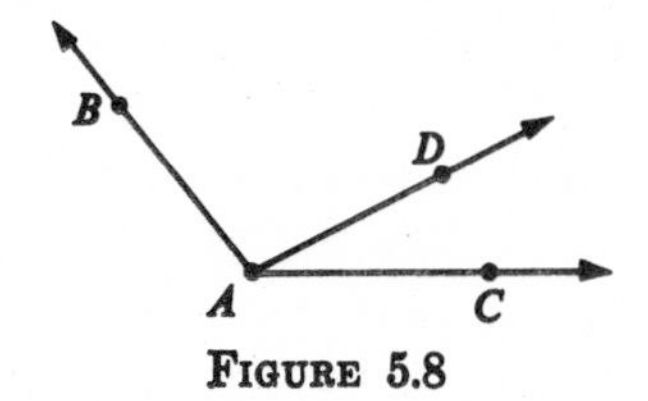

FIGURE 5.8

The analogy is incomplete, because it is *not* true that given any three rays with the same end point, one of them is between the other two. (Example?) We do have, however, the following theorems.

Theorem 1. For angles, congruence is an equivalence relation.

Theorem 2. *The Angle-Construction Theorem.* Let $\angle ABC$ be an angle, let $\overrightarrow{B'C'}$ be a ray, and let H be a half plane whose edge contains $\overleftrightarrow{B'C'}$. Then there is exactly one ray $\overrightarrow{B'A'}$, with A' in H, such that

$$\angle ABC \cong \angle A'B'C'.$$

Theorem 3. *The Angle-Addition Theorem.* If (1) D is in the interior of $\angle BAC$, (2) D' is in the interior of $\angle B'A'C'$, (3) $\angle BAD \cong \angle B'A'D'$, and (4) $\angle DAC \cong \angle D'A'C'$, then (5) $\angle BAC \cong \angle B'A'C'$.

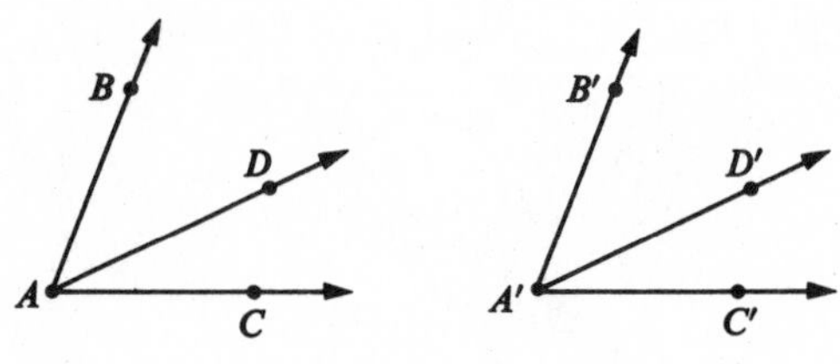

FIGURE 5.9

Theorem 4. *The Angle-Subtraction Theorem.* If (1) D is in the interior of $\angle BAC$, (2) D' is in the interior of $\angle B'A'C'$, (3) $\angle BAD \cong \angle B'A'D'$, and (4) $\angle BAC \cong \angle B'A'C'$, then (5) $\angle DAC \cong \angle D'A'C'$.

If we translate these theorems into the language of angular measure, using our definition of congruence for angles, they are seen to be trivial consequences of the postulates for m.

Two rays are called *perpendicular* if their union is a right angle. If $\overrightarrow{AB}$ and $\overrightarrow{AC}$ are perpendicular, then we write

$$\overrightarrow{AB} \perp \overrightarrow{AC}.$$

In this case we also say that the lines $\overleftrightarrow{AB}$ and $\overleftrightarrow{AC}$ are perpendicular, and write

$$\overleftrightarrow{AB} \perp \overleftrightarrow{AC}.$$

Two segments $\overline{AB}$, $\overline{BC}$ are perpendicular if the lines containing them are perpendicular. We use the same term and the same notation for a segment and a line, a line and a ray, and so on. Thus $\overleftrightarrow{AB} \perp \overline{PQ}$ means that $\overleftrightarrow{AB} \perp \overleftrightarrow{PQ}$; this in turn means that the union of the two lines contains a right angle.

An angle with measure less than 90 is called *acute*, and an angle with measure greater than 90 is called *obtuse*. Two angles are called *complementary* if the sum of their measures is 90.

If $m\angle BAC < m\angle B'A'C'$, then we say that $\angle BAC$ is *smaller than* $\angle B'A'C'$, and we write

$$\angle BAC < \angle B'A'C'.$$

Note that the relation "is smaller than" is not an order relation; it is quite possible for two angles to be different, without either of them being smaller than the other. In fact, this happens whenever the angles are congruent.

Two angles form a *vertical pair* if their sides form pairs of opposite rays, like this:

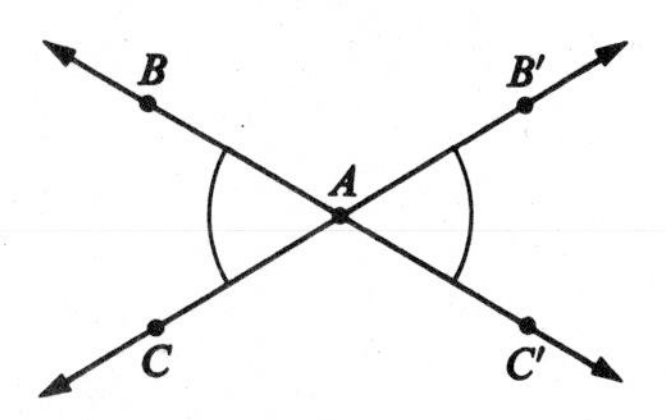

FIGURE 5.10

Here $\angle BAC$ and $\angle B'AC'$ form a vertical pair. To be more precise, if B-A-C', C-A-B', and the lines $\overleftrightarrow{AB}$ and $\overleftrightarrow{AB'}$ are different, then $\angle BAC$ and $\angle B'AC'$ form a *vertical pair*.

Theorem 5. *The Vertical Angle Theorem.* If two angles form a vertical pair, then they are congruent.

RESTATEMENT. If B-A-C', C-A-B', and the lines $\overleftrightarrow{AB}$ and $\overleftrightarrow{AB'}$ are different, then $\angle BAC \cong \angle B'AC'$.

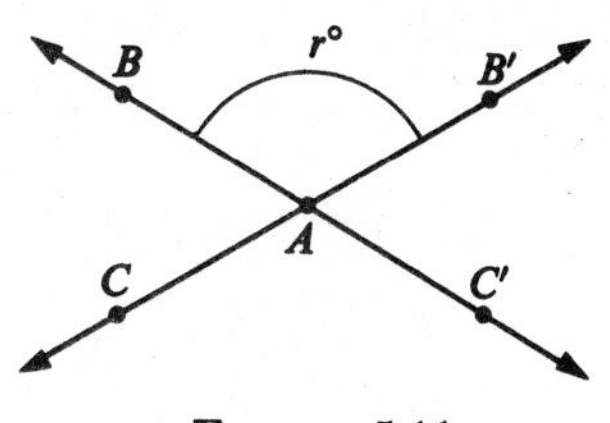

FIGURE 5.11

Proof. Let

$$r = m\angle BAB'.$$

Since B-A-C', it follows that $\overrightarrow{AB}$ and $\overrightarrow{AC'}$ are opposite rays. Therefore $\angle BAB'$ and $\angle B'AC'$ form a linear pair. Therefore these angles are supplementary, and we have

$$m\angle B'AC' = 180 - r.$$

Similarly, C-A-B', $\angle BAC$, and $\angle BAB'$ form a linear pair; these angles are therefore supplementary; and we have

$$m\angle BAC = 180 - r.$$

Therefore $m\angle BAC = m\angle B'AC'$, and $\angle BAC \cong \angle B'AC'$, which was to be proved.

If you suspect that the apparatus with which we stated the theorem and worked out the proof was superfluous, try "proving" the theorem while gazing at the following figure:

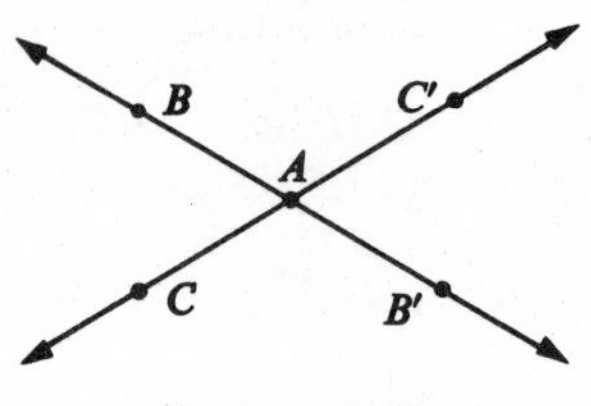

FIGURE 5.12

The point is that to prove anything at all about vertical angles, we have to have a definition of a vertical pair that is sufficiently exact to be usable.

Theorem 6. If two intersecting lines form one right angle, then they form four right angles.

Proof?

Chapter 6
Congruences between Triangles

6.1 THE IDEA OF A CONGRUENCE

As we explained in Chapter 3, the intuitive idea of congruence is the same for all types of figures. It means in every case that the first figure can be moved without changing its size or shape, so as to coincide with the second figure. There are two possible approaches to the problem of treating the idea of congruence mathematically. One way is to take it as undefined, state enough postulates to describe its essential properties, and then go on to prove whatever theorems turn out to be true. In later chapters, we shall show how such a postulational treatment works. For the present, however, we shall use a different scheme: we shall *define* congruence, in terms of distance and angular measure, and then proceed to prove our theorems on the basis of only one additional postulate.

The first of these approaches to congruence is called the *synthetic* approach, and the second, which we shall be using for some time to come, is called the *metric* method. We have already applied the metric method to the simplest cases, where the figures we are dealing with are segments or angles. Our basic definitions were as follows.

(1) $\overline{AB} \cong \overline{CD}$ means, by definition, that $AB = CD$.

(2) By definition, $\angle BAC \cong \angle PQR$ means that $m\angle BAC = m\angle PQR$.

We now proceed to the case where the figures involved are triangles. It is obvious, in the figures below, that all three of the triangles are congruent.

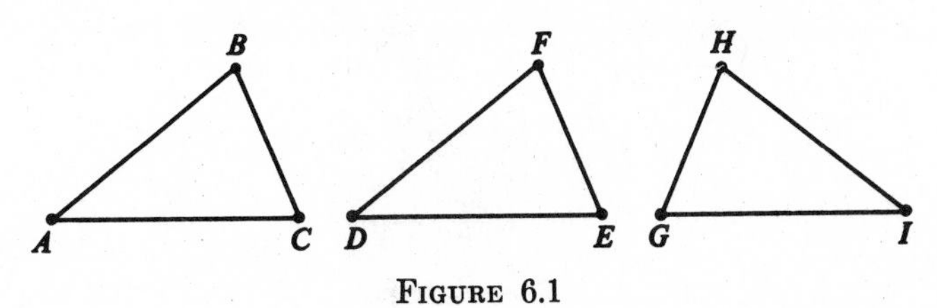

FIGURE 6.1

That is, any one of them can be moved onto any other one in such a way that it fits exactly. Thus, to move the first onto the second, we should put A on D, B on F, and C on E. These directions describe the motion by means of a one-to-one correspondence between the vertices of the first triangle and those of the second:

$$A \leftrightarrow D,$$
$$B \leftrightarrow F,$$
$$C \leftrightarrow E.$$

Similarly, the second triangle can be matched with the third by the correspondence,

$$D \leftrightarrow I,$$
$$E \leftrightarrow G,$$
$$F \leftrightarrow H.$$

Notice that there is no particular point in giving names to these functions, or in using functional notation to describe them; there are only three elements in each of the two sets, and we can therefore describe the function quite conveniently, simply by writing down all three of the matching pairs. There is a shorthand which is even briefer. We can describe our first correspondence in one line, like this:

$$ABC \leftrightarrow DFE.$$

Here it should be understood that the first letter on the left is matched with the first on the right, the second with the second, and the third with the third, like this:

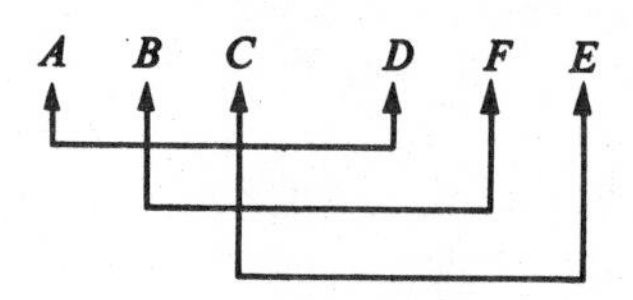

FIGURE 6.2

Given a correspondence between the vertices of two triangles, there is a naturally induced correspondence between the sides and the angles. Thus, given the correspondence

$$ABC \leftrightarrow DFE,$$

the induced correspondence between the sides is

$$\overline{AB} \leftrightarrow \overline{DF},$$
$$\overline{BC} \leftrightarrow \overline{FE},$$
$$\overline{AC} \leftrightarrow \overline{DE},$$

and the induced correspondence between the angles is

$$\angle A \leftrightarrow \angle D,$$
$$\angle B \leftrightarrow \angle F,$$
$$\angle C \leftrightarrow \angle E.$$

(Here, as usual, we are using the shorthand $\angle A$ for $\angle BAC$, $\angle B$ for $\angle ABC$, and so on.) When a correspondence between the vertices is given, then whenever we speak of *corresponding sides* or *corresponding angles*, we shall always be referring to the correspondence induced in the way shown above.

Of course, not every one-to-one correspondence between the vertices of two triangles describes a workable scheme for moving the first triangle onto the second, even if the triangles happen to be congruent. For example, the correspondence

$$ABC \leftrightarrow FED$$

is unworkable; the triangles can't be made to fit in this way. To test whether a correspondence between the vertices is "workable," we need to check whether the matching sides and angles are congruent. In fact, this is our official definition of a congruence.

DEFINITION. Given $\triangle ABC$, $\triangle DEF$, and a one-to-one correspondence

$$ABC \leftrightarrow DEF$$

between their vertices. If every pair of corresponding sides are congruent, and every pair of corresponding angles are congruent, then the correspondence is a *congruence.*

That is, the correspondence

$$ABC \leftrightarrow DEF$$

is a congruence if all six of the following conditions hold:

$$\begin{array}{ll} \overline{AB} \cong \overline{DE}, & \angle A \cong \angle D, \\ \overline{AC} \cong \overline{DF}, & \angle B \cong \angle E, \\ \overline{BC} \cong \overline{EF}, & \angle C \cong \angle F. \end{array}$$

If $ABC \leftrightarrow DEF$ is a congruence, then we write

$$\triangle ABC \cong \triangle DEF.$$

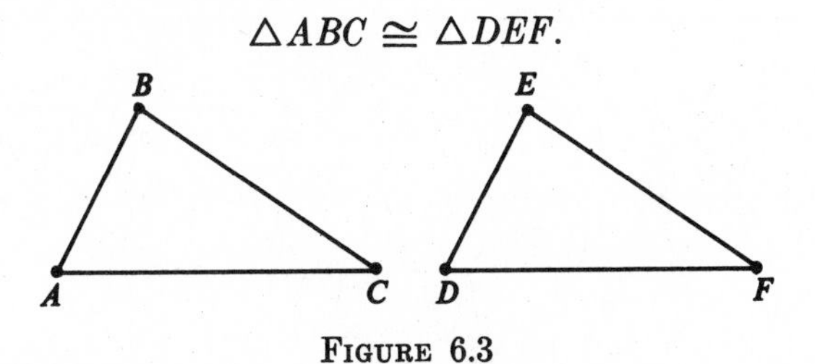

FIGURE 6.3

Two triangles are called *congruent* if there is *some* correspondence between their vertices which satisfies the six conditions for a congruence. Note that the expression $\triangle ABC \cong \triangle DEF$ says not merely that $\triangle ABC$ and $\triangle DEF$ are congruent, but also that they are congruent in a particular way, that is, under the correspondence $ABC \leftrightarrow DEF$.

Therefore, if we want to say merely that two triangles are congruent, we have to say this in words; we can't use the shorthand. It turns out, however, that this is no handicap, because in the geometry of triangles, the idea of congruence in the abstract hardly ever occurs. Nearly always, when we talk about congruent triangles, we go on to draw conclusions about "corresponding sides" or "corresponding angles"; and this means that what we really had in mind was a correspondence. The basic idea here is not the idea of *congruence*, but the idea of *a congruence*.

If you want to check that a given correspondence is a congruence, you don't have to check all six pairs of corresponding parts. For example, suppose that two sides and the included angle of the first triangle are congruent to the corresponding parts of the second, as the markings in the following figure indicate.

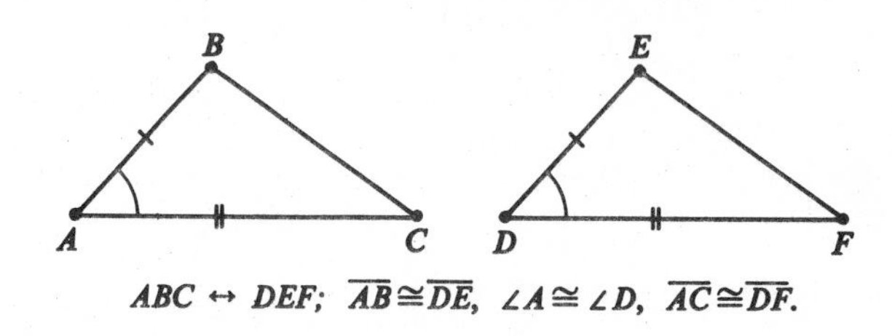

$ABC \leftrightarrow DEF;\ \overline{AB} \cong \overline{DE},\ \angle A \cong \angle D,\ \overline{AC} \cong \overline{DF}.$

FIGURE 6.4

It ought to follow that $ABC \leftrightarrow DEF$ is a congruence. In fact, this is our basic congruence postulate.

SAS. Given a correspondence between two triangles (or between a triangle and itself). If two sides and the included angle of the first triangle are congruent to the corresponding parts of the second triangle, then the correspondence is a congruence.

Here SAS stands for Side Angle Side. We shall refer to this postulate, hereafter, as the SAS postulate, or simply as SAS.

PROBLEM SET 6.1

These problems are not stated, and are not supposed to be solved, in terms of deductive geometry. You may take for granted that correspondences that look like congruences really are congruences.

1. Write down all of the congruences between an equilateral triangle and itself.

2. The figure on the right is a five-pointed star. Write down all of the congruences between the star and itself. Let us agree that a congruence is simply a matching scheme that "works," and that such a congruence is sufficiently described if we explain, in the one-line short notation, where the points A, B, C, D, and E of the star are supposed to go. Thus one of the congruences that we are looking for is $ABCDE \leftrightarrow CDEAB$.

3. Given a triangle $\triangle ABC$ which is isosceles but not equilateral. That is, $AB = AC$, but $AB \neq BC$. How many congruences are there, between $\triangle ABC$ and itself?

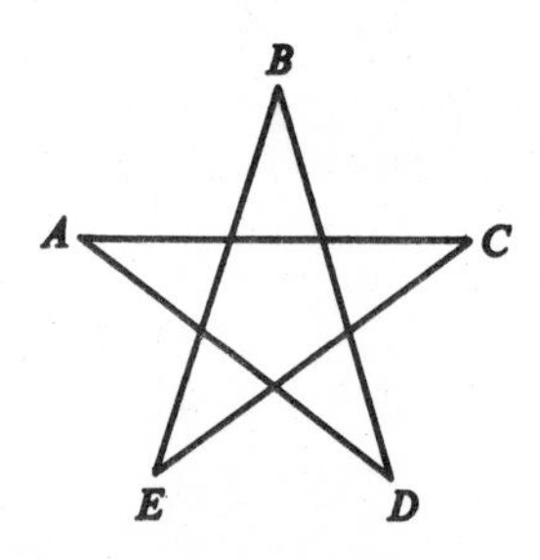

FIGURE 6.5

6.2 THE BASIC CONGRUENCE THEOREMS

An isosceles triangle is a triangle at least two of whose sides are congruent. A triangle which is not isosceles is called *scalene*. If all three sides are congruent, then the triangle is *equilateral*. The first and easiest consequence of the SAS postulate follows.

Theorem 1. *The Isosceles Triangle Theorem.* If two sides of a triangle are congruent, then the angles opposite them are congruent.

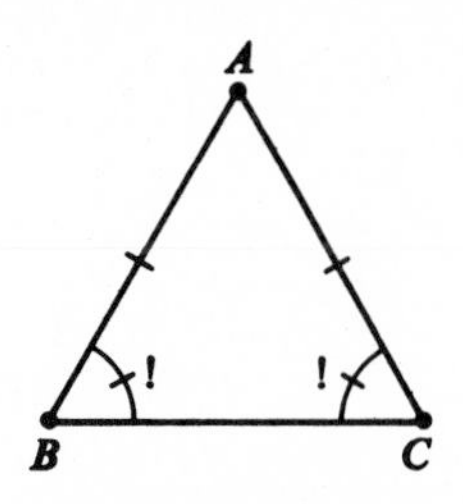

FIGURE 6.6

That is, *the base angles of an isosceles triangle are congruent.* Here by the base angles we mean the angles opposite the two congruent sides.

The marking of the figure gives a complete picture of the theorem. The marks on the sides $\overline{AB}$ and $\overline{AC}$ indicate that these sides are congruent by hypothesis. The marks for $\angle B$ and $\angle C$, with exclamation points, indicate the *conclusion* that these angles are congruent. Throughout this book, exclamation points will be used in figures in this way, to indicate conclusions.

RESTATEMENT. Given $\triangle ABC$. If $\overline{AB} \cong \overline{AC}$, then $\angle B \cong \angle C$.

Proof. Consider the correspondence $ABC \leftrightarrow ACB$.

Under this correspondence, $\overline{AB} \leftrightarrow \overline{AC}$, $\overline{AC} \leftrightarrow \overline{AB}$, and $\angle A \leftrightarrow \angle A$. Thus two sides and the included angle are congruent to the parts that correspond to them. Therefore, by SAS, the correspondence is a congruence: $\triangle ABC \cong \triangle ACB$. By definition of a congruence, this means that $\angle B \cong \angle C$, which was to be proved.

This is the famous *pons asinorum* theorem. The phrase pons asinorum means asses' bridge, and was suggested by the figure which accompanied Euclid's proof.

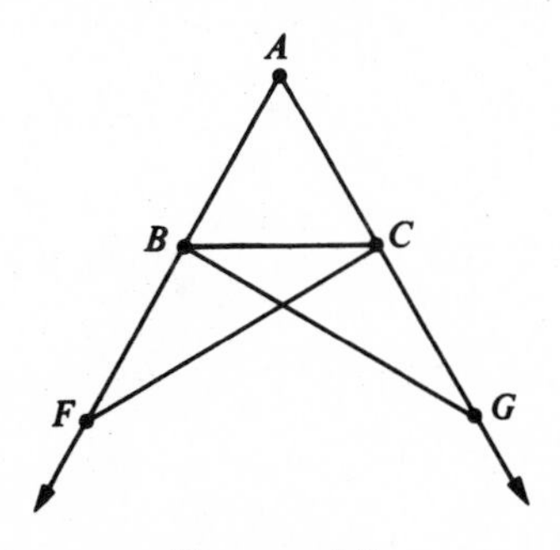

FIGURE 6.7

Euclid's proof was long; it takes over a page in print. The proof given above is due, essentially, to Pappus, although Pappus did not use the formulation of SAS that we have been using here. Not many years ago—or so the story goes—an electronic computing machine was programmed to look for proofs of elementary geometric theorems. When the *pons asinorum* theorem was fed into the machine, it promptly printed Pappus's proof on the tape. This is said to have been a surprise to the people who had coded the problem; Pappus' proof was new to them. What had happened, of course, was that the SAS postulate had been coded in some such form as the following:

"If (1) A, B, and C are noncollinear, (2) D, E, and F are noncollinear; (3) $\overline{AB} \cong \overline{DE}$; (4) $\overline{BC} \cong \overline{EF}$; and (5) $\angle ABC \cong \angle DEF$, then (6) $\overline{AC} \cong \overline{DF}$; (7) $\angle ACB \cong \angle DFE$; and (8) $\angle BAC \cong \angle EDF$."

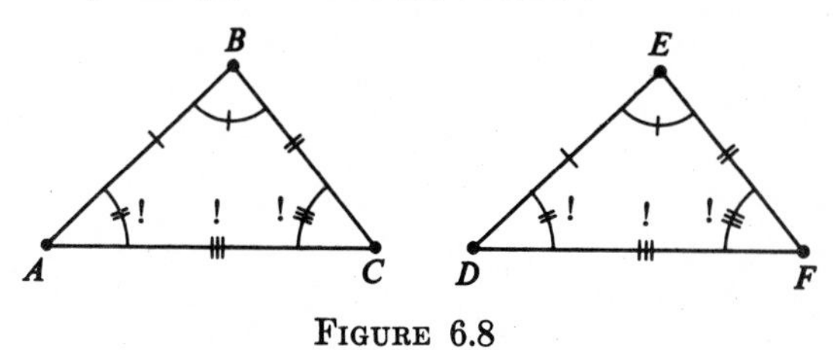

FIGURE 6.8

This is the sort of language in which people talk to transistors; you can't indoctrinate them with preconceptions and prejudices; and so, if you want the transistors to get the idea that the triangles (Fig. 6.8) in the SAS postulate are supposed to be different, you have to say so explicitly. It didn't occur to anybody to do this, and so the machine proceeded, in its simple-minded way, to produce the most elegant proof.

Corollary 1-1. Every equilateral triangle is equiangular.
That is, in an equilateral triangle, all three angles are congruent. Proof?

Theorem 2. *The ASA Theorem.* Given a correspondence between two triangles (or between a triangle and itself). If two angles and the included side of the first triangle are congruent to the corresponding parts of the second, then the correspondence is a congruence.

RESTATEMENT. Given $\triangle ABC$, $\triangle DEF$, and a correspondence $ABC \leftrightarrow DEF$. If $\angle A \cong \angle D$, $\angle C \cong \angle F$, and $\overline{AC} \cong \overline{DF}$, then $\triangle ABC \cong \triangle DEF$.

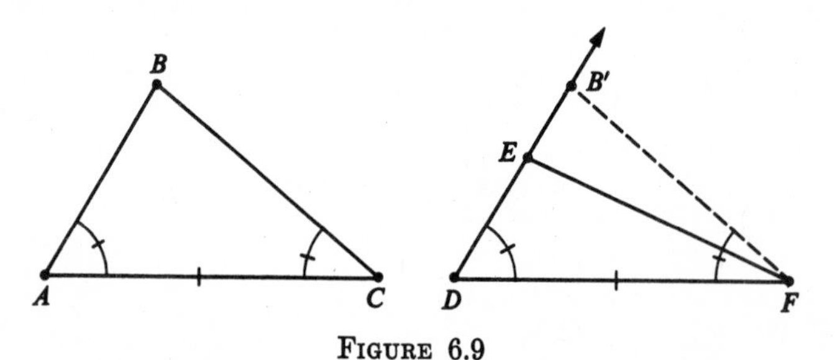

FIGURE 6.9

Proof. (1) By Theorem C-2 of Section 3.6 there is a point B' on the ray $\overrightarrow{DE}$, such that $\overline{DB'} \cong \overline{AB}$.

(2) By SAS, we have $\triangle ABC \cong \triangle DB'F$.

(3) By definition of a congruence, we have $\angle DFB' \cong \angle ACB$.

(4) By Theorem 2, Section 5.1, it follows that $\overrightarrow{FB'} = \overrightarrow{FE}$.

(5) Therefore $B' = E$, because the lines $\overleftrightarrow{DE}$ and $\overleftrightarrow{FE}$ intersect in only one point.

(6) Therefore, by (2), we have $\triangle ABC \cong \triangle DEF$, which was to be proved.

From this we get a corollary which is a converse of Theorem 1.

Corollary 2-1. If two angles of a triangle are congruent, then the sides opposite them are congruent.

The following corollary is a converse of Corollary 1-1.

Corollary 2-2. Every equiangular triangle is equilateral.

Proof?

The third of the basic congruence theorems is harder to prove.

Theorem 3. *The SSS Theorem.* Given a correspondence between two triangles (or between a triangle and itself). If all three pairs of corresponding sides are congruent, then the correspondence is a congruence.

RESTATEMENT. Given $\triangle ABC$, $\triangle DEF$, and a correspondence $ABC \leftrightarrow DEF$. If $\overline{AB} \cong \overline{DE}$, $\overline{BC} \cong \overline{EF}$, and $\overline{AC} \cong \overline{DF}$, then the correspondence is a congruence.

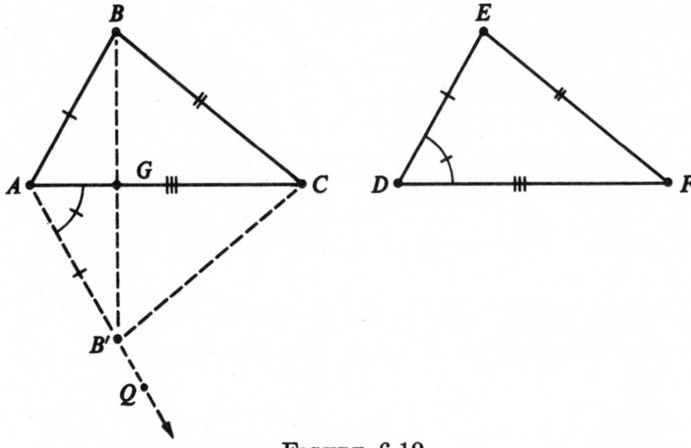

FIGURE 6.10

Proof. Before proceeding with details, let us explain what the idea of the proof is going to be. First we are going to copy $\triangle DEF$ on the under side of $\triangle ABC$; that is, we are going to set up $\triangle AB'C$, with B' on the opposite side of $\overleftrightarrow{AC}$ from B, so that $\triangle AB'C \cong \triangle DEF$. Second, we shall show that $\triangle ABC \cong \triangle AB'C$. It will follow that $\triangle ABC \cong \triangle DEF$, which was to be proved. In full, the proof is as follows.

(1) By the angle-construction theorem (Theorem 2, Section 5.1) there is a ray $\overrightarrow{AQ}$, with Q on the opposite side of $\overleftrightarrow{AC}$ from B, such that

$$\angle CAQ \cong \angle EDF.$$

(2) By the segment-construction theorem (Theorem C-2 of Section 3.6), there is a point B' of $\overrightarrow{AQ}$ such that

$$\overline{AB'} \cong \overline{DE}.$$

(3) Since we already know that $\overline{AC} \cong \overline{DF}$, it follows by SAS that

$$\triangle AB'C \cong \triangle DEF.$$

(Thus we have completed the first part of our program.)

(4) $\overline{B'B}$ intersects $\overleftrightarrow{AC}$ in a point G. (Because B and B' are on opposite sides of $\overleftrightarrow{AC}$.)

The proof now splits up into a number of cases. (i) A-G-C, as in the figure (Fig. 6.10). (ii) $A = G$. (iii) G-A-C. Strictly speaking, there are two more cases $G = C$ and A-C-G, but these are essentially the same as (ii) and (iii).

We proceed with the proof for Case (i).

(5) $\angle ABG \cong \angle AB'G$ (by the isosceles triangle theorem).

(6) $\angle CBG \cong \angle CB'G$ (for the same reason).

(7) G is in the interior of $\angle ABC$. [Since A-G-C in Case (i), this follows from Theorem 4, Section 4.2.]

(8) G is in the interior of $\angle AB'C$. (For the same reason.)

(9) By (5), (6), (7), and (8), together with Theorem 3, Section 5.1, it follows that

$$\angle ABC \cong \angle AB'C.$$

(10) By SAS, it follows that

$$\triangle ABC \cong \triangle AB'C.$$

(11) Therefore, by (10) and (3),

$$\triangle ABC \cong \triangle DEF.$$

(Proof?)

For Cases (ii) and (iii), the figures look like this:

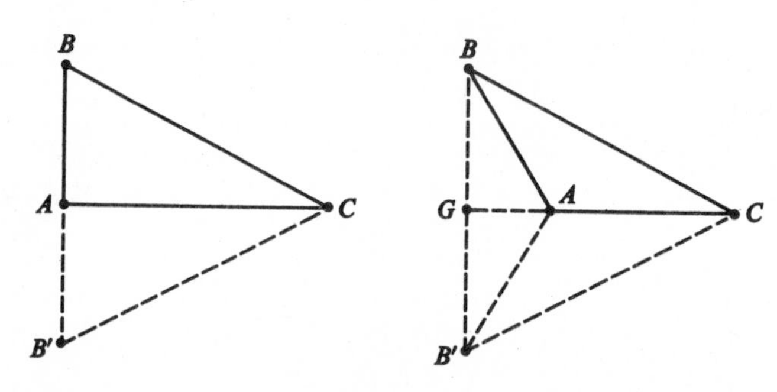

FIGURE 6.11

We leave these cases as problems.

Roughly speaking, a *bisector* of an angle is a ray in the interior which splits the angle into two congruent parts, like this.

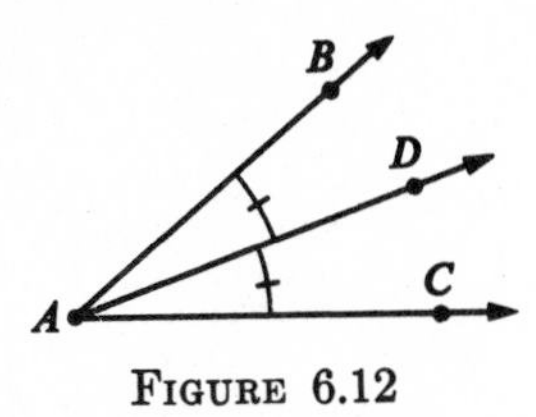

FIGURE 6.12

That is, $\overrightarrow{AD}$ *bisects* $\angle BAC$ if (1) D is in the interior of $\angle BAC$, and (2) $\angle BAD \cong \angle DAC$.

Theorem 4. Every angle has exactly one bisector.

Proof. Given $\angle BAC$. We lose no generality by supposing that $\overline{AB} \cong \overline{AC}$. We can always pick two points on different sides of the angle, equidistant from the vertex.

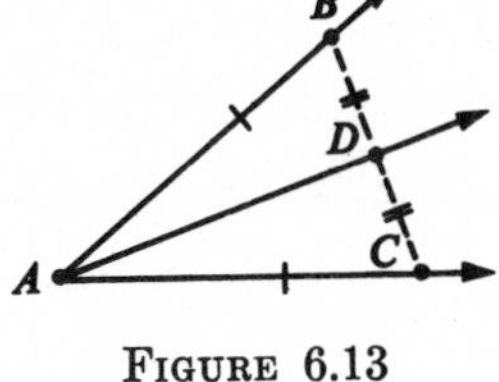

FIGURE 6.13

Let D be the midpoint of $\overline{BC}$. Then D is in the interior of $\angle BAC$. (Why?) And by the SSS theorem, $\triangle ADC \cong \triangle ADB$. Therefore $\angle BAD \cong \angle CAD$; and so $\overrightarrow{AD}$ bisects $\angle BAC$.

Thus we have shown that every angle has at least one bisector. This is half of our theorem. We need next to show that $\angle BAC$ has at most one bisector. To do this, it will be sufficient to show that every bisector of $\angle BAC$ passes through the midpoint D of $\overline{BC}$.

Suppose that $\overrightarrow{AE}$ bisects $\angle BAC$. Then automatically E is in the interior of $\angle BAC$. By the crossbar theorem (Theorem 3 of Section 4.3), it follows that $\overrightarrow{AE}$ intersects $\overline{BC}$ in some point D', between B and C. By the SAS postulate we have $\triangle AD'B \cong \triangle AD'C$. Therefore $\overline{D'B} \cong \overline{D'C}$, and D' is the midpoint of $\overline{BC}$. Since $\overline{BC}$ has only one midpoint, it follows that $\angle BAC$ has only one bisector, which was to be proved.

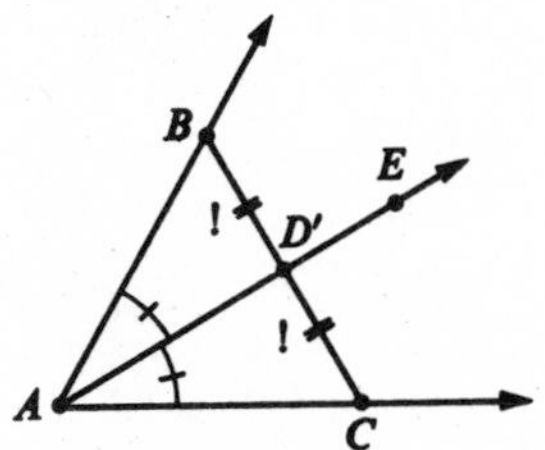

FIGURE 6.14

Problem Set 6.2

1. Complete the proof of the SSS theorem.

*2. Let the "special SAS postulate" be the statement that you get if you add to the SAS postulate the condition that the triangles must be different. Show that the general form of the SAS postulate can be proved as a theorem, on the basis of the special SAS postulate.

6.3 SOME REMARKS ON TERMINOLOGY

The language in which we have been discussing congruence in this book is different from the language used in much of the literature; it may be worthwhile to discuss the reasons for the differences.

We have already explained the reason for speaking about *congruences* between triangles, in the sense of correspondences having certain properties, rather than speaking of the relation of *congruence* in the abstract. The reason, briefly, is that the former is what we mean and what we need.

Our use of the word *congruent* in connection with segments and angles is a slightly different matter. Suppose that we have a pair of segments, a pair of angles, and a pair of circles which "match up" in the sense suggested by the following figures:

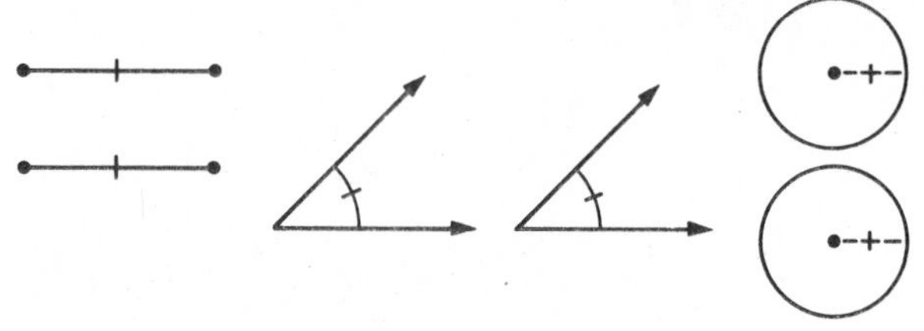

Figure 6.15

The situation can then be described in two ways. (1) We can say (following widespread usage) that the segments are *equal;* and similarly for the angles and the circles. (2) We can say (following the usage of the present book) that in each of the three cases, the figures are *congruent.*

The same people who call two segments *equal* if they have the same length also say that two triangles are equal if they have the same area.

There are two difficulties with the loose use of the word *equals* to describe equality of length, angular measure, and area. The first difficulty is that if the word *equals* is used in this way, there is no word left in the language with which we can say that A is—without ifs, buts, qualification, or fudging—the same as B. This latter relation is called the *logical identity*. It may seem a peculiar idea, at first, because if two things are exactly the same, there can't be two of them. But as soon as mathematics had begun to make heavy use of symbolism, the logical identity became important. For example, each of the expressions

$$\frac{1}{2\sqrt{3}-1}, \qquad \frac{2\sqrt{3}+1}{11}$$

describes a number. The *descriptions* are obviously different, but it is easy to check that they describe the same number; and this is what we mean when we write

$$\frac{1}{2\sqrt{3} - 1} = \frac{2\sqrt{3} + 1}{11}.$$

The relation denoted by the symbol =, in the above equation, is the logical identity.

The concept of the logical identity $A = B$ is so important, and comes up so often, that it is entitled to have a word to itself. For this reason, in nearly all modern mathematics, the word *equals* and the symbol = are used in only one sense: they mean *is exactly the same as.*

The second difficulty with the loose use of the word *equals* is that it puts us in the position of using two words to describe the same idea, when one word would do. Congruence is the basic equivalence relation in geometry. We may use different technical definitions for it, in connection with different types of figure, but the underlying idea is always the same: two figures are congruent if one can be placed on the other by a rigid motion. The basic equivalence relation of geometry is entitled to have a word all to itself; and the word *congruence* appears to be elected.

It was, no doubt, for these reasons that Hilbert adopted, in his *Foundations of Geometry*, the terminology that we are using in the present book. The problem is not logical but expository. A good terminology matches up the words with the ideas in the simplest possible way, so that the basic words are in one-to-one correspondence with the basic ideas.

It should be borne in mind that the strict mathematical interpretation of the word *equals*, in the sense of "is exactly the same as," is a technical usage. In ordinary literary English, the word is used even more loosely than in Euclid. For example, when Thomas Jefferson wrote, in the Declaration of Independence, that all men are created equal, he did not mean that there is only one man in the world, or that all men are congruent replicas of one another. He meant merely that all men have a certain property in common, namely, the property of being endowed by their Creator with certain unalienable rights.

In fact, only mathematics and logic need a word and a symbol for the idea of "is the same as," and the need did not appear even in mathematics and logic until the heavy use of symbolism developed. This development came along after Euclid; and sheer force of habit preserved Euclid's terminology long past the time when it had become awkward.

6.4 THE INDEPENDENCE OF THE SAS POSTULATE

We have seen that the ASA theorem and the SSS theorem can be proved on the basis of the SAS postulate. The question may arise whether the SAS postulate itself can be turned into a theorem; that is, *proved* on the basis of the postulates that precede it.

Certain general considerations suggest the contrary. If you reconsider the ruler postulate, you will see that it deals with lines one at a time. It does not seem to claim that there is any connection between distances measured along one line and distances measured along another. The first postulate that describes such a connection is SAS; SAS tells us, among other things, that if $\overline{AB} \cong \overline{DE}$, $\angle A \cong \angle D$ and $\overline{AC} \cong \overline{DF}$, then $\overline{BC} \cong \overline{EF}$.

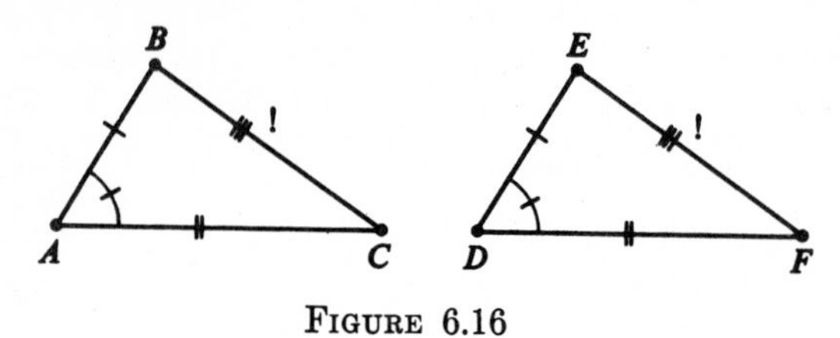

FIGURE 6.16

Here, except in very special cases, the distances BC and EF are measured along different lines.

We observe, moreover, that the postulates for the angular-measure function m do not mention distance at all.

These considerations suggest that SAS gives genuinely new information.

Given a set of postulates, say, $P_1, P_2, \ldots, P_n$, for a mathematical structure. We say that P_n is *independent* of the other postulates $P_1, P_2, \ldots, P_{n-1}$ if there is a mathematical system which satisfies $P_1, P_2, \ldots, P_{n-1}$ but does not satisfy P_n. For example, the postulate which says that every $a \neq 0$ has a reciprocal a^{-1} is independent of the other field postulates. The easiest way to see this is to observe that the integers satisfy all of the field postulates, with the sole exception of this one.

We shall give an example of the same sort to show that SAS is independent of the preceding postulates of metric geometry.

Consider a structure $[S, \mathcal{P}, \mathcal{L}, d, m]$ satisfying all of the postulates of metric geometry. We may think of this system as coordinate three-dimensional space, with the usual definitions of distance and angular measure. We shall define a new distance function d', by making a slight change in the "normal" distance function d. We do this in the following way. We choose a particular line L at random. We agree that $d'(P, Q)$ is to be the old $d(P, Q)$, except when P and Q both lie on L, in which case $d'(P, Q) = 2d(P, Q)$. It is plain that d' satisfies the ruler postulate

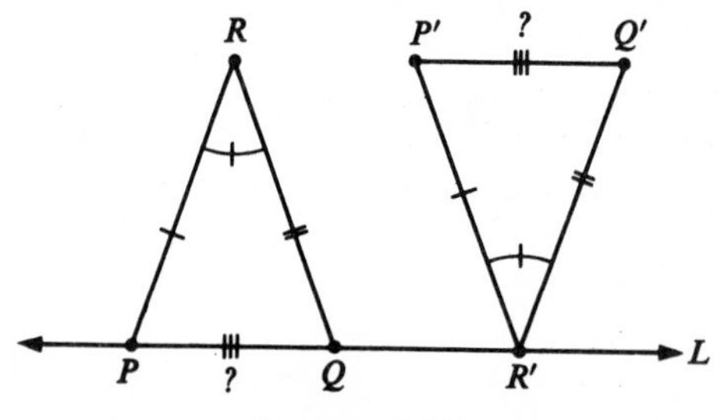

FIGURE 6.17

on all lines except perhaps for L. And on L, the ruler postulate also holds. Given a coordinate system f which works for the old d, we merely set $f'(P) = 2f(P)$, and the new coordinate system now works for d'.

The SAS postulate now fails for the old m and the new d' (Fig. 6.17). We ought to have $\triangle PQR \cong \triangle P'Q'R'$, because $d'(R, P) = d'(R', P')$, $\angle R \cong \angle R'$ and $d'(R, Q) = d'(R', Q')$. But the congruence between the triangles does not hold, because $d'(P, Q) = 2d'(P', Q')$.

It is possible also to show the independence of SAS by leaving d unchanged and defining a peculiar angular measure m' for angles with a certain point P as vertex. But examples of the latter sort are harder to describe and to check.

6.5 EXISTENCE OF PERPENDICULARS

You may have noted that in the proof of the SSS theorem in Case (i), we had $\overleftrightarrow{GB} \perp \overleftrightarrow{AC}$. The reason is as follows. By SAS, we have $\triangle AGB \cong \triangle AGB'$. Therefore $\angle AGB \cong \angle AGB'$. Since these two angles form a linear pair, they are supplementary. Thus $\angle AGB$ is congruent to a supplement of itself, and hence is a right angle. Therefore our proof of the SSS theorem included, implicitly, a proof that "a perpendicular can always be drawn, to a given line, through a given external point." We make this explicit in the following theorem.

Theorem 1. Given a line and a point not on the line, then there is a line which passes through the given point and is perpendicular to the given line.

Proof. Let L be the line, and let B be a point not on L. Let A and C be any two points of L. By the angle-construction theorem, there is a point Q such that (1) B and Q are on opposite sides of L, and (2) $\angle BAC \cong \angle QAC$.

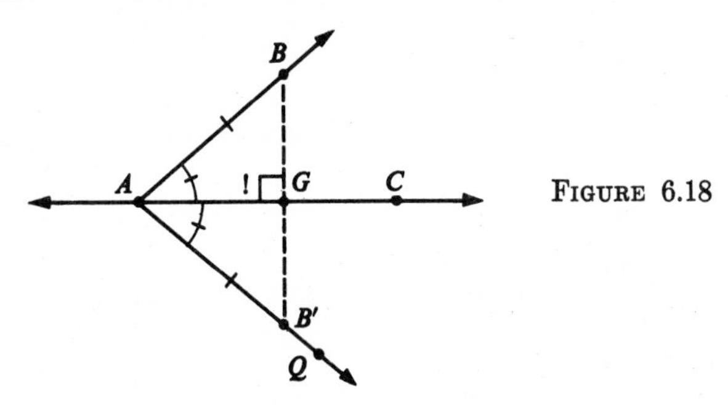

FIGURE 6.18

By the segment-construction theorem, there is a point B' of $\overrightarrow{AQ}$ such that $\overline{AB} \cong \overline{AB'}$. Since B and B' are on opposite sides of $L = \overleftrightarrow{AC}$, $\overline{BB'}$ intersects L in a point G. Now there are two possibilities:

(1) $G \neq A$. In this case it follows by SAS that $\triangle AGB \cong \triangle AGB'$. Therefore $\angle AGB \cong \angle AGB'$. Since these two angles form a linear pair, they are supplementary. Therefore each of them is a right angle. Therefore $\overleftrightarrow{BG} \perp \overleftrightarrow{AC}$, as desired.

(2) $G = A$. In this case $\angle BGC \cong \angle BAC$ and $\angle B'GC \cong \angle QAC$. Therefore $\angle BGC \cong \angle B'GC$; and it follows, as in Case (1), that $\overleftrightarrow{BG} \perp \overleftrightarrow{AC}$.

Note that Case (2) is "unlikely," because A and C were chosen at random on the line. But it is possible that we happened to pick the foot of the perpendicular.

Note also that if you use the full force of the angle-construction postulate (M-3), then you can prove easily that given a line L in a plane E, and a point A which is *on* the line, there is always a line in E which contains A and is perpendicular to L. So far, however, we have been making it a point not to use the "protractor postulates," but to use instead the first few theorems based on them; and so we do not yet state the above theorem officially.

Chapter 7
Geometric Inequalities

Up to now, in our study of the geometry of the triangle, we have been dealing only with congruence; our theorems have stated that under certain conditions we can infer that two segments (or two angles) are congruent. We shall now investigate conditions under which we can say that one segment is larger than another, or that one angle is larger than another.

Initially, we defined inequalities between angles by means of measure. That is, $\angle ABC < \angle DEF$ if $m\angle ABC < m\angle DEF$. It is clear, of course, that the same idea can be described simply in terms of congruence, without any regard to the source of our concept of congruence. We can say that $\angle ABC < \angle DEF$ if there is a point G, in the interior of $\angle DEF$ such that $\angle ABC \cong \angle GEF$.

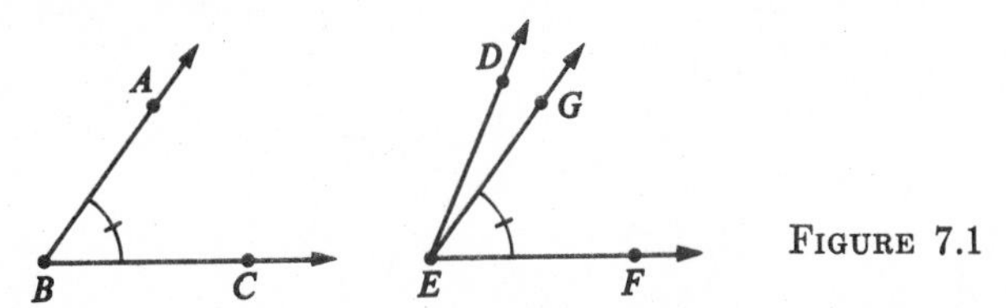

FIGURE 7.1

Similarly, for segments, we may say that $\overline{AB}$ is shorter than $\overline{CD}$ ($\overline{AB} < \overline{CD}$) if the distance AB is less than the distance CD. Or we may forget where the idea of congruence came from, and say that if there is a point E, between C and D, such that $\overline{AB} \cong \overline{CE}$, then $\overline{AB} < \overline{CD}$.

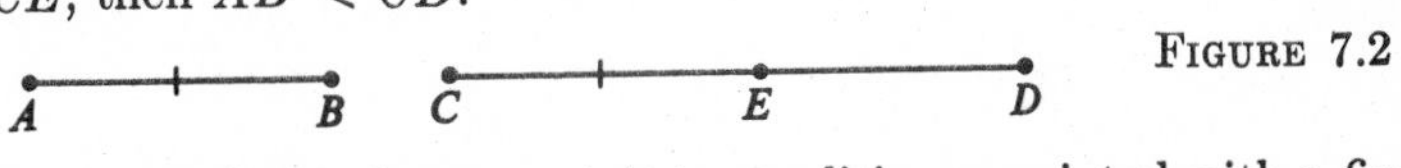

FIGURE 7.2

We now proceed to investigate the geometric inequalities associated with a fixed triangle.

In the figure below, the angle $\angle BCD$ is called an *exterior angle* of $\triangle ABC$. More precisely, if A-C-D, then $\angle BCD$ is an *exterior angle* of $\triangle ABC$.

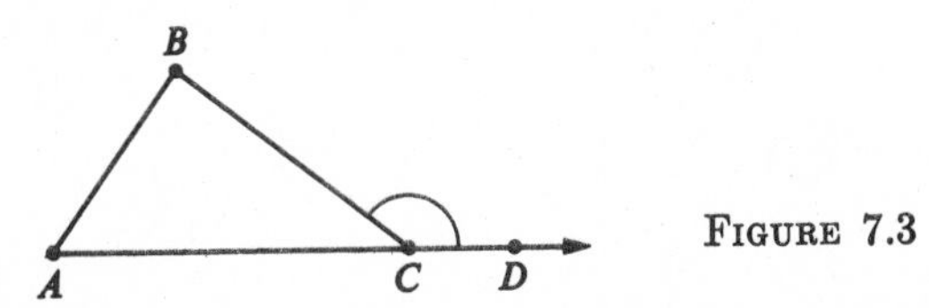

FIGURE 7.3

Every triangle has six exterior angles, as indicated in the following figure, and these six angles form three vertical pairs.

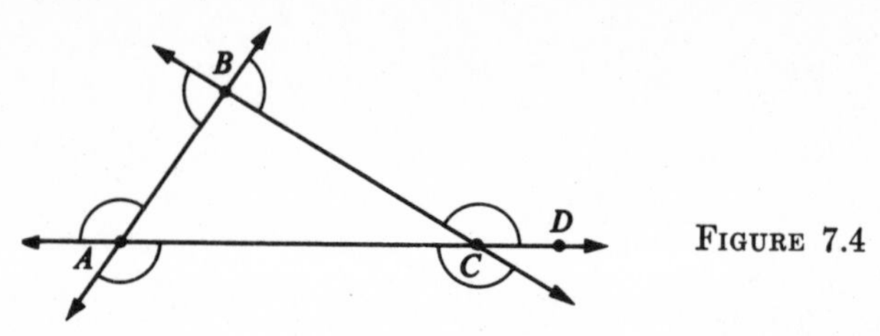

FIGURE 7.4

It follows, of course, that the two exterior angles at a given vertex are always congruent. The angles $\angle A$ and $\angle B$ of $\triangle ABC$ are called the *remote interior angles* of the exterior angles with vertex at C; similarly, $\angle A$ and $\angle C$ are the remote interior angles of the exterior angles with vertex at B; and so also for the third case.

Theorem 1. Any exterior angle of a triangle is greater than each of its remote interior angles.

RESTATEMENT. Given $\triangle ABC$. If A-C-D, then $\angle BCD > \angle B$.

First we observe that the restatement really does convey the entire content of the theorem.

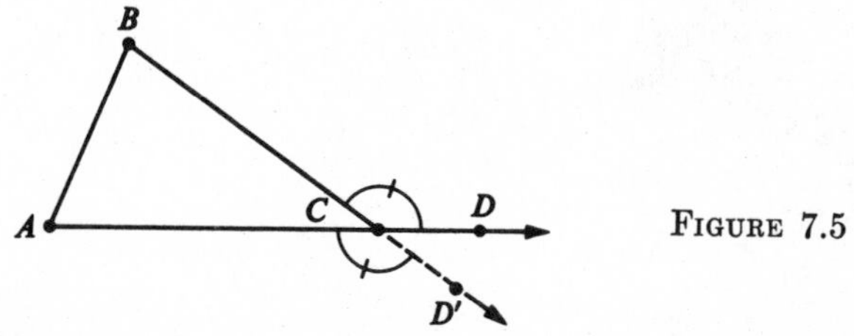

FIGURE 7.5

If we prove that the restatement holds true, then we will also know that the other exterior angle at C is greater than $\angle B$, because the two exterior angles at C are congruent. It will also follow, merely by a change of notation, that $\angle ACD' > \angle A$; and this means that the exterior angles at C are greater than *each* of their remote interior angles.

We now proceed to the proof of the restatement.

Let E be the midpoint of $\overline{BC}$.

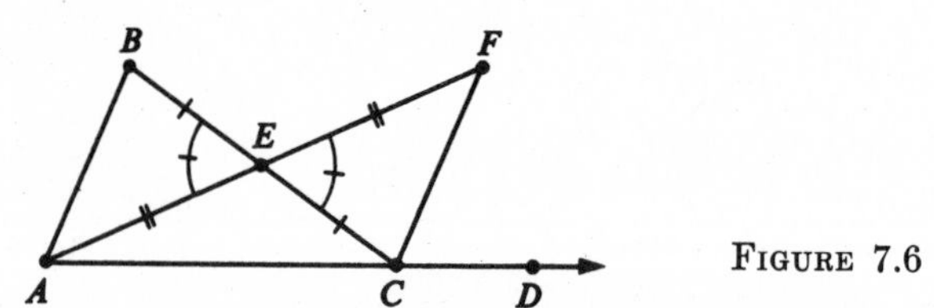

FIGURE 7.6

By Theorem C-2, there is a point F such that A-E-F and $\overline{EA} \cong \overline{EF}$.

Consider $\triangle AEB$ and $\triangle FEC$. Now $\overline{EB} \cong \overline{EC}$, by hypothesis for E; $\overline{EA} \cong \overline{EF}$, by hypothesis for F; and $\angle AEB \cong \angle FEC$, because vertical angles are congruent. By SAS it follows that the correspondence $AEB \leftrightarrow FEC$ is a congruence; that is, $\triangle AEB \cong \triangle FEC$. Therefore, $\angle B \cong \angle BCF$.

By Theorem 7, Section 4.2, we know that F is in the interior of $\angle BCD$. Therefore $\angle BCF < \angle BCD$; and therefore $\angle B < \angle BCD$, which was to be proved.

Corollary 1-1. The perpendicular to a given line, through a given external point, is unique.

RESTATEMENT. Let L be a line, and let P be a point not on L. Then there is only one line through P, perpendicular to L.

Proof. Suppose that there are two perpendiculars to L through P, intersecting L in points Q and R. We shall show that this is impossible.

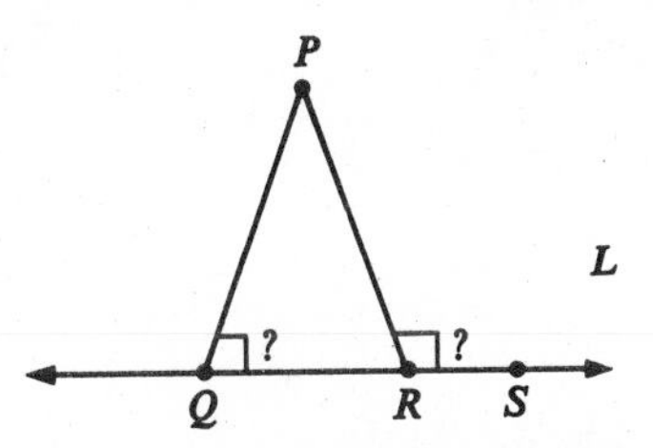

FIGURE 7.7

Let S be a point of L such that Q-R-S. Then $\angle PRS$ is an exterior angle of $\triangle PQR$; and $\angle PQR$ is one of its remote interior angles. This is impossible, because both $\angle PQR$ and $\angle PRS$ are right angles. (See Theorem 6 of Section 5.1.)

Theorem 2. If two sides of a triangle are not congruent, then the angles opposite them are not congruent, and the greater angle is opposite the longer side.

RESTATEMENT. Given $\triangle ABC$. If $\overline{AB} > \overline{AC}$, then $\angle C > \angle B$.

Proof. Let D be a point of $\overrightarrow{AC}$, such that $\overline{AD} \cong \overline{AB}$. Then A-C-D, as the figure indicates, because $\overline{AD} \cong \overline{AB} > \overline{AC}$. Since the base angles of an isosceles triangle are congruent, we have

$$\angle ABD \cong \angle D. \tag{1}$$

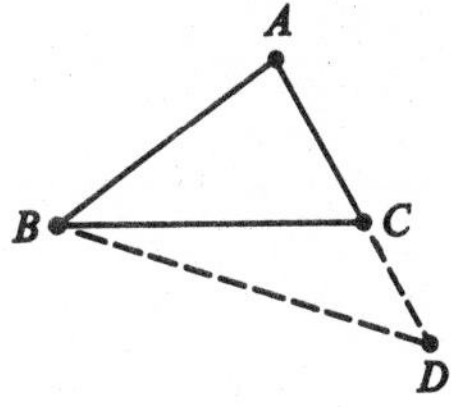

FIGURE 7.8

Since A-C-D, it follows by Theorem 5, Section 4.2, that C is in the interior of $\angle ABD$. Therefore

$$\angle ABC < \angle ABD; \tag{2}$$

therefore

$$\angle ABC < \angle D. \tag{3}$$

Since $\angle ACB$ is an exterior angle of $\triangle BCD$, we have

$$\angle D < \angle ACB. \tag{4}$$

By (3) and (4), $\angle ABC < \angle ACB$. Thus, in $\triangle ABC$ we have $\angle B < \angle C$, which was to be proved.

Theorem 3. If two angles of a triangle are not congruent, then the sides opposite them are not congruent, and the larger side is opposite the larger angle.

RESTATEMENT. Given $\triangle ABC$. If $\angle B < \angle C$, then $\overline{AC} < \overline{AB}$.

Proof. (1) If $\overline{AC} \cong \overline{AB}$, then by the isosceles triangle theorem it would follow that $\angle B \cong \angle C$; and this is false.

(2) If $\overline{AC} > \overline{AB}$, then by Theorem 2 it would follow that $\angle B > \angle C$; and this is false.

The only remaining possibility is $\overline{AC} < \overline{AB}$, which was to be proved.

Theorem 4. The shortest segment joining a point to a line is the perpendicular segment.

RESTATEMENT. Let L be a line, let P be a point not on L, let Q be the foot of the perpendicular to L through P, and let R be any other point of L. Then $\overline{PQ} < \overline{PR}$.

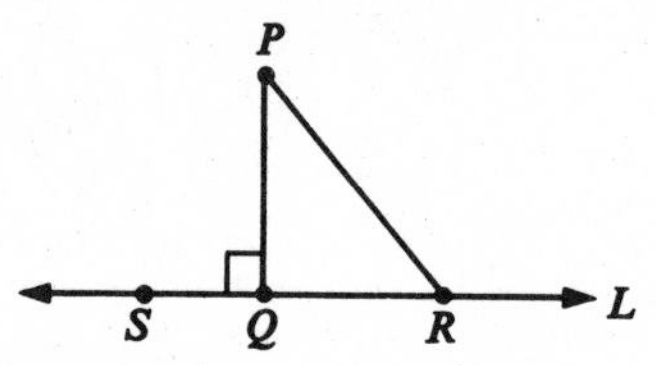

FIGURE 7.9

Proof. Let S be a point of L such that S-Q-R. Then $\angle PQS$ is an exterior angle of $\triangle PQR$. Therefore $\angle PQS > \angle PRQ$. Since $\overline{PQ} \perp L$, we know that $\angle PQS \cong \angle PQR$; therefore $\angle PQR > \angle PRQ$. By the preceding theorem, it follows that $\overline{PR} > \overline{PQ}$, which was to be proved.

Theorem 5. *The Triangular Inequality.* In any triangle, the sum of the lengths of any two sides is greater than the length of the third side.

RESTATEMENT. If A, B, and C are noncollinear, then $AB + BC > AC$.

Proof. Let D be a point of $\overrightarrow{CB}$ such that C-B-D and $BD = BA$. Then

$$CD = AB + BC. \quad (1)$$

Now B is in the interior of $\angle DAC$, by Theorem 5 of Section 4.2; therefore

$$\angle DAB < \angle DAC. \quad (2)$$

Since $\triangle BAD$ is isosceles, with $BA = BD$, it follows that $\angle D \cong \angle BAD$; therefore

$$\angle D < \angle DAC. \quad (3)$$

By applying Theorem 3 to $\triangle ADC$, we get $\overline{CD} > \overline{AC}$; and this may be expressed in terms of distance, as follows:

$$CD > AC. \quad (4)$$

By (1) and (4), we have

$$AB + BC > AC,$$

which was to be proved.

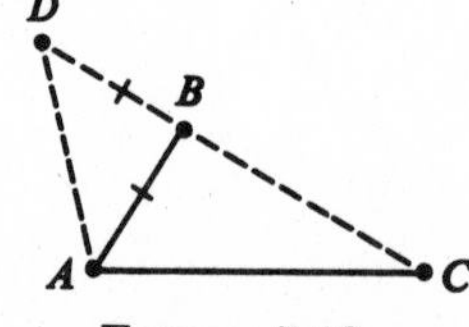

FIGURE 7.10

Note that this theorem was stated and proved in terms of distance, rather than in terms of the relation of congruence between segments. Here we have departed from the style of the preceding few chapters. The reasons for this departure will be explained in the following section. (The questions involved here are more complicated than you might think.)

Theorem 6. *The Hinge Theorem.* If two sides of one triangle are congruent, respectively, to two sides of a second triangle, and the included angle of the first triangle is larger than the included angle of the second triangle, then the opposite side of the first triangle is larger than the opposite side of the second triangle.

RESTATEMENT. Given $\triangle ABC$ and $\triangle DEF$. If $\overline{AB} \cong \overline{DE}$, $\overline{AC} \cong \overline{DF}$, and $\angle A > \angle D$, then $\overline{BC} > \overline{EF}$.

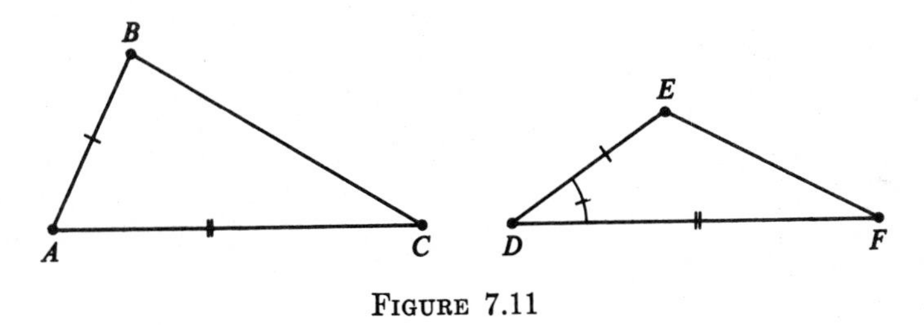

FIGURE 7.11

Proof. (1) We assert that there is a point K, in the interior of $\angle BAC$, such that $\triangle AKC \cong \triangle DEF$.

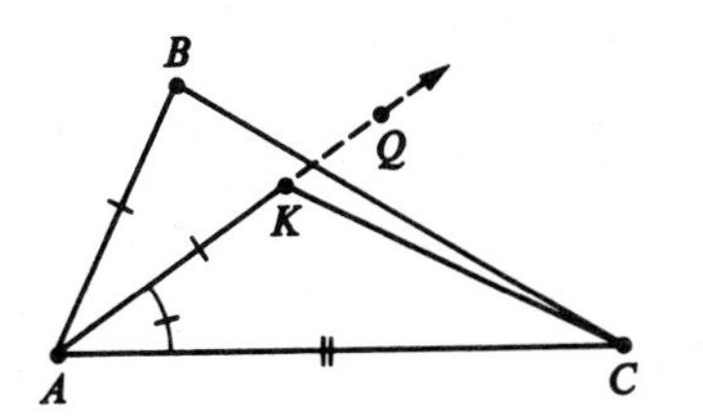

FIGURE 7.12

To show this, we first take a point Q, on the same side of $\overleftrightarrow{AC}$ as B, such that $\angle QAC \cong \angle EDF$. (This is by the angle-construction theorem.) Since $\angle A > \angle D$, Q is in the interior of $\angle BAC$; in fact, every point of $\overrightarrow{AQ} - A$ is in the interior of $\angle BAC$. Let K be the point of $\overrightarrow{AQ}$ such that $\overline{AK} \cong \overline{DE}$. By SAS, we have $\triangle AKC \cong \triangle DEF$, which is what we wanted.

(2) Next, let $\overrightarrow{AR}$ be the bisector of $\angle BAK$.

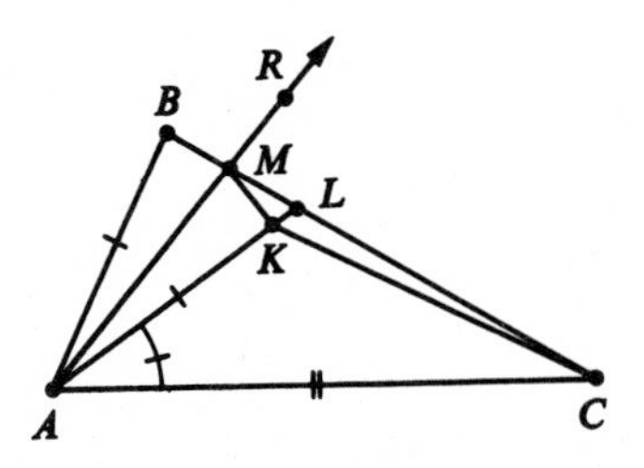

FIGURE 7.13

(See Theorem 4 of Section 6.2, which asserts that every angle has exactly one bisector.) By the crossbar theorem, $\overrightarrow{AK}$ intersects $\overline{BC}$ in a point L; and by another application of the crossbar theorem, $\overrightarrow{AR}$ intersects $\overline{BL}$ in a point M.

(3) By SAS, we have

$$\triangle ABM \cong \triangle AKM.$$

Therefore $MB = MK$. By Theorem 5, we know that

$$CK < CM + MK.$$

Therefore

$$CK < CM + MB,$$

because $MB = MK$. Now $CM + MB = CB$, because C-M-B. And $CK = EF$, because $\triangle AKC \cong \triangle DEF$. Therefore, we finally get

$$EF < CB,$$

which was what we wanted.

The above proof is correct and complete as it stands, but it works for reasons that are a little trickier than one might suspect. The last of the figures given does

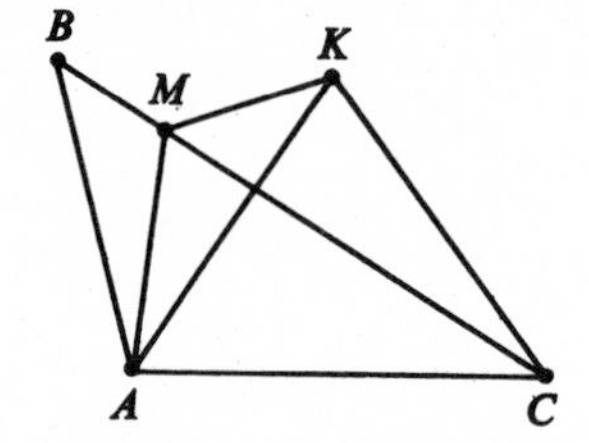

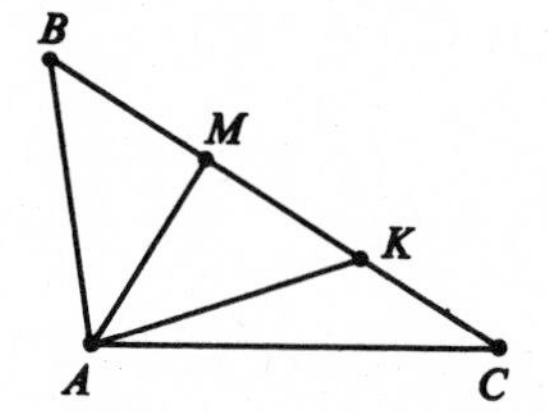

Figure 7.14

not indicate all of the possibilities. The figures might look like either of those above. The proof that we have given applies word for word, to the figure on the left, and for the figure on the right we merely need to give a different reason for the inequality $CK < CM + MK$.

Finally, we observe that the SAA theorem can be proved merely on the basis of the exterior angle theorem, without the use of the parallel postulate or theorems based on it.

Theorem 7. *The SAA Theorem.* Given a correspondence between two triangles. If two angles and a side of the first triangle are congruent to the corresponding parts of the second, then the correspondence is a congruence.

Restatement. Given $\triangle ABC$, $\triangle DEF$, and $ABC \leftrightarrow DEF$. If $\overline{AB} \cong \overline{DE}$, $\angle A \cong \angle D$, and $\angle C \cong \angle F$, then $\triangle ABC \cong \triangle DEF$.

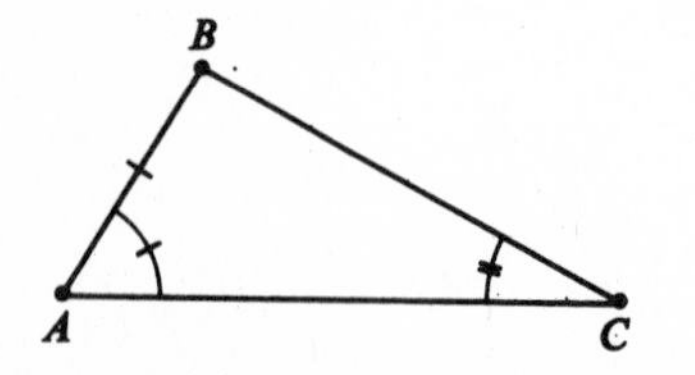

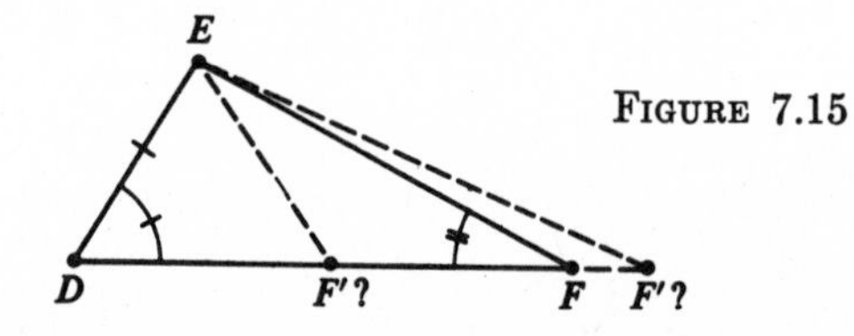

Figure 7.15

(Note that the case where the given side is included between the two angles has already been taken care of by the ASA theorem.)

Proof. Let F' be a point of $\overrightarrow{DF}$, such that $\overline{DF'} \cong \overline{AC}$. By SAS we have

$$\triangle ABC \cong \triangle DEF'.$$

Therefore $\angle F' \cong \angle C \cong \angle F$. But we must have (1) D-F-F', (2) D-F'-F or (3) $F = F'$. If D-F-F', then $\angle F$ is an exterior angle of $\triangle EFF'$, so that $\angle F > \angle F'$, which is false. If D-F'-F, then $\angle F'$ is an exterior angle of $\triangle EFF'$, and $\angle F' > \angle F$, which is also false. Therefore $F' = F$, and $\triangle ABC \cong \triangle DEF$, which was to be proved.

A triangle is called a *right* triangle if one of its angles is a right angle. By Corollary 1-1 we know that every triangle has at most one right angle. In a right triangle, the side opposite the right angle is called the *hypotenuse*, and the other two sides are called the *legs*. The following does not depend on the parallel postulate.

Theorem 8. *The Hypotenuse-Leg Theorem.* Given a correspondence between two right triangles. If the hypotenuse and one leg of one of the triangles are congruent to the corresponding parts of the other triangle, then the correspondence is a congruence.

RESTATEMENT. Given $\triangle ABC$ and $\triangle DEF$, such that $m\angle A = m\angle D = 90$, $\overline{AB} \cong \overline{DE}$, and $\overline{BC} \cong \overline{EF}$. Then $\triangle ABC \cong \triangle DEF$.

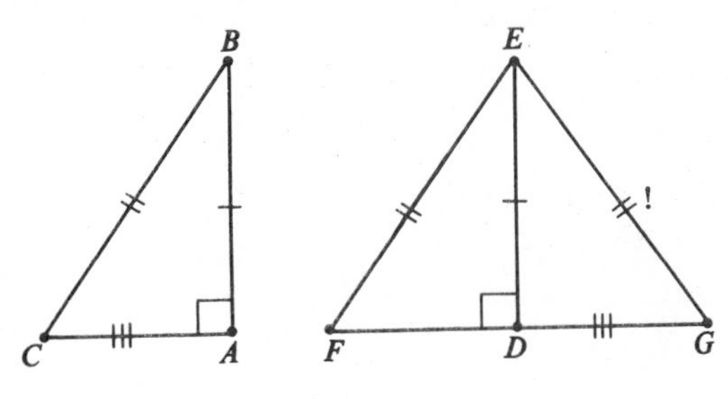

FIGURE 7.16

Proof. Let G be the point such that F-D-G and $\overline{DG} \cong \overline{AC}$. By the Supplement Postulate, $\angle EDG$ is a right angle, and $\angle EDG \cong \angle BAC$. By SAS we have $\triangle ABC \cong \triangle DEG$. It follows that $\overline{EG} \cong \overline{BC}$. Therefore $\overline{EG} \cong \overline{EF}$. By the Isosceles Triangle Theorem (Theorem 1 of Section 6.2) we have $\angle F \cong \angle G$. By the SAA Theorem, $\triangle DEG \cong \triangle DEF$. Therefore $\triangle ABC \cong \triangle DEF$, which was to be proved.

Chapter 8
The Euclidean Program: Congruence without Distance

8.1 THE SYNTHETIC POSTULATES

So far in this book, the real numbers have played a central role. We recall that the structure is

$$[S, \mathcal{L}, \mathcal{P}, d, m],$$

where d and m are real-valued functions, defined for point pairs and angles, respectively. The ideas of congruence for segments, betweenness for points on a line, and congruence for angles were defined in terms of distance and angular measure, in the following way.

(1) A-B-C means (by definition) that A, B, and C are different points of the same line, and

$$AB + BC = AC$$

(where PQ is the distance $d(P, Q)$ between P and Q).

(2) $\overline{AB}$ is defined as the union of A, B, and all points between A and B.

(3) $\overline{AB} \cong \overline{CD}$ means (by definition) that $AB = CD$.

(4) $\angle A \cong \angle B$ means (by definition) that $m\angle A = m\angle B$.

Under this scheme, nearly all the basic properties of betweenness and congruence for segments and angles could be proved as theorems; the only exception was the SAS postulate.

This scheme of presentation for geometry is not the classical one. It was proposed in the early 1930's by G. D. Birkhoff, and has only recently become popular. The classical scheme, as one finds it in Euclid or in David Hilbert's *Foundations of Geometry,* is different; the basic difference is that in Euclid the real numbers do not appear at all at the beginning, and appear only in a disguised form even at the end (see Chapter 20). Euclid's treatment of geometry is called the *synthetic* treatment. The Birkhoff scheme is called *metric,* because it uses measurement.

We now give a sketch of how the Euclid-Hilbert treatment works. At the beginning we have the system

$$[S, \mathcal{L}, \mathcal{P}].$$

The treatment of incidence theorems is exactly like the one given in Chapter 2. Immediately thereafter, however, a difference appears. Instead of adding to our structure the real-valued functions d and m, we add the following things.

(1) An undefined relation of *betweenness*, for triplets of points. This is regarded as given, in the same way that d and m were regarded as given, subject to certain postulates to be stated soon.

Segments are defined, as before, in terms of betweenness. Similarly for rays and angles. We then add to our structure the following:

(2) An undefined relation of *congruence* for segments, and

(3) An undefined relation, also called *congruence*, for angles.

It does no harm to denote the two undefined congruence relations by the same symbol, $\cong$. If we represent the betweenness relation as $\mathcal{B}$, then the structure becomes

$$[S, \mathcal{L}, \mathcal{P}, \mathcal{B}, \cong].$$

It is now out of the question to prove theorems about betweenness and congruence until we have postulates which describe their properties. These are in three groups, as follows.

Betweenness Postulates

B-1. If A-B-C, then C-B-A.

B-2. Of any three points of a line, exactly one is between the other two.

B-3. Any four points of a line can be named in an order A, B, C, D, in such a way that A-B-C-D.

B-4. If A and B are any two points, then (1) there is a point C such that A-B-C, and (2) there is a point D such that A-D-B.

B-5. If A-B-C, then A, B, and C are three different points of the same line.

These statements are the same as the theorems that were proved on the basis of the ruler postulate in Section 3.4, where betweenness was defined in terms of distance.

Congruence Postulates for Segments

C-1. For segments, congruence is an equivalence relation.

C-2. *The Segment-Construction Postulate.* Given a segment $\overline{AB}$ and a ray $\overrightarrow{CD}$. There is exactly one point E of $\overrightarrow{CD}$ such that $\overline{AB} \cong \overline{CE}$.

C-3. *The Segment-Addition Postulate.* If (1) A-B-C, (2) A'-B'-C', (3) $\overline{AB} \cong \overline{A'B'}$, and (4) $\overline{BC} \cong \overline{B'C'}$, then (5) $\overline{AC} \cong \overline{A'C'}$.

C-4. *The Segment-Subtraction Postulate.* If (1) A-B-C, (2) A'-B'-C', (3) $\overline{AB} \cong \overline{A'B'}$, and (4) $\overline{AC} \cong \overline{A'C'}$, then (5) $\overline{BC} \cong \overline{B'C'}$.

C-5. Every segment has exactly one midpoint; that is, for every $\overline{AB}$ there is exactly one point C such that A-C-B and $\overline{AC} \cong \overline{CB}$.

These statements are the theorems that were proved in Section 3.6, where congruence for segments was defined in terms of distance.

The separation postulates and the treatment of sides of a line, interiors of angles, and so on, are exactly as in Chapter 4.

Congruence Postulates for Angles

C-6. For angles, congruence is an equivalence relation.

C-7. *The Angle-Construction Postulate.* Let $\angle ABC$ be an angle, let $\overrightarrow{B'C'}$ be a ray, and let H be a half plane whose edge contains $\overrightarrow{B'C'}$. Then there is exactly one ray $\overrightarrow{B'A'}$, with A' in H, such that $\angle ABC \cong \angle A'B'C'$.

C-8. *The Angle-Addition Postulate.* If (1) D is in the interior of $\angle BAC$, (2) D' is in the interior of $\angle B'A'C'$, (3) $\angle BAD \cong \angle B'A'D'$, and (4) $\angle DAC \cong \angle D'A'C'$, then (5) $\angle BAC \cong \angle B'A'C'$.

C-9. *The Angle-Subtraction Postulate.* If (1) D is in the interior of $\angle BAC$, (2) D' is in the interior of $\angle B'A'C'$, (3) $\angle BAD \cong \angle B'A'D'$, and (4) $\angle BAC \cong \angle B'A'C'$, then (5) $\angle DAC \cong \angle D'A'C'$.

These statements are the ones that were proved in Section 5.1, where congruence for angles was defined in terms of the angular-measure function m.

Finally, we state the SAS postulate exactly as before.

Let us now consider the question of how much of our work in the preceding sections needs to be done over again if we decide, for some reason, that it would be good to get along without using the real-valued functions d and m. The answer is that *hardly any of it needs to be done over again,* because until we got to the preceding chapter, on geometric inequalities, *we almost never cited the definitions of betweenness and congruence except at the very beginning in each case, when we were proving the basic theorems that we now propose to take as postulates.* The next time you look at these early chapters, you might try checking this statement, a line at a time. The first place where you will get into trouble is at the end of Chapter 5, where we used angular measure in the proof of the vertical angle theorem. Theorem 1 of Section 8.3 can be used to replace angular measure in this proof. The second difficulty is at the end of Chapter 6, where we pointed out that we were postponing half of the theorem that one would naturally expect on the existence of perpendiculars to a line L, through a point P, in a given plane. We showed (Theorem 1, Section 6.5) that when P is not on L, such a perpendicular always exists. We still need to prove, on the basis of the postulates in this chapter, that when P lies on L, the same conclusion follows; and we still need to prove, on the same postulational basis, that all right angles are congruent.

The synthetic treatment of congruence begins with a synthetic treatment of inequalities. See the following sections.

8.2 THE LAWS OF INEQUALITY FOR SEGMENTS

At the start we have to explain, without mentioning distances, what it means to say that one segment is shorter than another or that one angle is smaller than another. The appropriate definitions have already been suggested in Chapter 7.

DEFINITION. $\overline{AB} < \overline{CD}$ if there is a point B', between C and D, such that $\overline{AB} \cong \overline{CB'}$.

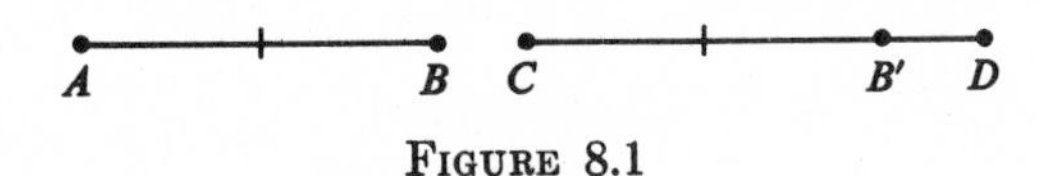

FIGURE 8.1

The basic properties of the relation $<$ are

(I) For every pair of segments $\overline{AB}$, $\overline{CD}$, exactly one of the following conditions holds:

$$\overline{AB} < \overline{CD}, \qquad \overline{AB} \cong \overline{CD}, \qquad \overline{CD} < \overline{AB}.$$

(II) If $\overline{AB} < \overline{CD}$ and $\overline{CD} < \overline{EF}$, then $\overline{AB} < \overline{EF}$.

(III) If $\overline{AB} \cong \overline{A'B'}$, $\overline{CD} \cong \overline{C'D'}$ and $\overline{AB} < \overline{CD}$, then $\overline{A'B'} < \overline{C'D'}$.

Note that (I) and (II) are very much like Conditions O-1 and O-2 for an order relation, with $=$ replaced by $\cong$, throughout. We shall now verify (I), (II), and (III).

Theorem 1. If A-B-C and A-C-D, then A-B-C-D.

Proof. We know that A, B, C, and D can be arranged in an order W, X, Y, Z, so that W-X-Y-Z. Here W cannot be B or C, because W is not between any two of the other three points. For the same reason, Z cannot be B or C. Therefore W and Z must be A and D in some order. Since W-X-Y-Z and Z-Y-X-W say the same thing, we can assume $W = A$, and $Z = D$. This gives

$$A\text{-}X\text{-}Y\text{-}D,$$

where X and Y are B and C. We cannot have A-C-B-D, because A-B-C. Therefore we have A-B-C-D, which was to be proved.

Theorem 2. If A-B-C and $\overline{AC} \cong \overline{A'C'}$, then there is a point B' such that A'-B'-C' and $\overline{AB} \cong \overline{A'B'}$.

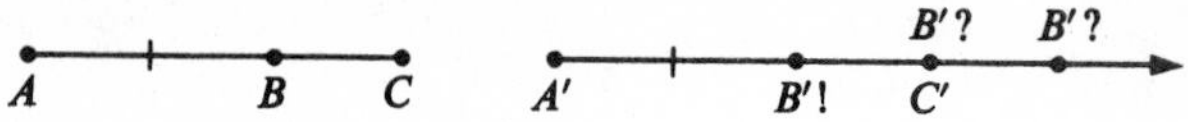

FIGURE 8.2

Proof. By the segment-construction postulate, C-2, we know that there is exactly one point B' on the ray $\overrightarrow{A'C'}$ such that $\overline{A'B'} \cong \overline{AB}$. There are now three possibilities:

$$(1)\ B' = C', \qquad (2)\ A'\text{-}C'\text{-}B', \qquad (3)\ A'\text{-}B'\text{-}C'.$$

We shall show that both (1) and (2) are impossible. It will follow that (3) holds true.

(1) Suppose that $B' = C'$. Then the ray $\overrightarrow{AC}$ contains *two* points X (namely, $X = B$ and $X = C$) such that $\overline{AX} \cong \overline{A'C'}$. This contradicts the segment-construction postulate, C-2.

(2) Suppose that $A'\text{-}C'\text{-}B'$. By the segment-construction postulate there is a point D on the ray opposite to $\overrightarrow{CA}$ such that $\overline{CD} \cong \overline{C'B'}$.

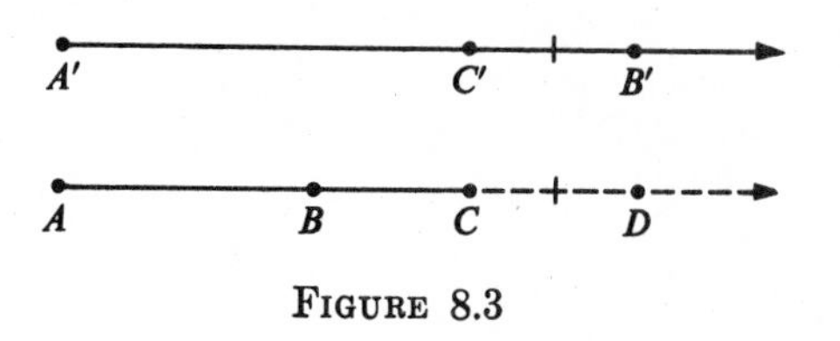

FIGURE 8.3

Thus $A\text{-}C\text{-}D$, $A'\text{-}C'\text{-}B'$, $\overline{AC} \cong \overline{A'C'}$, and $\overline{CD} \cong \overline{C'B'}$. Therefore, by the segment-addition postulate, we have $\overline{AD} \dot{\cong} \overline{A'B'}$. Since $\overline{A'B'} \cong \overline{AB}$, this gives $\overline{AD} \cong \overline{AB}$, which contradicts the uniqueness condition in the segment-construction postulate.

Theorem 3. If $\overline{AB} < \overline{CD}$, and $\overline{CD} \cong \overline{C'D'}$, then $\overline{AB} < \overline{C'D'}$.

Proof. We have a point B' such that $C\text{-}B'\text{-}D$ and $\overline{CB'} \cong \overline{AB}$. By Theorem 2, there is a point B'' such that $C'\text{-}B''\text{-}D'$ and $\overline{C'B''} \cong \overline{CB'}$. But $\overline{AB} \cong \overline{CB'}$. Therefore

$$\overline{AB} \cong \overline{C'B''}, \qquad \text{and} \qquad \overline{AB} < \overline{C'D'},$$

which was to be proved.

Theorem 4. If $\overline{AB} < \overline{CD}$, and $\overline{A'B'} \cong \overline{AB}$, then $\overline{A'B'} < \overline{CD}$.

This is easy even without the preceding theorems. Proof? Fitting these together, we get:

Theorem 5. If $\overline{AB} \cong \overline{A'B'}$, $\overline{CD} \cong \overline{C'D'}$, and $\overline{AB} < \overline{CD}$, then $\overline{A'B'} < \overline{C'D'}$.

This, of course, is Condition (III).

Theorem 6. The relation $\overline{AB} < \overline{AB}$ never holds, for any segment $\overline{AB}$.

Proof. If $\overline{AB} \cong \overline{AB'}$, for some point B' between A and B, then this contradicts the segment-construction postulate C-2.

Theorem 7. If $\overline{AB} < \overline{CD}$ and $\overline{CD} < \overline{EF}$, then $\overline{AB} < \overline{EF}$.

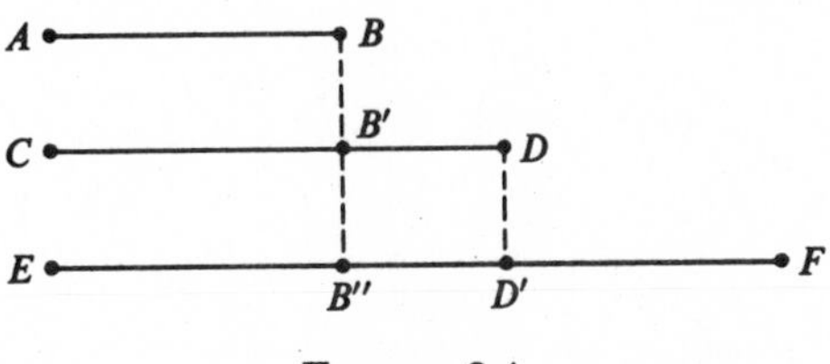

FIGURE 8.4

Proof. Take D' so that $E\text{-}D'\text{-}F$ and $\overline{ED'} \cong \overline{CD}$ (Fig. 8.4). By Theorem 3, there is a point B'' such that $E\text{-}B''\text{-}D'$ and $\overline{AB} \cong \overline{EB''}$. Since $E\text{-}B''\text{-}D'$ and $E\text{-}D'\text{-}F$, it follows by Theorem 1 that $E\text{-}B''\text{-}D'\text{-}F$, so that $E\text{-}B''\text{-}F$. Therefore $\overline{AB} < \overline{EF}$, which was to be proved. This is Condition (II).

Theorem 8. For every pair of segments $\overline{AB}$ and $\overline{CD}$, exactly one of the following conditions holds:

$$\overline{AB} < \overline{CD}, \qquad \overline{AB} \cong \overline{CD}, \qquad \overline{CD} < \overline{AB}.$$

Proof. There is a point B', on $\overrightarrow{CD}$, such that $\overline{CB'} \cong \overline{AB}$.

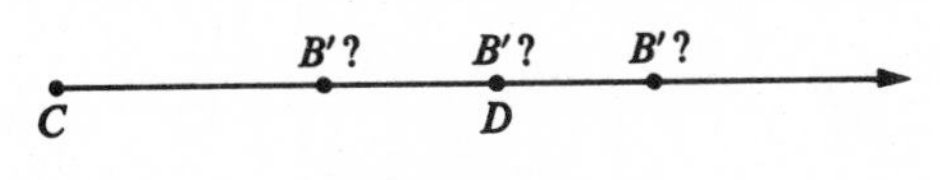

FIGURE 8.5

If $C\text{-}B'\text{-}D$, then (1) $\overline{AB} < \overline{CD}$. If $B' = D$, then (2) $\overline{AB} \cong \overline{CD}$. If $C\text{-}D\text{-}B'$, then by Theorem 2 it follows that there is a point D', between A and B, such that $\overline{AD'} \cong \overline{CD}$. Therefore (3) $\overline{CD} < \overline{AB}$. Thus at least one of the stated conditions holds.

And no two of these conditions can hold. If $\overline{AB} < \overline{CD}$ and $\overline{AB} \cong \overline{CD}$, then it follows by Theorem 4 that $\overline{CD} < \overline{CD}$, which contradicts Theorem 6. Similarly if $\overline{AB} \cong \overline{CD}$ and $\overline{CD} < \overline{AB}$. Finally, if $\overline{AB} < \overline{CD}$ and $\overline{CD} < \overline{AB}$, then it follows by Theorem 7 that $\overline{AB} < \overline{AB}$, which contradicts Theorem 6.

We have now verified (I), (II), and (III).

The synthetic treatment of inequalities for angles is very similar, and we shall not go through it in detail. It begins as follows.

DEFINITION. Given $\angle ABC$ and $\angle DEF$. If there is a point G in the interior of $\angle DEF$, such that $\angle ABC \cong \angle GEF$, then $\angle ABC < \angle DEF$.

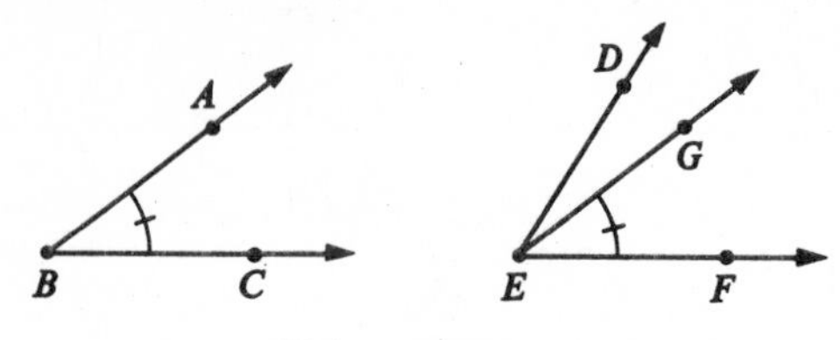

FIGURE 8.6

The basic properties of this relation are precisely analogous to (I), (II), and (III) above.

(IV) For every pair of angles $\angle A$, $\angle B$, exactly one of the following conditions holds:

$$\angle A < \angle B, \qquad \angle A \cong \angle B, \qquad \angle B < \angle A.$$

(V) If $\angle A < \angle B$ and $\angle B < \angle C$, then $\angle A < \angle C$.

(VI) If $\angle A \cong \angle A'$, $\angle B \cong \angle B'$ and $\angle A < \angle B$, then $\angle A' < \angle B'$.

The proofs are not trivial.

8.3 RIGHT ANGLES, SYNTHETICALLY CONSIDERED

The first step in handling right angles is to prove the following theorem, which is closely related to C-8.

Theorem 1. If (1) $\angle BAC$ and $\angle CAD$ form a linear pair, (2) $\angle B'A'C'$ and $\angle C'A'D'$ form a linear pair, and (3) $\angle CAD \cong \angle C'A'D'$, then (4) $\angle BAC \cong \angle B'A'C'$.

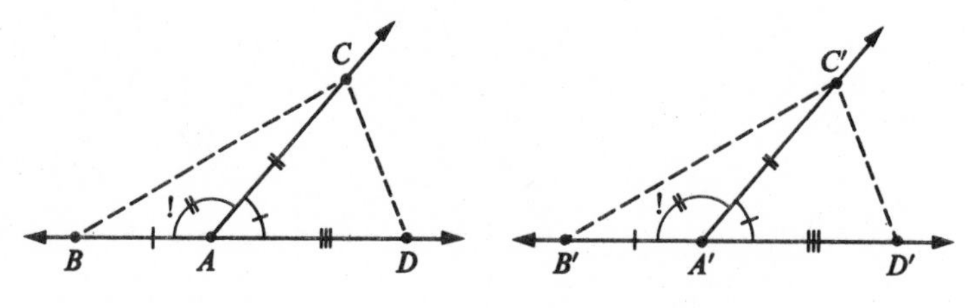

FIGURE 8.7

Proof. Evidently there is no harm in assuming that

$$\overline{AB} \cong \overline{A'B'}, \qquad \overline{AC} \cong \overline{A'C'}, \qquad \overline{AD} \cong \overline{A'D'},$$

since all six of our rays are equally well described if the points B', C', D' are chosen in this way. By SAS, we have $\triangle ADC \cong \triangle A'D'C'$. Therefore $\angle ADC \cong$

$\angle A'D'C'$; and so $\angle BDC \cong \angle B'D'C'$. By SAS, this means that $\triangle BDC \cong \triangle B'D'C'$. Therefore $\overline{BC} \cong \overline{B'C'}$. By the SSS theorem, $\triangle BAC \cong \triangle B'A'C'$. Therefore $\angle BAC \cong \angle B'A'C'$, which was to be proved.

Theorem 2. Any angle congruent to a right angle is also a right angle.

RESTATEMENT. Suppose that (1) $\angle BAC$ and $\angle CAD$ form a linear pair, and (2) $\angle BAC \cong \angle CAD$. Suppose that (3) $\angle B'A'C'$ and $\angle C'A'D'$ form a linear pair, and (4) $\angle B'A'C' \cong \angle BAC$. Then (5) $\angle B'A'C' \cong \angle C'A'D'$.

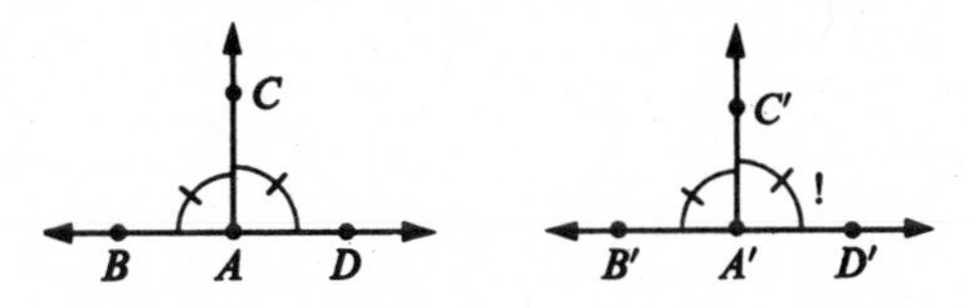

FIGURE 8.8

You should analyze this restatement carefully to see that it really is a restatement of Theorem 2. Once we have formulated the theorem in this way, the proof is trivial. By the preceding theorem, $\angle C'A'D' \cong \angle CAD$. By (2), $\angle CAD \cong \angle BAC$. Therefore

$$\angle C'A'D' \cong \angle BAC \cong \angle B'A'C', \qquad \text{and} \qquad \angle C'A'D' \cong \angle B'A'C',$$

which was to be proved.

Theorem 3. All right angles are congruent.

RESTATEMENT. Suppose that $\angle BAC$ and $\angle CAD$ form a linear pair and are congruent. Suppose that $\angle B'A'C'$ and $\angle C'A'D'$ form a linear pair and are congruent. Then $\angle BAC \cong \angle B'A'C'$.

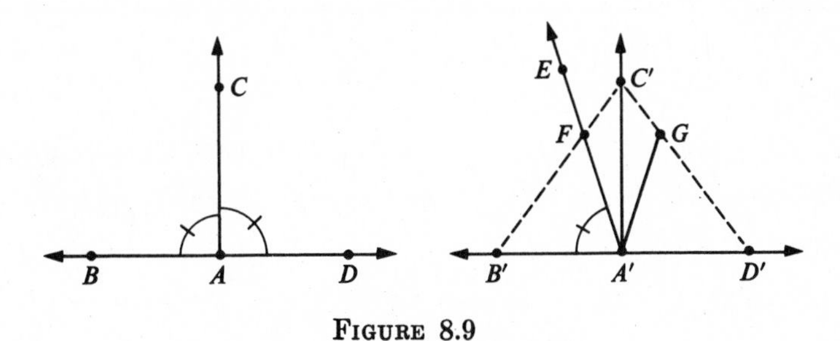

FIGURE 8.9

Proof. First we observe that the points B', and D' can be chosen so that

$$\overline{A'B'} \cong \overline{A'D'},$$

as indicated in the figure. It follows, by SAS, that $\triangle A'B'C' \cong \triangle A'D'C'$, so that we have also

$$\angle B' \cong \angle D'.$$

Let $\overrightarrow{A'E}$ be a ray, with E on the same side of $\overleftrightarrow{A'D'}$ as C', such that

$$\angle B'A'E \cong \angle BAC.$$

We need to prove that $\overrightarrow{A'E} = \overrightarrow{A'C'}$; it will then follow that $\angle BAC \cong \angle B'A'C'$.

If $\overrightarrow{A'E} \neq \overrightarrow{A'C'}$, then E lies in the interior of one of the angles, $\angle B'A'C'$, Then $\overrightarrow{A'E}$ intersects $\overline{B'C'}$ in a point F, such that B'-F-C'. By definition of $<$ (for segments), $\overline{B'F'} < \overline{B'C'}$. Since $\overline{B'C'} \cong \overline{D'C'}$, it follows by (III) of Section 8.2 that $\overline{B'F} < \overline{D'C'}$, which means that there is a point G such that D'-G-C' and $\overline{D'G} \cong \overline{B'F}$. By SAS we have

$$\triangle B'A'F \cong \triangle D'A'G.$$

Therefore

$$\angle B'A'F \cong \angle D'A'G.$$

But by Theorem 1, we know that $\angle D'A'E \cong \angle CAD$. Therefore $\angle D'A'E \cong \angle CAD \cong \angle BAC \cong \angle B'A'E \cong \angle D'A'G$.

We now have the following situation:

(1) G and E are both on that side of $\overleftrightarrow{A'D'}$ which contains C'.

(2) $\angle D'A'E \cong \angle D'A'G$.

(3) $\overrightarrow{A'E}$ and $\overrightarrow{A'G}$ are different, because E and G are on opposite sides of $\overleftrightarrow{A'C'}$ (the sides that contain B' and D', respectively).

This is impossible, by C-7, because that postulate says that conditions (1) and (2) determine a unique ray.

This proof is taken from Hilbert's *Foundations*, with many details added. The complications that arise in the proof are typical of what happens when one undertakes to reduce postulates to a minimum, in a subject as complicated as geometry. (Euclid postulated that all right angles are congruent, and Hilbert showed that the postulate was unnecessary.) Hilbert's postulates, in *Foundations*, were considerably weaker than those given in Section 8.1. In this book, our purpose is not to give the weakest postulates that can be made to work, but merely to give a valid and workable scheme.

Theorem 4. Given a plane E, a line L lying in E, and a point P of E. There is exactly one line in E which contains P and is perpendicular to L.

Proof. We already know that this is true for the case in which P is not on L. (See Theorem 1, Section 6.5; the proof of this theorem was essentially synthetic, based merely on C-7 and not on distance or angular measure.) We also know, by Theorem 1, Section 6.5, that *some* angles are right angles. Let Q be any point of L other than P. By C-7, there is a point R on one side of L in E, such that $\angle QPR$ is *congruent* to a right angle. By Theorem 2, this means that $\angle QPR$ *is* a right angle, and so $\overleftrightarrow{PR} \perp L$. If there were two such rays $\overrightarrow{PR}$, $\overrightarrow{PR'}$, then we would have $\angle QPR \cong \angle QPR'$, because all right angles are congruent. This is impossible, by C-7.

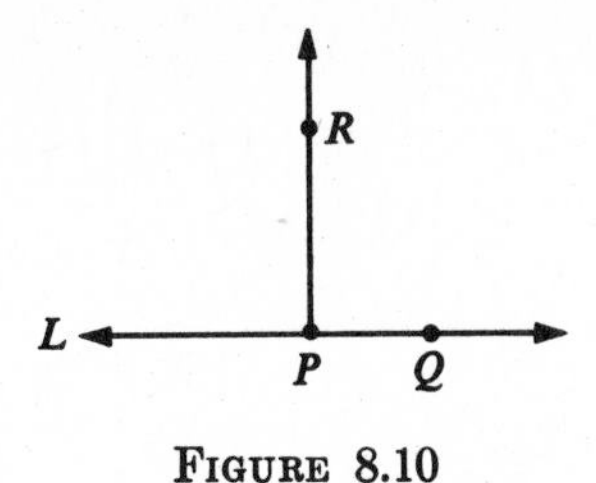

FIGURE 8.10

Note that in a metric treatment we see immediately that $\angle A$ is a right angle if and only if $m\angle A = 90$. Thus, in a metric treatment all the theorems in this section are almost too trivial to be worthy of explicit statement.

8.4 THE SYNTHETIC FORM OF THE TRIANGULAR INEQUALITY. ADDITION OF CONGRUENCE CLASSES

If we adopt the synthetic definitions of inequalities, for angles and segments, and reexamine Chapter 7 from this viewpoint, we find that the theory works in essentially the same way as before, until we get to the Triangular Inequality (Theorem 5 of Section 7.1). Our problems do not begin with the proof; in fact, our first problem is to *state* the theorem without mentioning distances. The following is logically correct, but not very natural.

Theorem 1. Given $\triangle ABC$, there is a point A' such that $\overline{A'B} \cong \overline{AB}$, A'-B-C, and $\overline{A'C} > \overline{AC}$.

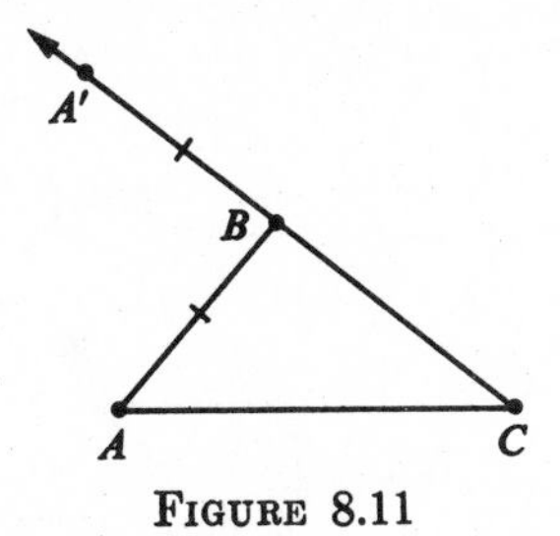

FIGURE 8.11

This really does convey the idea, because it says, intuitively speaking, that if $\overline{AB}$ and $\overline{BC}$ are laid end to end, they form a segment which is longer than $\overline{AC}$ (Fig. 8.11).

Note that the difficulty in handling this theorem synthetically is that it deals with the idea of addition. In metric geometry, the situation is simple, because the addition is performed with numbers. In synthetic geometry, we need to do a lot more talking, if we want to make sense, because the "sum" of two segments can be regarded as a segment only when the segments are end to end, like this:

FIGURE 8.12

Here it is reasonable to call $\overline{AC}$ the "sum" of $\overline{AB}$ and $\overline{BC}$. But if the segments look like this:

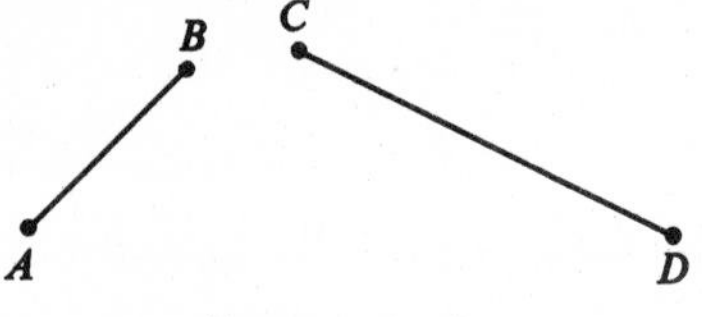

FIGURE 8.13

it is not plain what segment their sum ought to be.

The easiest way around this is as follows. Given $\overline{AB}$, let $[\overline{AB}]$ be *the set of all segments that are congruent to* $\overline{AB}$. Obviously, if $\overline{AB} \cong \overline{CD}$, then $[\overline{AB}] = [\overline{CD}]$. The sets $[\overline{AB}]$ are called *congruence classes.* What we have just pointed out is that a congruence class is equally well described by any of its members.

Suppose now that two segments $\overline{AB}$ and $\overline{CD}$ are given. Then there are always points X, Y, Z such that

$$X\text{-}Y\text{-}Z, \qquad \overline{XY} \cong \overline{AB}, \qquad \overline{YZ} \cong \overline{CD}.$$

The following observations are easy to check, on the basis of the congruence postulates.

(1) If X', Y', Z' are any other three points satisfying the same conditions, then it follows by the segment-addition postulate that $\overline{X'Z'} \cong \overline{XZ}$. That is, *the congruence class* $[\overline{XZ}]$ *is independent of the choice of* X, Y *and* Z.

(2) Suppose that $\overline{AB} \cong \overline{A'B'}$ and $\overline{CD} \cong \overline{C'D'}$. Let X, Y, Z be chosen for $\overline{AB}$ and $\overline{CD}$, as above; and let X', Y', Z' be chosen for $\overline{A'B'}$ and $\overline{C'D'}$. Then $\overline{XZ} \cong \overline{X'Z'}$. That is, *the congruence class* $[\overline{XZ}]$ *depends only on the congruence classes* $[\overline{AB}]$ *and* $[\overline{CD}]$; it is independent of the choice of representative segments $\overline{AB}$ and $\overline{CD}$.

Addition can now be defined, not between segments, of course, but between congruence classes. Given $[\overline{AB}]$ and $[\overline{CD}]$, we take X, Y, and Z such that

$$X\text{-}Y\text{-}Z, \qquad \overline{XY} \cong \overline{AB}, \qquad \overline{YZ} \cong \overline{CD}.$$

Then, by definition,

$$[\overline{AB}] + [\overline{CD}] = [\overline{XZ}].$$

Our remarks (1) and (2) show that this definition makes sense. The congruence class $[\overline{XZ}]$ is independent of the choice of $\overline{AB}$, $\overline{CD}$, X, Y, and Z; it depends only on the congruence classes $[\overline{AB}]$ and $[\overline{CD}]$.

Finally, we recall that (III) of Section 8.2 tells us that if $\overline{AB} < \overline{CD}$, then any segment in $[\overline{AB}]$ is less than every segment in $[\overline{CD}]$. We can therefore define

$$[\overline{AB}] < [\overline{CD}]$$

to mean that every segment congruent to $\overline{AB}$ is less than every segment congruent to $\overline{CD}$.

We can now give a more natural statement of the triangular inequality.

Theorem 2. For any triangle $\triangle ABC$, we have

$$[\overline{AB}] + [\overline{BC}] > [\overline{AC}].$$

The purely synthetic treatment is in a way more elegant. But if we write exact definitions, and really prove our theorems, then synthetic elegance must be bought at the price of technical complication. Under the metric scheme, a logically complete treatment is much easier.

8.5 A SUMMARY OF THE DIFFERENCES BETWEEN THE METRIC AND SYNTHETIC APPROACHES

We have now described two fundamentally different approaches to geometry. It may be helpful, by way of outline and review, to give a table indicating how these two approaches differ (Table 8.1). The basic ideas and statements appear in the left-most column, and in the next two columns we indicate how they are treated.

TABLE 8–1

	Metric approach	Synthetic approach
1. The given structure	$[S, \mathcal{L}, \mathcal{P}, d, m]$	$[S, \mathcal{L}, \mathcal{P}, \mathcal{B}, \cong]$
2. Distance	Given, in the structure	Never mentioned
3. Measure for angles	Given, in the structure	Never mentioned
4. Congruence for segments	Defined in terms of distance	Given, in the structure
5. Congruence for angles	Defined in terms of degree measure	Given, in the structure
6. Properties of congruence	Stated in theorems	Stated in postulates
7. Addition	Performed with numbers AB	Performed with congruence classes $[\overline{AB}]$
8. Inequalities	Defined between numbers, $AB < CD$	Defined between congruence classes, $[\overline{AB}] < [\overline{CD}]$

Henceforth, except in Chapter 20, we shall use the metric structure

$$[S, \mathcal{L}, \mathcal{P}, d, m].$$

Chapter 9
Three Geometries

9.1 INTRODUCTION

This chapter is purely informal; it does not form a part of the deductive sequence of the rest of the book. In fact, in this chapter we shall not prove anything at all; but everything that we discuss will be taken up more fully later. It may be of some help, however, to sketch in advance the kinds of geometry to which our theorems are going to apply.

For the sake of simplicity, we shall limit ourselves to geometry in a plane. The ideas that we shall discuss can be generalized to three dimensions, but only at the cost of considerable labor.

Two lines are called *parallel* if they lie in the same plane but do not intersect. In a *Euclidean* plane, the familiar parallel postulate holds.

The Euclidean Parallel Postulate. Given a line L and a point P not on L, there is one and only one line L' which contains P and is parallel to L.

This says that parallels always exist and are always unique.

For quite a while—for a couple of millennia, in fact—this proposition was regarded as a law of nature. In the nineteenth century, however, it was discovered by Lobachevski, Bolyai, and Gauss that you could get a consistent mathematical theory by starting with a postulate which states that parallels always exist, but denies that they are unique.

The Lobachevskian Parallel Postulate. Given a line L and a point P not on L, there are at least two lines L', L'' which contain P and are parallel to L.

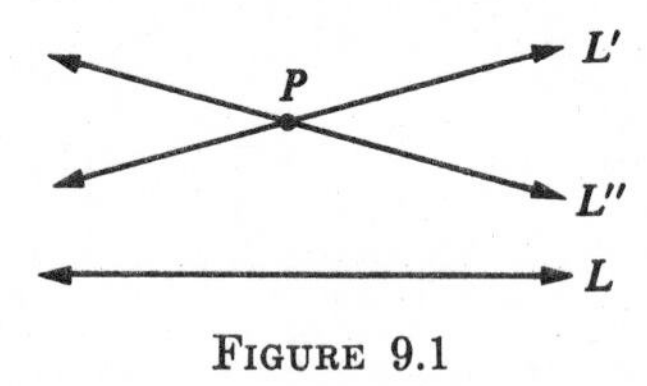

FIGURE 9.1

The picture looks implausible, because we are accustomed to thinking of the plane of the paper as Euclidean. But it is a fact, as we shall see, that a mathematical theory can be based on Lobachevski's postulate. And such a theory has applications in physics.

There is yet a third alternative. We can deny not the uniqueness of parallels but their existence.

The Riemannian Parallel Postulate. No two lines in the same plane are ever parallel.

These postulates give us three kinds of "plane geometry," the Euclidean, the Lobachevskian, and the Riemannian. In each of the three theories, of course, many other postulates are needed; we have merely been singling out their crucial difference. In this book, we shall be concerned mainly with the first of these geometries, incidentally with the second, and hardly at all with the third. In the following sections, we give concrete examples, or *models*, of these kinds of geometry, and indicate the most striking differences between them. In going through the rest of this book, you should have one of these models in mind most of the time; and at some points you should be thinking about two of them.

9.2 THE POINCARÉ MODEL FOR LOBACHEVSKIAN GEOMETRY

In this section we shall assume that there is a mathematical system satisfying the postulates of Euclidean plane geometry, and we shall use Euclidean geometry to describe a mathematical system in which the Euclidean parallel postulate fails, but in which the other postulates of Euclidean geometry hold.

Consider a fixed circle C in a Euclidean plane. We assume, merely for the sake of convenience, that the radius of C is 1. Let E be the interior of C.

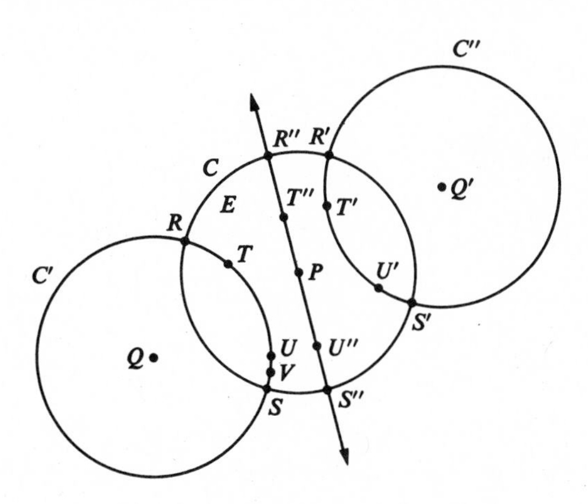

FIGURE 9.2

By an *L-circle* (L for Lobachevski) we mean a circle C' which is orthogonal to C. When we say that two circles are orthogonal, we mean that their tangents at each intersection point are perpendicular. If this happens at one intersection point R, then it happens at the other intersection point S. But we shall not stop to prove this; this chapter is purely descriptive and proofs will come later.

The *points* of our L-plane will be the points of the interior E of C. By an *L-line* we mean (1) the intersection of E and an L-circle, or (2) the intersection of E and a diameter of C.

It is a fact that

I-1. Every two points of E lie on exactly one L-line.

We are going to define a kind of "plane geometry," in which the "plane" is the set E and the lines are the L-lines. In our new geometry we already know what is meant by *point* and *line*. We need next to define distance and angular measure.

For each pair of points X, Y, either on C or in the interior of C, let XY be the usual Euclidean distance. Notice that if R, S, T, and U are as in the figure, then R and S are *not* points of our L-plane, but they *are* points of the Euclidean plane that we started with. Therefore, all of the distances TS, TR, US, UR are defined, and I-1 tells us that R and S are determined when T and U are named. There is one and only one L-line through T and U, and this L-line cuts the circle C in the points R and S. We shall use these four distances TS, TR, US, UR to define a new distance $d(T, U)$ in our "plane" E, by the following formula:

$$d(T, U) = \left|\log_e \frac{TR/TS}{UR/US}\right|.$$

Evidently we have the following postulate.

D-0. d is a function

$$d: E \times E \to \mathbf{R}.$$

Let us now look at the ruler postulate D-4. On any L-line L, take a point U and regard this point as fixed. For every point T of L, let

$$f(T) = \log_e \frac{TR/TS}{UR/US}.$$

That is, $f(T)$ is what we get by omitting the absolute value signs in the formula for $d(T, U)$. We now have a function,

$$f: L \to \mathbf{R}.$$

We shall show that f is a coordinate system for L.

If V is any other point of L, then

$$f(V) = \log_e \frac{VR/VS}{UR/US}.$$

Let $x = f(T)$ and $y = f(V)$. Then

$$|x - y| = \left|\log_e \frac{TR/TS}{UR/US} - \log_e \frac{VR/VS}{UR/US}\right| = \left|\log_e \frac{TR/TS}{VR/VS}\right|,$$

because the difference of the logarithms is the logarithm of the quotient. Therefore

$$|x - y| = d(T, V),$$

which means that our new distance function satisfies the ruler postulate.

Since D-4 holds, the other distance postulates automatically hold. (See Problem 1, Section 3.3.)

We define betweenness, segments, rays, and so on, exactly as in Chapter 3. All of the theorems of Chapter 3 hold in our new geometry, because the new geometry satisfies the postulates on which the proofs of the theorems were based. The same is true of Chapter 4; it is rather easy to convince yourself that the plane-separation postulate holds in E.

To discuss congruence of angles, we need to define an angular-measure function. Given an "L-angle" in our new geometry, we form an angle in the old geometry by using the two tangent rays:

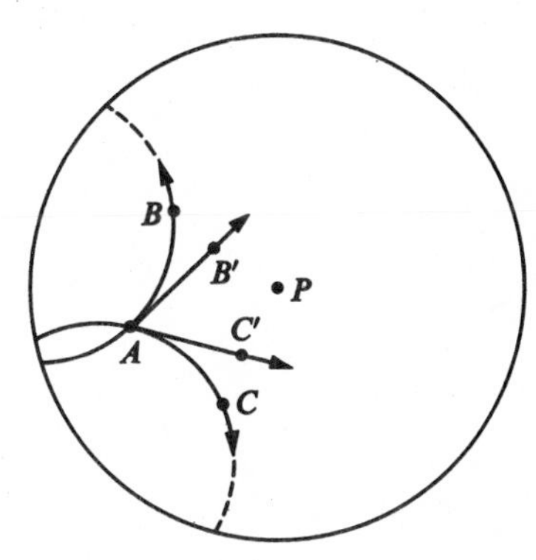

FIGURE 9.3

We then define the measure $m\angle BAC$ of $\angle BAC$ to be the measure (in the old sense) of the Euclidean angle $\angle B'AC'$.

It is a fact that the resulting structure

$$[E, L, d, m]$$

satisfies all the postulates of Chapters 2 through 6, including the SAS postulate. The proof of this takes time, however, and it requires the use of more Euclidean geometry than we know so far. Granted that the postulates hold, it follows that the theorems also hold. Therefore, the whole theory of congruence, and of geometric inequalities, applies to the Poincaré model of Lobachevskian geometry.

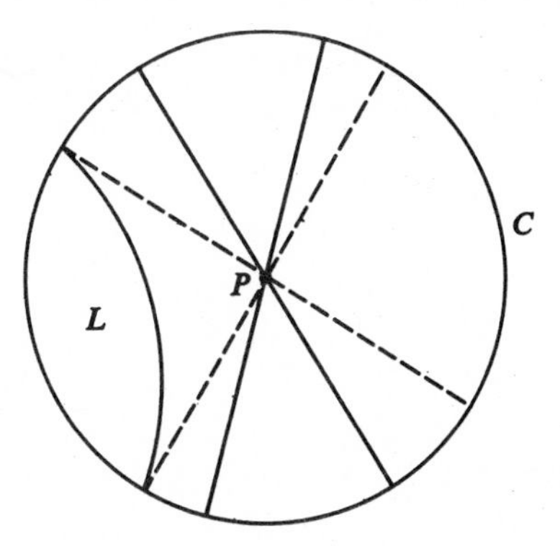

FIGURE 9.4

On the other hand, the Euclidean parallel postulate obviously does not hold for the Poincaré model. Consider, for example, an L-line L which does not pass through the center P of C (Fig. 9.4). Through P there are infinitely many L-lines which are parallel to L.

Lobachevskian geometry (also called *hyperbolic geometry*) is the kind represented by the Poincaré model. In such a geometry, when the familiar parallel postulate fails, it pulls down a great many familiar theorems with it. A few samples of theorems in hyperbolic geometry which are quite different from the analogous theorems of Euclidean geometry follow.

(1) No quadrilateral is a rectangle. In fact, if a quadrilateral has three right angles, the fourth angle is always acute.

(2) For any triangle, the sum of the measures of the angles is always *strictly less* than 180.

(3) No two triangles are ever similar, except in the case where they are also congruent.

The third of these theorems means that two figures cannot have exactly the same shape, unless they also have exactly the same size. Thus, in hyperbolic geometry, *exact scale models are impossible.*

In fact, each of the above three theorems characterizes hyperbolic geometry. If the angle-sum inequality,

$$m\angle A + m\angle B + m\angle C < 180,$$

holds, even for one triangle, then the geometry is hyperbolic; if the angle-sum equality holds, even for one triangle, then the geometry is Euclidean; and similarly for (1) and (3).

This has a curious consequence in connection with our knowledge of physical space. If physical space is hyperbolic, which it may be, it is theoretically possible for the fact to be demonstrated by measurement. For example, suppose that you measure the angles of a triangle, with an error less than $0.0001''$ for each angle. Suppose that the sum of the measures turns out to be $179°59'59.999''$. The difference between this and $180°$ is $0.001''$. This discrepancy could not be due to errors in measurement, because the greatest possible cumulative error is only $0.0003''$. Our experiment would therefore prove that the space that we live in is hyperbolic. (Granted, of course, that it satisfies the other postulates.)

On the other hand, no measurement however exact can prove that space is Euclidean. The point is that every physical measurement involves *some* possible error. Therefore we can never show by measurement that an equation,

$$r + s + t = 180,$$

holds exactly; and this is what we would have to do to prove that the space we live in is Euclidean.

Thus there are two possibilities: (1) The Euclidean parallel postulate does not hold in physical space, or (2) The truth about physical space will never be known.

9.3 THE SPHERICAL MODEL FOR RIEMANNIAN GEOMETRY

Let V be the surface of a sphere in space. We assume that the radius of V is $= 1$. A *great circle* is a circle which is the intersection of V with a plane through its center. If T and U are any points of V, then the shortest path on the surface joining T to U is an arc of a great circle.

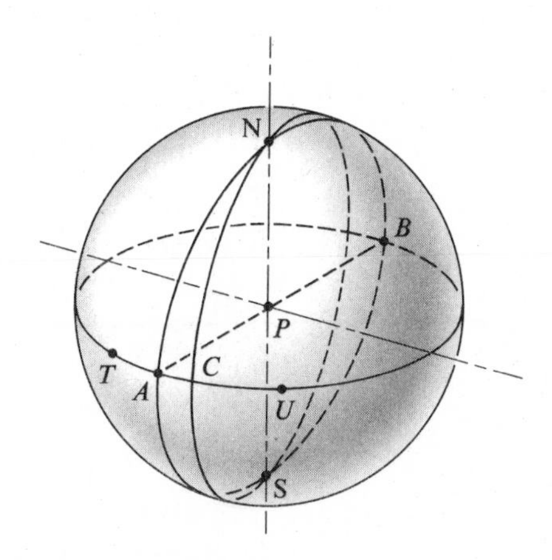

FIGURE 9.5

We might start to define a kind of "plane geometry" on V by taking the great circles as our lines. In this scheme we would take the length of the shortest path between each pair of points as the distance between the two points. The resulting system has some of the properties that we expect in plane geometry. For example, every "line" separates our "plane" into two "half planes," each of which is convex. But the Euclidean parallel postulate fails, because every two lines intersect. Our "geometry" has other peculiar properties.

(1) Two points do not necessarily determine a "line." For example, the north and south poles N and S lie on infinitely many great circles.

The same is true for the end points of any diameter of the sphere V. Such points are called *antipodal.* (More precisely, two points A and B of V are *antipodal* if the segment $\overline{AB}$ passes through the center of V.)

(2) While our "lines" never come to an end at any point, they are nevertheless finite in extent. In fact, since the radius of V is $= 1$, the maximum possible distance between any two points is π. Thus the ruler postulate cannot hold.

(3) Betweenness, in the form in which we are accustomed to it, collapses. In fact, given three points of a line it is not necessarily true that one of them is between the other two. We may have $AB = BC = AC$.

(4) The perpendicular to a line, from an external point, always exists, but is not necessarily unique. For example, any line joining the North Pole to a point of the equator is perpendicular to the equator.

(5) Some triangles have two right angles. (In the Fig. 9.5 at the start of this section, $\triangle ANC$ has right angles at both A and C.)

(6) The exterior angle theorem fails. (See the same example.)

The only one of these peculiarities that we can avoid is the first. We do this by altering the model in the following way. If two points A, B are antipodal, then we shall regard them as being the same. To be more precise, a *point* of our new geometry will be a pair of antipodal points of the sphere V. If A is a point of the sphere V, then $\overline{A}$ denotes the pair $\{A, A'\}$, where A' is the other end of the diameter that contains A. The *points* of our Riemannian plane E will be the pairs $\overline{A}$.

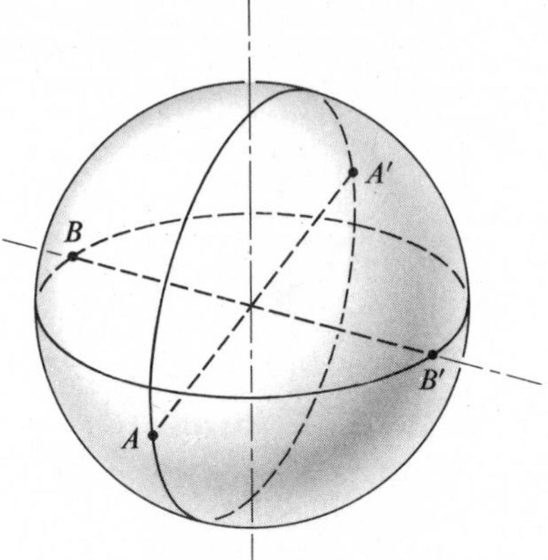

FIGURE 9.6

If L is a great circle on V, then $\overline{L}$ is the set of all points $\overline{A}$ for which A is on L. The sets $\overline{L}$ will be the *lines* in E.

The *distance* $d(\overline{A}, \overline{B})$ between two points $\overline{A}$ and $\overline{B}$ is the length of the shortest arc from A (or A') to B (or B'). Notice that this may be less than the length of the shortest arc from A to B.

In our new geometry, two points $\overline{A}$, $\overline{B}$ always determine a unique "line." The reason is that if A and B were antipodal on the sphere, $\overline{A}$ and $\overline{B}$ would be the same.

The Euclidean parallel postulate still fails, of course; two of our new lines always intersect in exactly one point. Lines are still of finite extent; the maximum possible distance between two points is now $\pi/2$. Betweenness still does not work. Perpendiculars still are not unique; we still have triangles with two right angles, and the exterior angle theorem still fails.

In fact, in arranging for two points to determine a line, we have introduced a new peculiarity: no line separates our Riemannian plane. In fact, if $\overline{L}$ is a line, and $\overline{A}$ and $\overline{B}$ are any two points not on $\overline{L}$, then there is always an arc, lying in a "line" $\overline{L}'$, which goes from $\overline{A}$ to $\overline{B}$ without intersecting $\overline{L}$.

In this book, we shall be concerned mainly with Euclidean geometry, but we shall devote considerable attention to hyperbolic geometry, mainly because it throws light on Euclidean geometry. The point is that these two kinds of geometry have so much in common that at the points where they differ, the differences are instructive. On the other hand, the differences between Riemannian and Euclidean geometry are so fundamental that the former is a technical specialty, remote from our main purpose. We shall not be concerned with it hereafter in this book.

9.4 SOME QUESTIONS FOR LATER INVESTIGATION

In this chapter, we have raised more questions than we have answered.

(1) We have said that the Poincaré model for hyperbolic geometry satisfies all the postulates of Euclidean geometry, with the sole exception of the Euclidean parallel postulate. This needs to be proved, and we haven't proved it with our conversational discussion in Section 9.2.

To check these postulates is a lengthy job. The reader is warned that this sort of verification is dismissed almost casually in much of the literature. If the models for hyperbolic geometry had in common with Euclidean geometry merely the trivial properties that are discussed in semipopular books, they would not have the significance which is commonly and rightly attributed to them.

(2) When the postulates are checked, for the Poincaré model, we will know that hyperbolic geometry is just as good, logically, as Euclidean geometry. We constructed the model on the basis of Euclidean geometry. Therefore, *if* there is a mathematical system satisfying the Euclidean postulates, it follows that there is a system satisfying the Lobachevskian postulates.

(3) There remains the *if* in (2). Is there a system satisfying the Euclidean postulates? To prove this, we need to set up a model. We shall see that this can be done, assuming that the real number system is given.

Chapter 10
Absolute Plane Geometry

10.1 SUFFICIENT CONDITIONS FOR PARALLELISM

Two lines are *parallel* if they lie in the same plane but do not intersect. We shall use the abbreviation $L_1 \parallel L_2$ to mean that the lines L_1 and L_2 are parallel. Later, as a matter of convenience, we shall say that two segments are parallel if the lines that contain them are parallel. We shall apply the same term to a line and a segment, a segment and a ray, and so on, just as we did somewhat earlier in our discussion of perpendicularity.

The Euclidean parallel postulate will be introduced in the next chapter, and used thereafter, except in the chapter on non-Euclidean geometry. The postulate, in the form in which it is usually stated, says that given a line and a point not on the line, there is exactly one line which passes through the given point and is parallel to the given line.

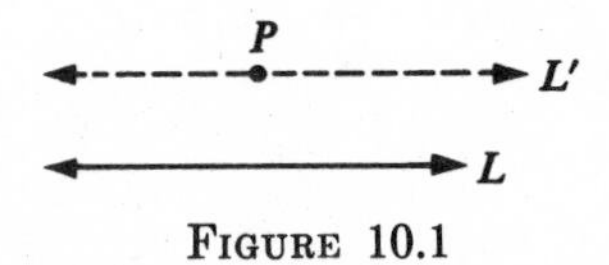

FIGURE 10.1

We shall see, however, from Theorems 1 and 2, that half of this statement can be proved on the basis of the postulates that we already have.

Theorem 1. If two lines lie in the same plane, and are perpendicular to the same line, then they are parallel.

RESTATEMENT. Let L_1, L_2, and T be three lines, lying in a plane E, such that $L_1 \perp T$ and $L_2 \perp T$. Then $L_1 \parallel L_2$.

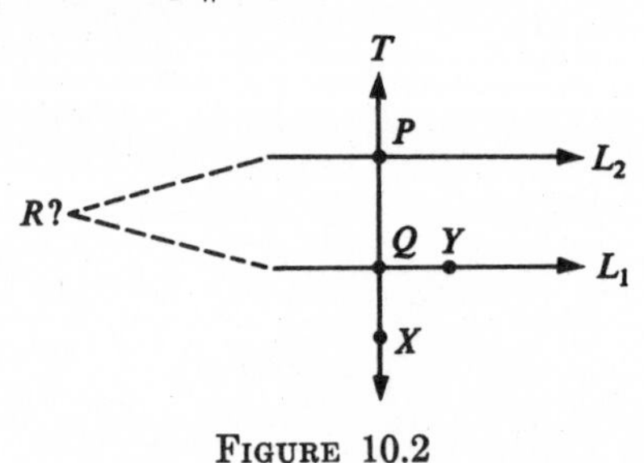

FIGURE 10.2

Proof. Suppose that L_1 and L_2 intersect T at points Q and P, respectively. Suppose that L_1 and L_2 are not parallel, and let R be the point at which they intersect. Then there are two perpendiculars to T through R; and this contradicts Corollary 1–1 of Chapter 7.

Theorem 2. Given a line and a point not on the line, there is always at least one line which passes through the given point and is parallel to the given line.

Proof. Let L be the line, let P be the point, and let E be the plane which contains them. By Theorem 1 of Section 6.5, there is a line T in E which passes through P and is perpendicular to L. By Theorem 4 of Section 8.3, there is a line L' in E which passes through P and is perpendicular to T. By the preceding theorem it follows that $L \parallel L'$, which was to be proved.

There is an easy generalization of Theorem 1, which we shall get to presently. In the figure below, T is a *transversal* to the lines L_1 and L_2.

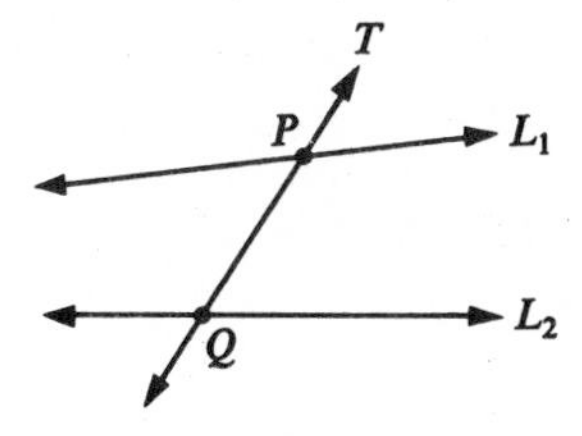

FIGURE 10.3

More precisely, if L_1, L_2, and T are three lines in the same plane, and T intersects L_1 and L_2 in two (different) points P and Q, respectively, then T is a *transversal* to L_1 and L_2.

In the figure below $\angle 1$ and $\angle 2$ are alternate interior angles; and $\angle 3$ and $\angle 4$ are alternate interior angles.

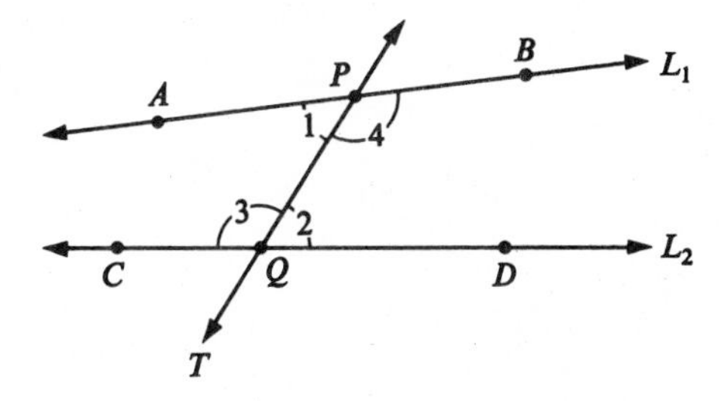

FIGURE 10.4

More precisely, (1) if T is a transversal to L_1 and L_2, intersecting L_1 and L_2 in P and Q, respectively, and (2) A and D are points of L_1 and L_2, respectively, lying on opposite sides of T, then $\angle APQ$ and $\angle PQD$ are *alternate interior angles.* (Under a change of notation, this definition says also that $\angle CQP$ and $\angle QPB$ are alternate interior angles.)

Theorem 3. Given two lines and a transversal. If a pair of alternate interior angles are congruent, then the lines are parallel.

The proof uses the exterior angle theorem.

In the figure below, $\angle 1$ and $\angle 1'$ are corresponding angles, $\angle 2$ and $\angle 2'$ are corresponding angles, and so on.

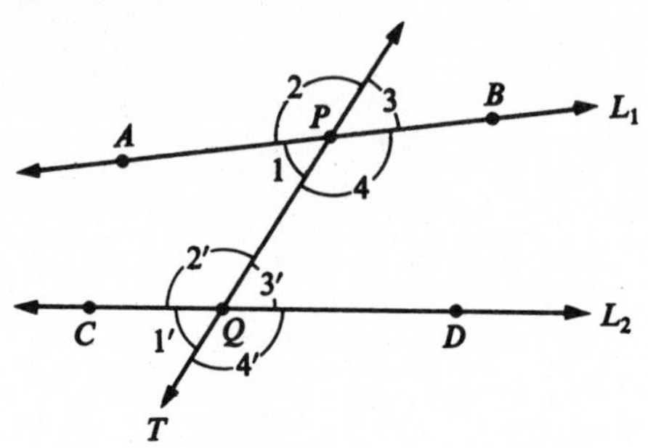

FIGURE 10.5

DEFINITION. If $\angle x$ and $\angle y$ are alternate interior angles, and $\angle y$ and $\angle z$ are vertical angles, then $\angle x$ and $\angle z$ are corresponding angles.

Theorem 4. Given two lines and a transversal. If a pair of corresponding angles are congruent, then a pair of alternate interior angles are congruent.

Theorem 5. Given two lines and a transversal. If a pair of corresponding angles are congruent, then the lines are parallel.

10.2 THE POLYGONAL INEQUALITY

The triangular inequality states that for any triangle ABC we have

$$AB + BC > AC.$$

If A, B, and C are not required to be noncollinear or even different, we get a weaker result.

Theorem 1. For any points $A, B, C,$

$$AB + BC \geqq AC.$$

Proof. If A, B, and C are noncollinear, this follows from the triangular inequality. If A, B, and C are collinear, we take a coordinate system on the line that contains them, and let their coordinates be x, y, and z. Let

$$a = x - y, \qquad b = y - z.$$

By Theorem 13, Section 1.4, we know that

$$|a| + |b| \geqq |a + b|,$$

therefore

$$|x - y| + |y - z| \geqq |x - z|.$$

Hence

$$AB + BC \geqq AC,$$

which was to be proved. From this we get the following theorem.

Theorem 2. *The Polygonal Inequality.* If $A_1, A_2, \ldots, A_n$ are any points $(n > 1)$, then

$$A_1A_2 + A_2A_3 + \cdots + A_{n-1}A_n \geqq A_1A_n.$$

The proof is by induction.

We shall need this result in the following section. For the first time, we are also about to use the Archimedean postulate for the real number system, given in Section 1.8. This says that if $e > 0$ and $M > 0$, then $ne > M$ for some positive integer n.

10.3 SACCHERI QUADRILATERALS

We recall, from Section 4.4, the definition of a quadrilateral. Given four points A, B, C, and D, such that they all lie in the same plane, but no three of them are collinear. If the segments $\overline{AB}$, $\overline{BC}$, $\overline{CD}$, $\overline{DA}$ intersect only at their end points, then their union is called a *quadrilateral*, and is denoted by $\square ABCD$. The segments $\overline{AB}$, $\overline{BC}$, $\overline{CD}$, $\overline{DA}$ are the *sides* of $\square ABCD$, and the segments $\overline{AC}$, $\overline{BD}$ are the *diagonals*. The *angles of* $\square ABCD$ are $\angle ABC$, $\angle BCD$, $\angle CDA$, and $\angle DAB$; they are often denoted briefly as $\angle B$, $\angle C$, $\angle D$, $\angle A$. If all four of the angles are right angles, then the quadrilateral is a *rectangle*.

On the basis of the postulates that we have so far, without the use of the parallel postulate, it is impossible to prove that any rectangles exist. If we try, in a plausible fashion, to construct a rectangle, we get what is called a *Saccheri quadrilateral*.

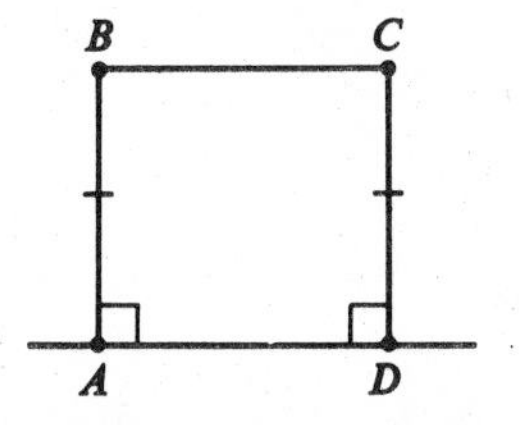

FIGURE 10.6

The definition is suggested by the markings on the figure above. To be precise, $\square ABCD$ is a *Saccheri quadrilateral* if $\angle A$ and $\angle D$ are right angles, B and C are on the same side of $\overleftrightarrow{AD}$, and $AB = CD$. The segment $\overline{AD}$ is called the *lower base;* and $\overline{BC}$ is called the *upper base.* The *lower base angles* are $\angle A$ and $\angle D$; and $\angle B$ and $\angle C$ are the *upper base angles.*

Theorem 1. The diagonals of a Saccheri quadrilateral are congruent.

Proof. By SAS, we have $\triangle BAD \cong \triangle CDA$. Therefore $\overline{BD} \cong \overline{AC}$.

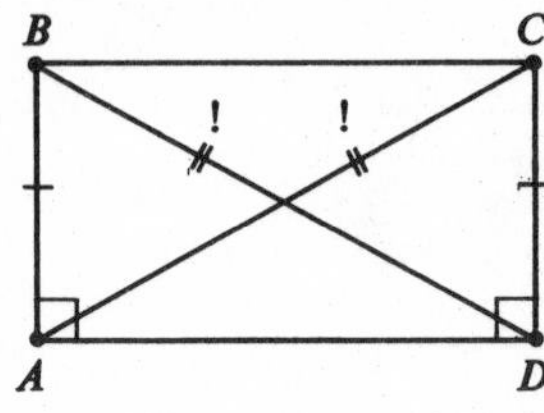

FIGURE 10.7

Roughly speaking, the following theorem states that a Saccheri quadrilateral is completely described, geometrically, by the distances AD and AB.

Theorem 2. Let $\square ABCD$ and $\square A'B'C'D'$ be Saccheri quadrilaterals, with lower bases $\overline{AD}$ and $\overline{A'D'}$. If $\overline{A'D'} \cong \overline{AD}$ and $\overline{A'B'} \cong \overline{AB}$, then $\overline{BC} \cong \overline{B'C'}$, $\angle B' \cong \angle B$, and $\angle C' \cong \angle C$.

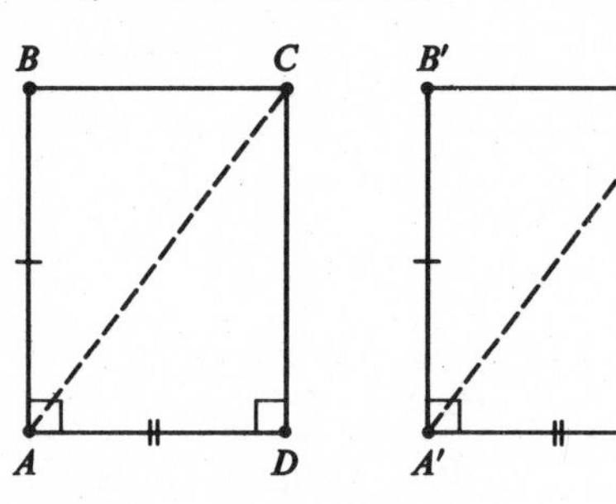

FIGURE 10.8

Proof. The main steps in the proof are as follows.
(1) $\triangle ACD \cong \triangle A'C'D'$ (by SAS).
(2) $\angle A \cong \angle A'$ (all right angles are congruent).
(3) C is in the interior of $\angle BAD$, and C' is in the interior of $\angle B'A'D'$.
(4) $\angle BAC \cong \angle B'A'C'$.
(5) $\overline{AC} \cong \overline{A'C'}$.
(6) $\triangle ABC \cong \triangle A'B'C'$.
(7) $\angle B \cong \angle B'$.
(8) $\overline{BC} \cong \overline{B'C'}$.
(9) $\angle C \cong \angle C'$.
Applying this theorem to the Saccheri quadrilaterals $\square ABCD$, $\square DCBA$, we get $\angle B \cong \angle C$. Thus we have the following theorem.

Theorem 3. In any Saccheri quadrilateral, the upper base angles are congruent.

Theorem 4. In any Saccheri quadrilateral, the upper base is congruent to or longer than the lower base.

RESTATEMENT. Given a Saccheri quadrilateral $\square A_1B_1B_2A_2$, with lower base $\overline{A_1A_2}$. Then $B_1B_2 \geqq A_1A_2$.

Proof. Let us set up a sequence of n Saccheri quadrilaterals, end to end, starting with the given one, like this:

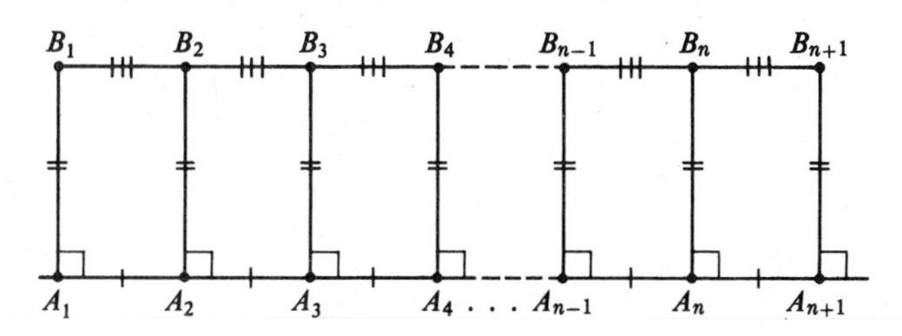

FIGURE 10.9

That is, $A_3, A_4, \ldots, A_{n+1}$ are points of the line $\overleftrightarrow{A_1A_2}$, appearing in the stated order on $\overleftrightarrow{A_1A_2}$; the angles $\angle B_2A_2A_3$, $\angle B_3A_3A_4, \ldots$ and so on are right angles;

$$A_1A_2 = A_2A_3 = A_3A_4 = \cdots = A_{n-1}A_n = A_nA_{n+1},$$

and

$$A_2B_2 = A_3B_3 = \cdots = A_nB_n = A_{n+1}B_{n+1}.$$

By Theorem 2, we have

$$B_1B_2 = B_2B_3 = \cdots = B_{n-1}B_n = B_nB_{n+1}.$$

We don't know whether the points $B_1, B_2, \ldots, B_{n+1}$ are collinear. But we know by the polygonal inequality that

$$B_1B_{n+1} \leqq B_1B_2 + B_2B_3 + \cdots + B_{n-1}B_n + B_nB_{n+1}.$$

Since all of the distances on the right are $=B_1B_2$, we have

$$B_1B_{n+1} \leqq n \cdot B_1B_2.$$

By the same principle, we get

$$A_1A_{n+1} \leqq A_1B_1 + B_1B_{n+1} + B_{n+1}A_{n+1} \leqq A_1B_1 + nB_1B_2 + A_1B_1.$$

Since $A_1A_{n+1} = nA_1A_2$, we have

$$nA_1A_2 \leqq nB_1B_2 + 2A_1B_1,$$

and this conclusion holds *for every n*.

Now suppose that our theorem is false. Then $A_1A_2 > B_1B_2$, so that $A_1A_2 - B_1B_2$ is a positive number. Obviously, $2A_1B_1$ is a positive number. Let

$$e = A_1A_2 - B_1B_2, \qquad \text{and} \qquad M = 2A_1B_1.$$

Then $e > 0$ and $M > 0$, but $ne \leqq M$ for every positive integer n. This contradicts the Archimedean postulate, and so completes the proof.

Problem Set 10.3

1. Show that every Saccheri quadrilateral is convex.

2. Let $\square ABCD$ be a quadrilateral, and let E be a point such that A-D-E. Suppose that B and C are on the same side of $\overleftrightarrow{AD}$, $AB = DC$, and $\angle BAD \cong \angle CDE$. Show that the quadrilateral is convex.

10.4 THE BASIC INEQUALITY FOR ANGLE-SUMS IN A TRIANGLE

A well-known theorem of Euclidean geometry asserts that, in any triangle, the sum of the degree measures of the angles is 180. Without the parallel postulate, we shall show that this sum is always *less than or equal to* 180. We shall need some preliminaries.

Theorem 1. In any Saccheri quadrilateral $\square ABCD$ (with lower base $\overline{AD}$), we have $\angle BDC \geqq \angle ABD$.

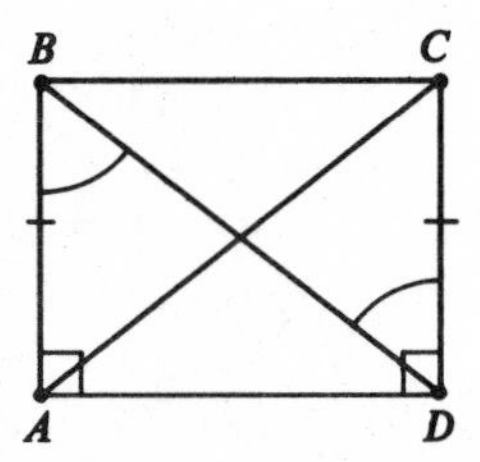

Figure 10.10

Proof. We know that $BA = DC$ and $BD = BD$. If it were true that $\angle ABD > \angle BDC$, then it would follow by Theorem 6, Section 7.1, that $AD > BC$; and this would contradict Theorem 4, Section 10.3. Therefore $\angle ABD \leqq \angle BDC$, which was to be proved.

From this we get an immediate consequence for right triangles.

Theorem 2. If $\triangle ABD$ has a right angle at A, then

$$m\angle B + m\angle D \leqq 90.$$

Proof. Let C be a point such that $\square ABCD$ is a Saccheri quadrilateral:

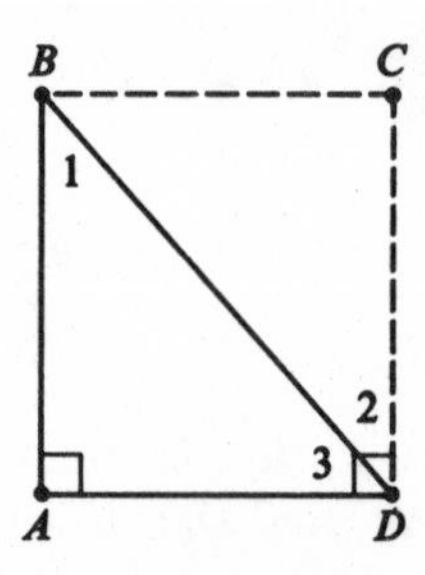

Figure 10.11

Then
$$m\angle 3 + m\angle 2 = 90,$$

because $\angle ADC$ is a right angle. By the preceding theorem, $m\angle 2 \geqq m\angle 1$. Therefore
$$90 - m\angle 3 \geqq m\angle 1, \quad \text{and} \quad m\angle 1 + m\angle 3 \leqq 90,$$

which was to be proved. Thus we have:

Theorem 3. Every right triangle has only one right angle; and its other two angles are acute.

Of course we observed this long ago, at the end of Chapter 7, using the exterior angle theorem. But we shall need Theorem 2 for other purposes anyway.

As in Chapter 7, we define the *hypotenuse* of a right triangle as the side opposite the (unique) right angle. The other two sides are called the *legs*.

Theorem 4. The hypotenuse of a right triangle is longer than either of the legs.

Because the angle opposite the hypotenuse is larger. (See Theorem 3 of Chapter 7.)

In a triangle, two sides can easily be congruent. Therefore we cannot always speak of *the* longest side. We can, however, always speak of *a* longest side; this means a side at least as long as any other side.

Theorem 5. In $\triangle ABC$, let D be the foot of the perpendicular from B to $\overleftrightarrow{AC}$. If $\overline{AC}$ is a longest side of $\triangle ABC$, then A-D-C.

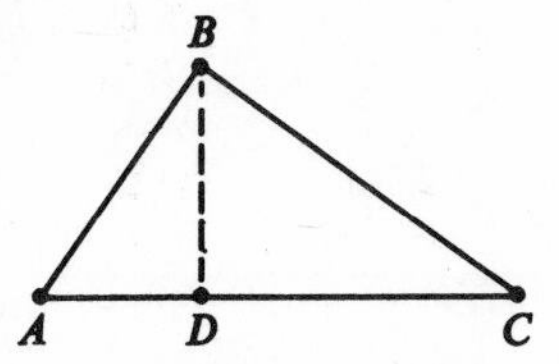

FIGURE 10.12

(This theorem, if available, would have simplified the proof of the SSS theorem.)

Proof. Suppose that the theorem is false. Then we have $D = A$, D-A-C, $D = C$ or A-C-D. We need to show that all of these cases are impossible. Since the latter two cases are essentially the same as the former two, it will suffice to show that $D = A$ and D-A-C are impossible.

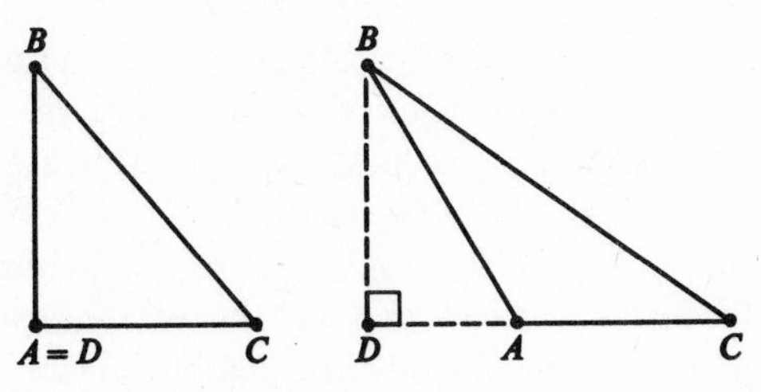

FIGURE 10.13

If $A = D$, then $\triangle ABC$ is a right triangle with right angle at A; therefore $BC > AC$, and $\overline{AC}$ is not a longest side.

If $D\text{-}A\text{-}C$, then $AC < DC$. Also $DC < BC$, because $\overline{BC}$ is the hypotenuse of $\triangle BDC$, and $\overline{DC}$ is one of the legs. Therefore $AC < BC$, and $\overline{AC}$ is not a longest side of $\triangle ABC$.

Finally, we can prove the theorem that we were working toward.

Theorem 6. In any triangle ABC, we have

$$m\angle A + m\angle B + m\angle C \leqq 180.$$

Proof. Suppose, without loss of generality, that $\overline{AC}$ is a longest side of $\triangle ABC$; and let $\overline{BD}$ be the perpendicular segment from B to the line $\overleftrightarrow{AC}$. By the preceding theorem we have $A\text{-}D\text{-}C$; this means that D is in the interior of $\angle ABC$.

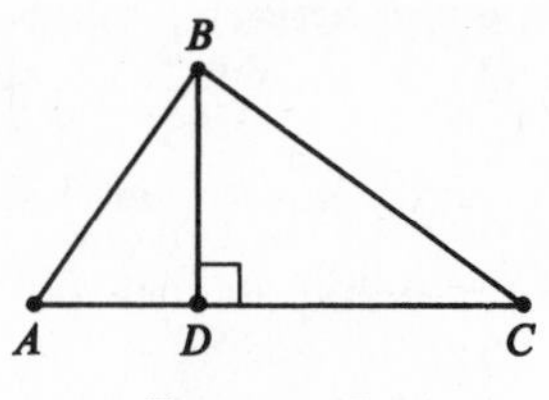

FIGURE 10.14

We now apply Theorem 2 to each of the right triangles $\triangle ADB$ and $\triangle BDC$. Thus

$$m\angle A + m\angle ABD \leqq 90$$

and

$$m\angle DBC + m\angle C \leqq 90.$$

Therefore

$$m\angle A + m\angle ABD + m\angle DBC + m\angle C \leqq 180.$$

Since D is in the interior of $\angle ABC$, we have

$$m\angle ABD + m\angle DBC = m\angle B.$$

Therefore

$$m\angle A + m\angle B + m\angle C \leqq 180,$$

which was to be proved.

10.5 A HISTORICAL COMEDY

Saccheri quadrilaterals are named after the Italian geometer Gerolamo Saccheri (1667–1733). Like most geometers of his time, Saccheri was dissatisfied with the situation of the parallel postulate; he believed that this statement ought to be proved as a theorem. He wrote a book, entitled *Euclides ab omne naevo vindicatus*, in which he undertook to "vindicate Euclid of every blemish" by showing that the parallel postulate was a consequence of the other postulates of synthetic geometry. On certain occasions, the development of mathematics involves high comedy. This was one of them.

In the first place, Saccheri's "proof" of the postulate was fallacious. However, the early stages of the proof were correct, and the preliminary theorems were new and important. If you omit the erroneous part of his book, what you get is the first treatise on what is now called *absolute geometry*. That is, Saccheri developed an extensive geometric theory which was independent of the whole question of the parallel postulate. (The preceding portions of this chapter are a sample of this sort of theory.) For this achievement, Saccheri is highly honored and justly so.

The final irony is that if Saccheri's enterprise had succeeded in the way he thought it had, no modern mathematician would have regarded his book as a vindication of Euclid. From a modern viewpoint, a proof of the parallel postulate would merely show that the postulate was redundant; and redundancy is not thought of as a virtue in a set of postulates. There are three main things that a modern mathematician wants to know about a postulate set.

(1) The postulates ought, by all means, to be *consistent*, in the sense that none of them contradicts the others. If this condition does not hold, then any mathematical theory based on the postulates is, quite literally, much ado about nothing, because, in this case, there isn't any mathematical system that satisfies the postulates. The only way to show that a set of postulates is consistent is to show that there is a system in which all of the postulates are satisfied. Such a system is called a *model* for the postulate set.

(2) The existence of a model is not enough. We use postulates as descriptions of the mathematical systems that satisfy them; and this means that a set of postulates is worth stating only if there is a mathematical system which satisfies them and is important enough to be worthy of study.

(3) The postulates ought, if possible, to be *independent*, in the sense that no one of them is deducible from the others as a theorem. Any postulate which is so deducible is called *redundant*. To show that a particular postulate is redundant, you have to prove it on the basis of the others; and this is what Saccheri tried to do for the parallel postulate. To show that a particular postulate is independent of the others, you have to show that there is a mathematical system in which all of the other postulates are satisfied, but in which this particular one is not. (See, for example, Section 6.4, in which we showed that the SAS postulate is independent of the postulates that precede it.)

In the nineteenth century, two fundamental questions were settled. First, it was shown that the postulates of synthetic geometry, including the parallel postulate, were consistent—granted, of course, that the real number system is consistent. It was shown further that the parallel postulate is independent of the others. This was done, in the only way that it could be done, by the discovery of "geometries" in which all the synthetic postulates except the parallel postulate were satisfied.

These two developments were the real vindication of Euclid from a modern viewpoint. The vindication of his vindicators lay in the fact that hyperbolic geometry turned out to be an important subject in its own right.

Chapter 11
The Parallel Postulate and Parallel Projection

11.1 THE UNIQUENESS OF PARALLELS

The Euclidean parallel postulate is as follows.

P-1. Given a line and an external point, there is only one line which passes through the given point and is parallel to the given line.

This gives us immediately a converse of Theorem 3, Section 10.1.

Theorem 1. Given two lines and a transversal. If the lines are parallel, then each pair of alternate interior angles are congruent.

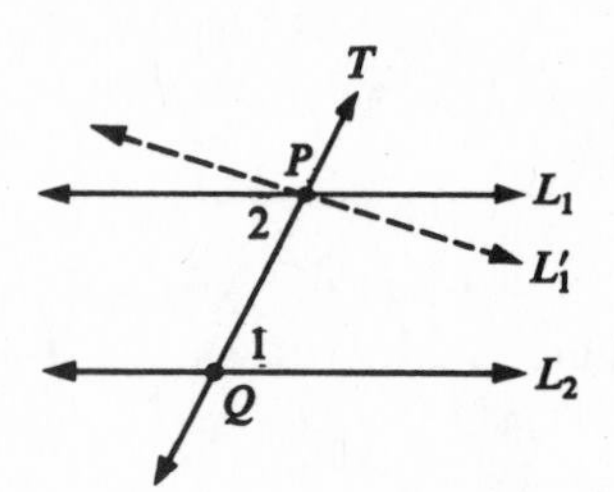

FIGURE 11.1

Proof. There is exactly one line L_1', through P, for which the alternate interior angles are congruent, and by Theorem 3, Section 10.1, we have $L_1' \parallel L_2$. Since there is only one such parallel, we have $L_1' = L_1$. Therefore $\angle 1 \cong \angle 2$, which was to be proved.

The proof of the following theorem is entirely analogous.

Theorem 2. Given two lines and a transversal. If the lines are parallel, then each pair of corresponding angles is congruent.

The inequality $m\angle A + m\angle B + m\angle C \leqq 180$ now becomes an equation.

Theorem 3. In any triangle $\triangle ABC$ we have

$$m\angle A + m\angle B + m\angle C = 180.$$

Proof. Let L be the parallel to $\overleftrightarrow{AC}$ through B. Let D and E be points of L, such that D-B-E, and such that D and A are on the same side of $\overleftrightarrow{BC}$. Then

$$m\angle 2 + m\angle B = m\angle DBC,$$

and

$$m\angle DBC + m\angle 1 = 180.$$

Therefore

$$m\angle 1 + m\angle B + m\angle 2 = 180.$$

By Theorem 1,

$$m\angle 1 = m\angle C \quad \text{and} \quad m\angle 2 = m\angle A,$$

therefore

$$m\angle A + m\angle B + m\angle C = 180,$$

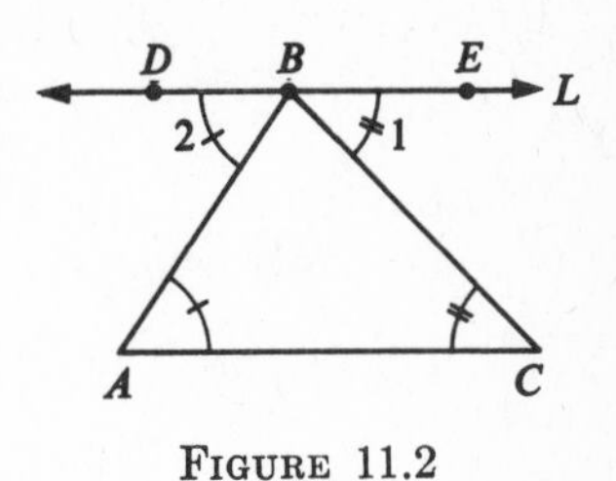

FIGURE 11.2

which was to be proved. The following theorems are an immediate consequence.

Theorem 4. The acute angles of a right triangle are complementary.

Theorem 5. Every Saccheri quadrilateral is a rectangle.

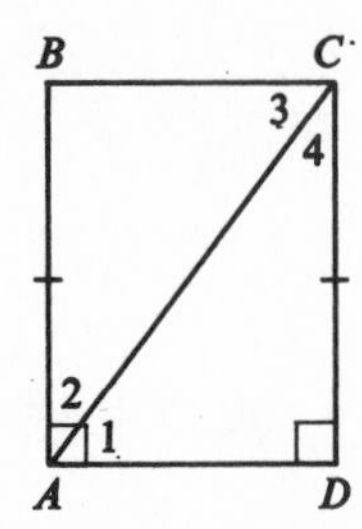

FIGURE 11.3

Proof. By Theorem 1, $\angle 2 \cong \angle 4$. Since $AB = DC$ and $AC = AC$, it follows that $\triangle BAC \cong \triangle DCA$. Therefore $\angle B \cong \angle D$, and $\angle B$ is a right angle. The proof that $\angle C$ is a right angle is obtained merely by permuting the notation. Thus we have finally shown that rectangles exist.

Note that in this proof we are using a figure to explain the notation. If the reader (or the writer) sees no other way to explain, say, the idea of alternate interior angles, then it is worthwhile to fight our way through the problem as we did in the previous chapter. But once we have done this, we have earned the right to speak in the abbreviated language of pictures.

A quadrilateral is a *trapezoid* if at least one pair of opposite sides are parallel. (It is sometimes required that the other pair of opposite sides be *nonparallel*, but this is artificial, just as it would be artificial to require that an isosceles triangle be nonequilateral.) If both pairs of opposite sides of a quadrilateral are parallel, then the quadrilateral is a *parallelogram*. If two adjacent sides of a parallelogram are congruent, then the quadrilateral is a *rhombus*. The proofs of the following theorems are omitted. (They are not much harder to write than to read.)

Theorem 6. For any triangle, the measure of an exterior angle is the sum of the measures of its two remote interior angles.

Theorem 7. In a plane, any two lines parallel to a third line are parallel to each other.

Theorem 8. If a transversal is perpendicular to one of two parallel lines, it is perpendicular to the other.

Theorem 9. Either diagonal divides a parallelogram into two congruent triangles.

More precisely: If $\square ABCD$ is a parallelogram, then $\triangle ABC \cong \triangle CDA$.

Theorem 10. In a parallelogram, each pair of opposite sides are congruent.

Theorem 11. The diagonals of a parallelogram bisect each other.

That is, they intersect at a point which is the bisector of each of them. Thus the proof must begin with a proof that the diagonals intersect each other (see Theorem 1, Section 4.4).

Theorem 12. Every trapezoid is a convex quadrilateral.

11.2 PARALLEL PROJECTIONS

We know, by Theorem 4, Section 8.2, that the perpendicular from a point to a line always exists and is unique.

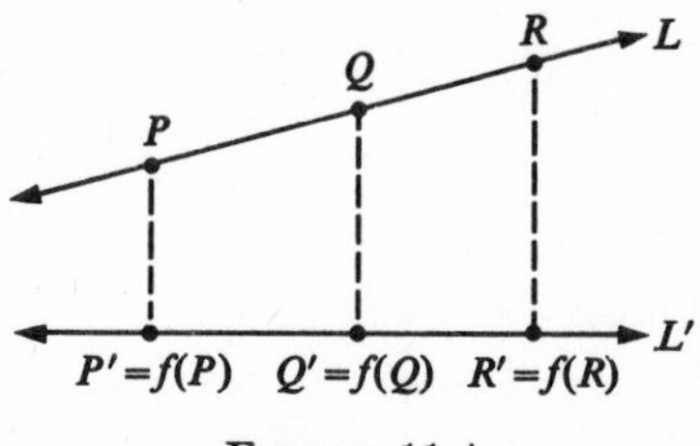

FIGURE 11.4

Thus, given two lines L, L' in the same plane, we can define the *vertical projection* of L into L'. This is the function

$$f: L \to L'$$

under which to each point P of L there corresponds the foot $P' = f(P)$ of the perpendicular from P to L'. In fact, the vertical projection can be defined equally well for the case where the lines are not necessarily coplanar; and the definition is exactly the same. This degree of generality, however, will not concern us. Note also that the existence and uniqueness of the vertical projection do not depend on the parallel postulate. We do, however, need this postulate to define and investigate the more general notion of *parallel* projection. Under this more general scheme, instead of following the perpendicular, to get from P to $P' = f(P)$, we proceed in any direction we want, providing, however, that we always go in the *same* direction for every point P of L. More precisely, the definition of parallel projections is as follows.

Given two lines L, L' and a transversal T. (By definition of a transversal, this means that all three of our lines are coplanar.) Let T intersect L and L' in points Q and Q', respectively.

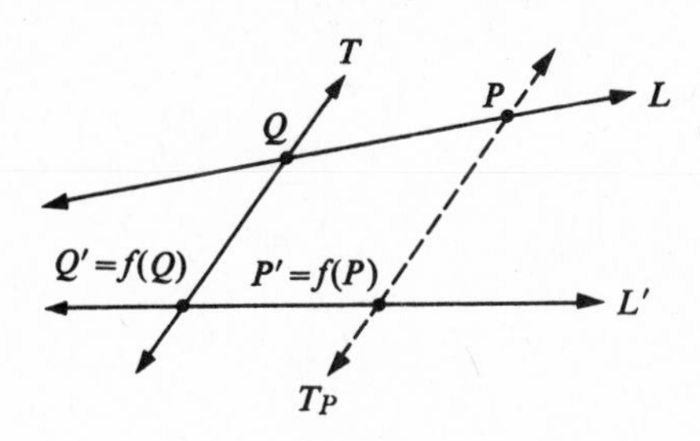

FIGURE 11.5

Let $f(Q)$ be Q'. For every other point P of L, let T_P be the line through P, parallel to T.

(1) If T_P were parallel to L', then it would follow that $T \parallel L'$, which is false, because T is a transversal to L and L'. Therefore T_P intersects L' in *at least* one point P'.

(2) If $T_P = L'$, it follows that $L' \parallel T$, which is false. Therefore T_P intersects L' in *at most* one point P'.

For each point P of L, let $f(P)$ be the unique point P' in which T_P intersects L'. This defines a function

$$f: L \to L'.$$

The function f is called the *projection of L onto L' in the direction T*.

Theorem 1. Every parallel projection is a one-to-one correspondence.

Proof. Given

$$f: L \to L',$$

the projection of L onto L' in the direction T. (See Fig. 11.5.) Let g be the projection of L' onto L in the direction T. Obviously g reverses the action of f; that is, if $P = g(P')$, then $P' = f(P)$. Therefore f has an inverse

$$f^{-1} = g: L' \to L.$$

Therefore f is a one-to-one correspondence $L \leftrightarrow L'$, which was to be proved.

(Another way of putting it is to say that every point P' of L' is $= f(P)$ for one and only one point P of L. It may be worthwhile, at this stage, to review the discussion of functions in Section 3.1.)

Theorem 2. Parallel projections preserve betweenness.

RESTATEMENT. Let $f: L \leftrightarrow L'$ be a parallel projection. If P-Q-R on L, then P'-Q'-R' on L'.

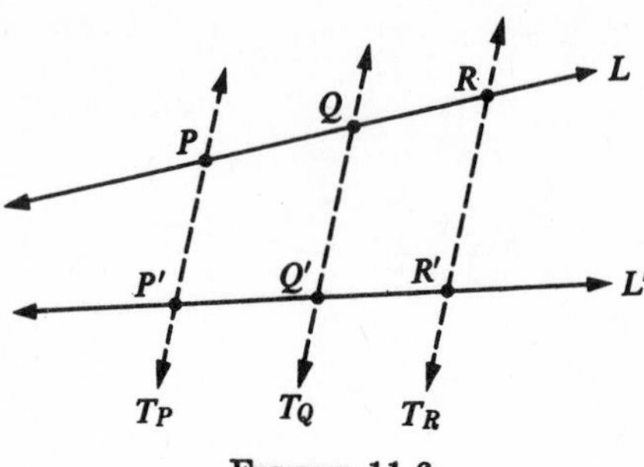

FIGURE 11.6

Here, of course, $P' = f(P)$, $Q' = f(Q)$, and $R' = f(R)$.

Proof. Let T_P, T_Q, T_R be as in the definition of a parallel projection, so that

$$T_P \parallel T_Q \parallel T_R.$$

Then R and R' are on the same side of T_Q, because $\overline{RR'}$ does not intersect T_Q. Similarly, P and P' are on the same side of T_Q. But P and R are on opposite sides of T_Q, because P-Q-R. By two applications of Theorem 2, Section 4.2, P' and R' are on opposite sides of T_Q. Therefore $\overline{P'R'}$ intersects T_Q in a point X. Since $T_Q \neq L'$, there is only one such point of intersection. Therefore $X = Q'$. Therefore Q' lies on $\overline{P'R'}$, and P'-Q'-R', which was to be proved.

Theorem 3. Parallel projections preserve congruence.

RESTATEMENT. Let $f: L \leftrightarrow L'$ be a parallel projection. If $\overline{AB} \cong \overline{CD}$ on L, then $\overline{A'B'} \cong \overline{C'D'}$ on L'.

Proof. (1) If $L \parallel L'$, then $\overline{AB}$ and $\overline{A'B'}$ are opposite sides of a parallelogram. By Theorem 10, Section 11.1, it follows that $\overline{AB} \cong \overline{A'B'}$. Similarly, $\overline{CD} \cong \overline{C'D'}$. Therefore $\overline{A'B'} \cong \overline{C'D'}$, as desired.

(2) Suppose that L and L' are not parallel, as in Fig. 11.7. Let V be the line through A, parallel to L', intersecting T_B at E. Let W be the line through C, parallel to L', intersecting T_D at F.

Now $V \parallel W$, and L is a transversal. For appropriate choice of the notation for C and D, $\angle 1 = \angle BAE$ and $\angle 1' = \angle DCF$ are corresponding angles. (This is true for the case shown in the figure; if it isn't true, we interchange the letters C and D.) By Theorem 2, Section 11.1, we have

$$\angle 1 \cong \angle 1'.$$

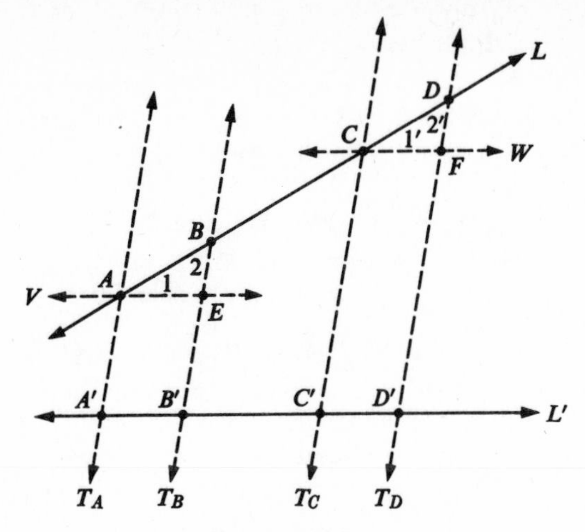

FIGURE 11.7

For the same reasons,

$$\angle 2 \cong \angle 2'.$$

Since $\overline{AB} \cong \overline{CD}$ by hypothesis, it follows by ASA that

$$\triangle ABE \cong \triangle CDF.$$

Therefore $\overline{AE} \cong \overline{CF}$. But $\overline{AE} \cong \overline{A'B'}$ and $\overline{CF} \cong \overline{C'D'}$, because these segments are opposite sides of parallelograms. Therefore $\overline{A'B'} \cong \overline{C'D'}$, which was to be proved.

Consider now three parallel lines with two common transversals, like this:

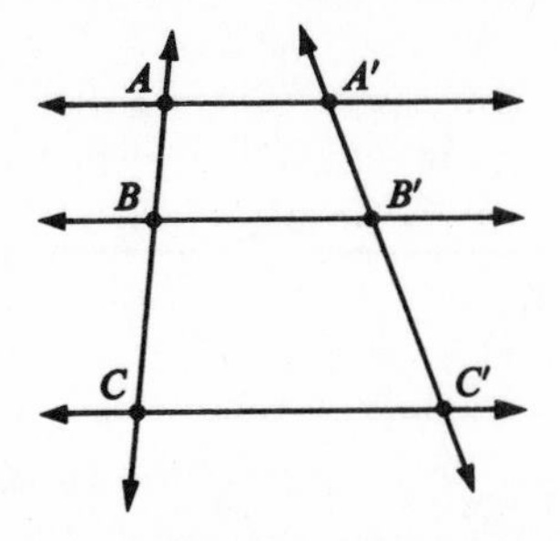

FIGURE 11.8

It ought to be true that

$$\frac{A'B'}{B'C'} = \frac{AB}{BC}.$$

In the style of our previous theorems, we could convey this by saying that *parallel projections preserve ratios.* In fact, this is true, but the proof is hard, and will be given in the following two sections. The theorem is worth working for; it is the foundation of the whole theory of similarity for triangles. The proof will depend on the Euclidean parallel postulate, as one might expect: if parallels are not unique, then parallel projections are not even well defined.

11.3 THE COMPARISON THEOREM

The algebraic method that we shall use, to prove that parallel projections preserve ratios, will be based on the following theorem.

Theorem 1. *The Comparison Theorem.* Let x and y be real numbers. Suppose (1) that every rational number less than x is also less than y, and (2) every rational number less than y is also less than x. Then $x = y$.

The proof is easy on the basis of Theorem 2, Section 1.8. Suppose that $x < y$. Then there is a rational number p/q, between x and y. Thus

$$x < \frac{p}{q} < y,$$

and p/q is less than y, but not less than x. This contradicts (2). Similarly, if $y < x$, we have

$$y < \frac{p}{q} < x$$

for some rational number p/q; this contradicts (1). Therefore $x = y$, which was to be proved.

This seemingly trivial observation turns out to be powerful, as we shall see.

11.4 THE BASIC SIMILARITY THEOREM

The purpose of this section is to show that parallel projections preserve ratios. Let us first consider the special case indicated by the following figure, and treated in the following theorem.

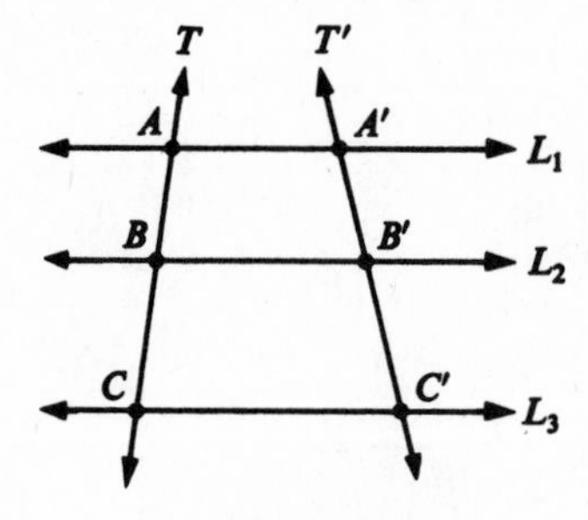

FIGURE 11.9

Here L_1, L_2, and L_3 are three parallel lines, with common transversals T and T'. We want to prove that

$$\frac{BC}{AB} = \frac{B'C'}{A'B'}.$$

Theorem 1. *The Basic Similarity Theorem.* Let L_1, L_2, and L_3 be three parallel lines, with common transversals T and T' intersecting them in points A, B, C and A', B', C'. If A-B-C (and A'-B'-C'), then

$$\frac{BC}{AB} = \frac{B'C'}{A'B'}.$$

Proof. Let

$$x = \frac{BC}{AB}, \qquad y = \frac{B'C'}{A'B'}.$$

Let p and q be any two positive integers.

(1) First we divide $\overline{AB}$ into q congruent segments, end to end, as in the figure:

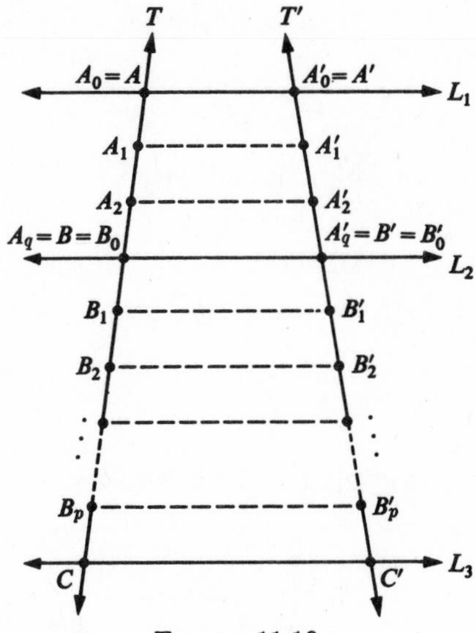

FIGURE 11.10

That is, we take a sequence of points

$$A = A_0, A_1, \ldots, A_q = B$$

in the stated order on the ray $\overrightarrow{AB}$, so that the length of each of the resulting segments is AB/q.

(2) Next we lay off, on the ray $\overrightarrow{BC}$, a sequence of p segments, of the same length AB/q. The end points of these segments are

$$B = B_0, B_1, \ldots, B_p.$$

(3) Now we project each of the points A_i, B_j onto T', in the direction L_1, thus getting the points A'_i, B'_j on T'.

Since all of the little segments on T are congruent, we have

$$\frac{BB_p}{AB} = \frac{p}{q}.$$

Since parallel projections preserve congruences, all the little segments on T' are congruent. Therefore

$$\frac{B'B'_p}{A'B'} = \frac{p}{q}.$$

We can now complete the proof easily, in two steps.

(4) Suppose that

$$\frac{p}{q} < x = \frac{BC}{AB}.$$

Then

$$p \cdot \frac{AB}{q} < BC.$$

Therefore

$$BB_p < BC.$$

Hence

$$B\text{-}B_p\text{-}C,$$

as indicated in Fig. 11.10, and

$$B'\text{-}B'_p\text{-}C',$$

because parallel projections preserve betweenness. Therefore

$$B'B'_p < B'C', \qquad p \cdot \frac{A'B'}{q} < B'C'$$

and

$$\frac{p}{q} < \frac{B'C'}{A'B'}.$$

(Here we have merely reversed the steps that led from

$$\frac{p}{q} < \frac{BC}{AB}$$

to $B\text{-}B_p\text{-}C$.) Thus we have proved that if $p/q < x$, then $p/q < y$.

(5) By exactly the same sort of reasoning, we conclude that if $p/q < y$, then $p/q < x$.

It follows by the comparison theorem that $x = y$, which was to be proved.

We extend Theorem 1, by various devices, to get the general case. Consider any four points on T, and the corresponding points on T', under a parallel projection:

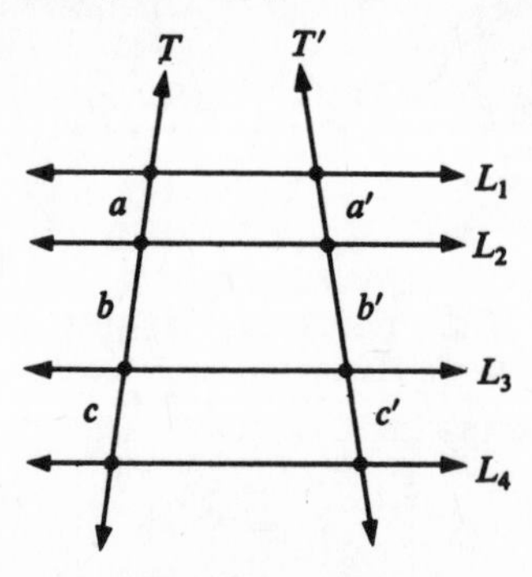

FIGURE 11.11

Here a, b, c and so on are the *lengths* of the indicated segments. By two applications of Theorem 1, we have

$$\frac{a}{b} = \frac{a'}{b'}, \qquad \frac{b}{c} = \frac{b'}{c'}.$$

Therefore we have

$$\frac{a}{a'} = \frac{b}{b'}, \qquad \frac{b}{b'} = \frac{c}{c'},$$

so that

$$\frac{a}{a'} = \frac{c}{c'}, \qquad \text{and} \qquad \frac{a}{c} = \frac{a'}{c'}.$$

Stated in words, our result is as follows.

Theorem 2. If two segments on the same line have no point in common, then the ratio of their lengths is preserved under every parallel projection.

From this it is easy to prove our main theorem.

Theorem 3. Parallel projection preserves ratios.

RESTATEMENT. Let T and T' be lines. Let A, B, C and D be any points of T, and let A', B', C', and D' be the corresponding points of T', under a parallel projection. Then

$$\frac{AB}{CD} = \frac{A'B'}{C'D'}, \qquad \text{and} \qquad \frac{AB}{A'B'} = \frac{CD}{C'D'}.$$

Proof. Let $\overline{XY}$ be a segment on T which has no point in common with $\overline{AB}$ or $\overline{CD}$, and let $\overline{X'Y'}$ be the corresponding segment on T', under the parallel projection. Then

$$\frac{AB}{A'B'} = \frac{XY}{X'Y'} = \frac{CD}{C'D'},$$

by Theorem 2. Therefore

$$\frac{AB}{A'B'} = \frac{CD}{C'D'},$$

and

$$\frac{AB}{CD} = \frac{A'B'}{C'D'},$$

which was to be proved.

Chapter 12
Similarities between Triangles

12.1 PROPORTIONALITIES

Given two sequences

$$a, b, c, \ldots; \qquad a', b', c', \ldots$$

of positive numbers. If it is true that

$$\frac{a'}{a} = \frac{b'}{b} = \frac{c'}{c} = \cdots,$$

then we say that the two sequences are *proportional*, and we write

$$a, b, c, \ldots \sim a', b', c', \ldots.$$

The constant ratio,

$$k = \frac{a'}{a} = \frac{b'}{b} = \cdots,$$

is called the *proportionality constant*. Note that proportionality is a symmetric relation. That is, if

$$a, b, c, \ldots \sim a', b', c', \ldots,$$

then

$$a', b', c', \ldots \sim a, b, c, \ldots,$$

and conversely. Note, however, that the proportionality constant depends on the order in which the sequences are written. If we reverse the order, we get a new constant which is the reciprocal of the old one.

To work with proportionalities, we merely express them in terms of equations between fractions, and then use the ordinary rules of algebra.
A sample theorem follows.

Theorem. If $a, b \sim c, d$, then $a, c \sim b, d$, and conversely.

Proof. The first proportionality means that

$$\frac{c}{a} = \frac{d}{b},$$

and the second means that

$$\frac{b}{a} = \frac{d}{c}.$$

Trivially, these equations are equivalent.

This being the case, the question arises why the notation deserves to be introduced at all. The reason is that the "$\sim$" relation is often easy to read off from a figure. Moving a little ahead of ourselves, for the sake of illustration, consider a pair of similar triangles, like this:

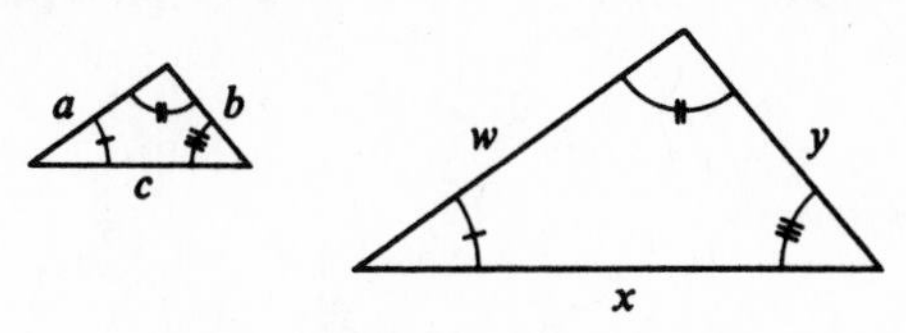

FIGURE 12.1

Writing the lengths of the sides in the proper order, we get the proportionality

$$a, b, c \sim w, y, x.$$

We have now made the transition from geometry to algebra, and can proceed to work algebraically with fractions, starting with the equations

$$\frac{w}{a} = \frac{y}{b} = \frac{x}{c}.$$

12.2 SIMILARITIES BETWEEN TRIANGLES

Given $\triangle ABC$, $\triangle DEF$ and a correspondence

$$ABC \leftrightarrow DEF.$$

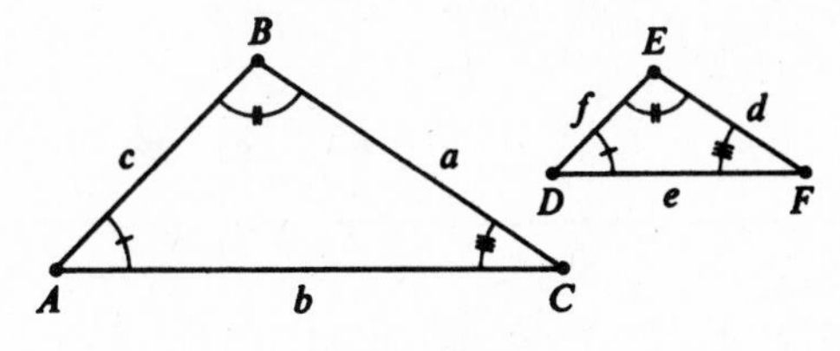

FIGURE 12.2

We use the familiar convention, under which a is the length of the side opposite $\angle A$, and so on. If

$$a, b, c \sim d, e, f,$$

then we say that *corresponding sides are proportional.* If corresponding sides are proportional, and every pair of corresponding angles are congruent, then we say that the correspondence is a *similarity*, and we write

$$\triangle ABC \sim \triangle DEF.$$

If there is a similarity between two triangles, then we say that the triangles are *similar*. (As in the case of congruences and congruence, the occasions when this is what we really mean are very rare.)

We remember that from the expression

$$\triangle ABC \cong \triangle DEF,$$

we could read off—without further reference to a figure—three angle-congruences

$$\angle A \cong \angle D, \qquad \angle B \cong \angle E, \qquad \angle C \cong \angle F$$

and three segment-congruences

$$\overline{AB} \cong \overline{DE}, \qquad \overline{AC} \cong \overline{DF}, \qquad \overline{BC} \cong \overline{EF}.$$

In the same way, from the expression

$$\triangle ABC \sim \triangle DEF,$$

we can read off the same three angle-congruences, and the proportionality

$$AB, AC, BC \sim DE, DF, EF.$$

To get the right-hand half of this expression, we simply replace each letter A, B, or C on the left by the corresponding letter D, E, or F.

Intuitively speaking, two triangles are similar if they have the same shape, although not necessarily the same size. It looks as if the shape ought to be determined by the angles alone, and this is true.

Theorem 1. *The AAA Similarity Theorem.* Given a correspondence between two triangles. If corresponding angles are congruent, then the correspondence is a similarity.

RESTATEMENT. Given $\triangle ABC$, $\triangle DEF$ and a correspondence

$$ABC \leftrightarrow DEF.$$

If $\angle A \cong \angle D$, $\angle B \cong \angle E$, and $\angle C \cong \angle F$, then

$$\triangle ABC \sim \triangle DEF.$$

Proof. Let E' and F' be points of $\overrightarrow{AB}$ and $\overrightarrow{AC}$, such that $AE' = f$ and $AF' = e$, as shown in Fig. 12.3. By SAS, we have

$$\triangle AE'F' \cong \triangle DEF.$$

Therefore $\angle AE'F' \cong \angle E$. Since $\angle E \cong \angle B$, we have $\angle AE'F' \cong \angle B$; thus $\overleftrightarrow{E'F'} \parallel \overleftrightarrow{BC}$, and A, F', and C correspond to A, E', and B under a parallel projection. Since parallel projections preserve ratios, we have

$$\frac{f}{AB} = \frac{e}{AC}.$$

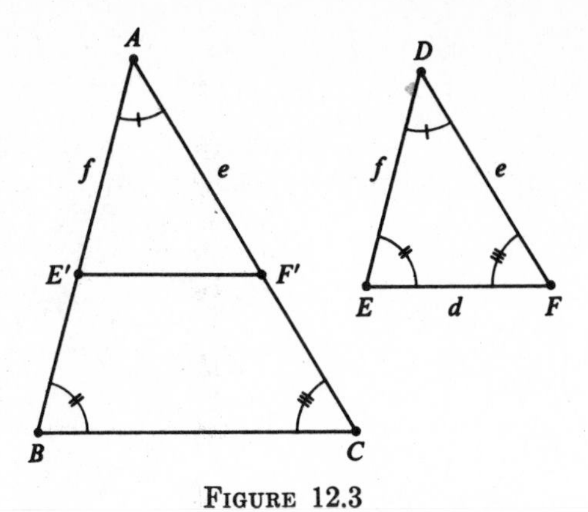

FIGURE 12.3

In exactly the same way, merely changing the notation, we can show that

$$\frac{e}{AC} = \frac{d}{BC};$$

therefore

$$d, e, f \sim BC, AC, AB,$$

and

$$d, e, f \sim a, b, c.$$

Hence corresponding sides are proportional, and the correspondence $ABC \leftrightarrow DEF$ is a similarity, which was to be proved.

Of course it follows from our angle-sum theorem that if two pairs of corresponding angles are congruent, so also is the third pair. Thus we have the following theorem.

Theorem 2. *The AA Similarity Theorem.* Given a correspondence between two triangles. If two pairs of corresponding angles are congruent, then the correspondence is a similarity.

We also have a sort of converse of Theorem 1.

Theorem 3. *The SSS Similarity Theorem.* Given two triangles and a correspondence between them. If corresponding sides are proportional, then corresponding angles are congruent, and the correspondence is a similarity.

RESTATEMENT. Given $\triangle ABC$, $\triangle DEF$, and a correspondence $ABC \leftrightarrow DEF$. If

$$a, b, c \sim d, e, f,$$

then

$$\triangle ABC \sim \triangle DEF.$$

Proof. Let E' be the point of $\overrightarrow{AB}$ for which $AE' = f$ (Fig. 12.4). Let L be the line through E', parallel to $\overleftrightarrow{BC}$. If $L \parallel \overleftrightarrow{AC}$, then $\overleftrightarrow{BC} \parallel \overleftrightarrow{AC}$, which is false. Therefore L intersects $\overleftrightarrow{AC}$ at a point F'. (In the figure, $b = AC$, not AF'.)

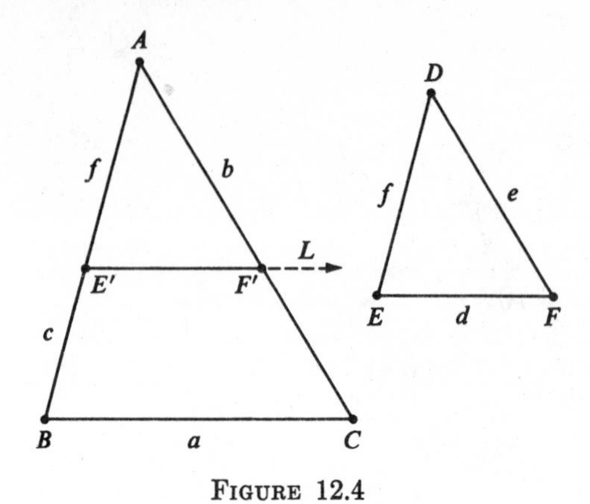

FIGURE 12.4

Now $\angle AE'F' \cong \angle B$, because these are corresponding angles; and $\angle A \cong \angle A$. Therefore

$$\triangle AE'F' \sim \triangle ABC;$$

hence

$$f, AF', E'F' \sim c, b, a.$$

Therefore

$$\frac{c}{f} = \frac{b}{AF'} = \frac{a}{E'F'}, \qquad \text{and} \qquad AF' = \frac{bf}{c}, \qquad E'F' = \frac{af}{c}.$$

But

$$f, e, d \sim c, b, a.$$

Thus

$$\frac{c}{f} = \frac{b}{e} = \frac{a}{d}, \qquad \text{and} \qquad e = \frac{bf}{c}, \qquad d = \frac{af}{c}.$$

By SSS, we have

$$\triangle AE'F' \cong \triangle DEF.$$

Therefore

$$\triangle DEF \sim \triangle ABC,$$

which was to be proved.

Next we have an analogue of SAS.

Theorem 4. *The SAS Similarity Theorem.* Given a correspondence between two triangles. If two pairs of corresponding sides are proportional, and the included angles are congruent, then the correspondence is a similarity.

RESTATEMENT. Given $\triangle ABC$, $\triangle DEF$, and the correspondence $ABC \leftrightarrow DEF$ If $\angle A \cong \angle D$, and $b, c \sim e, f$, then $\triangle ABC \sim \triangle DEF$.

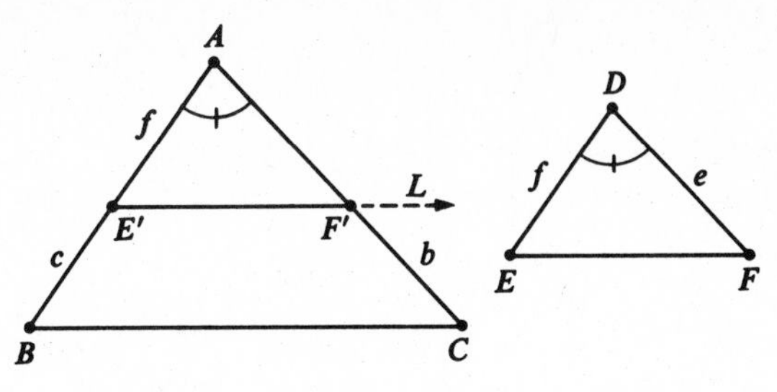

FIGURE 12.5

Proof. Let E' be the point of $\overrightarrow{AB}$ for which $AE' = f$. Let L be the line through E', parallel to $\overleftrightarrow{BC}$. Then L intersects $\overleftrightarrow{AC}$ in a point F'. The main steps in the proof, from here on, are as follows. You should be able to supply the reasons in each case.

(1) $\triangle AE'F' \sim \triangle ABC$.
(2) $b, c \sim AF', f$.
(3) $AF' = e$.
(4) $\triangle AE'F' \cong \triangle DEF$.
(5) $\triangle ABC \sim \triangle DEF$.

Theorem 5. Given a similarity between two triangles. If a pair of corresponding sides are congruent, then the correspondence is a congruence.

Proof? This was really a step in the proof of the preceding theorem.

An *altitude* of a triangle is a perpendicular segment from a vertex to the line containing the opposite side.

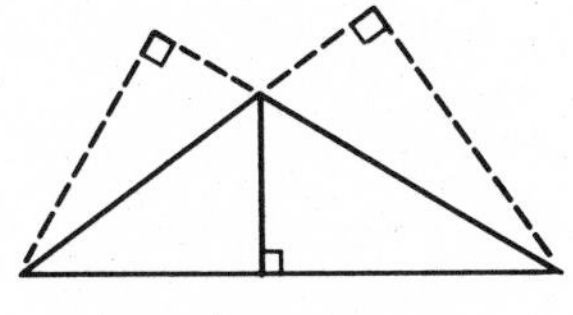

FIGURE 12.6

As the figure indicates, every triangle has three altitudes. We shall use the same word *altitude* for the length of such a perpendicular segment. In a right triangle, the altitude to the hypotenuse is always an "interior altitude," like this:

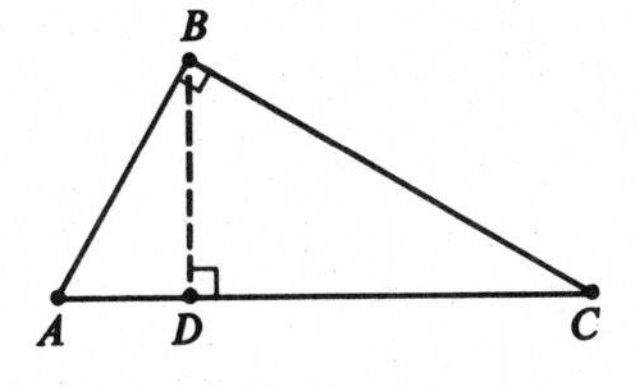

FIGURE 12.7

That is, if $\angle B$ is a right angle, and $\overline{BD} \perp \overline{AC}$, then A-D-C. (This follows from Theorem 5, Section 10.4).

12.3 THE PYTHAGOREAN THEOREM

To prove the Pythagorean Theorem, we need one preliminary result.

Theorem 1. The altitude to the hypotenuse of a right triangle divides it into two triangles each of which is similar to it.

RESTATEMENT. Let $\triangle ABC$ be a right triangle with its right angle at C, and let D be the foot of the perpendicular from C to $\overleftrightarrow{AB}$. Then $\triangle ACD \sim \triangle ABC \sim \triangle CBD$.

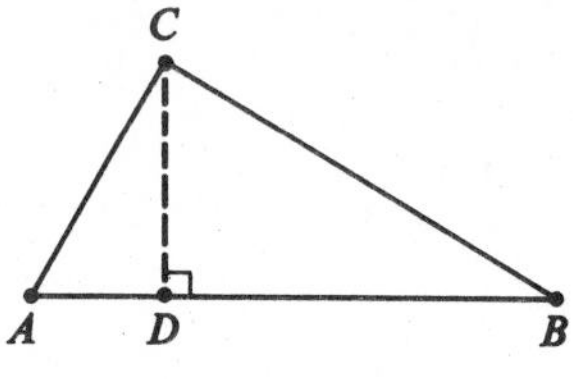

FIGURE 12.8

(To remember the way these similarities work, all you need to do is to observe that there is only one way that they *might* work. In the first correspondence, we must have $A \leftrightarrow A$, because $\angle A$ is common to $\triangle ACD$ and $\triangle ABC$. Also we must have $D \leftrightarrow C$ because these points are where the right angles are. Finally, we must have $C \leftrightarrow B$, because at this stage C has nowhere else left to go. Therefore the correspondence must be $ACD \leftrightarrow ABC$; and similarly for the second similarity.)

Proof. Obviously $\angle A \cong \angle A$. And $\angle ADC \cong \angle ACB$, because both of these are right angles. By the AA similarity theorem, we have

$$\triangle ACD \sim \triangle ABC.$$

The proof of the other half of the theorem is exactly the same.

Theorem 2. *The Pythagorean Theorem.* In any right triangle, the square of the length of the hypotenuse is the sum of the squares of the lengths of the other two sides.

RESTATEMENT. Let $\triangle ABC$ be a right triangle, with its right angle at C. Then

$$a^2 + b^2 = c^2.$$

Here we are using the usual notation for lengths of opposite sides.

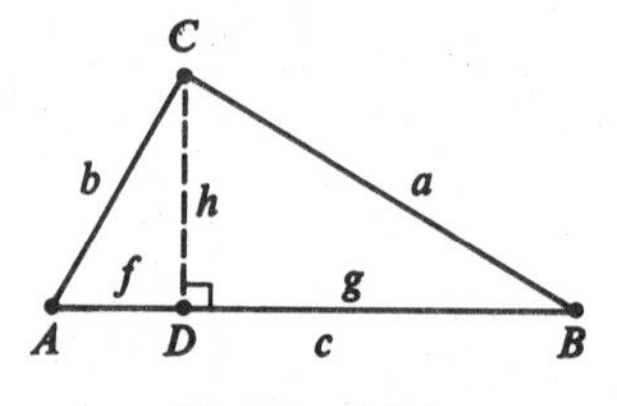

FIGURE 12.9

Proof. By the preceding theorem,

$$\triangle ACD \sim \triangle ABC \sim \triangle CBD.$$

Therefore

$$h, f, b \sim a, b, c \sim g, h, a.$$

We shall calculate f and g in terms of a, b, and c, and then use the fact that $f + g = c$. (Query: What theorem do we need to know, to conclude that $f + g = c$?)

Step 1. Since

$$\frac{f}{b} = \frac{b}{c},$$

we have

$$f = \frac{b^2}{c}.$$

And since

$$\frac{g}{a} = \frac{a}{c},$$

we have

$$g = \frac{a^2}{c}.$$

Step 2. Therefore

$$f + g = \frac{a^2 + b^2}{c} = c.$$

Therefore

$$a^2 + b^2 = c^2,$$

which was to be proved.

Legend has it that when Pythagoras discovered this theorem, he sacrificed an ox as a thank offering. (The legend is doubtful; and it is not even known whether the theorem was proved by Pythagoras personally.) The German poet, Heinrich Heine, remarked that ever since this sacrifice, the oxen have trembled whenever a great truth was discovered.

The proof given here is not the one given in Euclid. Euclid's proof (to be discussed later) made heavy use of the theory of area. During the last two thousand years or so, the literature of the Pythagorean theorem has become immense. Literally hundreds of proofs have been given. The converse of the Pythagorean theorem is also true, and its proof is easy.

Theorem 3. Given a triangle whose sides have lengths a, b, and c. If $a^2 + b^2 = c^2$, then the triangle is a right triangle, with its right angle opposite the side of length c.

Proof. Given $\triangle ABC$, with $a^2 + b^2 = c^2$. Let $\angle F$ be a right angle, and let D and E be the points on the sides of $\angle F$ such that $FE = a$ and $FD = b$ (Fig. 12.10). By the Pythagorean theorem,

$$DE^2 = a^2 + b^2.$$

Therefore

$$DE = \sqrt{a^2 + b^2} = c.$$

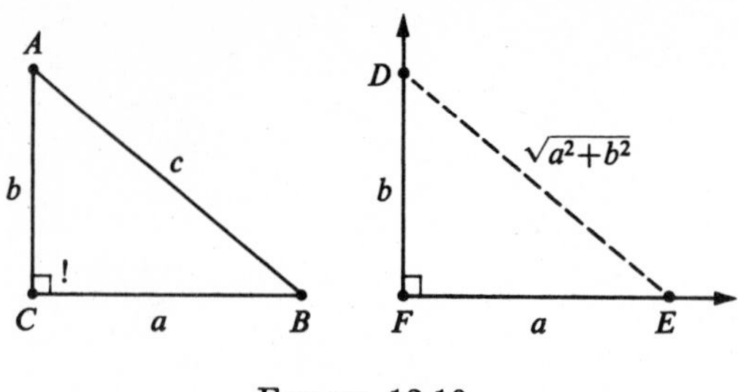

FIGURE 12.10

By SSS, $\triangle ABC \cong \triangle DEF$. Therefore $\triangle ABC$ is a right triangle, with its right angle at C.

Theorem 4. *The Hypotenuse-Leg Theorem.* If the hypotenuse and a leg of one right triangle are congruent to the hypotenuse and a leg of another, then the triangles are congruent.

RESTATEMENT. Given $\triangle ABC$, with a right angle at C, and $\triangle A'B'C'$, with a right angle at C'. If $a = a'$ and $c = c'$, then $\triangle ABC \cong \triangle A'B'C'$.

This was proved in Section 7.1. The proof belonged in Chapter 7, because it made no use of the parallel postulate, but the theorem itself belongs in the present context. Note that the Hypotenuse-Leg Theorem is an easy corollary of the Pythagorean Theorem.

We have already observed that every triangle has three altitudes, that is, one for each side considered as a base. A well-known formula asserts that the area of any triangle is equal to half the product of "the base" and "the altitude"; this means, of course, that the area is $= \frac{1}{2}bh$, where b is the length of any one of the three sides, and h is the corresponding altitude. Granted that this is right, which it is, it follows that the product bh must be independent of the choice of the base. That is, in the figure below, we must have

$$b_1h_1 = b_2h_2 = b_3h_3.$$

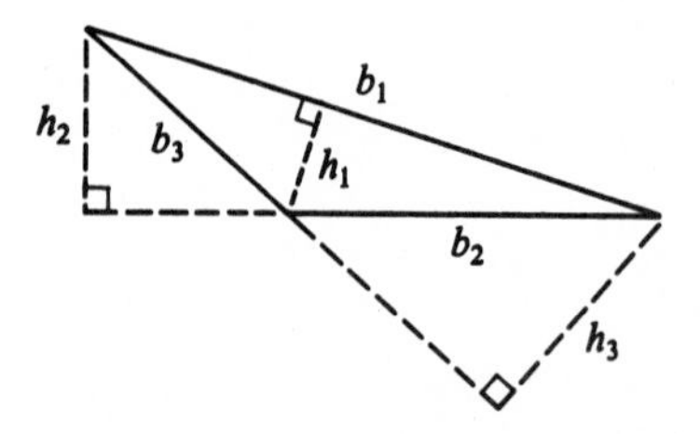

FIGURE 12.11

If the theory of plane area is set up by using postulates, then we can use area to prove that these equations hold. Later, however, we shall want to *prove* that simple figures such as triangles have areas, and that areas have the properties that we expect. For this purpose we want to prove the following theorem without using areas.

Theorem 5. In any triangle, the product of a base and the corresponding altitude is independent of the choice of the base.

RESTATEMENT. Given $\triangle ABC$. Let $\overline{AD}$ be the altitude from A to $\overleftrightarrow{BC}$, and let $\overline{BE}$ be the altitude from B to $\overline{AC}$. Then

$$AD \cdot BC = BE \cdot AC.$$

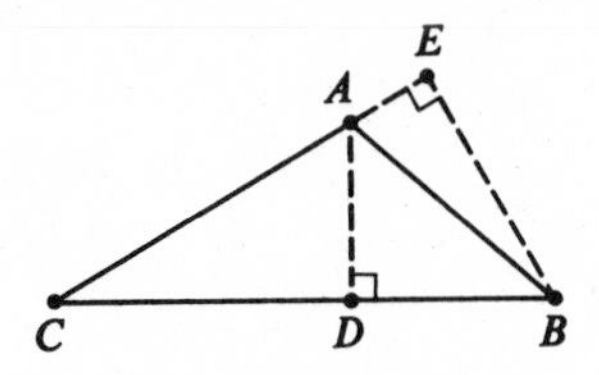

FIGURE 12.12

Proof. Suppose that $E \neq C$ and $D \neq C$, as shown in the figure. Then $\angle C = \angle C$, and $\angle BEC \cong \angle ADC$, because both are right angles. Therefore $\triangle BEC \sim \triangle ADC$. Hence

$$BE, BC \sim AD, AC.$$

Thus

$$\frac{AD}{BE} = \frac{AC}{BC},$$

and

$$AD \cdot BC = BE \cdot AC,$$

which was to be proved.

If $E = C$, then $\triangle ABC$ is a right triangle with its right angle at C and we also have $D = C$.

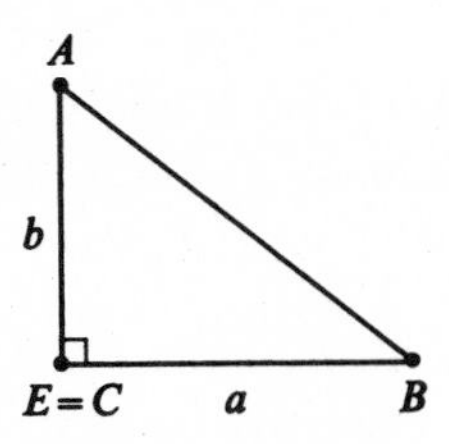

FIGURE 12.13

In this case, the theorem says trivially that $ab = ba$.

Theorem 6. For similar triangles, the ratio of any two corresponding altitudes is equal to the ratio of any two corresponding sides.

RESTATEMENT. Suppose that $\triangle ABC \sim \triangle A'B'C$. Let h be the altitude from A to $\overleftrightarrow{BC}$, and let h' be the altitude from A' to $\overleftrightarrow{B'C'}$. Then

$$\frac{h}{h'} = \frac{AB}{A'B'}.$$

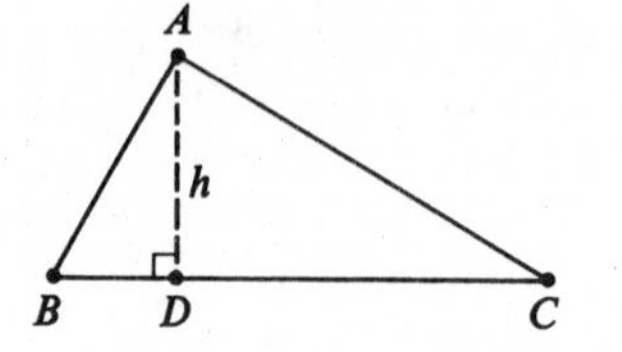

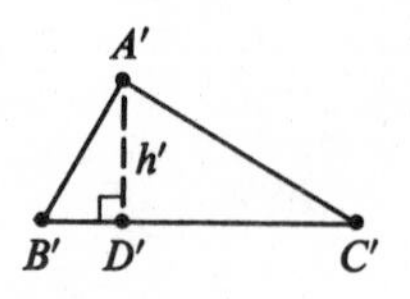

FIGURE 12.14

Proof. Let $\overline{AD}$ and $\overline{A'D'}$ be the altitudes whose lengths are h and h'. If $D = B$, then $D' = B'$, and there is nothing to prove. If not, $\triangle ABD \sim \triangle A'B'D'$, and the theorem follows.

Chapter 13
Polygonal Regions and Their Areas

13.1 THE AREA POSTULATES

A *triangular region* is a figure which is the union of a triangle and its interior like this:

FIGURE 13.1

The sides of the triangle are called *edges* of the region, and the vertices of the triangle are called *vertices* of the region.

A *polygonal region* is a figure like one of these:

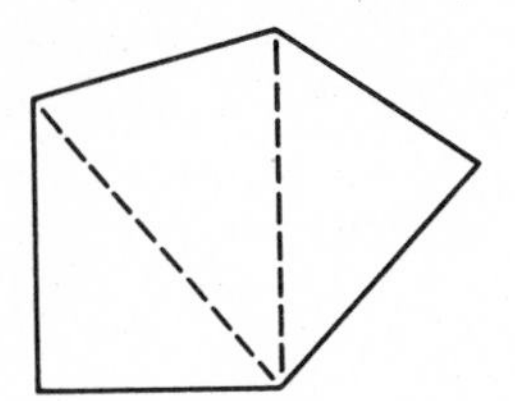

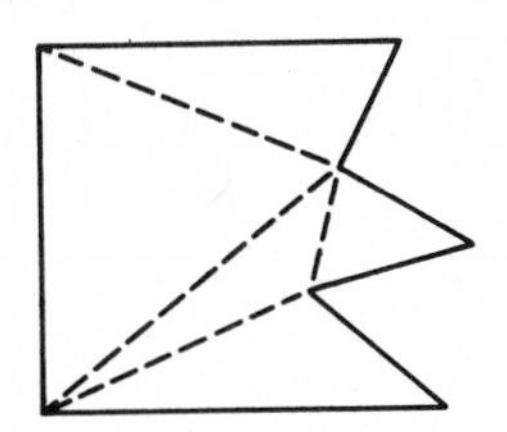

FIGURE 13.2

To be exact, a *polygonal region* is a plane figure which can be expressed as the union of a finite number of triangular regions, in such a way that if two of the triangular regions intersect, their intersection is an edge or a vertex of each of them. In each of the illustrations in Fig. 13.2, the dotted lines indicate how the regions can be cut up into triangular regions in such a way that the conditions of the definition are satisfied. Of course, there is nothing unique about the way in which a polygonal region can be cut up into triangular regions. In fact, if such a process can be carried out at all for a particular figure, it can be done in infinitely many ways. For example, a parallelogram plus its interior can be cut up in at least this many ways:

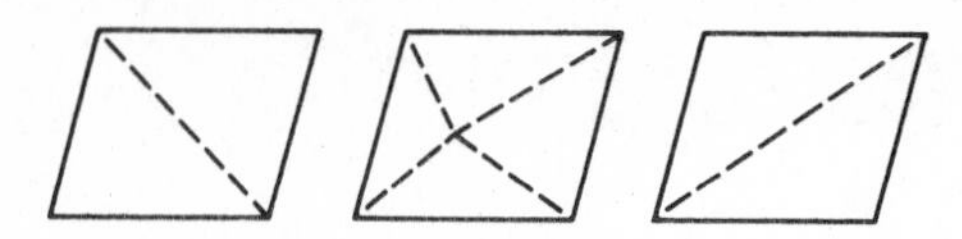

FIGURE 13.3

The theory of area is easiest to handle for the case where the figures are polygonal regions. And the easiest way to set up the theory is to suppose that an area function is given, under which to each polygonal region there corresponds a positive number called its area. Thus we let $\mathcal{R}$ be the set of all polygonal regions, and we add to our structure the function $\alpha : \mathcal{R} \to \mathbf{R}$.

Thus the total structure in our geometry is now

$$[S, \mathcal{L}, \mathcal{P}, d, m, \alpha],$$

and we need to state the postulates governing the area function α. (Here we are using the Greek letter alpha because the natural English letters have by now been used up for other purposes. We have used m for *measure*, and we want to go on using A and a to say, for example, that a is the length of the side opposite A in $\triangle ABC$.)

Our postulates are as follows.

A-1. α is a function $\mathcal{R} \to \mathbf{R}$, where $\mathcal{R}$ is the set of all polygonal regions and $\mathbf{R}$ is the set of all real numbers.

A-2. For every polygonal region R, $\alpha R > 0$.

A-3. *The Congruence Postulate.* If two triangular regions are congruent, then they have the same area.

A-4. *The Additivity Postulate.* If two polygonal regions intersect only in edges and vertices (or do not intersect at all), then the area of their union is the sum of their areas.

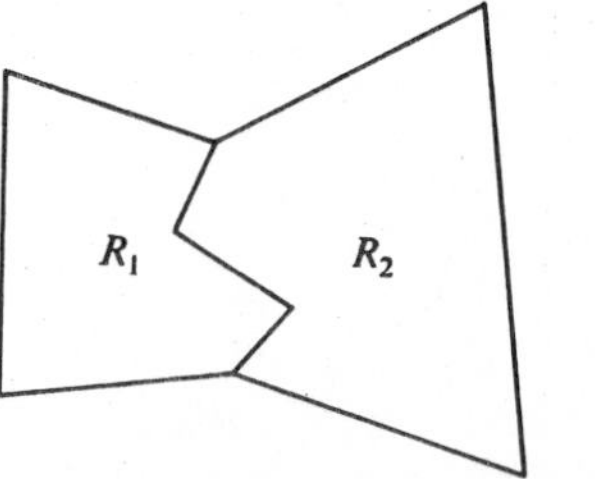

FIGURE 13.4

For example, $\alpha(R_1 \cup R_2) = \alpha R_1 + \alpha R_2$. Of course, areas cannot be added in this way if the intersection of the two regions contains a triangular region. Here $\alpha(T_1 \cup T_2)$ is obviously less than $\alpha T_1 + \alpha T_2$.

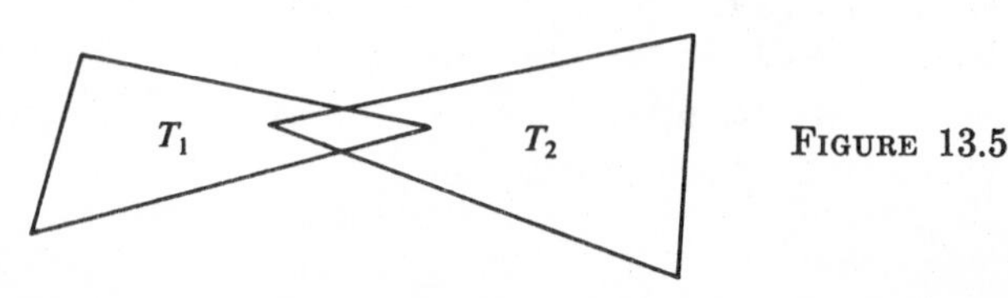

FIGURE 13.5

Note that if we have an area function that satisfies Postulates A-1 through A-4, and we then agree to multiply all areas by 2, then we get another area function that also satisfies A-1 through A-4. Intuitively speaking, area measured in square inches satisfies all our postulates so far, and so also does area measured in square cubits. We therefore need a postulate which, in effect, chooses a unit of measure for us by describing a connection between area and distance.

By a *square region* we mean the union of a square and its interior. Rectangular regions are defined in the same way. Either of the following statements, if taken as a postulate, is sufficient to determine the unit of area.

(1) If a square region has edges of length 1, then its area is 1.

(2) The area of a rectangular region is the product of its base and its altitude.

Statement (2) is an unreasonably strong postulate; we ought to be able to derive the formula for the area of a rectangular region from the formula for the area of a square region, and in fact the derivation is easy. (See below.) On the other hand, (1) is unreasonably weak: it leads to a difficult proof of the formula for the area of a square region of edge a, at the very beginning of the theory. (See Section 13.5.) We therefore split the difference, and use the following:

A-5. If a square region has edges of length a, then its area is a^2.

On this basis,we can prove (2) as a theorem.

Theorem 1. *The Rectangle Formula.* The area of a rectangular region is the product of its base and its altitude.

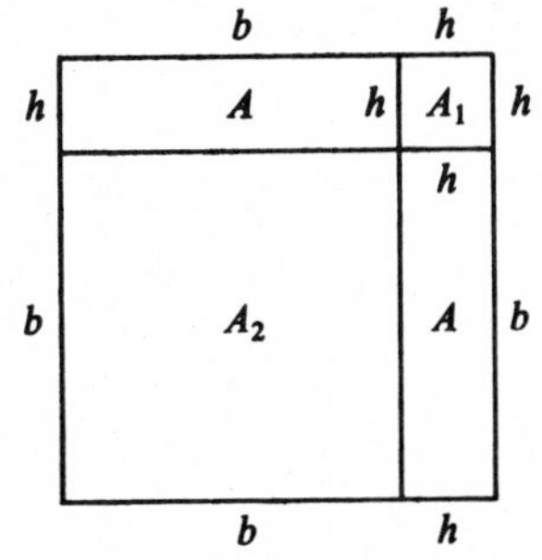

FIGURE 13.6

Proof. Given a rectangle of base b and altitude h, we construct a square of edge $b + h$, and decompose it into squares and rectangles as shown in the figure above. Then

$$\begin{aligned}(b + h)^2 &= 2A + A_1 + A_2,\\ b^2 + 2bh + h^2 &= 2A + h^2 + b^2,\\ 2bh &= 2A,\end{aligned}$$

and $A = bh$, which was to be proved.

Later we shall see that the theory of area, for polygonal regions, can be built up without using any new postulates at all. On the basis of the postulates stated earlier, it is possible to prove that there is an area function satisfying A-1 through A-5. Meanwhile, in the present chapter, we shall regard the area function as given, subject to A-1 through A-5, and show how it can be put to work on various geometric problems. The first step is to get formulas for the areas of the simplest figures.

Hereafter, for the sake of convenience, we shall speak of the areas of triangles, areas of rectangles, and so on; this is an abbreviation. Triangles and quadrilaterals are not polygonal regions, and obviously they are too thin to have areas greater than zero. Also we shall abbreviate $\alpha\triangle ABC$ as ABC, $\alpha\square ABCD$ as $ABCD$, and so on. This is consistent with our notation AB for the length of the segment $\overline{AB}$. In each case, when the letters appear without decoration, the resulting expression denotes a *number*, and this number measures, in some way, the corresponding geometric figure. To repeat:

$$ABC = \alpha\triangle ABC,$$

and

$$ABCD = \alpha\square ABCD,$$

by definition.

13.2 AREA THEOREMS FOR TRIANGLES AND QUADRILATERALS

Theorem 1. The area of a right triangle is half the product of the lengths of its legs.

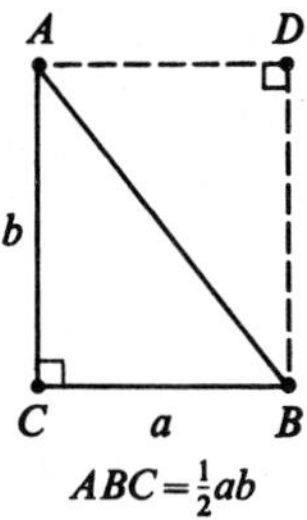

FIGURE 13.7

Proof. Given $\triangle ABC$, with a right angle at C. Let D be the point such that $\square ADBC$ is a rectangle. By the additivity postulate,

$$ADBC = ABC + ABD.$$

By the congruence postulate,

$$ABD = ABC.$$

Therefore

$$ADBC = 2ABC.$$

By the rectangle formula,

$$ADBC = ab.$$

Therefore $2ABC = ab$, and

$$ABC = \tfrac{1}{2}ab,$$

which was to be proved.

Theorem 2. The area of a triangle is half the product of any base and the corresponding altitude.

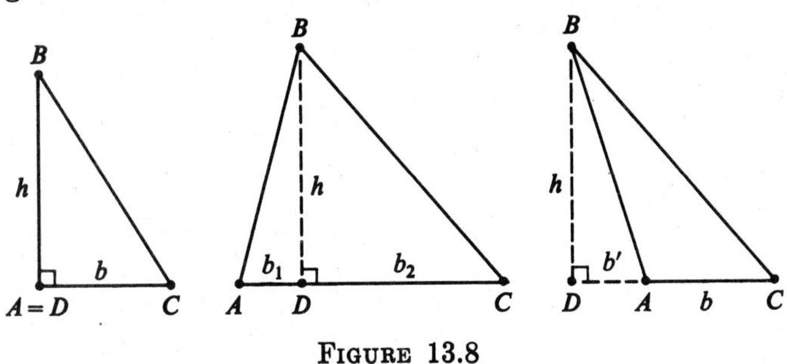

FIGURE 13.8

Proof. Given $\triangle ABC$. Let D be the foot of the perpendicular from B to $\overleftrightarrow{AC}$; let $AC = b$, and let $BD = h$ (as in each of the figures). There are, essentially, three cases to consider.

(1) If $A = D$, then $\triangle ABC$ is a right triangle and

$$ABC = \tfrac{1}{2}bh,$$

by Theorem 1.

(2) A-D-C. Let $AD = b_1$ and $DC = b_2$. By Theorem 1,

$$BDA = \tfrac{1}{2}b_1h \qquad \text{and} \qquad BDC = \tfrac{1}{2}b_2h.$$

By the additivity postulate,

$$ABC = BDA + BDC.$$

Therefore

$$\begin{aligned} ABC &= \tfrac{1}{2}b_1h + \tfrac{1}{2}b_2h = \tfrac{1}{2}(b_1 + b_2)h \\ &= \tfrac{1}{2}bh, \end{aligned}$$

which was to be proved.

(3) D-A-C. Let $b' = AD$. By Theorem 1,

$$BDC = \tfrac{1}{2}(b' + b)h.$$

Also by Theorem 1,

$$BDA = \tfrac{1}{2}b'h.$$

By the additivity postulate,

$$BDC = BDA + ABC.$$

Therefore

$$\begin{aligned} ABC &= BDC - BDA \\ &= \tfrac{1}{2}(b' + b)h - \tfrac{1}{2}b'h \\ &= \tfrac{1}{2}bh, \end{aligned}$$

which was to be proved.

Theorem 3. The area of a parallelogram is the product of any base and the corresponding altitude.

$$ABCD = bh.$$

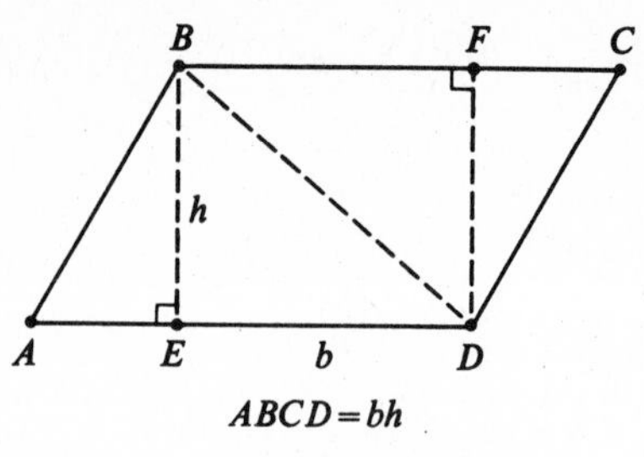

FIGURE 13.9

Proof. Given a parallelogram $\square ABCD$, with base $b = AD$ and corresponding altitude $h = BE$. By the additivity postulate,

$$ABCD = ABD + BDC.$$

By two applications of Theorem 2,

$$ABD = \tfrac{1}{2}bh \qquad \text{and} \qquad BDC = \tfrac{1}{2}bh.$$

(Details? We need to know that $BC = b$ and $DF = h$.) Therefore

$$ABCD = \tfrac{1}{2}bh + \tfrac{1}{2}bh = bh,$$

which was to be proved.

This is not the order of derivations that we most often see; usually we get the area formula for parallelograms first, and derive Theorem 2 from it. The "proof" of Theorem 3 then looks like this:

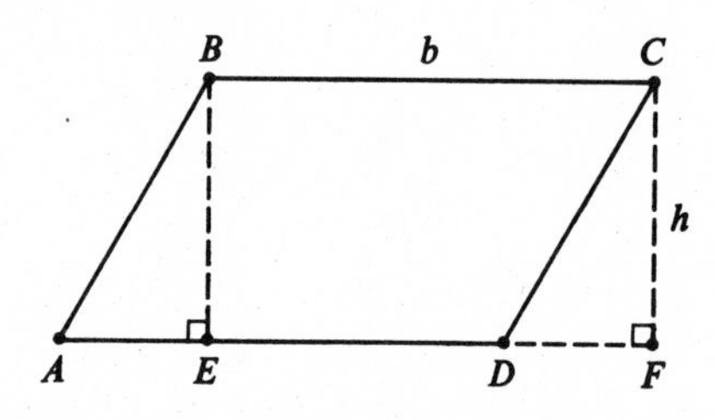

FIGURE 13.10

(1) $ABCD = ABE + BCDE$.
(2) $ABE = DCF$, because $\triangle ABE \cong \triangle DCF$.
(3) $ABCD = BCDE + DCF$.
(4) $BCDE + DCF = BCFE$, by the additivity postulate.
(5) $BCFE = bh$, by the unit postulate, because $BCFE$ is a rectangle.

This "proof" works only in the cases described by the figures that are drawn to illustrate it. Consider the following case. If the parallelogram looks like the figure below, then the above discussion becomes nonsense in the very first step, because there is no such thing as "the quadrilateral $\square BCDE$," and even if we allowed quadrilaterals to cross themselves, the equation $ABCD = ABE + BCDE$ would not hold for the areas of the corresponding polygonal regions.

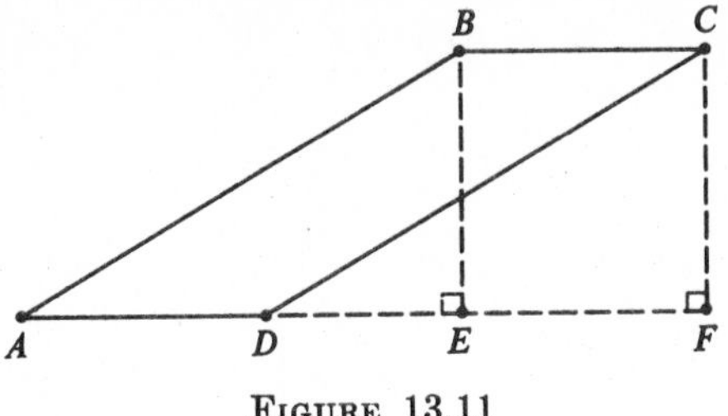

FIGURE 13.11

Theorem 4. The area of a trapezoid is half the product of the altitude and the sum of the bases.

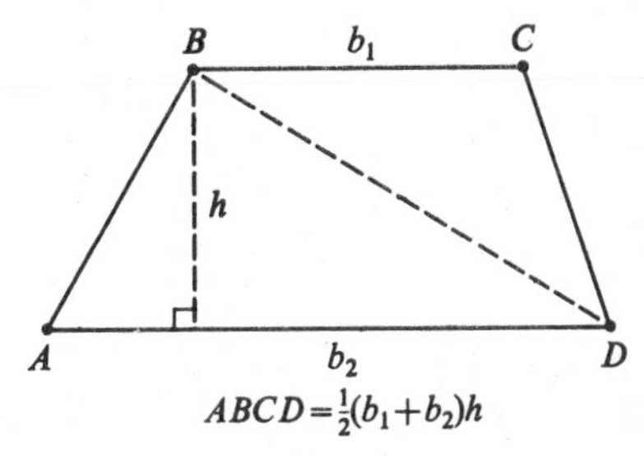

FIGURE 13.12

Proof. The main steps are as follows:

(1) $ABCD = ABD + BDC$,
(2) $ABD = \frac{1}{2}b_2h$,
(3) $BDC = \frac{1}{2}b_1h$,
(4) $ABCD = \frac{1}{2}(b_1 + b_2)h$.

Note that Theorem 4 includes Theorem 3 as a special case, because every parallelogram is a trapezoid. Note also that the proof of Theorem 4 is exactly the same as that of Theorem 3. The point is that although the invalid proof of Theorem 3 uses and needs the fact that $\overline{AB} \parallel \overline{CD}$, the valid proof does not.

Theorem 5. If two triangles have the same altitude, then the ratio of their areas is equal to the ratio of their bases.

This theorem follows immediately from the area formula. If the triangles $\triangle ABC$ and $\triangle DEF$ have bases b_1, b_2, and the corresponding altitude for each of them is h, then

$$\frac{ABC}{DEF} = \frac{\frac{1}{2}b_1h}{\frac{1}{2}b_2h} = \frac{b_1}{b_2},$$

which was to be proved. In much the same way, we get the following theorem.

Theorem 6. If two triangles have the same base, then the ratio of their areas is the ratio of their corresponding altitudes.

The following theorem is a corollary of each of the preceding theorems.

Theorem 7. If two triangles have the same base and the same altitude, then they have the same area.

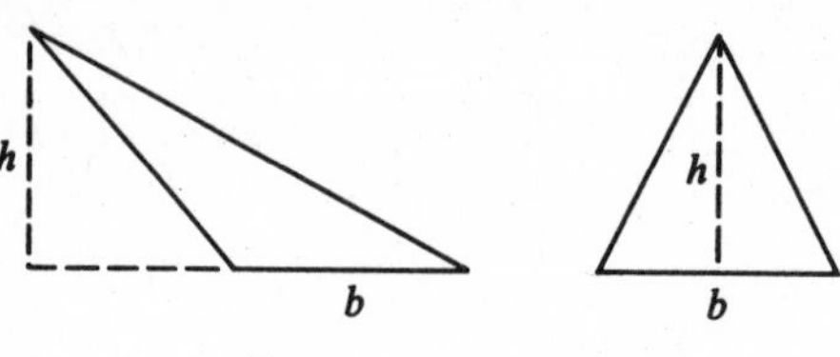

FIGURE 13.13

Theorem 8. If two triangles are similar, then the ratio of their areas is the square of the ratio of any two corresponding sides. That is, if

$$\triangle ABC \sim \triangle DEF,$$

then,

$$\frac{ABC}{DEF} = \left(\frac{a}{d}\right)^2 = \left(\frac{b}{e}\right)^2 = \left(\frac{c}{f}\right)^2.$$

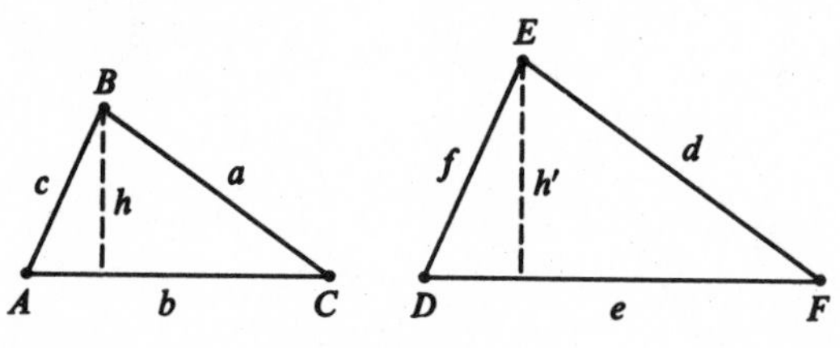

FIGURE 13.14

Proof. If the altitudes to $\overline{AC}$ and $\overline{DF}$ are h and h', as in the figure above, then we know by Theorem 6, Section 12.3, that

$$\frac{h}{h'} = \frac{a}{d} = \frac{b}{e} = \frac{c}{f}.$$

Now

$$\frac{ABC}{DEF} = \frac{\frac{1}{2}bh}{\frac{1}{2}eh'} = \left(\frac{b}{e}\right)\left(\frac{h}{h'}\right)$$
$$= \left(\frac{b}{e}\right)^2 = \left(\frac{a}{d}\right)^2 = \left(\frac{c}{f}\right)^2,$$

which was proved.

13.3 APPLICATIONS OF AREA THEORY: A SIMPLE PROOF OF THE BASIC SIMILARITY THEOREM

The theory of plane area is not merely an adjunct to the geometry on which it is based. If we use the area postulates A-1 through A-5 as part of our basic apparatus, then we can simplify some of our proofs considerably. Probably the most striking simplification of this kind is in the proof of the basic similarity theorem (Theorem 1, Section 11.4).

Consider first the case where the two transversals T and T' intersect in a point of L_1. (It is quite easy, as we shall see, to pass from this to the general case.) The picture then looks like this:

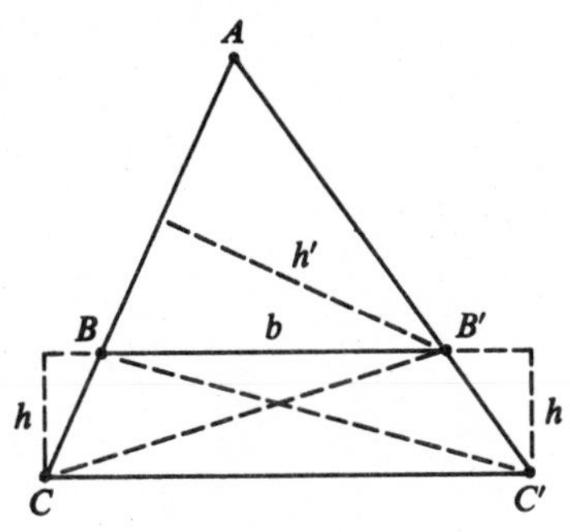

FIGURE 13.15

We have given $\overline{BB'} \parallel \overline{CC'}$, and we want to prove that

$$\frac{AB}{BC} = \frac{AB'}{B'C'}.$$

The steps in the area proof are as follows.

(1) $\triangle CBB'$ and $\triangle C'BB'$ have the same base $b = BB'$, and the same corresponding altitude h. Therefore

$$CBB' = C'BB'.$$

(2) Let us look at $\triangle ABB'$ and $\triangle CBB'$ sidewise, taking $\overline{AB}$ and $\overline{BC}$ as their bases. Then the corresponding altitudes are the same. In each case, the altitude is the length h' of the perpendicular from B' to $\overleftrightarrow{AC}$. Therefore

$$\frac{ABB'}{CBB'} = \frac{AB}{BC},$$

by Theorem 5, Section 13.2.

(3) For exactly the same sort of reason,

$$\frac{ABB'}{C'BB'} = \frac{AB'}{B'C'}.$$

[To get this, merely replace B by B', and C by C', in step (2).]

(4) Therefore

$$\frac{AB}{BC} = \frac{ABB'}{CBB'} = \frac{ABB'}{C'BB'} = \frac{AB'}{B'C'},$$

which was to be proved.

Consider now the general case. Let T'' be the line through A, parallel to T', intersecting L_2 and L_3 in B'' and C''. Then

$$\frac{AB}{BC} = \frac{AB''}{B''C''},$$

by the preceding proof. But we have

$$AB'' = A'B',$$
$$B''C'' = B'C',$$

because opposite sides of a parallelogram are congruent. Therefore

$$\frac{AB}{BC} = \frac{A'B'}{B'C'},$$

which was to be proved.

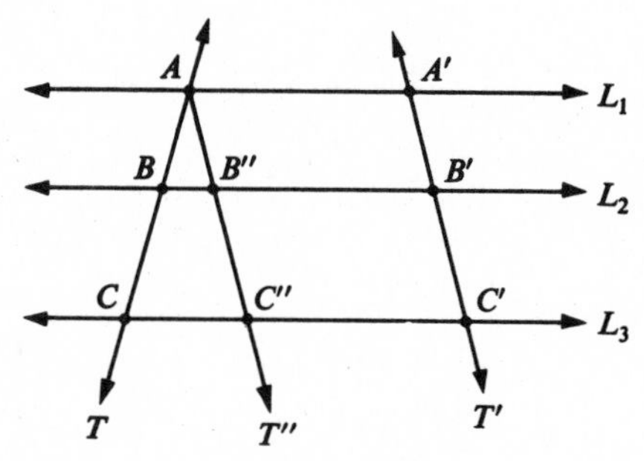

FIGURE 13.16

This proof of the basic similarity theorem is, in a way, inelegant because it creates the impression that the theorem depends on the theory of area. As we have seen, this impression is false. The area proof has, however, an important virtue; that is, it makes the theorem a part of elementary geometry.

The area proof, incidentally, is the one given in Euclid. This fact, and the proof itself, are not as widely known as they deserve to be.

13.4 FURTHER APPLICATIONS OF AREA THEORY: THE PYTHAGOREAN THEOREM

The Pythagorean theorem can be proved without using the concept of a similarity. Such a proof goes like this. Given a right triangle $\triangle ABC$, with legs of length a, b, and hypotenuse, c.

Take a square $\square DEFG$, of edge $a + b$. In the square, construct four congruent copies of $\triangle ABC$, as shown in the figure. (We construct them in the corners of the big square, using SAS.) Then $\angle KHI$ is a right angle, because $\angle DHK$ and $\angle EHI$ are complementary. For the same reason, all the angles of $\square HIJK$ are right angles, and $\square HIJK$ is a square.

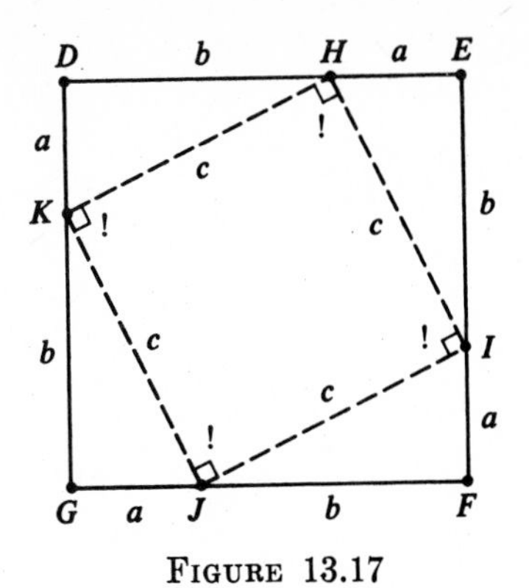

FIGURE 13.17

By the additivity postulate for area, the area is equal to the area of the inner square, plus the areas of the four right triangles. Since the right triangles are all congruent, we have

$$DEFG = HIJK + 4 \cdot KDH$$

or

$$(a + b)^2 = c^2 + 4 \cdot \tfrac{1}{2}ab$$

or

$$a^2 + 2ab + b^2 = c^2 + 2ab.$$

Therefore

$$a^2 + b^2 = c^2,$$

which was to be proved.

Euclid's proof also used areas, but it was different from the one above and considerably more complicated. Given a right triangle, he constructed squares on each of the sides, like this:

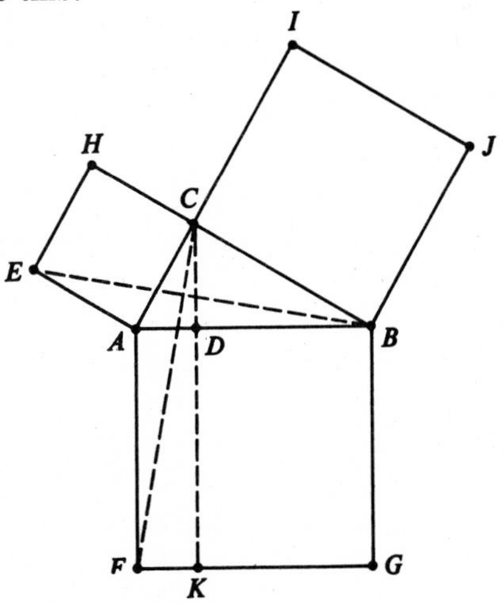

FIGURE 13.18

Here $\triangle ABC$ is a right triangle, with its right angle at C. The outer figures are squares, and $\overline{CK}$ is perpendicular to $\overline{AB}$ and $\overline{FG}$, intersecting then at D and K. The main steps in Euclid's proof were as follows.

(1) $ADKF = 2CAF$. (The rectangle and the triangle have the same base AF, and the same altitude AD.)

(2) $EHCA = 2EAB$. (The square and the triangle have the same base EA, and the same altitude AC.)

(3) $\angle EAC \cong \angle FAD$, because both are right angles, and $\angle CAD \cong \angle CAD$. Therefore $\angle EAB \cong \angle CAF$. But $EA = AC$, because $\square EHCA$ is a square; and $AB = AF$, because $\square ABGF$ is a square. By SAS, we have

$$\triangle EAB \cong \triangle CAF.$$

Therefore

$$EAB = CAF.$$

(4) From (1), (2), and (3), we get

$$ADKF = EHCA.$$

That is, *the square at the upper left has the same area as the rectangle at the lower left.* By exactly the same reasoning, we get the same conclusion for the square and rectangle on the right:

$$KDBG = BCIJ.$$

That is, *the square at the upper right has the same area as the rectangle at the lower right.*

(5) Since the area of the lower square $\square ABGF$ is the sum of the areas of the two rectangles $\square ADKF$ and $\square KDBG$, it follows that the area of the square on the hypotenuse is equal to the sum of the areas of the squares on the legs.

This proof is quite different in spirit, and in an important way, more elegant than the two that we have seen already. One way of putting it is that while our first two proofs depended on calculations, Euclid's reasoning is more geometric and more conceptual. The figure that goes with it is more than a reminder of the hypothesis and the notation; it is, in a sense, a picture of the Pythagorean phenomenon, so that if you understand the figure, you understand why the theorem is true.

In a way, the above presentation of Euclid's proof is misleading. It suggests that the Pythagorean theorem meant the same thing to Euclid, in the *Elements*, that it meant to us in Chapter 12 of this book; and this is far from true. Theorem 2, Section 12.3, stated that under certain conditions, the *numbers* a, b, c must satisfy the equation $a^2 + b^2 = c^2$; and the theorem appeared in a presentation of geometry in which the real numbers are given, independently of geometric concepts, and are *used* in the geometry to measure things. Thus we use the *number* AB as the measure of the length of the segment $\overline{AB}$; we use the number ABC as the measure of the area of the triangle $\triangle ABC$; and so on.

In Euclid, there are no numbers independent of geometric concepts, except, of course, for the positive integers, which are used to count things. To recast his theory, in a form meeting modern standards of explicitness and exactitude, is a formidable task. We have given a sample of this in Chapter 8. There we showed

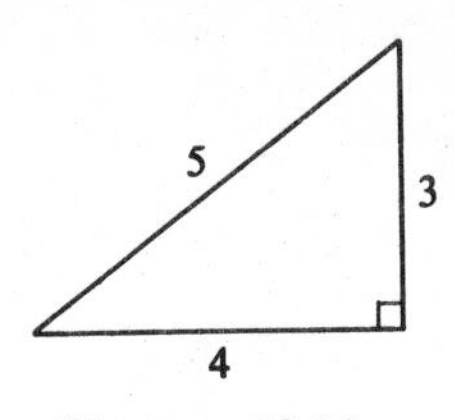

FIGURE 13.19

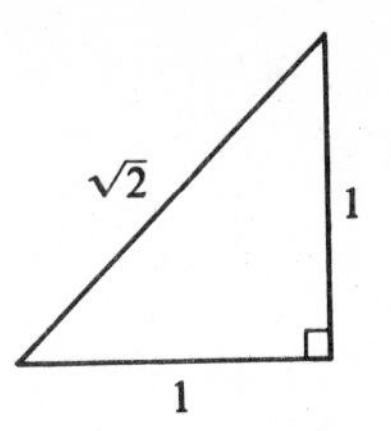

FIGURE 13.20

that instead of using the concept of *length* for segments (this length being a real number) you can use the concept of *same-length:* two segments have the same length if they are congruent, and congruence between segments is an undefined term, subject to certain postulates. Much the same thing can be done for the theory of area, but it is technically difficult. It is therefore customary, especially in elementary courses, to use the simpler apparatus of metric geometry. Often this is done unobtrusively, and it is not easy to see what is going on. A moment's reflection, however, will convince us that any time we label a figure as shown in the illustrations above, we are doing metric geometry, whether or not we say so.

13.5 A WEAKER FORM OF THE UNIT POSTULATE A-5

Our fifth postulate for area asserted that the area of a square is the square of the length of its edges. We remarked that the following postulate would have been sufficient.

A-5′. If a square has edges of length 1, then its area is 1.

This postulate gives us, in a minimal sense, a "unit of measure." We shall show that it implies A-5.

Theorem 1. If a square has edges of length $1/q$ (q a positive integer), then its area is $1/q^2$.

Proof. A unit square region can be decomposed into q^2 square regions, all with the same edge $1/q$ and with the same area A (Fig. 13.21). Therefore $1 = q^2A$, and $A = 1/q^2$.

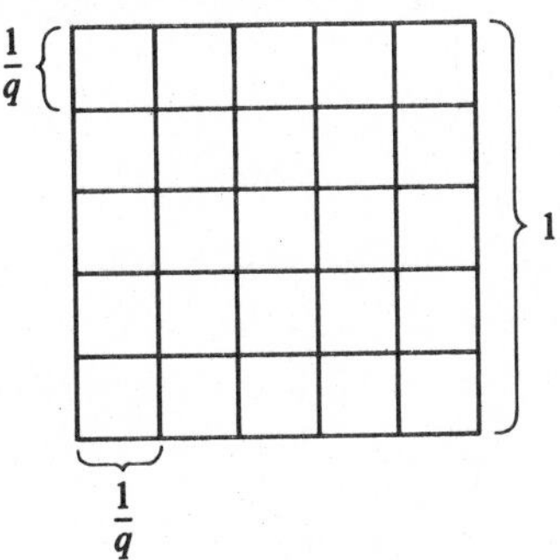

FIGURE 13.21

Theorem 2. If a square has edges of rational length p/q, then its area is p^2/q^2.

Proof. Such a square can be decomposed into p^2 squares, each of edge $1/q$.

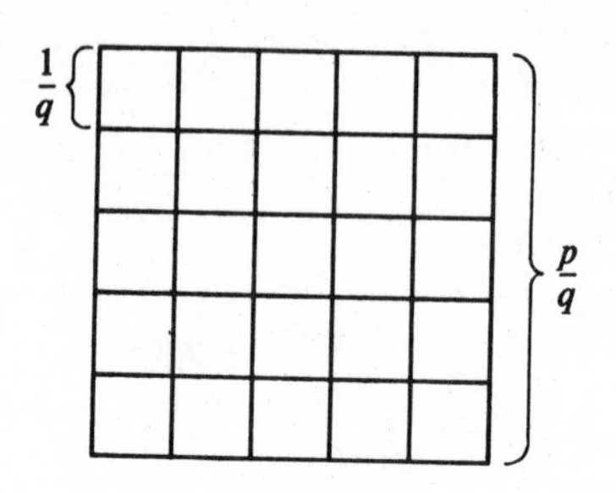

FIGURE 13.22

If A is its area, then

$$A = p^2 \cdot \frac{1}{q^2} = \frac{p^2}{q^2}.$$

Theorem 3. If a square has edges of length a, then its area is a^2.

Proof. Given a square S_a with edges of length a. Given any rational number p/q, let $S_{p/q}$ be a square of edge p/q, with an angle in common with S_a, like this:

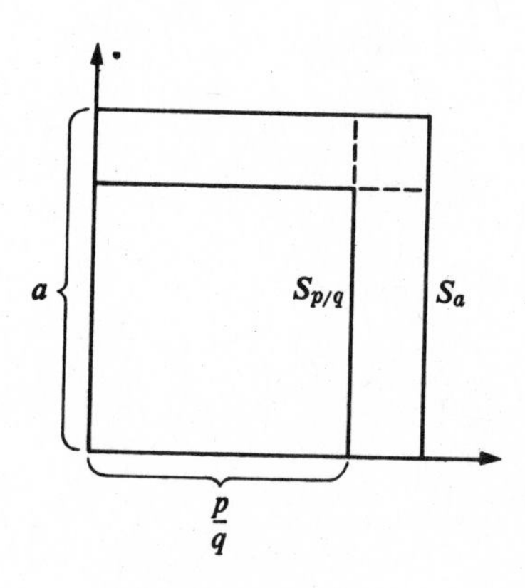

FIGURE 13.23

Each of the following statements is easily seen to be equivalent to the next.
(1) $p/q < a$
(2) $S_{p/q}$ lies in S_a (with two rectangular regions and a square region left over.)
(3) $\alpha S_{p/q} < \alpha S_a$.
(4) $p^2/q^2 < \alpha S_a$.
(5) $p/q < \sqrt{\alpha S_a}$.

Since (1) and (5) are equivalent, we have

$$a = \sqrt{\alpha S_a},$$

so that $a^2 = \alpha S_a$, which was to be proved.

The above treatment of A-5 is due to Peter Lawes, who was a student of the author at the time.

Chapter 14
The Construction of an Area Function

14.1 THE PROBLEM

In the preceding chapter we assumed that an area function was given, satisfying Postulates A-1 through A-5. This was reasonable; in fact, in elementary geometry, it is the only approach that is simple enough to be manageable.

It is natural, however, to ask whether this complicated set of assumptions was merely a matter of convenience or actually a logical necessity. The question is whether on the basis of our other postulates, we can *define* an area function for polygonal regions and *prove* that our function satisfies Postulates A-1 through A-5. The answer is yes.

In setting up such a function, we have to begin by assigning an area to *some* sort of figure. It seems hopeless to try to do everything at once, by assigning an area to every polygonal region at the very start. We ought to begin by defining areas for certain simple figures, and try to handle the more complicated figures in terms of the simple ones. For this purpose, rectangles are not promising, because they cannot be used as building blocks, except for very special figures. For example, a triangular region cannot be expressed as the union of a finite number of rectangular regions. It is true that we got the formula $\frac{1}{2}bh$, for the area of a right triangle, from the formula bh for the area of a rectangle. But to do this, we had to suppose that there was such a thing as the area of a right triangle; and in our present program, the latter question is the whole point at issue.

This suggests that our "point of entry" should be not rectangles, but triangles. We begin building our area function by stating, as a *definition*, that for any $\triangle ABC$, the area is

$$ABC = \tfrac{1}{2}bh,$$

where b is any base and h is the corresponding altitude. This makes sense, because we have shown (Theorem 5, Section 12.3) that the product bh depends only on the triangle, and does not depend on the choice of the base.

We now want to use triangles as building blocks. Given a polygonal region, we cut it up into a finite number of triangular regions intersecting only in edges and vertices (Fig. 14.1). For each of the triangles, the area is already defined, by the formula $\frac{1}{2}bh$. If the theory works, then the area of the region is the sum of the

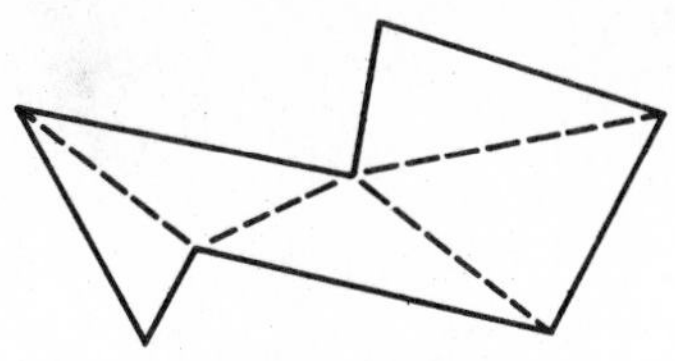

FIGURE 14.1

areas of the triangles. This suggests a tempting procedure: We might solve our problem at once by *defining* the area of the region as the sum of the areas of the triangles.

There is, however, a difficulty. Any polygonal region can be cut up into triangular regions in infinitely many ways:

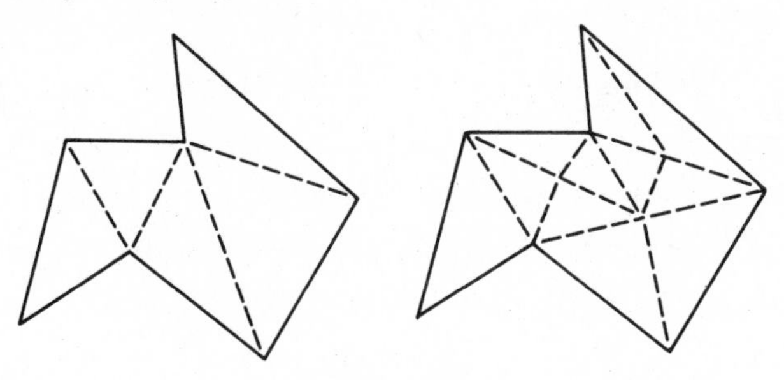

FIGURE 14.2

Therefore, before we can define the area of a polygonal region as the sum of the areas of the triangular regions, we need to prove that *this sum depends only on the region that we started with, and is independent of the way in which we cut it up.*

At first glance, this problem may seem trivial, and after a little reflection it may seem almost impossible. The truth is somewhere in between, as we shall see.

14.2 COMPLEXES AND THEIR FORMULA AREAS

Given a polygonal region R, expressed as the union of a finite number of triangular regions, intersecting only in edges and vertices. The set K whose elements are the triangular regions is called a *complex*, and is called a *triangulation* of R.

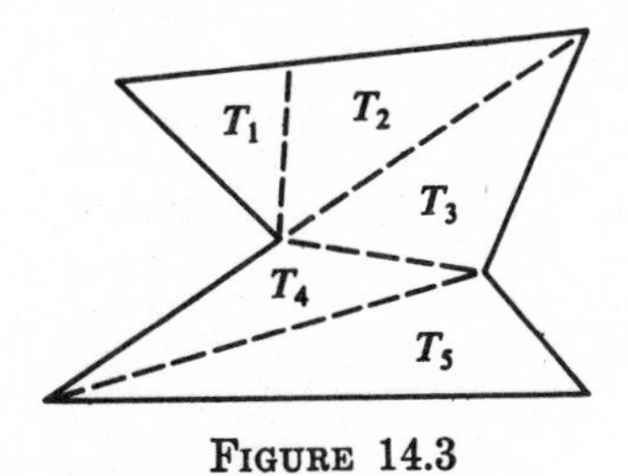

FIGURE 14.3

In the figure, the complex K is the set

$$\{T_1, T_2, T_3, T_4, T_5\}.$$

Thus R and K are objects of quite different kinds. R is an infinite set of points, and K is a finite set of triangular regions. The vertices and edges of the T_i's are also called vertices and edges of K.

We are going to define the area of a triangle by the formula $\frac{1}{2}bh$. To avoid confusion between the area function that we shall finally set up and the apparatus that we are using in the process, we shall call this the *formula area*, and state our first official definitions as follows.

DEFINITION 1. The formula area of a triangle is half the product of any base and the corresponding altitude.

DEFINITION 2. The formula area of a complex is the sum of the formula areas of its elements.

A *strip complex* is a complex that looks like either of the figures below:

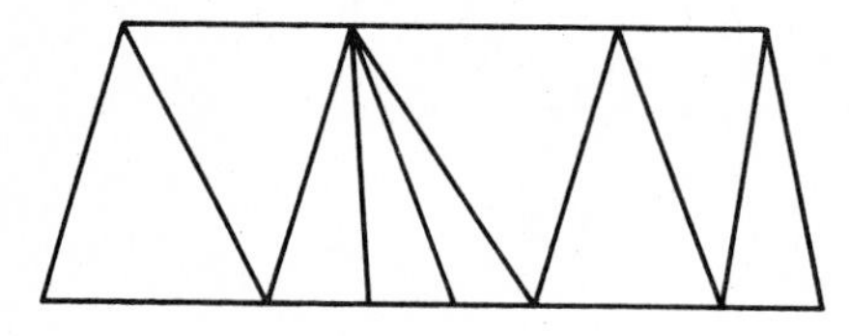

FIGURE 14.4

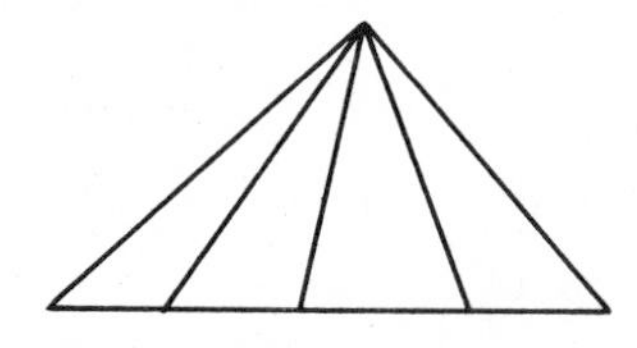

FIGURE 14.5

More precisely, a complex K is a *strip complex* (1) if K is a triangulation of a trapezoid or a triangle, (2) for the trapezoidal case, all vertices of K are on the upper or lower base, and (3) for the triangular case, all vertices of K are on the base or the opposite vertex.

Note that these cases are being handled together, as if the triangle were a "trapezoid with upper base = 0." The following theorem should also be interpreted in this way.

Theorem 1. The formula area of a strip complex is $\frac{1}{2}(b_1 + b_2)h$, where b_1 and b_2 are the bases and h is the altitude.

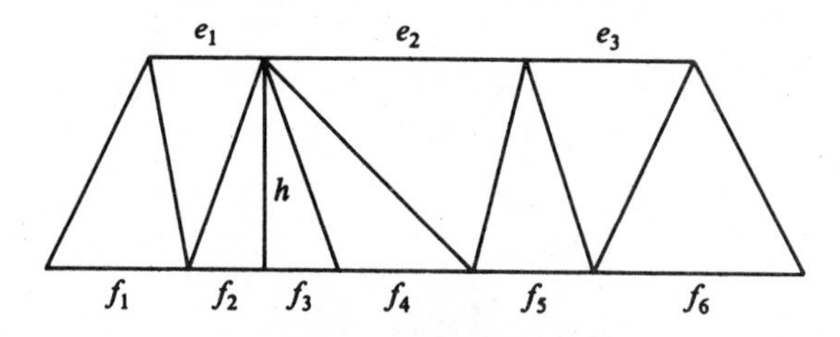

FIGURE 14.6

Proof. The sum of the formula areas of the triangles with two vertices on the upper base is

$$\tfrac{1}{2}e_1h + \tfrac{1}{2}e_2h + \cdots + \tfrac{1}{2}e_nh,$$

where $e_1, e_2, \ldots, e_n$ are the lengths of the segments into which the upper base is divided by the vertices. The sum of the formula areas of the other triangles is

$$\tfrac{1}{2}f_1h + \tfrac{1}{2}f_2h + \cdots + \tfrac{1}{2}f_mh.$$

Therefore the formula area of the complex is

$$\begin{aligned}\tfrac{1}{2}h(e_1 + e_2 + \cdots + e_n) + \tfrac{1}{2}h(f_1 + f_2 + \cdots + f_m) &= \tfrac{1}{2}hb_1 + \tfrac{1}{2}hb_2 \\ &= \tfrac{1}{2}(b_1 + b_2)h,\end{aligned}$$

which was to be proved.

It will be convenient to generalize this result slightly. By a *strip decomposition* of a triangle or trapezoid, we mean a decomposition into triangles and trapezoids, like this:

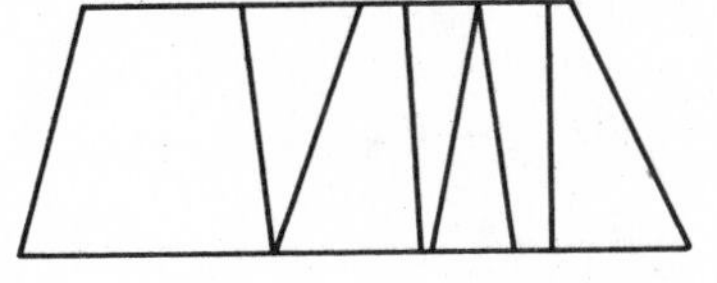

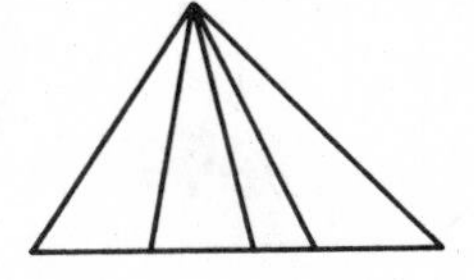

FIGURE 14.7

The *formula area* of a trapezoid is defined to be $\frac{1}{2}(b_1 + b_2)h$, where h is the altitude and b_1 and b_2 are the bases. The triangles and trapezoids in a strip decomposition are called its *parts.*

Theorem 2. For any strip decomposition of a triangle or trapezoid, the formula area of the original figure is the sum of the formula areas of the parts.

Proof. If none of the parts are trapezoids, this follows immediately from Theorem 1. If some trapezoids appear, we split each of them into two triangles by putting in a diagonal:

FIGURE 14.8

This does not change the sum of the formula areas of the parts. Therefore Theorem 2 follows from Theorem 1.

Theorem 3. If A-D-C, A-E-B, and $\overline{DE} \parallel \overline{BC}$, then the formula area of $\triangle ABC$ is the sum of the formula areas of $\triangle ADE$ and $\square DEBC$.

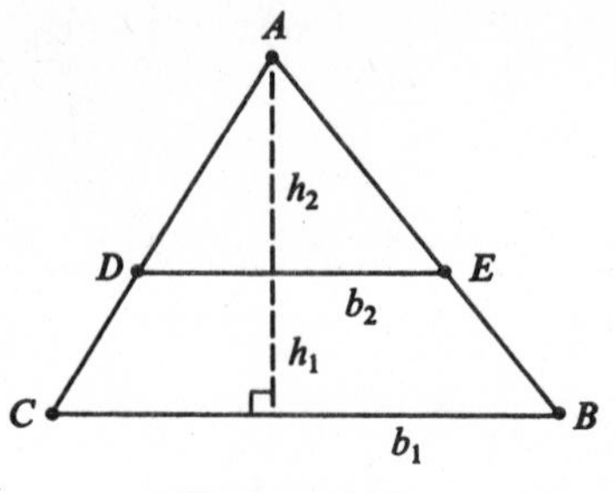

FIGURE 14.9

Proof. Let $b_1 = BC$ and $b_2 = DE$; let h_1 and h_2 be the altitudes of $\square DEBC$ and $\triangle ADE$. Our theorem then says that

$$\tfrac{1}{2}b_2h_2 + \tfrac{1}{2}(b_1 + b_2)h_1 = \tfrac{1}{2}b_1(h_1 + h_2)$$

or

$$b_2h_2 + b_1h_1 + b_2h_1 = b_1h_1 + b_1h_2$$

or

$$b_2(h_1 + h_2) = b_1h_2$$

or

$$\frac{h_1 + h_2}{h_2} = \frac{b_1}{b_2}.$$

This is true, by Theorem 6, Section 12.3, because $\triangle ADE \sim \triangle ACB$.

By a *parallel decomposition* of a triangular region, we mean a decomposition into one or two triangles and a finite number of trapezoids, like either of the following figures. (Let us spare ourselves the formal definition.)

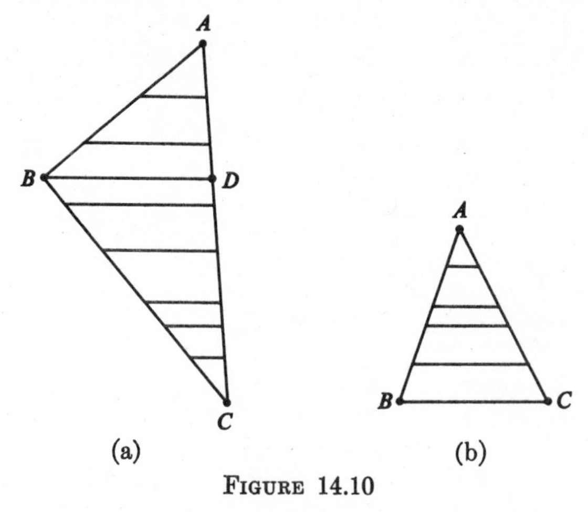

FIGURE 14.10

Theorem 4. For any parallel decomposition of a triangle, the formula area of the original triangle is the sum of the formula areas of the trapezoids and the triangles in the decomposition.

Proof. For Case 1 (Fig. 14.10b) this follows by repeated applications of the preceding theorem, working from the top downward. For Case 2, [Fig. 14.10(a)] we observe, first, that by Theorem 1, the formula area of $\triangle ABC$ is the sum of the formula areas of $\triangle BDC$ and $\triangle ABD$. (To apply Theorem 1, we must look at the figure sidewise.) Now apply the result of Case 1 to each of these. We are now ready to prove our main theorem.

Theorem 5. All triangulations of the same polygonal region have the same formula area.

Proof. Given two triangulations K_1 and K_2 of a polygonal region R.

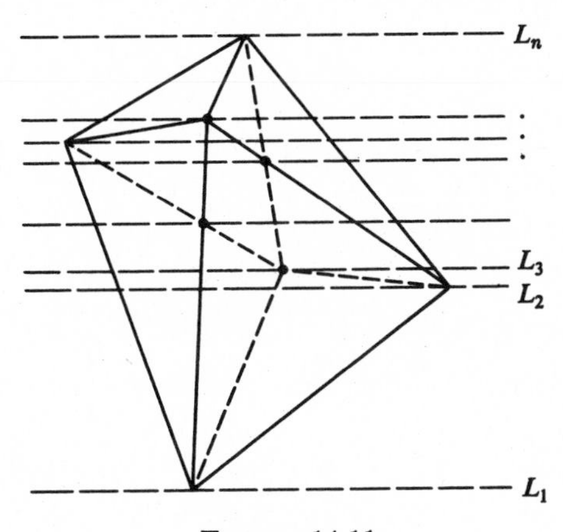

FIGURE 14.11

In the figure, the edges of K_1 are solid, and those of K_2 are dashed.

We take a family of parallel lines $L_1, L_2, \ldots, L_n$ passing through all vertices of K_1, all vertices of K_2, and all points where edges of K_1 intersect edges of K_2. (These are the horizontal lines, formed with long dashes, in the figure.) Now the lines L_i give parallel decompositions of each triangle of K_1:

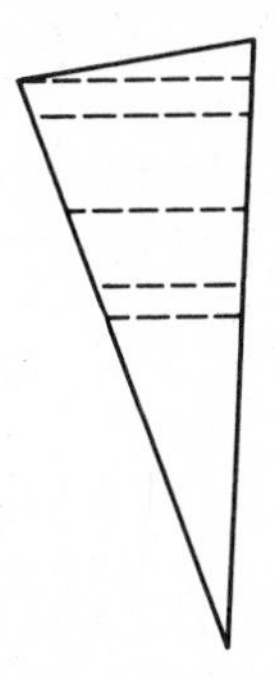

FIGURE 14.12

Let us call these triangles and trapezoids the *primary parts* of K_1. By definition, the formula area of K_1 is the sum of the formula areas of these triangles. We apply Theorem 4 to each triangle, and add the results. This gives:

(1) The formula area of K_1 is the sum of the formula areas of the primary parts of K_1.

The edges of K_2 give a strip decomposition of each primary part of K_1:

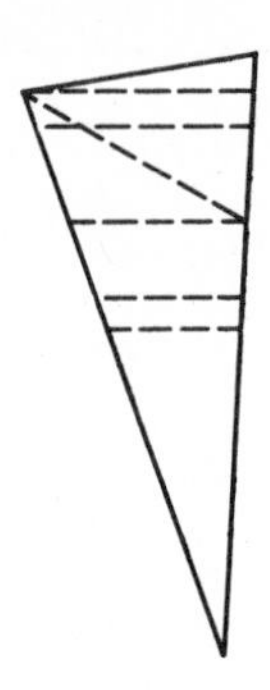

FIGURE 14.13

Let us call these smaller triangles and trapezoids the *secondary parts* of K_1. (In Fig. 14.13, there are a total of eight secondary parts, in the triangle shown. In Fig. 14.11 at the beginning of the proof of Theorem 5, there are a grand total of 31 secondary parts of K_1.)

We know, by Theorem 2, that:

(2) The formula area of each primary part of K_1 is the sum of the formula areas of the secondary parts that lie in it.

Combining (1) and (2), we get:

(3) The formula area of K_1 is the sum of the formula areas of the secondary parts of K_1.

Applying the same reasoning to K_2, we get:

(4) The formula area of K_2 is the sum of the formula areas of the secondary parts of K_2.

This tells us all that we need to know, because the secondary parts of K_2 are exactly the same as the secondary parts of K_1. Therefore, by (3) and (4), K_1 and K_2 have the same formula area, which was to be proved. Thus, we can state the following definition.

DEFINITION. The area αR of a polygonal region R is the number which is the formula area of every triangulation of R.

14.3 VERIFICATION OF THE AREA POSTULATES FOR THE FUNCTION α

Trivially, we know that

A-1. α is a function $\mathcal{R} \to \mathbf{R}$.

A-2. $\alpha R > 0$ for every R.

A-3. Any two congruent triangles have the same area.

These follow immediately from the definition of α.

A-5. If a square region has edges of length a, then its area is a^2.

The reason is that either diagonal divides the square region into two right triangular regions, each of which has formula area $a^2/2$. (This is very much like the proof of Theorem 1, Section 13.2, only now it works in reverse.)

We proceed to verify A-4. Given regions R_1 and R_2, with triangulations K_1 and K_2. Suppose that R_1 and R_2 intersect only in edges and vertices. It may

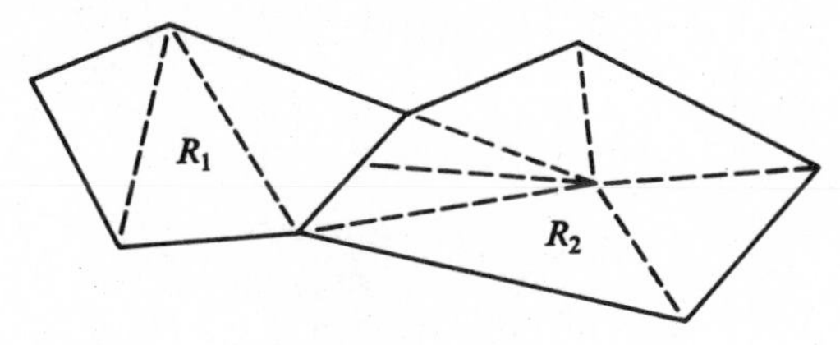

FIGURE 14.14

happen, as indicated in the figure, that some edges of K_1 (or K_2) contain more than one edge of K_2 (or K_1). If so, we split some triangles in K_1 (or K_2) into smaller triangles.

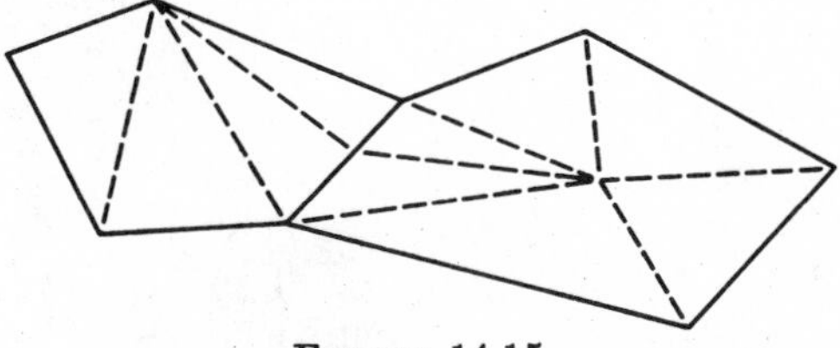

FIGURE 14.15

In this way we get new triangulations K_1', K_2' whose union K is a triangulation of $R_1 \cup R_2$. Now $\alpha(R_1 \cup R_2) = \alpha R_1 + \alpha R_2$, because the formula area of K is the sum of the formula areas of K_1 and K_2. Thus A-4 is satisfied by α.

Chapter 15
Perpendicular Lines and Planes in Space

15.1 THE BASIC THEOREMS

Given a line L and a plane E, intersecting in a point P.

If *every* line in E, passing through P, is perpendicular to L, then we say that L and E are *perpendicular*, and we write $L \perp E$, or $E \perp L$. In the figure, L is supposed

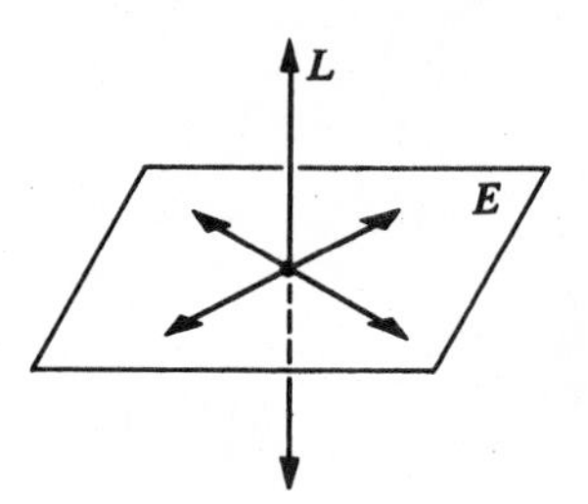

FIGURE 15.1

to be perpendicular to E. We have indicated two lines in E, passing through P. These are both perpendicular to L, although, in a perspective drawing, they don't look as if they were. Note that when we say that $L \perp E$, we are making a statement about an infinite collection of lines; that is, all of the lines that lie in E and contain P must be perpendicular to E. If we required merely that E contain *one* line perpendicular to L, this wouldn't mean anything. You can easily convince yourself that every plane that intersects L contains such a line. Soon we shall prove that if E contains two lines perpendicular to L, then $E \perp L$. The following two theorems are preliminaries.

A point A is *equidistant* from two points P and Q if $AP = AQ$ (Fig. 15.2).

Theorem 1. If A and B are equidistant from P and Q, then every point between A and B has the same property.

The main steps in the proof are as follows.

(1) $\triangle PAB \cong \triangle QAB$.

(2) $\angle PAB \cong \angle QAB$.

(3) $\triangle PAX \cong \triangle QAX$.

(4) $PX = QX$.

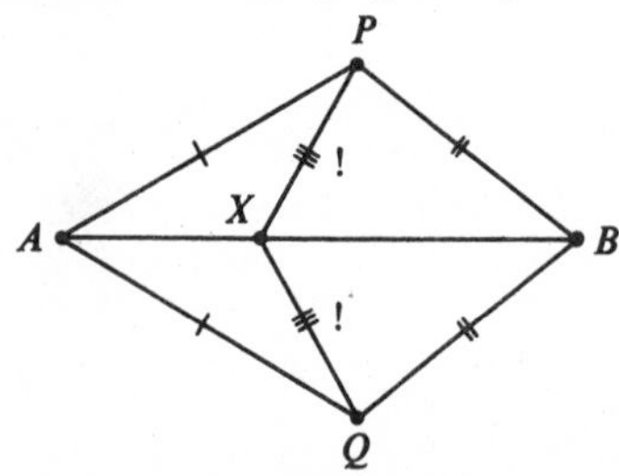

FIGURE 15.2

Theorem 2. If a line L contains the midpoint of $\overline{PQ}$, and contains another point which is equidistant from P and Q, then $L \perp \overline{PQ}$.

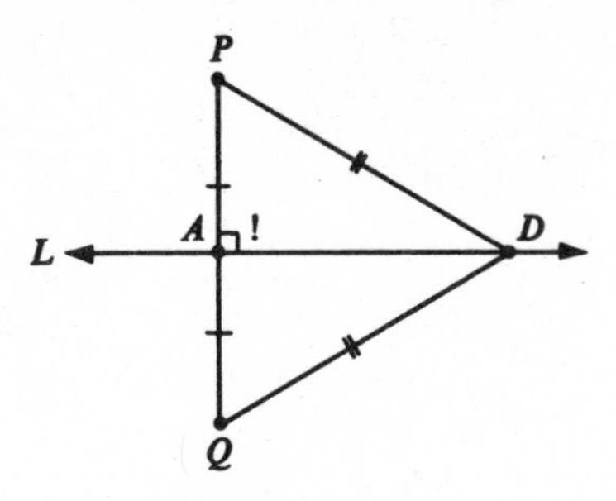

FIGURE 15.3

For $\triangle PAD \cong \triangle QAD$ by SSS. Therefore $\angle PAD \cong \angle QAD$, and each is a right angle.

Theorem 3. If a line is perpendicular to each of two intersecting lines at their point of intersection, then it is perpendicular to the plane that contains them.

RESTATEMENT. Let L_1 and L_2 be two lines intersecting at A, and let E be the plane that contains them. Let L be a line which is perpendicular to L_1 and L_2 at A. Then every line in E through A is perpendicular to L.

Proof. Let P and Q be points of L such that P-A-Q and $AP = AQ$. Let L_3 be any third line in E through A. Now each of the lines L_1 and L_2 contains points on each side of L_3 in E. Let B and C be points of L_1 and L_2, lying on opposite sides of L_3 in E. Then $\overline{BC}$ intersects L_3 in a point $D \neq A$. Now

(1) $\triangle PAB \cong \triangle QAB$, by SAS.
(2) $PB = QB$.
(3) Similarly, $PC = QC$.
(4) $PD = QD$, by Theorem 1.
(5) $\overleftrightarrow{PQ} \perp L_3$, by Theorem 2.

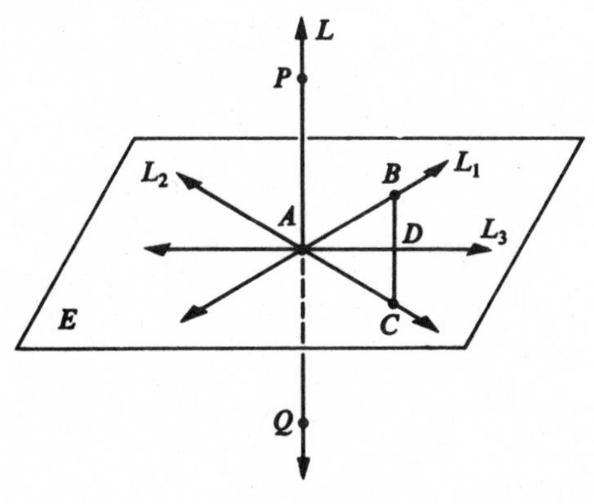

FIGURE 15.4

The perpendicular bisector of a segment in a plane is the line which is perpendicular to the segment at its midpoint. The following theorems are easy to prove, and serve as an introduction to the analogous theorems in space.

Theorem 4. If L is the perpendicular bisector of the segment $\overline{AB}$ (in a plane E), then all points of L are equidistant from A and B.

The converse is also true.

Theorem 5. Let A, B, and P be points of a plane E. If P is equidistant from A and B, then P lies on the perpendicular bisector of $\overline{AB}$.

Combining these two theorems, we get the following.

Theorem 6. The perpendicular bisector of a segment in a plane is the set of all points of the plane that are equidistant from the end points of the segment.

A theorem of this sort is called a *characterization* theorem. You have characterized a figure—that is, a set of points—if you state a condition that is satisfied by the points belonging to the given set, and by no other points. Usually the proof of a characterization theorem is in two parts. For example, let L be the perpendicular bisector of $\overline{AB}$ in the plane E and let

$$W = \{P | P \text{ lies in } E, \text{ and } PA = PB\}.$$

Theorem 4 says that every point of L lies in W; that is, $L \subset W$. Theorem 5 says that every point of W lies in L; that is, $W \subset L$. Taken together, these statements tell us that $L = W$; and this is precisely the content of Theorem 6.

Under a widespread usage, theorems like Theorem 6 are called *locus* theorems; people sometimes say that "the perpendicular bisector of a segment is the *locus* of all points of the plane that are equidistant from the end points." The Latin meaning of the word *locus* is no clue to this usage, because in Latin the word simply means *place*, which doesn't fit the context at all. The most straightforward way to describe the usage is to say that a set of points is referred to as a locus if the speaker is about to give a characterization of the set. For this reason, the word *locus* is superfluous; it has been superseded by the universally applicable word *set*.

Pursuing the analogy of perpendicular bisectors in a plane, we may now be tempted to say that *the perpendicular bisector of a segment (in space) is the plane which is perpendicular to the segment at its midpoint.* This definition tacitly assumes that through the midpoint of a segment there is one and only one perpendicular plane; and therefore we—or rather you—should prove these statements, in order to legitimize the definition.

Theorem 7. Given a line L and a point P of L. There is at least one plane which is perpendicular to L at P.

Theorem 8. Given a line L and a point P of L. There is only one plane which is perpendicular to L at P.

[*Hint:* L is the intersection of two planes E_1 and E_2.] We can now prove theorems analogous to Theorems 4, 5, and 6.

Theorem 9. Every point of the perpendicular bisecting plane of a segment is equidistant from the end points of the segment.

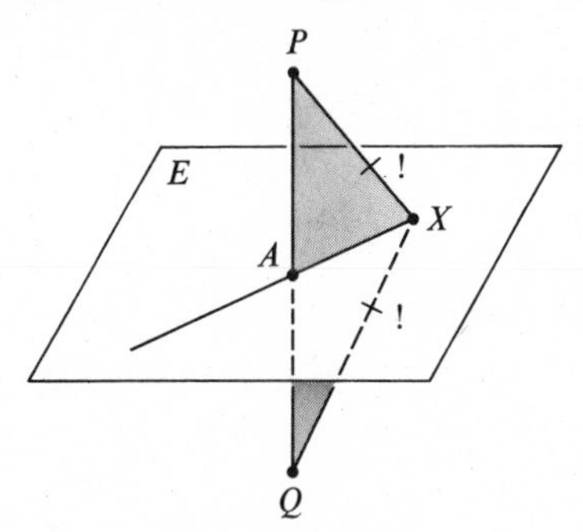

FIGURE 15.5

Proof. Let A be the midpoint of $\overline{PQ}$, let E be perpendicular to $\overline{PQ}$ at A, and let X be any point of E. If $X = A$, then X is equidistant from P and Q. If $X \neq A$, then by SAS we have $\triangle PAX \cong \triangle QAX$, so that $PX = QX$, which was to be proved.

Theorem 10. Every point equidistant from the end points of a segment lies in the perpendicular bisecting plane of the segment.

RESTATEMENT. Let M be the midpoint of $\overline{PQ}$, let E be the plane perpendicular to $\overline{PQ}$ at M, and let X be a point such that $PX = QX$. Then X lies in E.

Proof. By Theorem 2, $\overleftrightarrow{MX} \perp \overline{PQ}$ at M. Let F be the plane that contains $\overline{PQ}$ and X. Then F intersects E in a line L, and $L \perp \overline{PQ}$ at M. Therefore $L = \overleftrightarrow{MX}$, because perpendicular lines in a plane are unique. Therefore X lies in E, which was to be proved.

Fitting these theorems together, as in the planar discussion, we get a characterization theorem.

Theorem 11. The perpendicular bisecting plane of a segment is the set of all points that are equidistant from the end points of the segment.

We shall need the following theorem in discussing perpendiculars to planes through external points.

Theorem 12. Any two lines perpendicular to the same plane are coplanar.

Proof. Let L_1 and L_2 be perpendicular to the plane E at points A and B, respectively. Let M be the midpoint of $\overline{AB}$, let L be the perpendicular bisector of $\overline{AB}$ in E, and let P and Q be two points of L, equidistant from M.

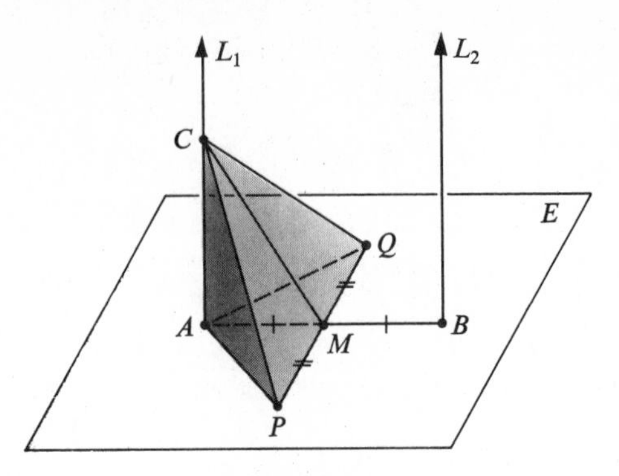

FIGURE 15.6

We shall show that *every point C of L_1 is equidistant from P and Q*. Now $AP = AQ$, because $\overleftrightarrow{AB}$ is the perpendicular bisector of $\overline{PQ}$. Since $L_1 \perp E$, it follows that $\triangle CAP \cong \triangle CAQ$. Therefore $CP = CQ$.

In the same way it follows that every point of L_2 is equidistant from P and Q. Therefore L_1 and L_2 lie in the same plane, namely, the perpendicular bisecting plane of the segment $\overline{PQ}$. (See Theorem 11.)

The rest of the theorems in this section are stated without proof; you should be able to furnish the proofs in each case. (But for Theorem 15 you will probably need the hints at the end of the section.)

Theorem 13. Through a given point in a given plane there is at least one line perpendicular to the given plane.

Theorem 14. Through a given point in a given plane there is at most one line perpendicular to the given plane.

Theorem 15. Through a given point not in a given plane there is at least one line perpendicular to the given plane.

Theorem 16. Through a given point not in a given plane there is at most one line perpendicular to the given plane.

The preceding four theorems fit together to give the following theorems.

Theorem 17. Given a point and a plane, there is exactly one line which passes through the given point and is perpendicular to the given plane.

Theorem 18. If a plane E and line L are perpendicular at a point P, then E contains every line that passes through P and is perpendicular to L.

The main stages in the proof of Theorem 15 are the following. We have given a plane E and an external point P.

(1) Let L_1 by *any* line in E.

(2) Let E_1 be the plane containing P and L_1.

(3) Let L_2 be the perpendicular from P to L_1, intersecting L_1 at Q.
(4) Let L_3 be the perpendicular to L_1 at Q, in E.
(5) Let E_2 be the plane containing P and L_3.
(6) Let L be the perpendicular from P to L_3, in E_2.
We then show that $L \perp E$.

PROBLEM SET 15.1

1. Prove Theorems 7, 8, 13, 14, 15, 16, and 18, above.

2. Show that if a line L contains two points equidistant from P and Q, then every point of L is equidistant from P and Q.

3. Show that if a plane E contains three noncollinear points which are equidistant from P and Q, then all points of E are equidistant from P and Q.

15.2 PARALLEL LINES AND PLANES IN SPACE

Two planes are called *parallel* if they have no point in common. If the planes E_1 and E_2 are parallel, then we write $E_1 \parallel E_2$. Similarly, a line L and a plane E are parallel if they have no point in common; we abbreviate this by writing $L \parallel E$, or $E \parallel L$.

The theory of parallelism in space is closely analogous to that of parallelism in a plane. No new postulates are needed, and the proofs are easy, as we shall see.

Theorem 1. If a plane intersects two parallel planes, then it intersects them in two parallel lines.

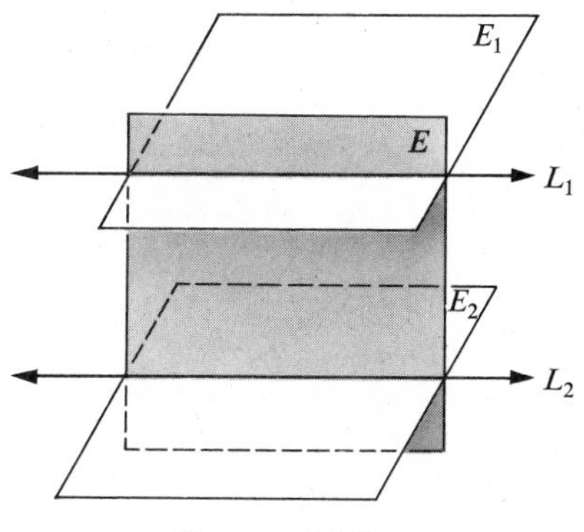

FIGURE 15.7

Proof. We have given a plane E, intersecting two parallel planes E_1 and E_2 in the nonempty sets L_1 and L_2. By Postulate I-4, the sets L_1 and L_2 are lines. They are coplanar, because E contains both of them, and they have no point in common, because L_1 lies in E_1 and L_2 lies in E_2. Therefore L_1 and L_2 are parallel lines, which was to be proved.

Theorem 2. Two lines perpendicular to the same plane are parallel.

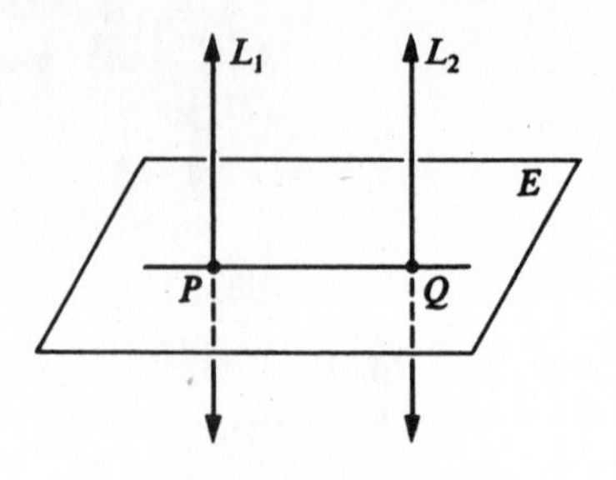

FIGURE 15.8

Proof. Given L_1 and L_2, perpendicular to a plane E at points P and Q. By Theorem 12, Section 15.1, L_1 and L_2 are coplanar. And each of them is perpendicular to the line $\overleftrightarrow{PQ}$, because $\overleftrightarrow{PQ}$ lies in E. Therefore $L_1 \parallel L_2$. (Why?)

Theorem 3. If a line is perpendicular to one of two parallel planes, it is perpendicular to the other.

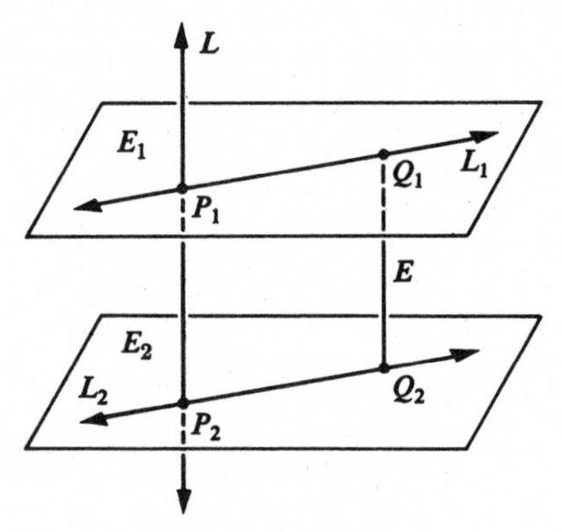

FIGURE 15.9

Proof. Let E_1 and E_2 be the two parallel planes, and let L be a line perpendicular to E_1 at P_1. Thus we have given that every line in E_1 that contains P_1 is perpendicular to L. We need to prove two things:

(1) L intersects E_2 (in a point P_2).

(2) Every line in E_2 that contains P_2 is perpendicular to L.

Proof of (1). Let Q_2 be any point of E_2, and let $\overline{Q_1Q_2}$ be the perpendicular segment from Q_2 to a point Q_1 of E_1. If $Q_1 = P_1$, there is nothing to prove, because, in this case, $\overleftrightarrow{Q_1Q_2} = L$, by Theorem 14, Section 15.1, and L therefore intersects E_2.

Suppose, then, that $Q_1 \neq P_1$. By the preceding theorem, it follows that $L \parallel \overleftrightarrow{Q_1Q_2}$. Let F be the plane that contains L and $\overleftrightarrow{Q_1Q_2}$. Since F intersects E_2, it follows that the intersection is a line L_2. If it were true that $L \parallel L_2$, then it would follow that $L_2 \parallel \overleftrightarrow{Q_1Q_2}$, which is false. Therefore L and L_2 are *not* parallel; L and L_2 intersect, at a point P_2. Then P_2 lies in both L and E_2, which was to be proved.

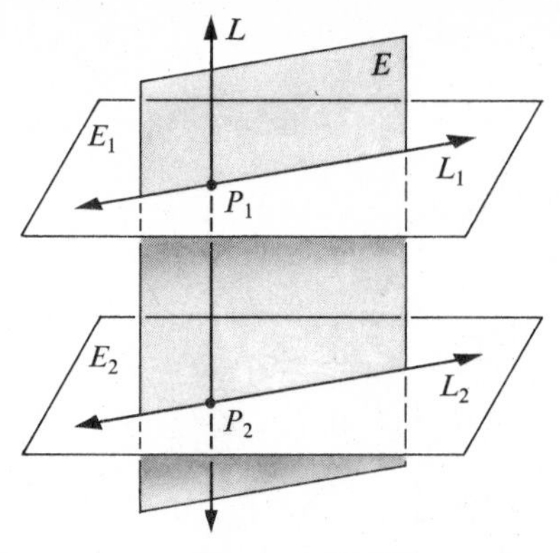

FIGURE 15.10

Proof of (2). We now have given that L intersects E_2 at P_2. Let L_2 be any line in E_2, passing through P_2. We need to prove that $L_2 \perp L$.

Let E be the plane that contains L and L_2. Then E intersects E_1 in a line L_1. By Theorem 1, $L_1 \parallel L_2$, and $L_1 \perp L$, because every line in E_1 through P_1 is perpendicular to L. Therefore $L_2 \perp L$, which was to be proved.

Theorem 4. Any two planes perpendicular to the same line are parallel.

Proof. Suppose that L is perpendicular to E_1 at P and perpendicular to E_2 at Q. If E_1 intersects E_2 at a point R, then $\triangle PQR$ has right angles at both P and Q, which is impossible.

PROBLEM SET 15.2

Prove the following theorems. They have fairly short proofs.

Theorem 5. A plane perpendicular to one of two parallel lines is perpendicular to the other.

Remember that you must begin by showing that the given plane intersects the second of the given lines.

Theorem 6. If two lines are each parallel to a third line, they are parallel to each other.

Theorem 7. Two parallel planes are everywhere equidistant. That is, all perpendicular segments from one of the two planes to the other are congruent.

Theorem 8. Let H be a half space with face E. Let e be a positive number. Let F be the set of all points Q of H whose perpendicular distance from E is $=e$. Then F is a plane.

[*Hint:* For the proof, let P be a point of E, and let $\overline{AP}$ be a segment perpendicular to E, such that $AP = e$. Let E' be the plane through A, perpendicular to $\overline{AP}$. Show that (1) $F \subset E'$, and (2) $E' \subset F$. It will follow that F is a plane, namely, the plane E'.]

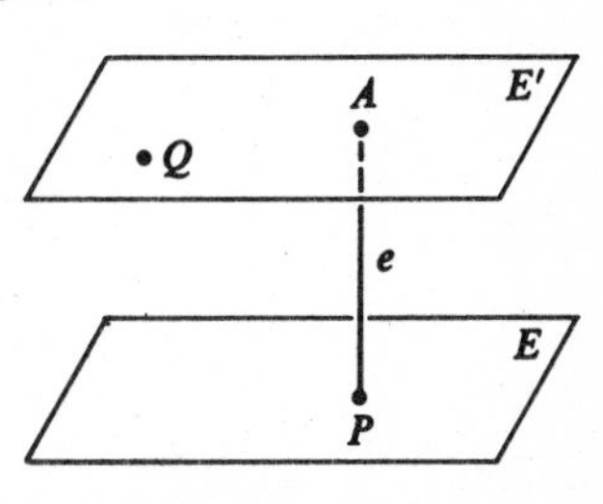

FIGURE 15.11

15.3 THE MEASURE OF A DIHEDRAL ANGLE; PERPENDICULAR PLANES.

We recall that a *dihedral angle* is the union of a line and two noncoplanar half planes having the line as their common edge. The line is called the *edge* of the dihedral angle, and the two half planes are called its *sides* or *faces.*

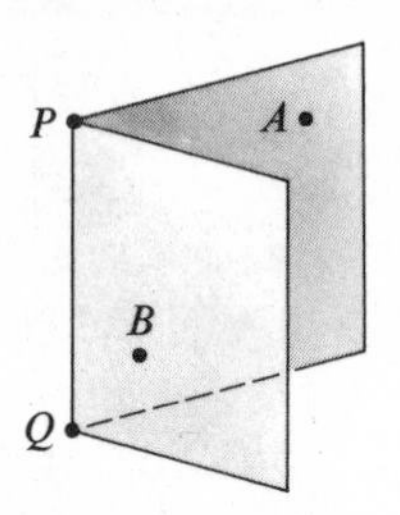

FIGURE 15.12

If $\overleftrightarrow{PQ}$ is the edge, and A and B are points on different sides, then the dihedral angle is denoted by $\angle A\text{-}PQ\text{-}B$. It is clear, of course, that the dihedral angle is completely determined when P, Q, A, and B are named.

It is not hard to see that if E is a plane which intersects the edge $\overleftrightarrow{PQ}$ in only one point C, then E intersects $\angle A\text{-}PQ\text{-}B$ in an angle:

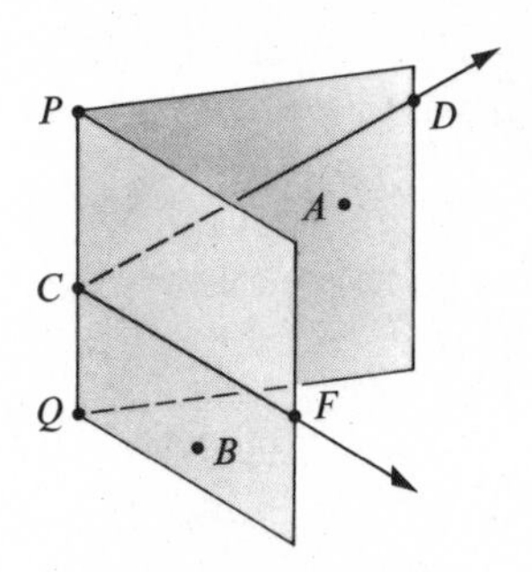

FIGURE 15.13

In the figure, E is the plane $\overleftrightarrow{CDF}$, intersecting the dihedral angle $\angle A\text{-}PQ\text{-}B$ in $\angle DCF$. The size of $\angle DCF$ depends, of course, on the position of the plane E. In the figure, $\angle DCF$ is "fairly large"; but if $\angle PCD$ and $\angle PCF$ are both very "small," then $\angle DCF$ will also be "very small."

If through a point C of the edge we pass a plane E, perpendicular to the edge, then the intersection of E with the dihedral angle is called a *plane angle* of the dihedral angle. We shall use the plane angles to define a degree-measure for dihedral angles. To do this, we need the following theorem.

Theorem 1. Any two plane angles of the same dihedral angle are congruent.

Proof. Let $\angle C$ and $\angle D$ be two plane angles of the given dihedral angle. Take points P, Q, R, S on the sides of $\angle C$ and $\angle D$, as in the figure, so that $CP = DQ$

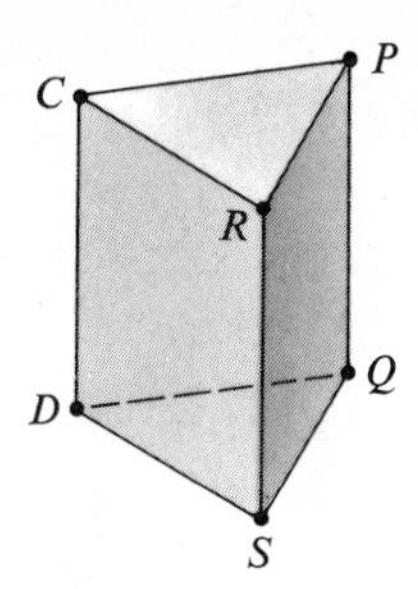

FIGURE 15.14

and $CR = DS$. (When we say "as in the figure," this means that P and Q are on one side of the dihedral angle; R and S are on the other side of the dihedral angle.)

The main steps in the proof are as follows.

(1) $\overline{CP} \parallel \overline{DQ}$. (These segments are coplanar, and perpendicular to the same line.)

(2) $\square CPQD$ is a parallelogram. (A pair of opposite sides are congruent and parallel.)

(3) $PQ = CD$. (Why?)

(4) $\overline{PQ} \parallel \overline{CD}$. (Why?)

(1′) $\overline{CR} \parallel \overline{DS}$.

(2′) $\square CRSD$ is a parallelogram.

(3′) $RS = CD$.

(4′) $\overline{RS} \parallel \overline{CD}$.

(5) $PQ = RS$ (by (3) and (3′)).

(6) $\overline{PQ} \parallel \overline{RS}$ (by (4) and (4′)).

(7) $PR = QS$. (Why?)

(8) $\triangle PCR \cong \triangle QDS$ (by SSS).

(9) $\angle PCR \cong \angle QDS$, which was to be proved.

We can now define the measure of a dihedral angle. The measure $m\angle A\text{-}PQ\text{-}B$ is the number which is the measure of all plane angles of $\angle A\text{-}PQ\text{-}B$.

The same theorem enables us to define right dihedral angles. A dihedral angle is a *right angle* if its plane angles are right angles. Two planes are *perpendicular* if their union contains a right dihedral angle.

PROBLEM SET 15.3

Prove the following theorems.

Theorem 2. If a line is perpendicular to a plane, then every plane that contains the given line is perpendicular to the given plane.

Theorem 3. If two planes are perpendicular, then any line in one of them, perpendicular to their line of intersection, is perpendicular to the other.

Chapter 16
Circles and Spheres

16.1 BASIC DEFINITIONS

Let P be a point of a plane E, and let r be a positive number. The *circle with center P and radius r* is the set of all points Q of E whose distance from P is equal to r. Two or more circles with the same center are called *concentric*.

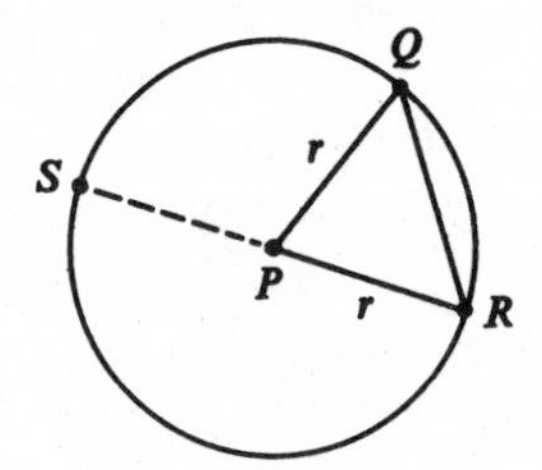

FIGURE 16.1

If Q is any point of the circle, then the segment $\overline{PQ}$ is a *radius* of the circle, and Q is called its *outer end*. If Q and R are any two points of the circle, then the segment $\overline{QR}$ is a *chord* of the circle. A chord that contains the center is called *a diameter* of the circle. Evidently the length of every diameter is the number $2r$. This number $2r$ is called *the diameter* of the circle. (Note that the word *radius* is used in two senses. It may mean either a number r or a segment $\overline{PQ}$. But it will always be easy to tell which is meant. When we speak of *the* radius, we mean the number r, and when we speak of *a* radius, we mean a segment. Similarly for the two uses of the word *diameter*.)

The *interior* of a circle is the set of all points of the plane whose distance from the center is less than the radius. The *exterior* of a circle is the set of all points of the plane whose distance from the center is greater than the radius.

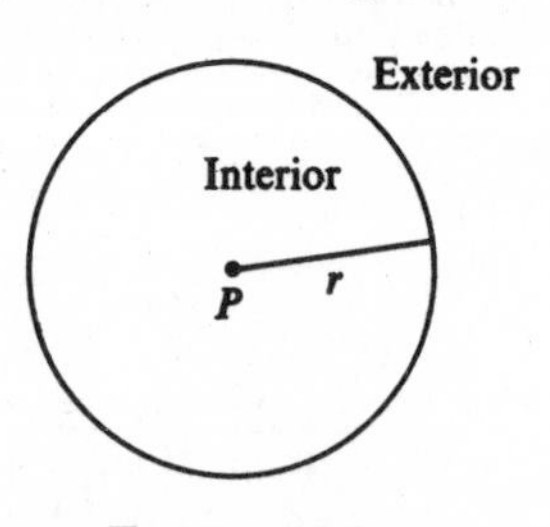

FIGURE 16.2

The corresponding definitions for spheres in space are precisely analogous. They are as follows.

Given a point P and a positive number r. The *sphere with center P and radius r* is the set of all points Q whose distance from P is equal to r. Two or more spheres with the same center are called *concentric*.

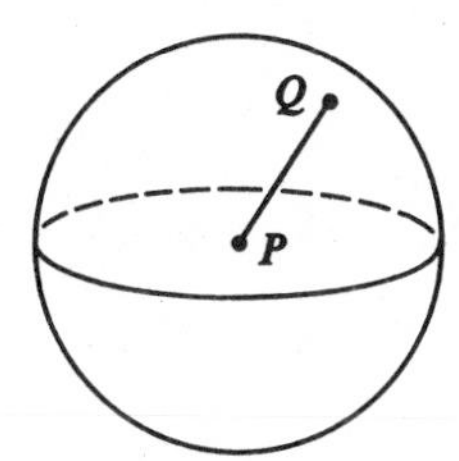

FIGURE 16.3

If Q is any point of the sphere, then the segment $\overline{PQ}$ is *a radius* of the sphere, and Q is called its *outer end*. If Q and R are any two points of the sphere, then the segment $\overline{QR}$ is called a *chord* of the sphere. A chord which contains the center is called *a diameter* of the sphere. Evidently the length of every diameter is the number $2r$. The number $2r$ is called *the diameter* of the sphere.

The *interior* of a sphere is the set of all points whose distance from the center is less than the radius. The *exterior* of a sphere is the set of all points whose distance from the center is greater than the radius.

PROBLEM SET 16.1

1. Show that every circle has only one center and only one radius. That is, if the circle with center P' and radius r' is the same as the circle with center P and radius r, then $P' = P$ and $r' = r$. [*Hint:* Suppose that $P \neq P'$, and consider the line $\overleftrightarrow{PP'}$.]

16.2 SECANT AND TANGENT LINES. THE LINE-CIRCLE THEOREM

Given a circle C and a line L in the same plane. If the line and the circle have one and only one point in common, then the line is called a *tangent* line, and the common point is called the point of *tangency*, or *point of contact*. If the line intersects the circle at more than one point, it is called a *secant* line. The following theorem is familiar, and is easy to prove.

Theorem 1. If a line is perpendicular to a radius of a circle at its outer end, then the line is a tangent.

Proof. Let C be a circle with center at P; let $\overline{PQ}$ be a radius, and let L be perpendicular to $\overline{PQ}$ at Q (Fig. 16.4). If R is any other point of L, then $PR > PQ$, because the shortest segment joining a point to a line is the perpendicular segment.

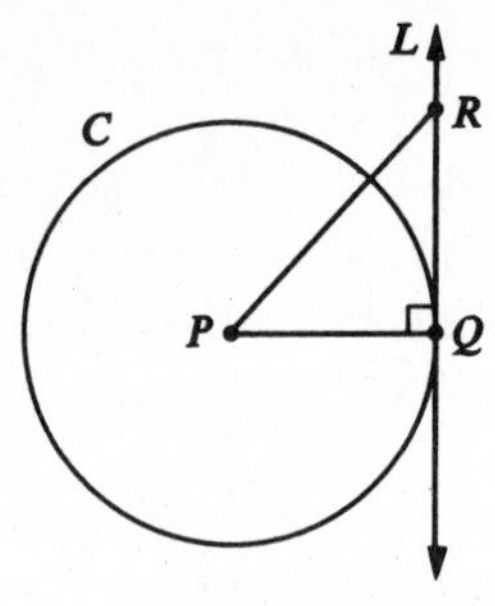

FIGURE 16.4

Therefore R is in the exterior of C. Therefore L intersects C only at Q and hence is a tangent line.

The converse is also true.

Theorem 2. Every tangent to a circle is perpendicular to the radius drawn to the point of contact.

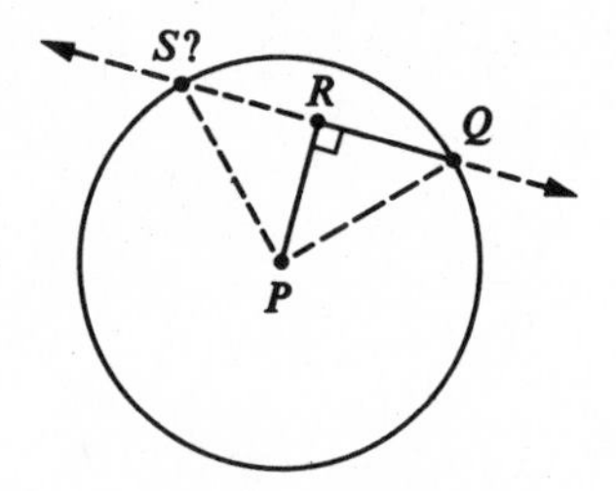

FIGURE 16.5

Proof. Let C be a circle with center at P, and let L be tangent to C at Q. Suppose that Q is not the foot of the perpendicular from P to L, and let R be the point which *is* the foot of the perpendicular. By the segment-construction postulate (or the segment-construction theorem, according to the chapter in which you refer to it), there is a point S of L such that Q-R-S and $RQ = RS$. By the Pythagorean theorem, applied twice, we have

$$PR^2 + RS^2 = PS^2, \qquad PR^2 + RQ^2 = PQ^2.$$

Therefore $PS = PQ$, S lies on the circle, and L is not a tangent line.

The proofs of the following theorems are fairly straightforward.

Theorem 3. Any perpendicular from the center of C to a chord bisects the chord.

Theorem 4. The segment joining the center to the midpoint of a chord is perpendicular to the chord.

Theorem 5. In E, the perpendicular bisector of a chord passes through the center.

Circles with the same radius r are called *congruent.* By the distance between the center of a circle and a chord, we mean, of course, the perpendicular distance; that is, the length of the perpendicular segment from the center to the chord. Two chords are *equidistant* from the center if their distances from the center are the same.

Theorem 6. In the same circle or in congruent circles, chords equidistant from the center are congruent.

Theorem 7. In the same circle or in congruent circles, any two congruent chords are equidistant from the center.

The following innocent looking theorem is of special interest.

Theorem 8. *The Line-circle Theorem.* If a line intersects the interior of a circle, then it intersects the circle in exactly two points.

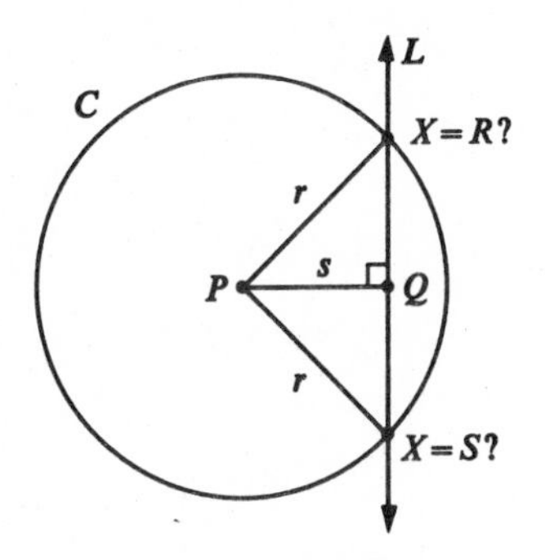

FIGURE 16.6

Proof. Let C be the circle with center at P and radius r, and let L be the line. Let Q be the foot of the perpendicular from P to L. Since $PZ < r$ for some point Z of L, it follows that $PQ < r$; that is, Q lies in the interior.

Let

$$PQ = s < r,$$

as indicated in the figure. We want to prove that C intersects L in exactly two points, R and S.

If X is a point where the line intersects the circle, then $\triangle PQX$ has a right angle at Q. Therefore

$$s^2 + QX^2 = r^2,$$

by the Pythagorean theorem. Hence

$$QX = \sqrt{r^2 - s^2}.$$

And conversely, if X lies on L, and $QX = \sqrt{r^2 - s^2}$, it follows that

$$\begin{aligned} PX^2 &= s^2 + (\sqrt{r^2 - s^2})^2 \\ &= s^2 + r^2 - s^2 \\ &= r^2. \end{aligned}$$

Now $r^2 - s^2 > 0$, because $s < r$. By the Euclidean completeness postulate, $r^2 - s^2$ has a positive square root $\sqrt{r^2 - s^2}$. By the ruler postulate, there are exactly two points X of L such that $QX = \sqrt{r^2 - s^2}$. Therefore exactly two points lie both on the line and on the circle.

Note that in the proof of this theorem, we have used the Euclidean completeness postulate, for the first time. This is not surprising, because the theorem itself describes a completeness property of the plane; it asserts the existence of points satisfying certain conditions. If we delete a few points, at random, from a plane, then we may get a "plane" in which the line-circle theorem fails.

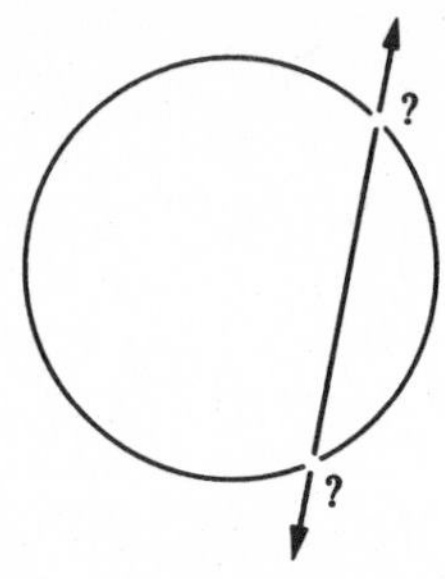

FIGURE 16.7

In the present treatment, this sort of thing is ruled out by the Euclidean completeness of the real number system. In a purely synthetic treatment, Theorem 8 should be taken as a postulate.

16.3 SECANT AND TANGENT PLANES

The analogy between the following discussion and the preceding one is fairly close. Given a sphere S and a plane E. If the plane and the sphere have one and only one point Q in common, then the plane is called a *tangent* plane, and the common point Q is called the *point of tangency*, or *point of contact*. If the plane intersects the sphere in more than one point, it is called a *secant* plane.

Theorem 1. Every plane perpendicular to a radius at its outer end is tangent to S.

Proof. If R is any point of the plane E, other than Q, then $PR > PQ$, because the perpendicular segment from P to E is the shortest.

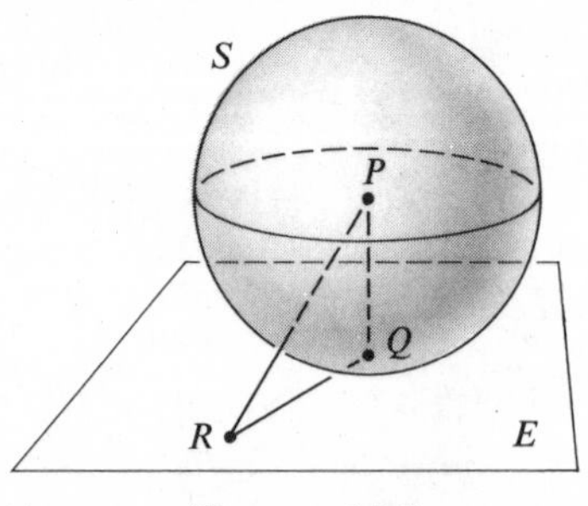

FIGURE 16.8

Theorem 2. Every tangent to a sphere is perpendicular to the radius drawn to the point of contact.

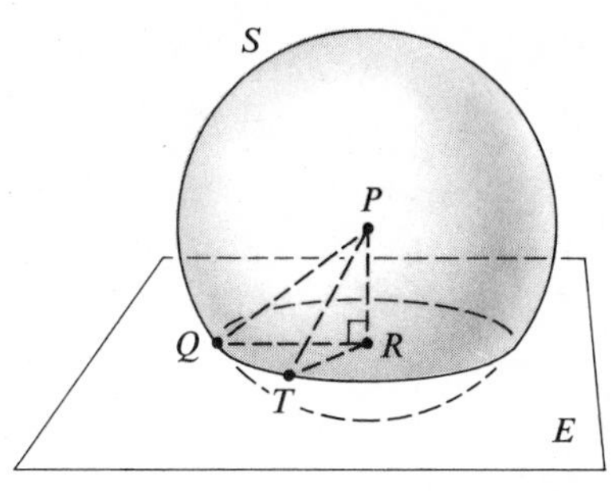

FIGURE 16.9

Proof. Let E be tangent to S at Q. Suppose that Q is not the foot of the perpendicular from P to E, and let R be the point which *is* the foot of the perpendicular. Let C be the circle with center at R, with radius RQ, in the plane E. If T is any point of C, then $\overline{RT} \perp \overline{PR}$. Therefore, by the Pythagorean theorem, we have

$$\begin{aligned} PT^2 &= PR^2 + RT^2 \\ &= PR^2 + RQ^2. \end{aligned}$$

But $PR^2 + RQ^2 = PQ^2$. Therefore T lies not only in E but also in S. Thus E intersects S in an entire circle; and this contradicts the hypothesis for E.

Theorem 3. If a plane E intersects the interior of S, then E intersects S in a circle.

Proof. Let Q be the foot of the perpendicular from the center P of S to the plane E, and let $s = PQ$. Then $s < r$. Therefore $r^2 - s^2 > 0$. Let

$$t = \sqrt{r^2 - s^2}.$$

(By Euclidean completeness, again.) It is now straightforward to check that the intersection of S and E is precisely the circle in E with center at Q and radius t. The proofs of the following theorems are straightforward, and are omitted.

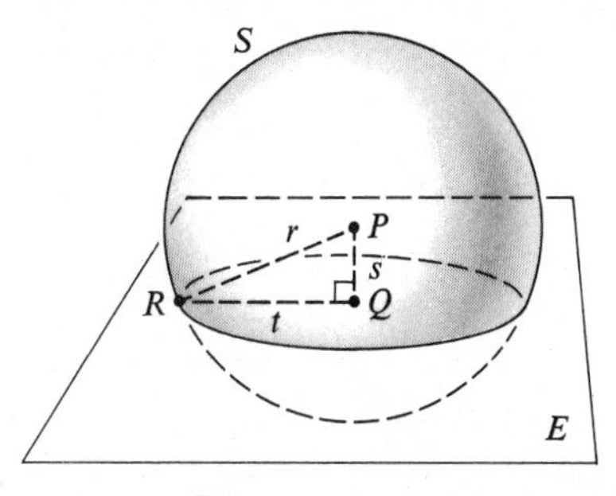

FIGURE 16.10

Theorem 4. The perpendicular from the center of S to a secant plane E passes through the center of the circle in which E intersects S.

Theorem 5. If the plane E intersects S in a circle C, then the segment between the center of S and the center of C is perpendicular to E.

Theorem 6. If E, S, and C are as in the preceding theorem, and L is a line perpendicular to E at the center of C, then L passes through the center of S.

16.4 ARCS OF CIRCLES

A *central angle* of a given circle is an angle whose vertex is the center of the circle.

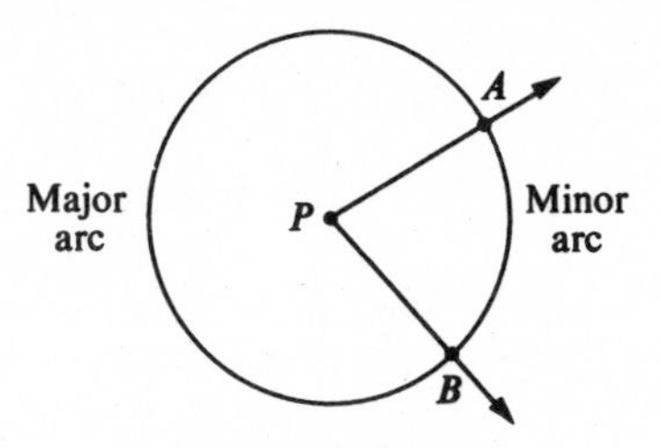

FIGURE 16.11

Let A and B be the points in which the sides of the central angle intersect the circle, so that the central angle is $\angle APB$. The *minor arc* $\widehat{AB}$ is the set consisting of A and B together with all points of the circle that are in the interior of $\angle APB$. The *major arc* $\widehat{AB}$ is the set consisting of A and B together with all points of the circle that lie in the exterior of $\angle APB$. In either case, the points A and B are called the *end points* of the arc.

If A and B are the end points of a diameter, then there are two arcs with A and B as end points. Each of these arcs $\widehat{AB}$ consists of A and B together with all points of the circle that lie on a given side of the line. These are called *semicircles.*

Of course the notation $\widehat{AB}$ for arcs is always ambiguous, because there are always two different arcs with A and B as end points. In cases where misunderstanding might occur, we remove the ambiguity by taking some third point X of the arc, and then denoting the arc by $\widehat{AXB}$. In the figure, $\widehat{AXB}$ is a minor arc, $\widehat{AYB}$ is the corresponding major arc, and $\widehat{CAB}$ and $\widehat{CYB}$ are semicircles.

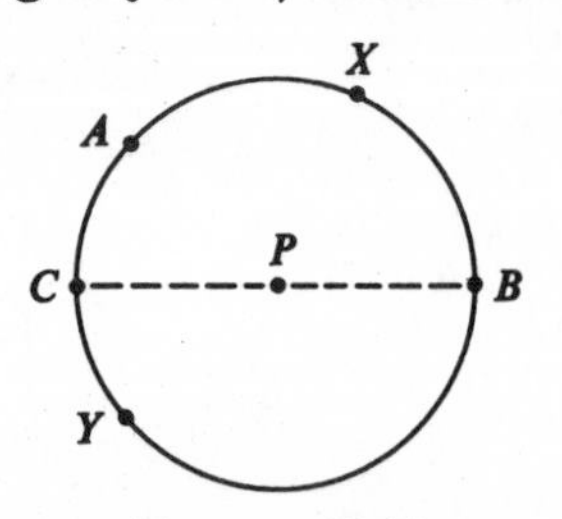

FIGURE 16.12

The *degree measure* $m\widehat{AXB}$ of an arc $\widehat{AXB}$ is defined in the following way.

(1) If $\widehat{AXB}$ is a minor arc, then $m\widehat{AXB}$ is the measure $m\angle APB$ of the corresponding central angle.

(2) If $\widehat{AXB}$ is a semicircle, then $m\widehat{AXB} = 180$.

(3) If $\widehat{AXB}$ is a major arc, then $m\widehat{AXB} = 360 - m\angle APB$.

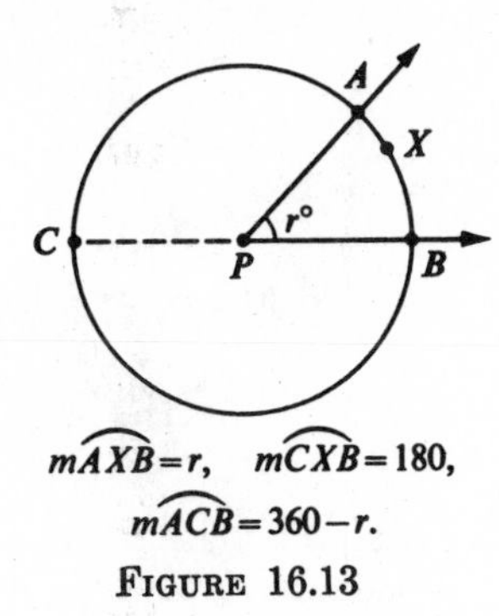

$m\widehat{AXB} = r$, $m\widehat{CXB} = 180$,

$m\widehat{ACB} = 360 - r$.

FIGURE 16.13

The following theorem says that degree measure for arcs is additive in the way that we might expect.

Theorem 1. If $\widehat{AB}$ and $\widehat{BC}$ are arcs of the same circle, having only the point B in common, and their union is an arc $\widehat{AC}$, then $m\widehat{AB} + m\widehat{BC} = m\widehat{AC}$.

That is, we always have

$$m\widehat{ABC} = m\widehat{AB} + m\widehat{BC}.$$

The proof is tedious, because we need to discuss five cases, but each of the five cases is easy. We describe them, give the figures, and leave the verifications to the reader.

CASE 1. $\widehat{ABC}$ is a minor arc.

CASE 2. $\widehat{ABC}$ is a semicircle.

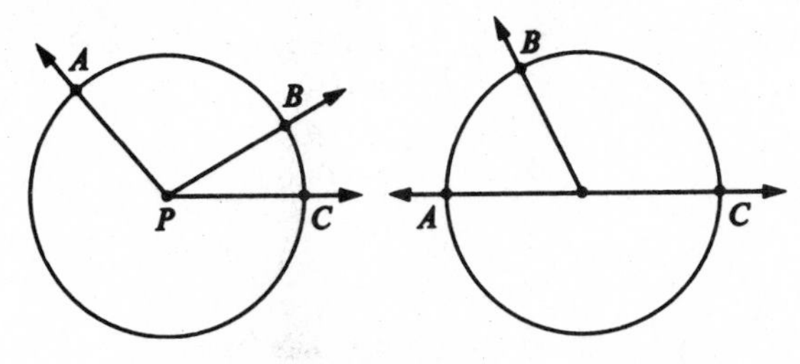

FIGURE 16.14

In these first two cases, $m\widehat{AB}$ and $m\widehat{BC}$ are simply $m\angle APB$ and $m\angle BPC$.

CASE 3. $\widehat{ABC}$ is a major arc, and A and C are on opposite sides of the diameter that contains B. (What are the equations relating r, u, s, and t?)

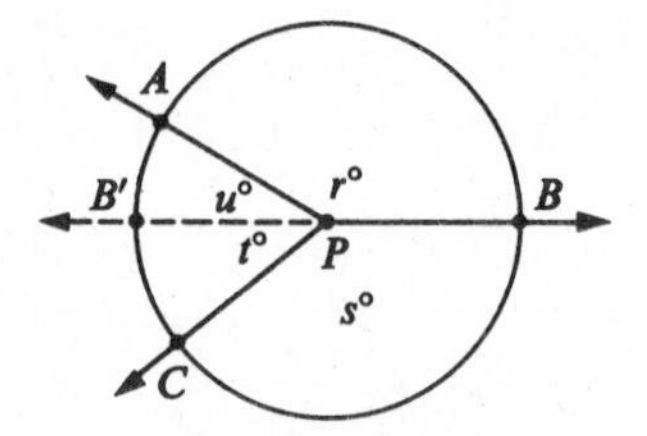

FIGURE 16.15

CASE 4. $\widehat{ABC}$ is a major arc, and A and C are on the same side of the diameter that contains B.

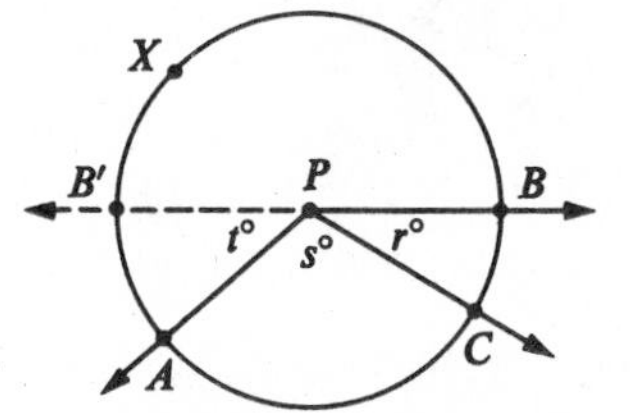

FIGURE 16.16

CASE 5. $\widehat{ABC}$ is a major arc, and one of the arcs $\widehat{AB}$, $\widehat{BC}$ is a semicircle. (Here $m\widehat{ABC} = 360 - t = 180 + 180 - t = 180 + s = m\widehat{AB} + m\widehat{BC}$.)

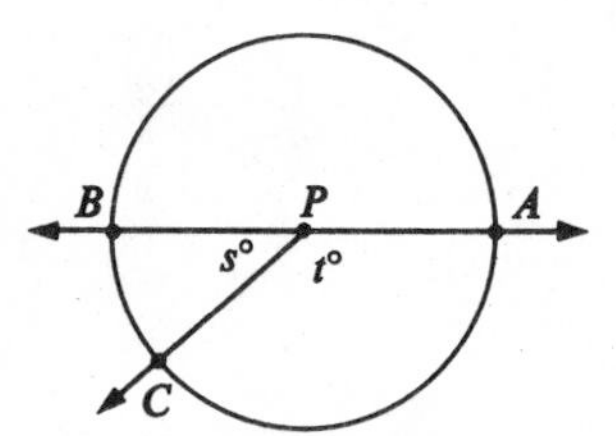

FIGURE 16.17

In the figures below, the angle $\angle ABC$ is *inscribed in* the dotted arc $\widehat{ABC}$. To be exact, an angle is *inscribed in* an arc of a circle if (1) the two end points of the arc lie on the two sides of the angle, and (2) the vertex of the angle is a point, but not an end point, of the arc. (We can put this more briefly: $\angle ABC$ is *inscribed in* $\widehat{ABC}$, by definition.)

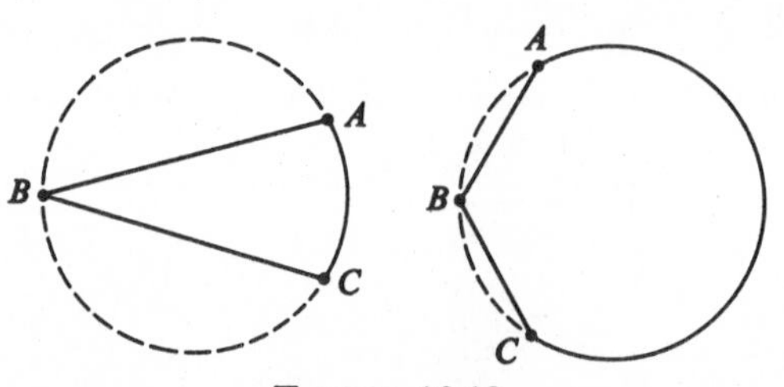

FIGURE 16.18

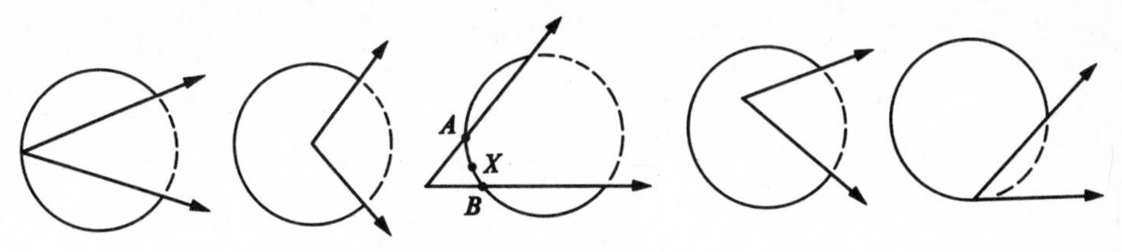

FIGURE 16.19

In the figures above, the indicated angle *intercepts* the dotted arc. In the third of these cases, the angle intercepts not only the dotted arc but also the arc $\widehat{AXB}$.

We shall now give a mathematical definition of the idea conveyed by the figures above. An angle *intercepts* an arc if (1) the end points of the arc lie on the angle, (2) each side of the angle contains at least one end point of the arc, and (3) except for its end points, the arc lies in the interior of the angle.

Theorem 2. The measure of an inscribed angle is half the measure of its intercepted arc.

RESTATEMENT. Let $\angle A$ be inscribed in an arc $\widehat{BAC}$ of a circle, intercepting the arc $\widehat{BC}$. Then $m\angle A = \frac{1}{2}m\widehat{BC}$.

Proof. CASE 1. Consider first the case where $\angle A$ contains a diameter of the circle. Let $\angle x = \angle ABP$, $\angle y = \angle BPC$, and $\angle z = \angle APB$, as in the figure.

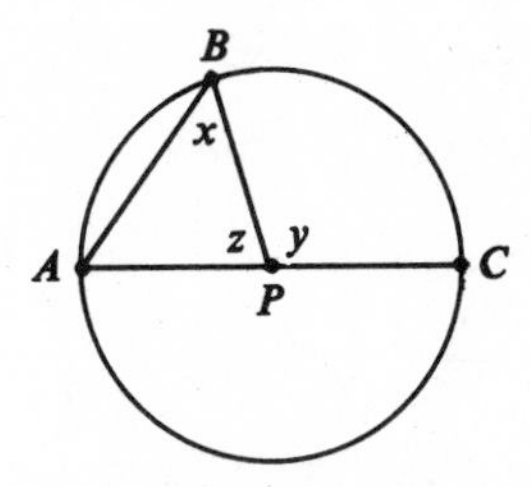

FIGURE 16.20

Then we have

$$m\angle A + m\angle x + m\angle z = 180 \quad \text{and} \quad m\angle z + m\angle y = 180.$$

Since A and B are on the circle, we have $PA = PB$. Therefore, by the isosceles triangle theorem, we have $m\angle A = m\angle x$, so that

$$\begin{aligned} 2m\angle A &= 180 - m\angle z \\ &= m\angle y \\ &= m\widehat{BC}. \end{aligned}$$

Therefore

$$m\angle A = \tfrac{1}{2}m\widehat{BC},$$

which was to be proved.

CASE 2. Suppose that B and C are on opposite sides of the diameter through A. Then

$$m\angle A = m\angle v + m\angle w.$$

Also, by Case 1,

$$m\angle v = \tfrac{1}{2}m\widehat{BD},$$

and

$$m\angle w = \tfrac{1}{2}m\widehat{DC}.$$

By Theorem 1,

$$m\widehat{BD} + m\widehat{DC} = m\widehat{BC}.$$

Therefore

$$m\angle A = \tfrac{1}{2}mBC,$$

which was to be proved.

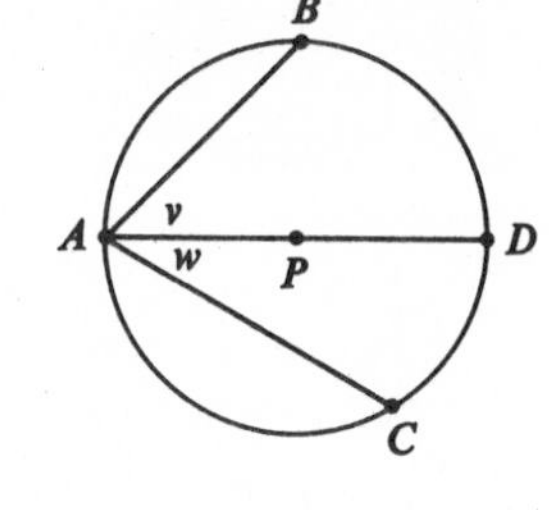

FIGURE 16.21

CASE 3. Suppose that B and C are on the same side of the diameter through A. Here

$$m\angle x + m\angle y = m\angle z,$$

and

$$m\widehat{BC} + m\widehat{CD} = m\widehat{BD}.$$

Therefore

$$\begin{aligned} m\angle A = m\angle y &= m\angle z - m\angle x \\ &= \tfrac{1}{2}mBD - \tfrac{1}{2}m\widehat{CD}, \end{aligned}$$

by Case 1. Therefore

$$m\angle A = \tfrac{1}{2}m\widehat{BC},$$

which was to be proved.

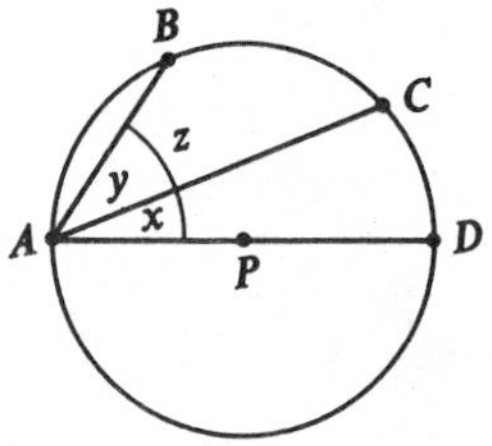

FIGURE 16.22

This theorem has two immediate consequences.

Theorem 3. An angle inscribed in a semicircle is a right angle.

Theorem 4. All angles inscribed in the same arc are congruent.

PROBLEM SET 16.4

1. Given two circles with a common tangent at a point A such that the second circle passes through the center of the first. Show that every chord of the first circle that passes through A is bisected by the second circle.

2. Three or more points are called *concyclic* if there is a circle that contains all of them. Show that every three noncollinear points (in a plane) are concyclic.

3. Show that three collinear points are never concyclic.

4. An *inscribed quadrilateral* is one whose vertices are concyclic. Prove that in an inscribed quadrilateral each pair of opposite angles are supplementary.

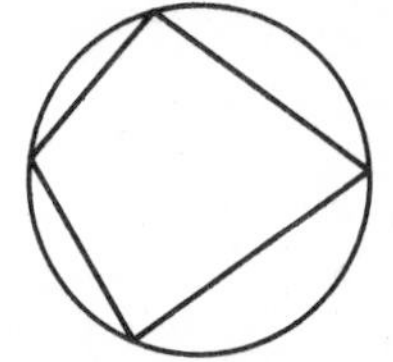

FIGURE 16.23

5. Show, conversely, that if a pair of opposite angles of a convex quadrilateral are supplementary, then the quadrilateral is inscribed.

6. A pair of parallel lines *intercept* the arc $\widehat{AB}$ of a circle if (1) the lines intersect the circle at A and B, and (2) every other point of $\widehat{AB}$ lies between the two lines.

Theorem 5. If two parallel lines intersect a circle, then they intercept congruent arcs. (There are three cases to be considered: two secants, two tangents, one secant, and one tangent.)

Theorem 6. In the same circle or in congruent circles, if two chords are congruent, then so also are the corresponding minor arcs.

Theorem 7. In the same circle or in congruent circles, if two arcs are congruent, then so are the corresponding chords.

Theorem 8. Given a circle, and an angle formed by a secant ray and a tangent ray with its vertex on the circle. Then the measure of the angle is half the measure of its intercepted arc.

Theorem 9. No two different circles intersect in more than two points.

16.5 THE TWO-CIRCLE THEOREM

We shall now discuss tangent lines to a circle through an external point. The fact is that given a circle C and a point Q of its exterior, there are always exactly two lines which pass through Q and are tangent to C:

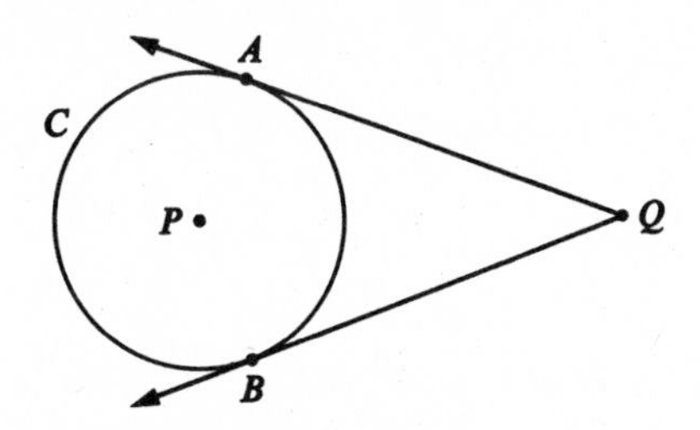

FIGURE 16.24

The natural way to try to prove this is as follows. Let M be the midpoint of the segment $\overline{PQ}$, where P is the center of C. Let C' be the circle with center M and radius $MP = MQ$.

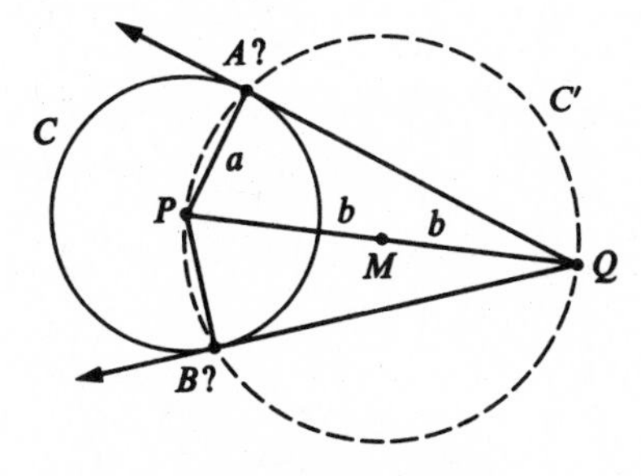

FIGURE 16.25

If C' intersects C in two points A and B, as the figure suggests, then $\overleftrightarrow{QA}$ and $\overleftrightarrow{QB}$ are tangent to C at A and B, respectively. The reason is that each of the angles $\angle PAQ$ and $\angle PBQ$ is inscribed in a semicircle and hence each is a right angle.

We can now apply Theorem 1, Section 16.2, which says that a line perpendicular to a radius at its outer end is tangent to the circle. In this reasoning there is a dangling *if*. To complete the above proof that there are two tangents through Q, we need to show that C and C' intersect in two points. For this purpose we need the following theorem.

Theorem 1. *The Two-Circle Theorem.* Let C and C' be circles of radius a and b, and let c be the distance between their centers. If each of the numbers a, b, c is less than the sum of the other two, then C and C' intersect in two points. And the two points of intersection lie on opposite sides of the line of centers.

Before proceeding to prove the theorem, let us see how it applies to our problem in connection with the external tangents through Q. In Fig. 16.25, let the radius of C be a, and let the radius of C' be $b = PM$. The distance between the centers is $c = MP = b$. Since Q is an external point, we have

$$PQ > a,$$

so that

$$a < 2b.$$

Then (1) $a < b + c$, because $b + c = 2b$, and $a < 2b$. Also (2) $b < a + c$, because $b = c$ and $a > 0$. Finally, (3) $c < a + b$, because $c = b$. Therefore the two-circle theorem applies, and so it follows that C and C' intersect in two points A and B. Therefore there are at least two tangents to C through Q. Later (in the next section) we shall show that there are exactly two tangents through Q. The rest of this section will be devoted to the proof of the two-circle theorem.

If a, b, and c are the lengths of the sides of a triangle, then each of the numbers a, b, c is less than the sum of the other two. (We know this by three applications of the triangular inequality.) We shall prove the converse.

Theorem 2. *The Triangle Theorem.* Given three positive numbers a, b, c. If each of these numbers is less than the sum of the other two, then there is a triangle whose sides have length a, b, c.

Proof. Without loss of generality, let us suppose that

$$a \geqq b \geqq c.$$

Take a segment $\overline{BC}$, of length a. We want to find a point A such that $AB = c$ and $AC = b$. We shall start by *assuming* that there is a triangle $\triangle ABC$, of the

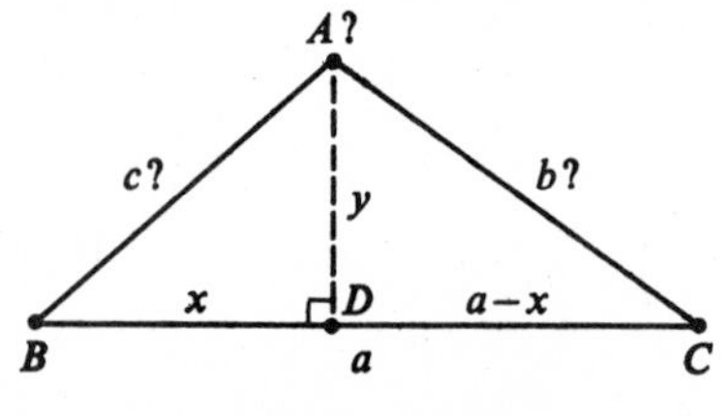

FIGURE 16.26

sort that we are looking for, and then find out where the point A must be. This procedure will not prove anything, because we start by assuming the very thing that we are supposed to prove. But once we have found the exact location of the points that *might* work, it will be easy to check that they *do* work.

Suppose, then, that $\triangle ABC$ is given, with sides of the desired length, as indicated in Fig. 16.26. Let D be the foot of the perpendicular from A to $\overleftrightarrow{BC}$. Then B-D-C, because $\overline{BC}$ is a longest side of $\triangle ABC$. Therefore, if $BD = x$, then $DC = a - x$. Let $AD = y$. Then by two applications of the Pythagorean theorem, we have

$$y^2 = c^2 - x^2, \tag{1}$$

$$y^2 = b^2 - (a - x)^2. \tag{2}$$

Therefore

$$c^2 - x^2 = b^2 - (a - x)^2,$$

so that

$$c^2 - x^2 = b^2 - a^2 + 2ax - x^2,$$

and

$$2ax = a^2 + c^2 - b^2.$$

Therefore

$$x = \frac{a^2 + c^2 - b^2}{2a}; \tag{3}$$

and from (1), we get

$$y = \sqrt{c^2 - x^2}. \tag{4}$$

What we have proved so far is that *if x and y satisfy* (1) *and* (2), *then x and y satisfy* (3) *and* (4). We shall check, conversely, that *if x and y satisfy* (3) *and* (4), *then x and y satisfy* (1) *and* (2). Half of this is trivial. If (4) holds, then so also does (1). Suppose then, that (3) is satisfied. Reversing the steps in the derivation, we get

$$c^2 - x^2 = b^2 - (a - x)^2.$$

Since (1) is known to hold, it follows that $y^2 = b^2 - (a - x)^2$, which is Eq. (2).

We can summarize this by writing

$$(1) \text{ and } (2) \Leftrightarrow (3) \text{ and } (4).$$

Now that we know what triangle to look for, let us start all over again. We have three positive numbers a, b, c, with

$$a \geqq b \geqq c.$$

Let

$$x = \frac{a^2 + c^2 - b^2}{2a}.$$

Then $x > 0$, because $a^2 \geqq b^2$ and $c^2 \geqq 0$. We want to set

$$y = \sqrt{c^2 - x^2},$$

but first we have to prove that $c > x$, to make sure that the radicand is positive.

Obviously it will be sufficient to show that $c - x > 0$. Now

$$\begin{aligned} c - x &= c - \frac{a^2 + c^2 - b^2}{2a} \\ &= \frac{2ac - a^2 - c^2 + b^2}{2a} \\ &= \frac{b^2 - (a^2 - 2ac + c^2)}{2a} \\ &= \frac{b^2 - (a - c)^2}{2a}. \end{aligned}$$

We know that $a < b + c$. Therefore $a - c < b$. Since both $a - c$ and b are $\geqq 0$, it follows that $(a - c)^2 < b^2$; and this means that $c - x > 0$, or $c > x$.

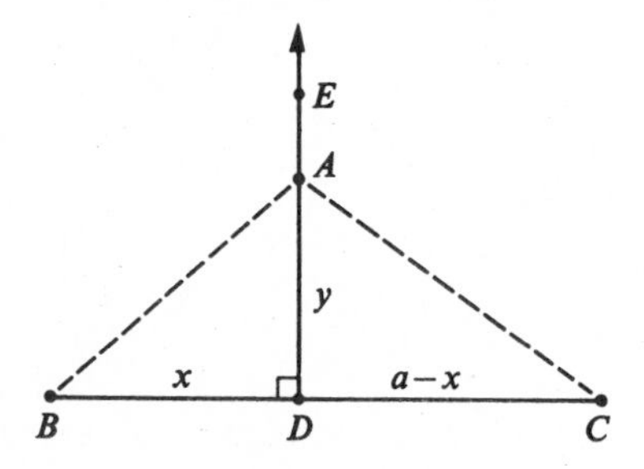

FIGURE 16.27

We are now ready to construct our triangle. Let $\overline{BC}$ be a segment of length a. Let D be a point of $\overline{BC}$ such that

$$BD = x = \frac{a^2 + c^2 - b^2}{2a}. \tag{3'}$$

Let $\overrightarrow{DE}$ be a ray starting at D, perpendicular to $\overline{BC}$, and let A be a point of $\overrightarrow{DE}$ such that

$$AD = y = \sqrt{c^2 - x^2}. \tag{4'}$$

Since x and y satisfy (3) and (4), it follows that x and y satisfy (1) and (2). Thus

$$x^2 + y^2 = c^2 \tag{1'}$$

$$(a - x)^2 + y^2 = b^2. \tag{2'}$$

But $x^2 + y^2 = AB^2$, and $(a - x)^2 + y^2 = AC^2$. Therefore $AB^2 = c^2$ and $AC^2 = b^2$. Since b and c are positive, this means that $AB = c$ and $AC = b$. Therefore $\triangle ABC$ is a triangle of the sort that we were looking for.

On the basis of the triangle theorem, it is easy to prove the two-circle theorem. We have given a circle C, with center P and radius a, and a circle C', with center M and radius b:

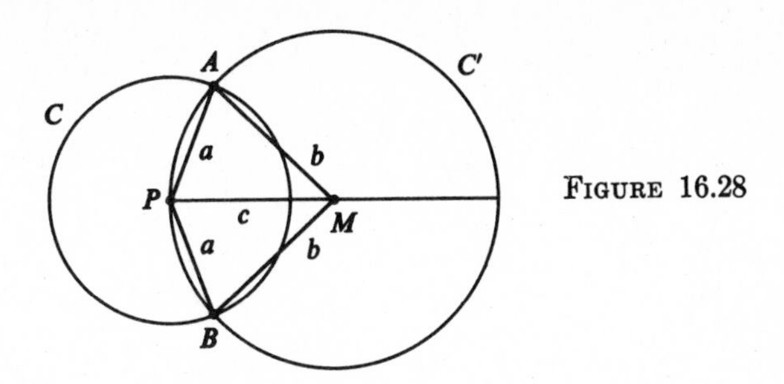

FIGURE 16.28

The distance PM between the centers is c, and each of the numbers a, b, c is less than the sum of the other two. Therefore there is a triangle $\triangle RST$, with $RS = a$, $ST = b$ and $RT = c$:

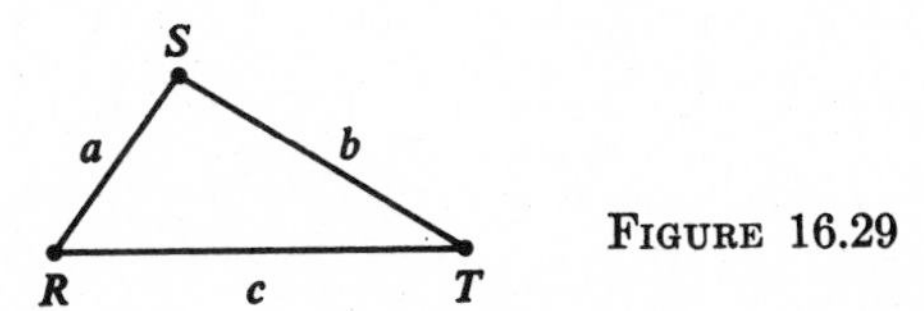

FIGURE 16.29

Let A be a point in the plane of our two circles such that $\angle APM \cong \angle R$ and $AP = a = RS$. By SAS, $\triangle RST \cong \triangle PAM$, so that $AM = ST = b$. Thus A is on both C and C'. Let B be a point on the opposite side of $\overleftrightarrow{PM}$ from A, such that $\angle BPM \cong \angle R$ and $BP = a = RS$. By SAS, $\triangle RST \cong \triangle PBM$, so that B is on both C and C'. It is not hard to check that these points A and B are the only points where the two circles intersect. (This was Theorem 9 in Problem Set 16.4.)

In a purely synthetic treatment, something equivalent to the two-circle theorem needs to be stated as a postulate. In effect, Euclid used such a postulate, without stating it. In modern terms, the required postulate takes the following form.

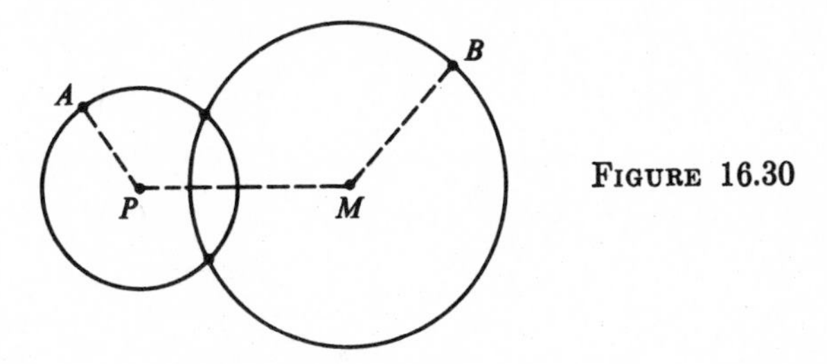

FIGURE 16.30

The Two-Circle Postulate. Let C_1 and C_2 be circles, with centers P and M. Let A and B be points of C_1 and C_2. If each of the congruence classes $[\overline{PA}]$, $[\overline{PM}]$, and $[\overline{MB}]$ is less than the sum of the other two, then C_1 and C_2 intersect in two points.

Here addition and inequalities, for congruence classes, are defined as in Section 8.4. The postulate does not say that the two intersection points lie on opposite sides of $\overleftrightarrow{PM}$, because this can easily be proved.

Chapter 17
Cartesian Coordinate Systems

17.1 COORDINATES, EQUATIONS, AND GRAPHS

Obviously, all readers of this book know about coordinate systems, from elementary analytic geometry. For the sake of completeness, however, we explain them here from the beginning. To achieve speed and simplicity, and reduce the amount of outright repetition, we have introduced various novelties in the derivations.

In a plane E we set up a Cartesian coordinate system in the following way. First we choose a line X, with a coordinate system as given by the ruler postulate. The zero point of X will be called the *origin*. We now take a line Y, perpendicular to X at the origin, with a coordinate system in which the origin has coordinate $= 0$.

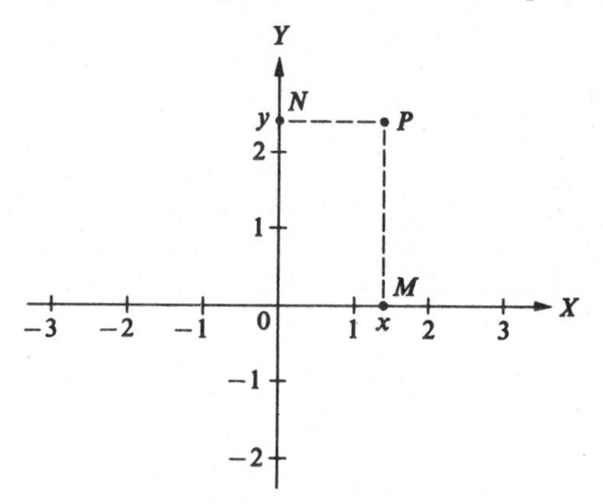

FIGURE 17.1

Given a point P of E, we drop a perpendicular to a point M of X. The coordinate x of M on X is called the *x-coordinate*, or the *abscissa*, of P. We drop a perpendicular from P to a point N of Y. The coordinate of N on Y is called the *y-coordinate*, or the *ordinate*, of P. Thus to every point P of E there corresponds an ordered pair (x, y) of real numbers, that is, an element of the product set $\mathbf{R} \times \mathbf{R}$. Clearly this is a one-to-one correspondence

$$E \leftrightarrow \mathbf{R} \times \mathbf{R}.$$

For short, we shall speak of "the point (x, y)," meaning, of course, the point corresponding to (x, y) in the coordinate system under discussion.

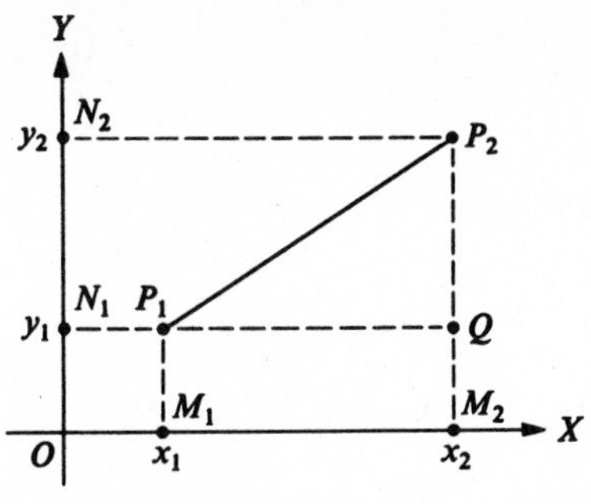

FIGURE 17.2

Theorem 1. The distance between the points $P_1 = (x_1, y_1)$ and $P_2 = (x_2, y_2)$ is given by the formula

$$P_1P_2 = \sqrt{(x_2 - x_1)^2 + (y_2 - y_1)^2}.$$

Proof. Let M_1, N_1, M_2, N_2 be the projections of P_1 and P_2 onto the axes, as in the definition of coordinates. If $x_1 = x_2$, then

$$\overleftrightarrow{P_1P_2} \parallel \overleftrightarrow{N_1N_2}, \qquad |x_2 - x_1| = 0,$$

and

$$\begin{aligned} P_1P_2 &= |y_2 - y_1| \\ &= \sqrt{(y_2 - y_1)^2} \\ &= \sqrt{(x_2 - x_1)^2 + (y_2 - y_1)^2}. \end{aligned}$$

(Here we are ignoring the trivial case where $P_1 = N_1$ and $P_2 = N_2$.) If $y_1 = y_2$, the same conclusion follows in a similar way. Suppose, then, that $x_1 \neq x_2$ and $y_1 \neq y_2$, as in the figure. Then the horizontal line through P_1 intersects the vertical line through P_2, in a point Q, and $\triangle P_1P_2Q$ has a right angle at Q. (Here, and hereafter, a horizontal line is X or a line parallel to X; and a vertical line is Y or a line parallel to Y.) Thus

$$P_1Q = M_1M_2,$$

and

$$P_2Q = N_2N_1,$$

either because the point pairs are the same or because opposite sides of a rectangle are congruent. By the Pythagorean theorem,

$$P_1P_2^2 = P_1Q^2 + P_2Q^2.$$

Therefore

$$\begin{aligned} P_1P_2^2 &= M_1M_2^2 + N_1N_2^2 \\ &= |x_2 - x_1|^2 + |y_2 - y_1|^2 \\ &= (x_2 - x_1)^2 + (y_2 - y_1)^2, \end{aligned}$$

and from this the distance formula follows.

By a *linear equation in x and y* we mean an equation of the form

$$Ax + By + C = 0,$$

where A, B, and C are real numbers, and A and B are not both $= 0$. By the *graph* of an equation, we mean the set of all points that satisfy the equation. More generally, by the graph of a *condition* we mean the set of all points that satisfy the given condition. Thus the interior of a circle with center Q and radius r is the *graph* of the condition $PQ < r$; and one of our theorems tells us that the perpendicular bisector of a segment $\overline{AB}$ is the *graph* of the condition $PA = PB$.

Theorem 2. Every line in E is the graph of a linear equation in x and y.

Proof. Let L be a line in E. Then L is the perpendicular bisector of some segment P_1P_2, where $P_1 = (x_1, y_1)$ and $P_2 = (x_2, y_2)$. Thus L is the graph of the condition

$$PP_1 = PP_2.$$

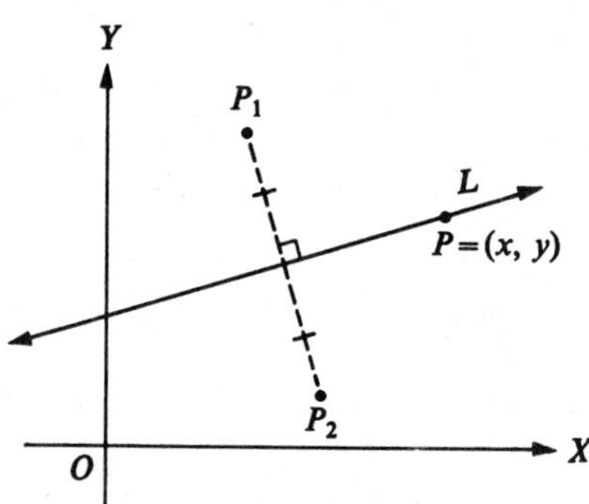

FIGURE 17.3

With $P = (x, y)$ this can be written algebraically in the form

$$\sqrt{(x - x_1)^2 + (y - y_1)^2} = \sqrt{(x - x_2)^2 + (y - y_2)^2},$$

or

$$x^2 - 2x_1x + x_1^2 + y^2 - 2y_1y + y_1^2 = x^2 - 2x_2x + x_2^2 + y^2 - 2y_2y + y_2^2$$

or

$$2(x_2 - x_1)x + 2(y_2 - y_1)y + (x_2^2 + y_2^2 + x_1^2 + y_1^2) = 0.$$

This has the form

$$Ax + By + C = 0.$$

And A and B cannot both be $= 0$, because then we would have $x_2 = x_1$ and $y_2 = y_1$; this is impossible, because $P_1 \neq P_2$.

Theorem 3. If L is not vertical, then L is the graph of an equation of the form

$$y = mx + k.$$

Proof. L is the graph of an equation

$$Ax + By + C = 0.$$

Here $B \neq 0$, because for $B = 0$ the equation takes the form $x = -C/A$; and the graph is then vertical. Therefore we can divide by B, getting the equivalent equation

$$y = -\frac{Ax}{b} - \frac{C}{B}.$$

This has the desired form, with

$$m = -\frac{A}{B}, \qquad k = -\frac{C}{B}.$$

Theorem 4. If L is the graph of $y = mx + b$, and (x_1, y_1), (x_2, y_2) are any two points of L, then

$$\frac{y_2 - y_1}{x_2 - x_1} = m.$$

Proof. Since both points are on the line, we have

$$y_2 = mx_2 + k, \qquad y_1 = mx_1 + k.$$

Therefore

$$y_2 - y_1 = m(x_2 - x_1),$$

and $x_2 \neq x_1$, because L is not vertical. Therefore

$$\frac{y_2 - y_1}{x_2 - x_1} = m.$$

Thus the number m is uniquely determined by the line. It is called the *slope* of the line.

Theorem 5. Let L and L' be two nonvertical lines, with slopes m and m'. If L and L' are perpendicular, then

$$m' = -\frac{1}{m}.$$

Proof. Let

$$P_1 = (x_1, y_1) \qquad \text{and} \qquad P_2 = (x_2, y_2)$$

be points of L', such that L is the perpendicular bisector of $\overline{P_1P_2}$. (See Fig. 18.3.) As in the proof of Theorem 2, L is the graph of the equation

$$2(x_2 - x_1)x + 2(y_2 - y_1)y + (x_2^2 + y_2^2 + x_1^2 + y_1^2) = 0.$$

This has the form

$$Ax + By + C = 0,$$

where

$$A = 2(x_2 - x_1), \qquad B = 2(y_2 - y_1).$$

Therefore

$$m = -\frac{A}{B} = -\frac{2(x_2 - x_1)}{2(y_2 - y_1)} = -\frac{x_2 - x_1}{y_2 - y_1}.$$

But, by Theorem 4, we have

$$m' = \frac{y_2 - y_1}{x_2 - x_1}.$$

Therefore $m' = -(1/m)$, which was to be proved.

Theorem 6. Every circle is the graph of an equation of the form

$$x^2 + y^2 + Ax + By + C = 0.$$

Proof. By the distance formula, the circle with center (a, b) and radius r is the graph of the equation

$$\sqrt{(x - a)^2 + (y - b)^2} = r,$$

or

$$x^2 - 2ax + a^2 + y^2 - 2by + b^2 - r^2 = 0.$$

This has the required form, with

$$A = -2a,$$
$$B = -2b,$$
$$C = a^2 + b^2 - r^2.$$

The converse of Theorem 6 is false, of course. The graph of

$$x^2 + y^2 = 0$$

is a point, and the graph of

$$x^2 + y^2 + 1 = 0$$

is the empty set.

Problem Set 17.1

In proving the following theorems, try to use as little geometry as possible, putting the main burden on the algebra and on the theorems of this section.

1. Show that the graph of an equation of the form

$$x^2 + y^2 + Ax + By + C = 0$$

is always a circle, a point, or the empty set.

2. Show that if the graphs of the equations

$$y = m_1x + k_1, \qquad y = m_2x + k_2$$

are two (different) intersecting lines, then $m_1 \neq m_2$.

3. Show that if $m_1 = m_2$, then the graphs are either parallel or identical.

4. In the chapter on similarity, we defined

$$A_1, B_1, C_1 \sim A_2, B_2, C_2$$

to mean that all the numbers in question were positive and that

$$\frac{A_2}{A_1} = \frac{B_2}{B_1} = \frac{C_2}{C_1}.$$

Let us generalize this in the following way. Given A_1, B_1, C_1, not all $= 0$. If there is a $k \neq 0$ such that

$$A_2 = kA_1, \qquad B_2 = kB_1, \qquad C_2 = kC_1,$$

then we say that the sequences A_1, B_1, C_1 and A_2, B_2, C_2 are *proportional*, and we write

$$A_1, B_1, C_1 \sim A_2, B_2, C_2.$$

With this understanding, show that if

$$A_1x + B_1y + C_1 = 0 \qquad \text{and} \qquad A_2x + B_2y + C_2 = 0$$

have the same line L as their graph, then

$$A_1, B_1, C_1 \sim A_2, B_2, C_2.$$

[*Hint:* Discuss first the case where L is vertical, and then the case where L is not vertical.]

5. Describe the graphs of the following equations.

(a) $x^2 + y^2 + 1 + 2x + 2y + 2xy = 0$ (b) $xy = 0$ (c) $x^3 + xy^2 - x = 0$

Chapter 18
Rigid Motion

18.1 THE MOST GENERAL CONCEPT OF CONGRUENCE

By now we have given five different definitions of the word *congruent*, for five different kinds of figures. Two segments are congruent if they have the same length. Two angles are congruent if they have the same measure. Two triangles are congruent if there is a one-to-one correspondence between their vertices, such that every pair of corresponding sides are congruent (that is, have the same length) and every pair of corresponding angles are congruent (that is, have the same measure). Two circles are congruent if they have the same radius. Finally, two circular arcs are congruent if (1) the circles in which the arcs lie are congruent, and (2) the arcs have the same degree measure. All this is logically correct, but in a number of ways it leaves things to be desired.

In the first place, it was promised at the outset that the intuitive meaning of the word *congruent* was always going to be the same: two figures would be called congruent if they had exactly the same size and shape, that is, if one could be moved so as to coincide with the other. This promise has, in a sense, been kept. It is not hard to convince yourself that all five of the technical definitions that we have just reviewed have this intuitive meaning. On the other hand, it is artificial to have five different definitions to convey the same idea in five different cases. It would be better to have one definition which applies in the same way to segments, angles, triangles, and so on.

In the second place, as a matter of common sense, most of us would agree that two squares with edges of the same length ought to be congruent:

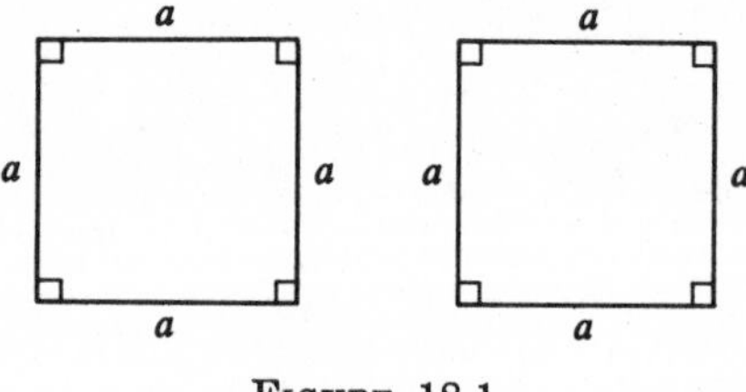

FIGURE 18.1

If the language of geometry doesn't allow us to say so, then the language of geometry must be inadequate.

Finally, it would be good to make some sort of contact with the old Euclidean idea of congruence. Euclid based all of his congruence proofs on a postulate that said that "things which coincide with one another are equal to one another." (See the Common Notions, in Book I of the *Elements*.) This was not adequate to account for the things that Euclid actually did. Strictly speaking, figures coincide only with themselves. And it is plain that the idea of motion, or superposition, is implicit in Euclid's congruence proofs. Some authors have attempted to make this idea explicit by stating a postulate to the effect that "geometric figures can be moved without changing their size or shape." But this still is not enough; it clarifies the difficulty without removing it. The difficulty is that while the term *figure* is plain enough (a figure is a set of points) the terms *moved*, *size*, and *shape* have an insecure status. They must be regarded as undefined terms, since no definitions have been given for them. But if they are undefined, then postulates must be given, conveying their essential properties; and this has not been done either. The general drift of the postulate is plain, but you cannot base a mathematical proof on a general drift.

It is possible, however, to formulate Euclid's idea in an exact mathematical way. We shall do this, by defining the general idea of *rigid motion*, or *isometry*.

The simplest example of this is as follows. Consider a rectangle $\square ABB'A'$. (The vertical sides of the rectangle are dotted in the figure because we are really interested in the two bases only.) Let f be the vertical projection,

$$f\colon \overline{AB} \leftrightarrow \overline{A'B'},$$

of the upper base onto the lower base. Thus, for each point P of $\overline{AB}$, $f(P)$ is the foot of the perpendicular from P to $\overline{A'B'}$. We know, of course, that f is a one-to-one correspondence between $\overline{AB}$ and $\overline{A'B'}$. That is, to every point P of $\overline{AB}$ there corresponds exactly one point $P' = f(P)$ of $\overline{A'B'}$, and to each point P' of $\overline{A'B'}$ there corresponds exactly one point $P = f^{-1}(P')$ of $\overline{AB}$. And this correspondence f has a special property: if P and Q are any two points of $\overline{AB}$, and P' and Q' are the corresponding points of $\overline{A'B'}$, as in the figure, then

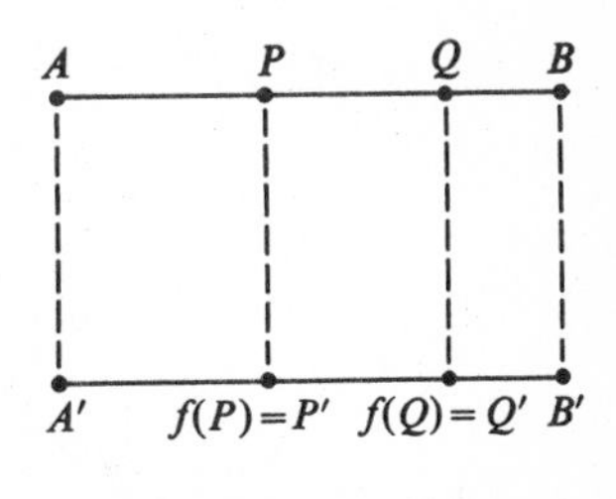

FIGURE 18.2

$$P'Q' = PQ,$$

because the segments $\overline{PQ}$ and $\overline{P'Q'}$ are opposite sides of a rectangle. Thus, *for any two points P, Q of $\overline{AB}$, the distance between $f(P)$ and $f(Q)$ is the same as the distance between P and Q.* More briefly, the correspondence f *preserves distances.*

The correspondence $f\colon \overline{AB} \leftrightarrow \overline{A'B'}$ is our first and simplest example of what is called a *rigid motion*, or an *isometry*. The general definition of this concept is as follows.

DEFINITION. Let M and N be sets of points, and let

$$f\colon M \leftrightarrow N$$

be a one-to-one correspondence between them. Suppose that for every two points P, Q of M we have

$$f(P)f(Q) = PQ.$$

Then f is called a *rigid motion*, or an *isometry*, between M and N. [Here $f(P)f(Q)$ denotes the distance between the points $f(P)$ and $f(Q)$.] If there is an isometry between M and N, then we say that M and N are *isometric*, and we write

$$M \approx N.$$

In this language, we can sum up our discussion of the vertical projection $f: \overline{AB} \leftrightarrow \overline{A'B'}$ in the form of the following theorem.

Theorem 1. Opposite sides of a rectangle are isometric.

Problem Set 18.1

1. Consider two triangles $\triangle ABC$ and $\triangle A'B'C'$, and suppose that

$$\triangle ABC \cong \triangle A'B'C'.$$

Let

$$V = \{A, B, C\}$$
$$V' = \{A', B', C'\}.$$

(Thus V and V' are *finite* sets of three elements each.) Does it follow that $V \approx V'$? That is, is there a rigid motion

$$f: V \leftrightarrow V'?$$

2. Let V be the set of vertices of a square of edge 1, and let V' be the set of vertices of another square of edge 1. Show that $V \approx V'$. (First you have to set up a one-to-one correspondence $f: V \leftrightarrow V'$, and then you have to show that f is an isometry.)

3. Do the same for the sets of vertices of the two parallelograms in the figure below.

4. Show that if V is a set of three collinear points, and V' is a set of three noncollinear points, then V and V' are *not* isometric.

5. Show that two segments of different lengths are never isometric.

6. Show that a line and an angle are never isometric.

7. Show that every two rays are isometric.

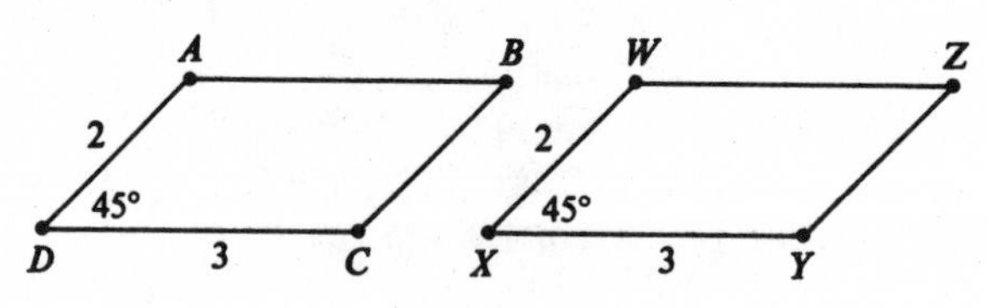

FIGURE 18.3

8. Show that two circles of different radius are never isometric.

9. Let L and L' be two lines in the same plane, and let $f: L \leftrightarrow L'$ be the vertical projection of L onto L'. Show that (1) if $L \parallel L'$, then f is an isometry, and conversely (2) if f is an isometry, then $L \parallel L'$.

10. Show that isometry is an equivalence relation. That is,

$M \approx M$ for every set M (Reflexivity).
If $M \approx N$, then $N \approx M$ (Symmetry).
If $M_1 \approx M_2$ and $M_2 \approx M_3$, then $M_1 \approx M_3$ (Transitivity).

18.2 ISOMETRIES BETWEEN TRIANGLES

Theorem 1 of the preceding section was, of course, more special than it needed to be. More generally, we have the following Theorem.

Theorem 1. If $\overline{AB} \cong \overline{CD}$, then there is an isometry $f: \overline{AB} \leftrightarrow \overline{CD}$, such that $f(A) = C$ and $f(B) = D$.

Proof. First we need to define a correspondence f between the two segments, and then we need to show that f preserves distances.

On $\overleftrightarrow{AB}$ let us set up a coordinate system, in such a way that the coordinate of A is 0 and the coordinate of B is positive. (It follows, of course, that the coordinate of B is the number AB.)

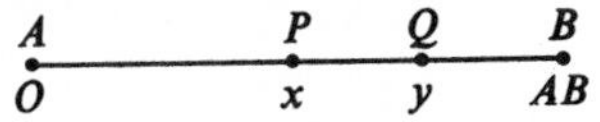

FIGURE 18.4

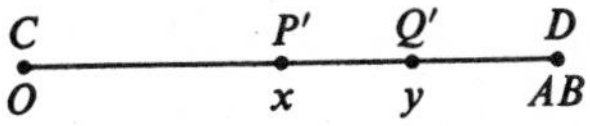

FIGURE 18.5

Similarly, we set up a coordinate system on $\overleftrightarrow{CD}$ in such a way that the coordinate of C is 0 and the coordinate of D is positive (and hence $= CD = AB$). The figures suggest how the correspondence f ought to be defined. Given a point P of $\overline{AB}$, the corresponding point $f(P)$ of $\overline{CD}$ is the point P' which has the same coordinate as P. Obviously this is a one-to-one correspondence between the two segments. And distances are preserved. Proof: Suppose that P and Q are points of $\overline{AB}$, with coordinates x and y. Then $P' = f(P)$ and $Q' = f(Q)$ have the same coordinates x and y, respectively. Since

$$PQ = |x - y|,$$

and

$$P'Q' = |x - y|,$$

it follows that

$$PQ = P'Q',$$

which was to be proved.

We can restate Theorem 1 in the following way.

Theorem. Given a correspondence

$$A \leftrightarrow C, \qquad B \leftrightarrow D$$

between the end points of two segments. If $\overline{AB} \cong \overline{CD}$, then there is an isometry $f: \overline{AB} \leftrightarrow \overline{CD}$ which agrees with the given correspondence at the end points.

Here f is called the isometry *induced by* the given correspondence. If we think of the theorem in these terms, then the extension to triangles is immediate.

Theorem 2. Given a correspondence

$$ABC \leftrightarrow DEF$$

between the vertices of two triangles. If

$$\triangle ABC \cong \triangle DEF,$$

then there is an isometry

$$f: \triangle ABC \leftrightarrow \triangle DEF,$$

such that $f(A) = D$, $f(B) = E$, and $f(C) = F$.

Proof. Let

$$f_1: \overline{AB} \leftrightarrow \overline{DE}$$

be the isometry induced by the correspondence $A \leftrightarrow D$, $B \leftrightarrow E$. Similarly, let

$$f_2: \overline{BC} \leftrightarrow \overline{EF}$$

be the isometry induced by the correspondence $B \leftrightarrow E$, $C \leftrightarrow F$; and let

$$f_3: \overline{AC} \leftrightarrow \overline{DF}$$

be the isometry induced by the correspondence $A \leftrightarrow D$, $C \leftrightarrow F$. Let f be the correspondence obtained by combining f_1, f_2, and f_3. That is, if P is on $\overline{AB}$, then $f(P) = f_1(P)$; if P is on BC, then $f(P) = f_2(P)$, and so on.

Since each f_i is an isometry, f preserves the distance between any two points that lie on the same side of $\triangle ABC$. Thus it remains only to show that f preserves

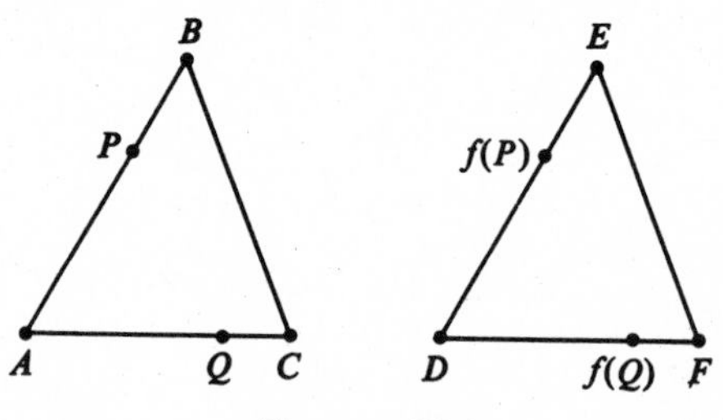

FIGURE 18.6

the distance between any two points P, Q on different sides of $\triangle ABC$. Suppose, without loss of generality, that P is on $\overline{AB}$ and Q is on $\overline{AC}$, as in the figure (Fig. 18.6). Let $P' = f(P)$ and let $Q' = f(Q)$. Then

$$\overline{AP} \cong \overline{DP'},$$

because f_1 is an isometry;

$$\overline{AQ} \cong \overline{DQ'},$$

because f_3 is an isometry; and

$$\angle PAQ \cong \angle P'DQ',$$

because $\triangle ABC \cong \angle DEF$. By SAS, we have

$$\triangle PAQ \cong \triangle P'DQ',$$

so that

$$PQ = P'Q',$$

which was to be proved.

The isometry f that was defined in the proof of Theorem 2 is called the isometry *induced by* the congruence $\triangle ABC \cong \triangle DEF$.

You may wonder, at this point, why we didn't define congruence by means of isometry in the first place. The reason is that in the sort of geometry that we have been discussing so far in this book, the elementary definitions based on distance, angular measure and correspondences between the *vertices* of triangles are the definitions that are convenient to work with. Thus, if we had defined $\angle A \cong \angle B$ to mean that $\angle A \approx \angle B$, the next thing that we should have done is to prove that $\angle A \approx \angle B$ if and only if $m\angle A = m\angle B$, so that we could work with the latter statement instead of the former. Similarly, we would have shown that $\triangle ABC \approx \triangle DEF$ if and only if the triangles are congruent in the elementary sense, and this would enable us to talk about correspondences between triplets of points, instead of talking about correspondences between infinite sets of points. In general, basic definitions in mathematics should be stated in such a way that they can be put to work quickly and easily.

Problem Set 18.2

1. Suppose that the correspondence

$$ABC \leftrightarrow A'B'C'$$

is an isometry. Show that if A-B-C, then A'-B'-C'.

2. Given an isometry

$$f: M \leftrightarrow N.$$

Let A and B be points of M. Show that if M contains the segment between A and B, then N contains the segment between $f(A)$ and $f(B)$.

3. Show that if M is convex, and $M \approx N$, then N is convex.

4. Given $M \approx N$. Show that if M is a segment, then so is N.

5. Given $M \approx N$. Show that if M is a ray, then so is N.

6. Suppose that M is a segment and N is a circular arc. Then M and N are not isometric.

18.3 GENERAL PROPERTIES OF ISOMETRIES. REFLECTIONS

Theorem 1. A-B-C if and only if (1) A, B, and C are all different and (2) $AB + BC = AC$.

Proof. A-B-C was originally defined to mean that (1) and (2) hold, and also (3) A, B, and C are collinear. By the Triangular Inequality (Theorem 5 of Section 7.1), (2) $\Rightarrow$ (3). The theorem follows.

Hereafter, if f is an isometry, and $A, B, C, \ldots$ are points, then $f(A)$, $f(B)$, $f(C), \ldots$ will be denoted by $A', B', C', \ldots$.

Theorem 2. Isometries preserve betweenness.

That is, A-B-$C \Rightarrow A'$-B'-C'. Proof: Isometries preserve conditions (1) and (2) of Theorem 1.

Theorem 3. Isometries preserve collinearity.

That is, if $f:M \leftrightarrow N$ is an isometry, and M lies on a line L, then N also lies on a line.

Proof. Suppose not. Then N contains three noncollinear points A', B', C'. The inverse-image points A, B, and C can be arranged in an order X, Y, Z, such that X-Y-Z. By Theorem 2, one of the points A', B', and C' is between the other two. This contradicts the hypothesis that A', B', and C' are noncollinear.

Theorem 4. If $f:M \leftrightarrow N$ is an isometry, then so also is $f^{-1}:N \leftrightarrow M$.

Proof. Obviously f^{-1} is a one-to-one correspondence $N \leftrightarrow M$. It remains to show that f^{-1} preserves distances. Let C and D be points of N. Then $f(A) = C$, $f(B) = D$, for some points A and B of M. Since f is an isometry, $CD = AB$. Therefore $f^{-1}(C)f^{-1}(D) = CD$, which was to be proved.

Hereafter, for the sake of simplicity, we shall discuss isometries $E \leftrightarrow E$, where E is a plane.

Theorem 5. Let f be an isometry $E \leftrightarrow E$. Then f preserves lines. That is, if L is a line, then so also is $f(L)$.

Proof. Let $L = \overleftrightarrow{AB}$. Since f preserves collinearity, we have

$$f(L) \subset \overleftrightarrow{A'B'}.$$

Since f^{-1} is an isometry, it follows in the same way that

$$f^{-1}(\overleftrightarrow{A'B'}) \subset \overleftrightarrow{AB}.$$

Therefore

$$f(f^{-1}(\overleftrightarrow{A'B'}) \subset f(\overleftrightarrow{AB}),$$

and

$$\overleftrightarrow{A'B'} \subset f(\overleftrightarrow{AB}).$$

Therefore $\overleftrightarrow{A'B'} = f(\overleftrightarrow{AB})$, which was to be proved.

Theorem 6. Let $f:E \leftrightarrow E$ be an isometry. Then f preserves segments. That is, $f(\overline{AB}) = \overline{A'B'}$.

Proof. Since f preserves betweenness, we have $f(\overline{AB}) \subset \overline{A'B'}$. (Details?) Since f^{-1} is an isometry, we have $f^{-1}(\overline{A'B'}) \subset \overline{AB}$. Applying f on both sides, in the preceding formula, we get $\overline{A'B'} \subset f(\overline{AB})$. Therefore $f(\overline{AB}) = \overline{A'B'}$.

Theorem 7. Let $f:E \leftrightarrow E$ be an isometry. Then f preserves triangles. That is, $f(\triangle ABC) = \triangle A'B'C'$.

Proof. If A', B', and C' were collinear, then A, B, and C would be collinear, by Theorems 3 and 4. Therefore A', B', and C' are not collinear. Theorem 7 now follows by three applications of Theorem 6.

Theorem 8. Let f and g be isometries $E \leftrightarrow E$. Then the composition $f(g)$ is also an isometry $E \leftrightarrow E$.

Obviously.

We now consider a special type of isometry. Let L be a line in a plane E. The *reflection* of E across L is a function r, defined as follows. (1) If P lies on L, then $r(P) = P$. (2) If P is not on L, then $r(P)$ is the point P' such that L is the perpendicular bisector of $\overline{PP'}$. Such a transformation r is called a *line-reflection*.

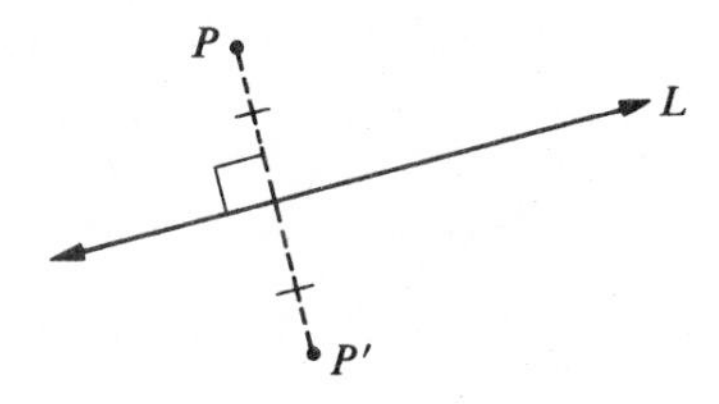

FIGURE 18.7

Theorem 9. Every line-reflection is an isometry.

Proof. Here it is convenient to introduce a coordinate system in which the x-axis is L. Then r has the form $(x, y) \mapsto (x, -y)$, and Theorem 9 now follows by the distance formula. (If we don't use a coordinate system, then we need to discuss various special cases.)

Theorem 10. Let P and P' be points. Then there is a line-reflection $P \mapsto P'$.

(Reflect across the perpendicular bisector of $\overline{PP'}$.)

Theorem 11. Let P, Q, and Q' be three points, such that $PQ = PQ'$. Then there is a line-reflection $r:P \mapsto P$, $Q \mapsto Q'$.

Proof. If the points are collinear, then P is the midpoint of $\overline{QQ'}$; we reflect across the perpendicular bisector of $\overline{QQ'}$. If not, we reflect across the line which contains the bisector of $\angle QPQ'$.

Theorem 12. Suppose that $\triangle ABC \cong \triangle PQR$, where the two triangles lie in the same plane E. Then there is an isometry

$$\begin{aligned} f&:E \leftrightarrow E,\\ &:A \mapsto P,\ B \mapsto Q,\ C \mapsto R,\\ &:\triangle ABC \leftrightarrow \triangle PQR, \end{aligned}$$

such that f is the composition of either two or three line-reflections.

Proof. By Theorem 10 there is a line-reflection $r_1: E \leftrightarrow E$, $A \mapsto P$. Then $r_1(\triangle ABC) = \triangle PB'C'$, and $PB' = PQ$. (Why?) By Theorem 11 there is a line-reflection $r_2: E \leftrightarrow E$, $P \mapsto P$, $B' \mapsto Q$. There are now two possibilities.

(1) If C'' $(= r_2(C'))$ and R are on the same side of $\overleftrightarrow{PQ}$, then $C'' = R$, because $\angle C''PQ \cong \angle RPQ$ and $PC'' = PR$. Therefore we are done: let $f = r_2(r_1)$.

(2) If C'' and R are on opposite sides of $\overleftrightarrow{PQ}$, then we still have $\angle C''PQ \cong \angle RPQ$ and $PC'' = PR$. Let r_3 be the reflection of E across $\overleftrightarrow{PQ}$. Then $r_3(P) = P$, $r_3(Q) = Q$, and $r_3(C'') = R$. Let $f = r_3(r_2(r_1))$.

Note that this theorem is stronger than Theorem 2 of Section 18.2. Now we know that every congruence between two triangles can be represented by an isometry of the entire plane onto itself.

Problem Set 18.3

1. Show that isometries preserve circles.

2. Let A, B, and C be three noncollinear points, and let f be an isometry $A \mapsto A$, $B \mapsto B$, $C \mapsto C$. Show that $P \mapsto P$ for every P.

3. Let $\triangle ABC$ and $\triangle PQR$ be triangles, with $\triangle ABC \cong \triangle PQR$. Show that there is only one isometry $f: E \leftrightarrow E$, $A \mapsto P$, $B \mapsto Q$, $C \mapsto R$. Is it true that there is only one isometry f such that $f(\triangle ABC) = \triangle PQR$?

4. Given $\overline{AB}$, show that there are four and only four isometries $f_i: E \leftrightarrow E$, $\overline{AB} \leftrightarrow \overline{AB}$.

5. Given $\overline{AB} \cong \overline{CD}$, show that there are four and only four isometries $g_i: E \leftrightarrow E$, $\overline{AB} \leftrightarrow \overline{CD}$.

6. Given a square $\square ABCD$, show that there are eight and only eight isometries $f_i: E \leftrightarrow E$, $\square ABCD \leftrightarrow \square ABCD$.

7. We showed (Theorem 6) that segments are preserved by isometries $E \leftrightarrow E$. Show that this holds for all isometries. That is, if f is an isometry $\overline{AB} \leftrightarrow N$, then N is a segment (namely, the segment $\overline{A'B'}$.)

8. Similarly, show that if L is a line, and f is an isometry $L \leftrightarrow N$, then N is a line.

9. Show that every isometry preserves triangles.

10. Find a way to deduce the results of the preceding three problems from the theorems of Section 18.3 (if you have not already done this).

18.4 DILATIONS AND SIMILARITIES

Let P_0 be a point of a plane E, let k be a positive number, and let

$$d: E \leftrightarrow E$$

be a transformation defined as follows. (1) $d(P_0) = P_0$. (2) If $P \neq P_0$, let $P' = d(P)$ be the point of $\overrightarrow{P_0P}$ such that $P_0P' = kP_0P$. Then d is called a *dilation*; P_0 is called its *center*, and k is called its *proportionality constant*. (Note that for $0 < k < 1$, it might seem more natural to call d a *contraction*.)

Theorem 1. Every dilation is a one-to-one correspondence $E \leftrightarrow E$.

Proof?

Theorem 2. Let $d: E \leftrightarrow E$ be a dilation, with proportionality constant k. Then for every pair of points A, B we have

$$A'B' = kAB.$$

Proof. Choose a coordinate system for E, with the center P_0 of d as the origin. Then

$$d(x, y) = (kx, ky).$$

Theorem 2 now follows by the distance formula.

Theorem 3. Dilations preserve betweenness.

The proof is like that of Theorem 2 of Section 18.3; the point is that

$$\begin{aligned} AB + BC &= AC \\ \Rightarrow kAB + kBC &= kAC \\ \Rightarrow A'B' + B'C' &= A'C'. \end{aligned}$$

Theorem 4. Dilations preserve collinearity.

The proof is like that of Theorem 3 of Section 18.3.

Theorem 5. The inverse of a dilation is a dilation.

(Use $k' = 1/k$.)

Theorem 6. Dilations preserve segments.

Theorem 7. Dilations preserve triangles.

The proofs are like those of Theorems 6 and 7 of Section 18.3.

Theorem 8. If $\triangle ABC \sim \triangle PQR$ (in a plane E), then there is an isometry $f:E \leftrightarrow E$ and a dilation $d:E \leftrightarrow E$ such that

$$d(f(A)) = P, \qquad d(f(B)) = Q, \qquad d(f(C)) = R,$$

and

$$d(f(\triangle ABC)) = \triangle PQR.$$

Proof. Let B' and C' be points of $\overrightarrow{PQ}$ and $\overrightarrow{PR}$, such that

$$PB' = AB, \qquad PC' = AC.$$

Then

$$\triangle ABC \cong \triangle PB'C',$$

and

$$\triangle PB'C' \sim \triangle PQR.$$

By Theorem 12 of Section 18.3, there is an isometry

$$\begin{aligned} f&:E \leftrightarrow E \\ &:A \mapsto P,\ B \mapsto B',\ C \mapsto C' \\ &:\triangle ABC \leftrightarrow \triangle PB'C'. \end{aligned}$$

Now let d be the dilation using $P_0 = P$ and

$$k = PQ/PA' = PR/PB'.$$

Then

$$d(P) = P, \qquad d(A') = Q, \qquad d(B') = R.$$

By Theorem 7, $d(\triangle PA'B') = \triangle PQR$. It follows that $d(f(A)) = P$, $d(f(B)) = Q$, $d(f(C)) = R$, and

$$d(f(\triangle ABC)) = \triangle PQR,$$

which was to be proved.

We have an easy converse.

Theorem 9. Let $\triangle ABC$ be a triangle, let f be an isometry, and let d be a dilation. Let $A' = d(f(A))$, $B' = d(f(B))$, $C' = d(f(C))$. Then

$$\triangle ABC \sim \triangle A'B'C'.$$

Note that we can now generalize the idea of a similarity in much the same way that we generalized the idea of a congruence. A *similarity* (between two plane figures of any kind) is a transformation which is equal to an isometry followed by a dilation; and two figures are *similar* if there is a similarity between them. In the case in which the figures are triangles, these definitions agree with the elementary definitions.

Problem Set 18.4

1. Write an explicit proof of Theorem 4.
2. Same, for Theorem 6.
3. Same, for Theorem 7.
4. Show that dilations preserve lines.
5. Show that dilations preserve circles.
6. Show that dilations preserve parabolas.
7. Prove Theorem 2 without using a coordinate system. (You will need to discuss various cases.)
8. Prove that dilations preserve half-planes.
9. Let P_1 and P_2 be parabolas. Show that there is an isometry f, and a dilation d, such that $d(f(P_1)) = P_2$.
10. Show that every two squares in the same plane are similar.

Chapter 19
Constructions with Ruler and Compass

19.1 UNMARKED RULERS AND COLLAPSIBLE COMPASSES

In the introduction to the foundations of geometry, given in the first few chapters of this book, we used postulates which fall into six groups: (1) the incidence postulates, (2) the separation postulates, (3) the ruler postulate, (4) the "protractor postulates," (5) the SAS postulate, and finally (6) the parallel postulate. (Here, of course, by the "protractor postulates" we mean the ones dealing with the measures of angles.) Our treatment of incidence and separation was standard; and so also was our treatment of parallelism. In fact, where these topics are concerned, the differences between one book and another are mainly in the style of exposition and the degree of explicitness.

The ruler and protractor postulates, however, are another matter. These were invented rather recently (by G. D. Birkhoff) and the mathematical spirit reflected by them is quite different from the mathematical spirit of the Greeks. The basic difference, roughly speaking, is that rulers and protractors are used for measuring things; we place them on geometric figures and we read off real numbers from the scales that are marked on them. Birkhoff's metric postulates tell us, in effect, that we have an "ideal ruler" and an "ideal protractor," with which we can measure segments and angles exactly. The Greeks, on the other hand, considered that measurement was merely one of the practical arts. It was considered unworthy of the attention of mathematicians and philosophers. Just as we have described the metric treatment in terms of two drawing instruments, the marked ruler and protractor, so we can describe Greek geometry in terms of two different drawing instruments—the *unmarked* ruler and the compass. The Greeks thought about geometry in terms of these two instruments; and they investigated at length the question of what figures could be constructed by means of them.

Before we consider problems of construction with ruler and compass, several warnings are in order.

(1) When we speak of a ruler and a compass, we mean an "ideal ruler" and an "ideal compass," which draw straight lines and circles exactly. The thickness of pencil marks and the approximations involved in draftsmanship will not concern us.

(2) The Euclidean ruler has no marks on it. We can use it to draw the line through two given points, but that is *all* we can use it for. We cannot use it to measure distances between points, or even to tell whether two segments are congruent.

(3) The Euclidean compass can be used in the following way. Given a point P and a point Q (in the plane), we can draw the circle that has center at P and contains Q. This is *all* that we can use the Euclidean compass for. That is, given a third point P', we are not allowed to move the spike of the compass to P' and then draw the circle with center at P' and radius PQ. For this reason, the Euclidean compass is called collapsible; you can't move the spike because "when you lift the spike off the paper the compass collapses." Another way of putting it is that you can't use the compass as a pair of dividers. Oddly enough, modern draftsmen feel the same way that Euclid did on this delicate question.

(4) In studying ruler and compass constructions, we shall not attempt to build the foundations of geometry all over again. In proving that our constructions work, we shall make free use of the theorems of metric plane geometry. In particular, we shall make continual use of the line-circle theorem and the two-circle theorem.

(5) In a construction problem, when we say that a line is "given," we mean that at least two points of the line are given.

Let us now try a few constructions.

Construction 1. To construct the perpendicular bisector of a given segment.

Given two points P, Q. First draw the circle C_1 with center at P, containing Q; and then draw the circle with center at Q, containing P:

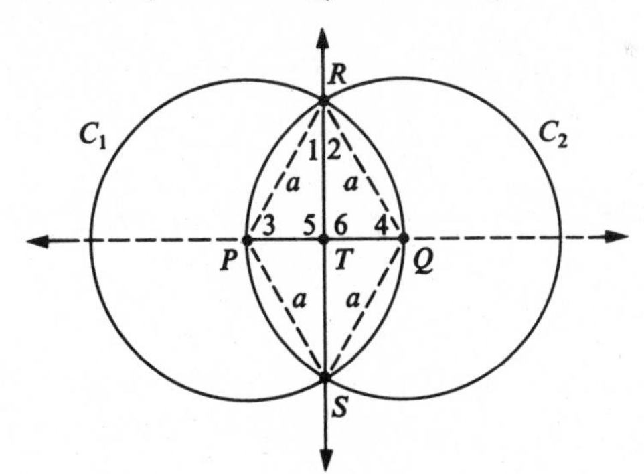

FIGURE 19.1

Let $PQ = a$. Since $a < 2a$, it follows that each of the numbers a, a, a is less than the sum of the other two. Therefore the hypothesis of the two-circle theorem is satisfied. (Most of the time, our verifications of the hypothesis of the two-circle theorem will be as trivial as this one.) Therefore C_1 and C_2 intersect in two points R and S, lying on opposite sides of $\overleftrightarrow{PQ}$. Therefore $\overline{RS}$ intersects $\overleftrightarrow{PQ}$ in a point T.

Since T lies on $\overline{RS}$, and $\overline{RS}$ is a chord of both circles, it follows that T is in the interior of both circles, and both TP and TQ are less than PQ. Therefore T-P-Q and P-Q-T are impossible. Therefore P-T-Q.

Now $\triangle SPR \cong \triangle SQR$, by SSS. Hence $\angle 1 \cong \angle 2$. By SAS we have $\triangle PRT \cong \triangle QRT$. Therefore $\angle 5 \cong \angle 6$, so that $\overleftrightarrow{RS} \perp \overline{PQ}$, and $\overline{PT} \cong \overline{TQ}$. Thus $\overleftrightarrow{RS}$ is the perpendicular bisector of $\overline{PQ}$. This gives us a sort of "corollary construction":

Construction 2. To bisect a given segment.

First construct the perpendicular bisector, as above. The point T is the bisector.

Construction 3. To construct the perpendicular to a given line, through a given point on the line.

Given a line L, and a point X of L. Since L was given, at least one other point P of L must be given. Draw the circle C which has center at X and contains P. Then C will intersect L in exactly one other point Q. Now construct the perpendicular bisector of $\overline{PQ}$; this will be perpendicular to L at X.

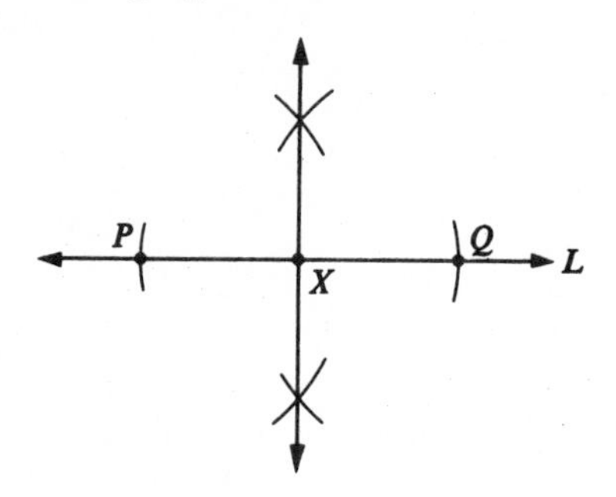

FIGURE 19.2

Construction 4. Given three points P, Q, R. To construct a rectangle $\square PQST$, such that $\overline{PT} \cong \overline{PR}$.

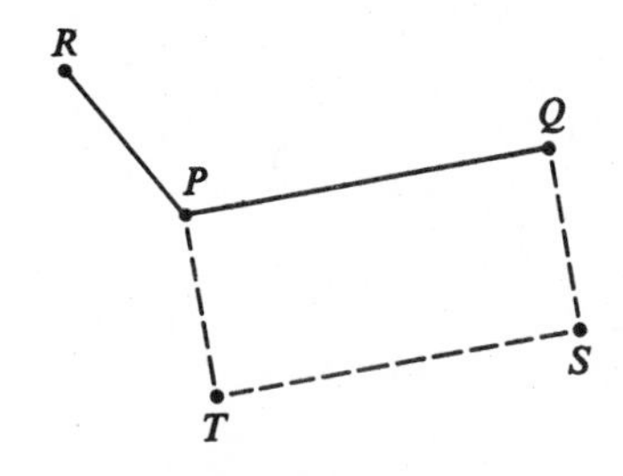

FIGURE 19.3

First construct the perpendicular L_1 to $\overline{PQ}$ at P. Then draw the circle C that has center at P and contains R. The circle C will intersect L_1 in points T and T'.

Now construct L_2 perpendicular to $\overline{PQ}$ at Q; and construct L_3 perpendicular to L_1 at T. If L_2 and L_3 were parallel, then $\overleftrightarrow{PT}$ and $\overleftrightarrow{PQ}$ would be parallel, which is false. Therefore L_2 intersects L_3 in a point S. We know that each pair of opposite sides of $\square PQST$ are parallel; and three of its angles are right angles. Therefore $\square PQST$ is a rectangle.

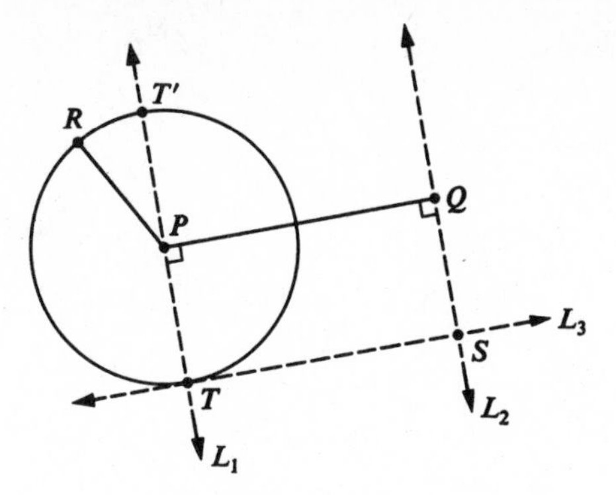

FIGURE 19.4

Construction 5. Given a segment $\overline{PQ}$ and a ray $\overrightarrow{AB}$. To construct a point C of $\overrightarrow{AB}$ such that $\overline{AC} \cong \overline{PQ}$.

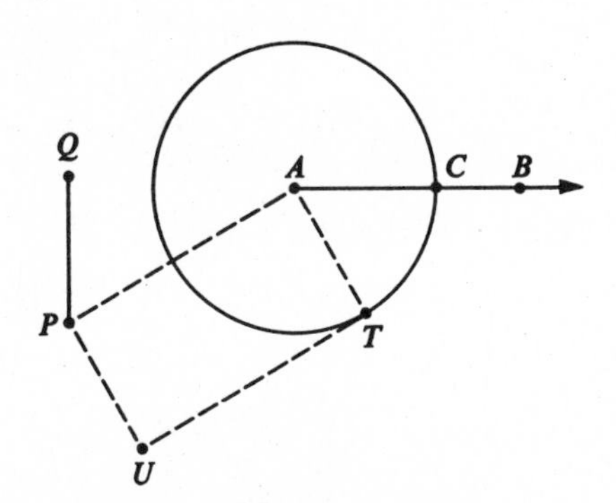

FIGURE 19.5

First we construct a rectangle $\square PATU$, with $\overline{PU} \cong \overline{PQ}$. Then $\overline{AT} \cong \overline{PQ}$. Draw the circle with center at A, containing T. This will intersect $\overrightarrow{AB}$ in a point C; and we will have $\overline{AC} \cong \overline{PQ}$, as desired.

Note that once we have Construction 5, we are free to forget about the collapsibility property of our Euclidean compass. The point is that the ruler and compass, in combination, furnish us in effect with a pair of dividers.

PROBLEM SET 19.1

1. Find a simpler method of doing Construction 5, which works whenever $PQ < PA$. [*Hint:* First construct an equilateral triangle $\triangle PAU$.]

2. Show that if Construction 5 can be done in the case where $PQ < PA$, then it can be done in the general case.

3. *Construction* 6. Given $\triangle ABC$, and $\overline{A'B'} \cong \overline{AB}$. To construct a point C' on a given side of $\overleftrightarrow{A'B'}$, such that $\triangle ABC \cong \triangle A'B'C'$.

4. *Construction* 7. Given $\angle ABC$, $\overrightarrow{PQ}$, and a side H of $\overleftrightarrow{PQ}$. To construct a ray $\overrightarrow{PR}$, with R in H, such that $\angle ABC \cong \angle RPQ$.

5. *Construction* 8. Given a line L and point P. To construct the line through P parallel to L.

6. *Construction* 9. Given $\overline{AB}$ and a positive integer n. To divide $\overline{AB}$ into n congruent segments.

7. Suppose that your Euclidean ruler is not a "theoretical ruler," of infinite extent both ways, but a ruler of finite length, say, one inch. Suppose also that the points of your compass cannot be spread apart more than an inch. Show that given any two points—no matter how far apart they may be—you can draw the segment between them.

8. In carrying out Construction 9, you probably found it convenient to draw, at random, a ray $\overrightarrow{AP}$ which was not collinear with $\overline{AB}$. Show that this process of random choice can be avoided.

Strictly speaking, random choices of points are not allowed in doing construction problems. The reason is curious: if they were allowed, then the so-called "impossible construction problems" to be discussed later in this chapter would be not quite impossible. For example, an infinitely lucky person might manage to pick, at random, a point on a trisector of any given angle.

19.2 HOW TO DO ALGEBRA WITH RULER AND COMPASS

Suppose that we have given a segment of length 1, and two segments of length a and b:

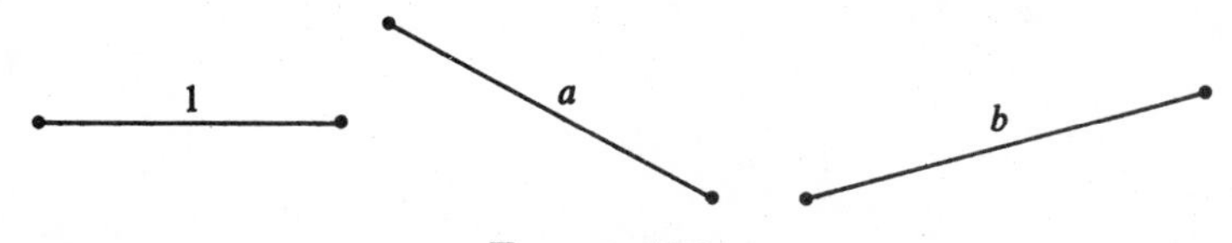

FIGURE 19.6

We shall show that all the elementary operations of algebra for the numbers a and b can be carried out with ruler and compass. That is, with ruler and compass we can construct segments whose lengths are

$$a + b, \qquad \frac{1}{a}, \qquad ab, \qquad \frac{b}{a}, \qquad \sqrt{a}.$$

(1) The first of these constructions is trivial. On any line L, we lay off a segment $\overline{PQ}$ of length a, and then lay off a segment $\overline{QR}$ of length b, in such a way that P-Q-R. The others require tricks.

(2) In Fig. 19.7, we have $\angle 1 \cong \angle 2$, so that $\triangle ABC \sim \triangle ADE$. Therefore

$$\frac{AC}{AB} = \frac{AE}{AD}.$$

If $AB = a$, $AC = 1$, and $AD = 1$, this says that

$$\frac{1}{a} = \frac{AE}{1}.$$

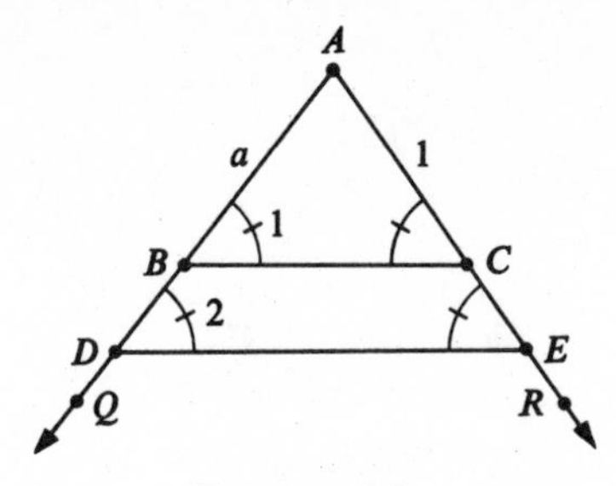

FIGURE 19.7

This indicates a way to construct a segment of length $1/a$. We start with any angle $\angle QAR$. On $\overrightarrow{AQ}$ we lay off $\overline{AB}$ with $AB = a$, and $\overline{AD}$ with $AD = 1$. On $\overrightarrow{AR}$ we lay off $\overline{AC}$, with $AC = 1$:

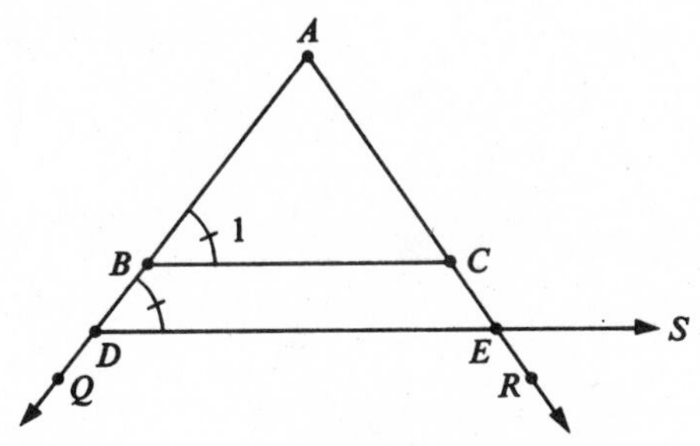

FIGURE 19.8

We now construct a ray $\overrightarrow{DS}$ so that $\angle ADS \cong \angle 1$. (Any angle can be copied with ruler and compass.) You ought to be able to prove, with no trouble, that $\overrightarrow{DS}$ intersects $\overrightarrow{AR}$ in a point E. The segment $\overline{AE}$ is the segment that we wanted.

(3) In the figure below, we have $\angle 1 \cong \angle 2$, so that $\triangle PXY \sim \triangle PZW$.

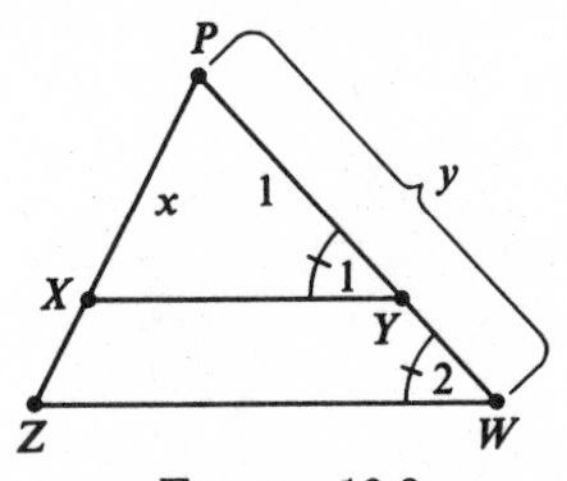

FIGURE 19.9

Therefore

$$\frac{PZ}{x} = \frac{y}{1},$$

so that $PZ = xy$. The construction of the figure is exactly as in the preceding case.

(4) To construct a segment of length b/a, we first find $1/a$, and then multiply the result by b.

(5) Finally, we want a segment of length $\sqrt{a}$. First we construct segments $\overline{PQ}$,

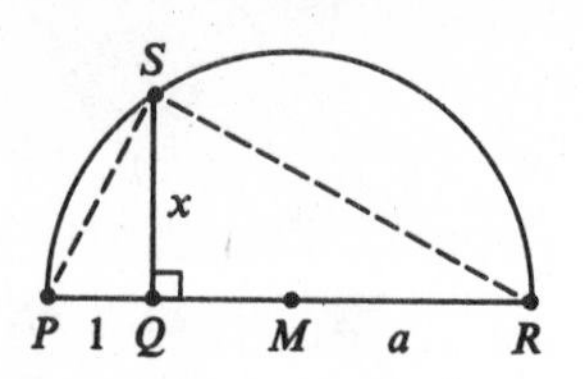

FIGURE 19.10

$\overline{QR}$, so that P-Q-R, $PQ = 1$ and $QR = a$. We bisect the segment. Let M be its bisector. With center M and radius

$$MP = MR = \frac{1 + a}{2},$$

we draw a circle. Next we construct a perpendicular to $\overline{PR}$ at Q. This line intersects the circle in two points, one of which is the S shown in the figure. Let $x = QS$. Since $\triangle PQS \sim \triangle SQR$, we have

$$\frac{QS}{PQ} = \frac{QR}{SQ}$$

or

$$\frac{x}{1} = \frac{a}{x}$$

or

$$x^2 = a.$$

Therefore $x = \sqrt{a}$, and $\overline{QS}$ is the segment that we were looking for.

19.3 SOLVING EQUATIONS WITH RULER AND COMPASS

We have found that with ruler and compass we can add, multiply, divide and extract square roots, starting with positive numbers. Let us now suppose that we have a coordinate system in the plane:

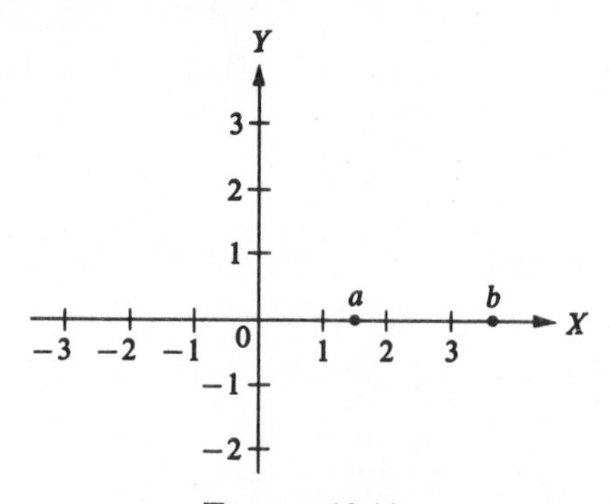

FIGURE 19.11

We want to know which points we can plot with ruler and compass, given the points with coordinates 1, a, and b on the x-axis. Of course, negative numbers have now entered the picture: a and b may easily be negative. But this causes no trouble; if we can plot the point with coordinate x, then surely we can plot the point with coordinate $-x$; given x and y, we can plot $x - y$, $y - x$, $(-x)y$, $x(-y)$, and so on.

This means that *with ruler and compass we can perform all the operations described in the postulates for a Euclidean ordered field*. That is, we can add, subtract, multiply, divide, and extract square roots, in all cases where these operations are algebraically possible. Hereafter, when we speak of "plotting a number," we shall mean, of course, plotting the corresponding point on the x-axis. Obviously, if we can plot both h and k, then we can plot the point (h, k) in the coordinate plane. We merely construct perpendiculars, as in the figure, thus getting their intersection. And conversely, if $P = (h, k)$ is given, we can plot h and k by *dropping* perpendiculars to the axes.

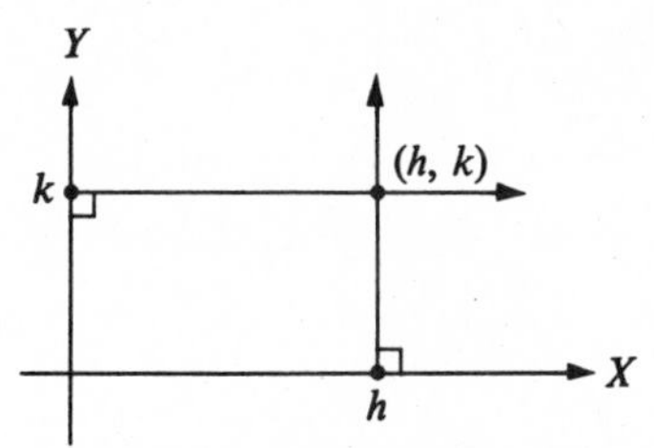

FIGURE 19.12

For this reason, in many cases we can solve algebraic problems by going through ruler-and-compass constructions. This process is not merely a stunt. We shall use it to solve some construction problems which would otherwise be very hard.

Problem 1. Given the points with coordinates a, b, c on the x-axis, with $b^2 - 4ac > 0$. We wish to plot, with ruler and compass, the roots of the equation

$$ax^2 + bx + c = 0.$$

These roots are the numbers

$$x_1 = \frac{-b + \sqrt{b^2 - 4ac}}{2a}, \quad \text{and} \quad x_2 = \frac{-b - \sqrt{b^2 - 4ac}}{2a}.$$

Each of them can be computed, starting from a, b, and c, by a finite number of additions, subtractions, multiplications, divisions and root-extractions. Each of these operations can be performed geometrically. Therefore the roots can be plotted.

Problem 2. Given the points on the x-axis with coordinates A, B, C, A', B', C'. We wish to plot the numbers x_1, y_1 which are the solution of the system

$$Ax + By + C = 0, \tag{1}$$
$$A'x + B'y + C' = 0. \tag{2}$$

We are interested in the case where the graphs of the equations are nonparallel lines intersecting in a single point (x_1, y_1). This occurs when

$$AB' - BA' \neq 0.$$

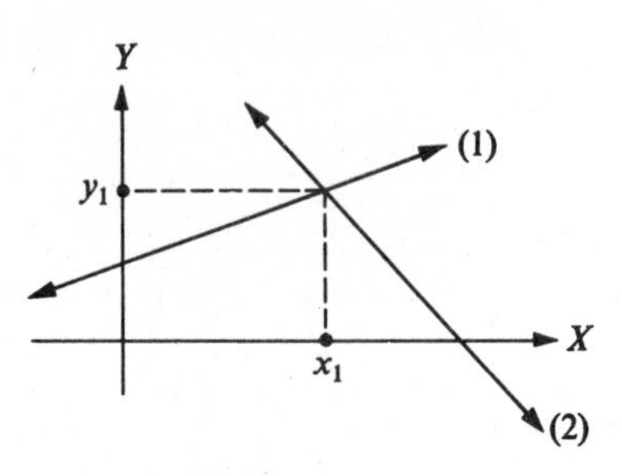

FIGURE 19.13

In this case, by the usual elementary methods, we get the solution in the form

$$x_1 = \frac{BC' - B'C}{AB' - BA'},$$

$$y_1 = \frac{A'C - AC'}{AB' - BA'}.$$

All the operations required here can be done with ruler and compass. Therefore x_1 and y_1 can be plotted. In fact, if you are actually going to plot x_1 and y_1, there is a much shorter method. We claim that if A, B, and C are plotted, then at least two points of the line

$$Ax + By + C = 0$$

can be plotted. If $B \neq 0$, we can set $x = 0$ and $x = 1$, getting the points $(0, -C/B)$, $(1, (-C - A)/B)$. If $B = 0$, then the corresponding line is vertical, and we can plot the points $((-C/A), 0)$, $((-C/A), 1)$. Once we have plotted these points, we can draw the line that contains them:

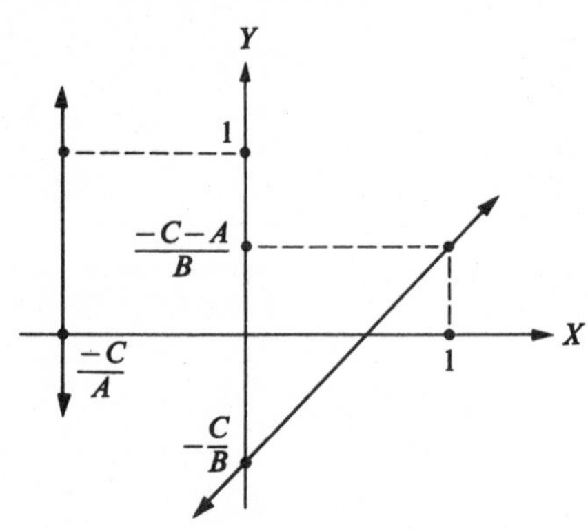

FIGURE 19.14

Thus, if the coefficients in the equation of a line can be plotted, the line itself can be drawn by a rule-and-compass construction. The short way to construct the solution of a pair of linear equations, therefore, is to draw the corresponding lines and see where they intersect. The same idea leads to an even greater economy in more difficult cases, as we shall see.

Problem 3. Suppose that a, b, r, A, B, and C are plotted, with $r > 0$. We want to plot the common solutions (x_1, y_1), (x_2, y_2) of the equations

$$(x - a)^2 + (y - b)^2 = r^2, \tag{1}$$

$$Ax + By + C = 0, \tag{2}$$

in the case in which such common solutions exist. We assume, as usual, that A and B are not both 0, so that the graph of (2) is a line.

This one is easy. First we draw the graph of (2) by the method used in the preceding problem. The graph of (1) is a circle with center at (a, b) and radius r. Since a, b, and r are given as plotted, we can draw this circle. We have now plotted whatever intersection points may happen to exist.

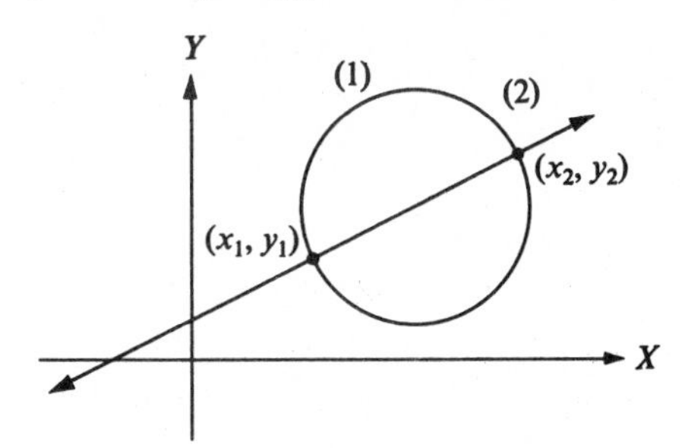

FIGURE 19.15

Problem 4. Suppose that A, B, C, D, E, and F are given as plotted. Suppose that the graphs of

$$x^2 + y^2 + Ax + By + C = 0, \tag{1}$$

$$Dx + Ey + F = 0, \tag{2}$$

are a line and a circle, respectively. We want to plot the common solutions (x_1, y_1), (x_2, y_2), in the cases where such solutions exist.

The first step is to draw the graph of (2). The remaining problem is to show that the graph of (1) can also be drawn.

Completing the square in the usual way, we convert (1) to the form

$$x^2 + Ax + \frac{A^2}{4} + y^2 + By + \frac{B^2}{4} = -C + \frac{A^2}{4} + \frac{B^2}{4},$$

or

$$\left(x + \frac{A}{2}\right)^2 + \left(y + \frac{B}{2}\right)^2 = \frac{A^2 + B^2 - 4C}{4}.$$

Thus the center of our circle is the point

$$(a, b) = \left(-\frac{A}{2}, -\frac{B}{2}\right),$$

and its radius is

$$r = \tfrac{1}{2}\sqrt{A^2 + B^2 - 4C}.$$

All three of the numbers a, b, and r can be plotted, since A, B, and C can be plotted. Therefore the circle can be drawn, exactly as in the preceding problem.

Problem 5. Given a system of equations

$$x^2 + y^2 + Ax + By + C = 0,$$
$$x^2 + y^2 + Dx + Ey + F = 0,$$

where the coefficients are given as plotted. Again, we want to plot the common solutions in the cases where there are any. To do this, we draw the two circles by the method used in the preceding problems.

We have now gotten involved in a lengthy investigation of relations between geometry and algebra. This may seem foreign to the spirit of geometry, but in fact it is foreign merely to the spirit of *Greek* geometry. We shall see that once algebra has been introduced, we can get easy solutions of problems that the Greeks found difficult, and we can get difficult solutions of problems which the Greeks found impossible.

19.4 THE PROBLEM OF APOLLONIUS

Given three circles C_1, C_2, and C_3 in the plane. The problem of Apollonius is to construct, with ruler and compass, all possible circles C which are tangent to all three of the circles C_1, C_2, and C_3. One such C is shown in the figure.

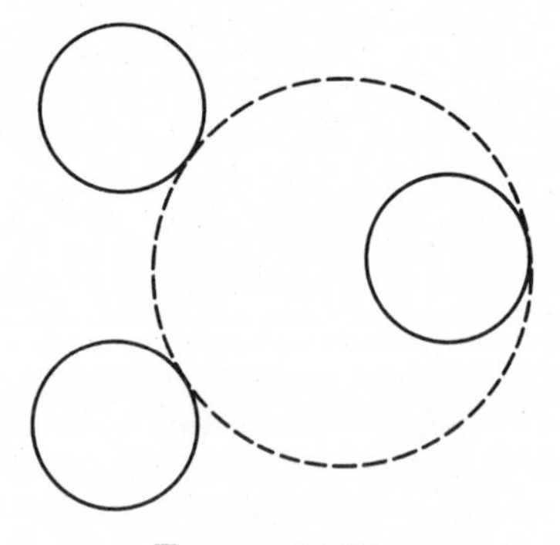

FIGURE 19.16

Using the methods of the preceding section, we shall show that all possible circles C are constructible with ruler and compass.

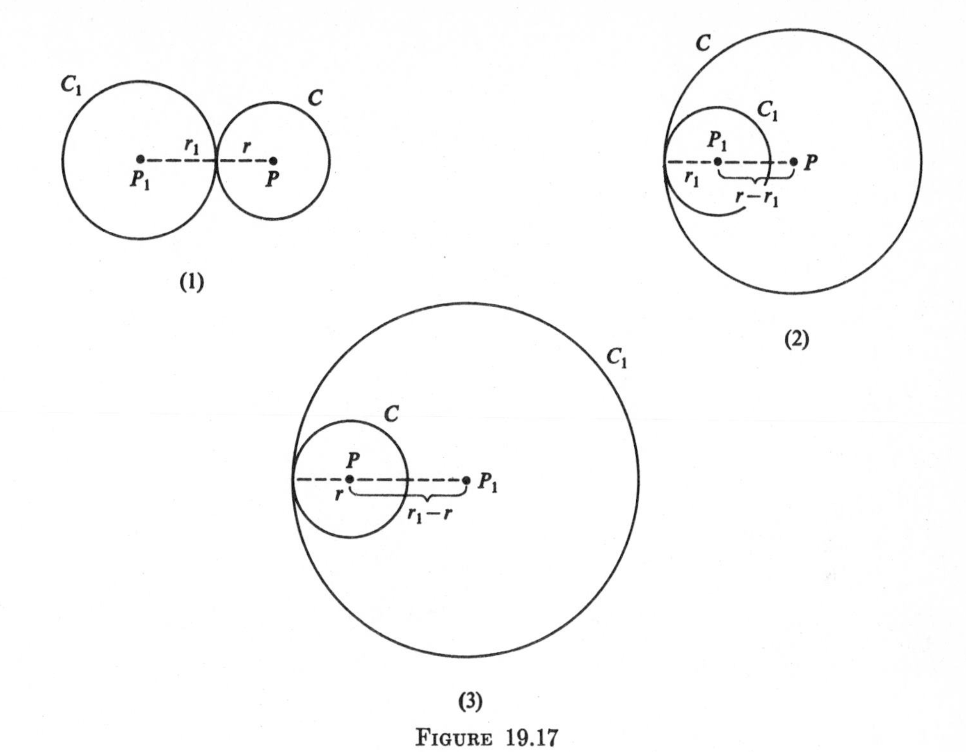

FIGURE 19.17

Given C_1, with center P_1 and radius r_1, and C, with center P and radius r, it is easy to find the conditions under which the two circles are tangent. If the circles are tangent and mutually exterior, then

$$PP_1 = r + r_1. \tag{1}$$

If they are tangent, and C_1 lies inside C, then

$$PP_1 = r - r_1. \tag{2}$$

Finally, if they are tangent, and C lies inside C_1, then

$$PP_1 = r_1 - r. \tag{3}$$

Conversely, each of these equations implies that the corresponding geometric condition holds. We can sum all this up by saying that *C and C_1 are tangent if and only if*

$$PP_1^2 = (r \pm r_1)^2.$$

To connect this up with our algebra, we let the centers be the points (a, b), (a_1, b_1). Our equations then take the form

$$(a - a_1)^2 + (b - b_1)^2 = (r \pm r_1)^2. \tag{4}$$

We can now restate the problem of Apollonius. C is tangent to all three of the circles C_1, C_2, C_3 if all three of the equations

$$(a - a_1)^2 + (b - b_1)^2 = (r \pm r_1)^2 \tag{4}$$

$$(a - a_2)^2 + (b - b_2)^2 = (r \pm r_2)^2 \tag{5}$$

$$(a - a_3)^2 + (b - b_3)^2 = (r \pm r_3)^2 \tag{6}$$

hold for some choice of $+$ or $-$ in each equation.

Here, for each of the eight possible choices of signs, we have a system of three equations in the three unknowns a, b, and r. Our problem is to solve these equations with ruler and compass, in the cases where solutions exist.

First we multiply out and collect terms, getting

$$a^2 + b^2 + r^2 - 2a_1a - 2b_1b \mp 2r_1r = r_1^2 - a_1^2 - b_1^2.$$

This has the form

$$a^2 + b^2 - r^2 + A_1a + B_1b + C_1r + D_1 = 0, \tag{7}$$

where the coefficients are numbers that can be plotted. Similarly we get

$$a^2 + b^2 - r^2 + A_2a + B_2b + C_2r + D_2 = 0, \tag{8}$$

$$a^2 + b^2 - r^2 + A_3a + B_3b + C_3r + D_3 = 0, \tag{9}$$

where all of the coefficients can be plotted. Subtracting (7), term by term, from (8) and (9), we get two equations of the form

$$E_2a + F_2b + G_2r + H_2 = 0, \tag{8'}$$

$$E_3a + F_3b + G_3r + H_3 = 0, \tag{9'}$$

where all of the coefficients can be plotted, being differences of plottable numbers.

We now propose to solve for a and b in terms of r. That is, we are going to regard $G_2r + H_2$ and $G_3r + H_3$ as constant terms. By the methods referred to in Problem 2 of the preceding section, we get

$$a = \frac{F_2(G_3r + H_3) - F_3(G_2r + H_2)}{E_2F_3 - F_2E_3},$$

$$b = \frac{F_3(G_2r + H_2) - F_2(G_3r + H_3)}{E_2F_3 - F_2E_3}.$$

Here a and b have the forms

$$a = J_1r + K_1, \qquad b = J_2r + K_2,$$

where all the coefficients can be plotted, because the previous coefficients were.

We now substitute in (7) these expressions for a and b. Equation (7) now becomes a quadratic equation in r alone, with coefficients all of which can be plotted:

$$(J_1r + K_1)^2 + (J_2r + K_2)^2 - r^2 + A_1(J_1r + K_1) \\ + B_1(J_2r + K_2) + C_1r + D_1 = 0.$$

We can solve such an equation with ruler and compass to plot the point r. We then plot a and b. This solves our problem.

Note that if any of these steps are algebraically impossible, this means that the geometric problem was impossible in the first place. This can easily happen. For example, if the three given circles are concentric, then no circle is tangent to all of them.

Various modifications of Apollonius' problem can be solved by the same method. Suppose, for example, that we want to construct all circles C which pass through a given point P_1 and are tangent to two given circles C_2 and C_3. To solve the

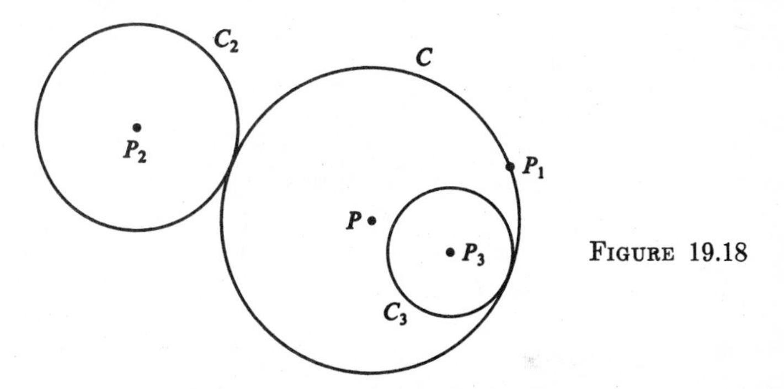

FIGURE 19.18

problem, we would simply set $r_1 = 0$, and then proceed exactly as before; similarly, if two points and a circle are given. You can even use the method to pass a circle through three given points, if you want to, but you surely don't want to. (Why?)

It should be understood that the sort of analysis that we have been going through does not lead to valid methods in mechanical drawing. In complicated construction procedures, there are so many steps that the cumulative error is likely to make the final result unrecognizable as a "right answer." To keep this cumulative error small, we need to do the algebra algebraically.

19.5 THE IMPOSSIBLE CONSTRUCTION PROBLEMS OF ANTIQUITY

The methods that we have been using were unknown to the Greeks. Some of the Greek mathematicians (notably Archimedes) were as good as any mathematicians who have ever lived. But some easily stated problems defeated them completely; and many centuries later the reasons for this became clear. It turned out that geometry, in the sense in which the Greeks understood it, is not a self-contained subject, and that some of its elementary problems require, for their solution, branches of mathematics that the Greeks did not discover.

Probably the most famous of these problems are the trisection of the angle and the duplication of the cube. Given an angle, we are asked to construct its trisectors with ruler and compass.

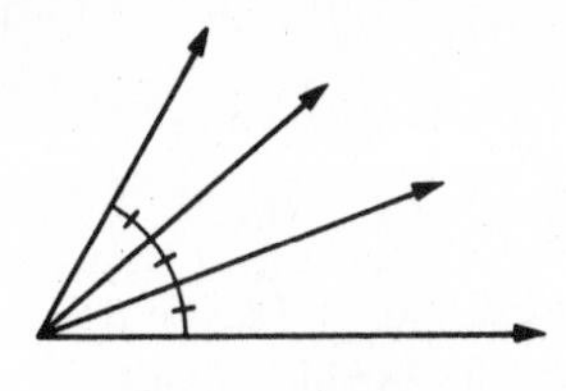

FIGURE 19.19

Given a segment of length a, it is required to construct a segment of length b, such that

$$b^3 = 2a^3.$$

If $a = 1$, then $b = \sqrt[3]{2}$. Thus our problem is to plot $\sqrt[3]{2}$ on the x-axis. The Greeks, of course, did not put it this way. In purely synthetic terms, given a segment $\overline{AB}$, we are to construct a segment $\overline{CD}$, such that a cube with $\overline{CD}$ as an edge has twice the volume of a cube with $\overline{AB}$ as an edge.

It turned out that both these problems are impossible: no such constructions exist. The rest of this chapter will be devoted to the proofs of these statements. In the following sections, it will seem that we are going far afield; we shall, because we have to. It was the need for going far afield that made the impossibility proofs so hard to discover.

19.6 THE SURD FIELD

A number x is called a *surd* if we can calculate x by a finite number of additions, subtractions, multiplications, divisions, and extractions of square roots, starting with 0 and 1.

For example, 2 is a surd, because $2 = 1 + 1$, and 1 is a surd *ex officio*. Given that n is a surd, it follows by addition that $n + 1$ is a surd, because the sum of two surds is a surd. By induction we conclude that every positive integer is a surd. By subtraction, every negative integer is also a surd. By division, we have the following:

Theorem 1. Every rational number is a surd.

And it is easy to show that:

Theorem 2. The surds form a Euclidean ordered field.

To check this, we note first that the associative, commutative, and distributive laws automatically hold for surds, since they hold for all real numbers. The same observation applies to the postulates for order. Since we are allowed to form surds from other surds by addition, subtraction, multiplication, and division, it follows that the set of surds contains sums, products, negatives, and reciprocals of all its elements. (Except, of course, that 0 has no reciprocal.) Therefore, the surds form an ordered field. Finally, since we are allowed to form surds by extracting square roots of positive surds, it follows that the surd field satisfies the Euclidean completeness postulate.

We denote the surd field by S.

Later we shall see that some numbers are not surds. Granted that there is one nonsurd, it follows that there are lots of others. For example, if x is not a surd, and p and q are nonzero integers, then $y = (p/q)x$ is not a surd. The reason is that if y were a surd, then x would be the product of the surds y and q/p. Therefore x would be a surd after all. In general, if a is a surd and x is not, and $a \neq 0$, then ax is not a surd.

We return now to the coordinate plane. By an S-point (S for *surd*), we mean a point both of whose coordinates are in S. By an S-line, we mean a line which contains at least two S-points. By an S-circle, we mean a circle whose center is an S-point and whose radius is in S. By an *S-equation* we mean an equation of the form $Ax + By + C = 0$ or $x^2 + y^2 + Dx + Ey + F = 0$, in which all of the coefficients are in S. These ideas are connected up by the following theorems.

Theorem 3. Every S-line is the graph of an S-equation.

Proof. Let L be a line containing the S-points (x_1, y_1), (x_2, y_2). If L is vertical, then L is the graph of the S-equation $x - x_1 = 0$. If L is not vertical, the slope of L is

$$m = \frac{y_2 - y_1}{x_2 - x_1}.$$

Here m is a surd, because the surds form a field; and L is the graph of the equation

$$y - y_1 = m(x - x_1),$$

or

$$mx - y - (-y_1 + mx_1) = 0,$$

which is an S-equation. The converse is also true.

Theorem 4. If a line L is the graph of an S-equation, then L is an S-line.

Proof. Suppose that L is the graph of the S-equation

$$Ax + By + C = 0.$$

If $B \neq 0$, then we can set $x = 0$ and $x = 1$, getting the S-points $(0, -(C/B))$, $(1, -(C + A)/B)$. If $B = 0$, then L contains the S-points $(-(C/A), 0)$ and $(-(C/A), 1)$.

Theorem 5. Every S-circle is the graph of an S-equation.

Proof. Let C be an S-circle with center (a, b) and radius r. Then C is the graph of

$$(x - a)^2 + (y - b)^2 = r^2,$$

or

$$x^2 + y^2 - 2ax - 2by + a^2 + b^2 - r^2 = 0,$$

which is an S-equation.

Theorem 6. If a circle C is the graph of an S-equation, then C is an S-circle.

Proof. Given that

$$x^2 + y^2 + Dx + Ey + F = 0$$

is an S-equation, and that its graph is a circle. We can then convert the equation to the form

$$\left(x + \frac{D}{2}\right)^2 + \left(y + \frac{E}{2}\right)^2 = \frac{D^2 + E^2 - 4F}{4}.$$

Therefore the center is the S-point $(-(D/2), -(E/2))$ and the radius is the surd

$$r = \tfrac{1}{2}\sqrt{D^2 + E^2 - 4F}.$$

This sort of elementary algebra is easier to write than to read. We therefore leave to you the verification of the following theorem.

Theorem 7. Let P be a point in the intersection of (1) two S-circles, (2) two S-lines, or (3) an S-circle and an S-line. Then P is an S-point.

The reason is roughly as follows. The coordinates x_1, y_1 of P are common solutions of two S-equations, each of which has the form

$$Ax + By + C = 0 \qquad \text{or} \qquad x^2 + y^2 + Dx + Ey + F = 0.$$

When we solve such a system by the usual algebraic methods, we find ourselves calculating x_1 and y_1 by starting with the coefficients and then performing the operations $+, \cdot, -, \div, \sqrt{\ }$. These operations keep us within the surd field at every stage. Therefore the final results x_1 and y_1 must be surds.

Problem Set 19.6

1. Let L be a nonvertical line. Show that if L contains an S-point, and the slope of L is a surd, then L is an S-line.

2. Let L be a nonvertical line. Show that if L is an S-line, then the slope of L is a surd.

3. Let C be a circle. Show that if the center of C is an S-point, and C contains an S-point, then C is an S-circle.

4. Let C be a circle. Show that if C contains three S-points, then C is an S-circle.

5. Show, conversely, that every S-circle contains three S-points.

6. Show that if x is a real number which is not a surd, and s is a surd different from 0, then xs is not a surd.

7. Assuming that S is not all of $\mathbf{R}$, show that every circle contains two non-S-points.

8. Assuming as before that $S \neq \mathbf{R}$, show that there is a circle which contains no S-points at all.

19.7 THE SURD PLANE

Let E be a coordinate plane. Let $\overline{E}$ be the set of all surd points in E. The set $\overline{E}$ will be called the *surd plane*. For each S-line L, let $\overline{L} = \overline{E} \cap L$.

The sets $\overline{L}$ are called *surd lines*. Similarly, if C is an S-circle, let

$$\overline{C} = \overline{E} \cap C.$$

The sets $\overline{C}$ are called *surd circles*.

As we have remarked before, it will turn out that not all real numbers are surds. Therefore $\overline{E}$ is full of holes. In fact, $\overline{E}$ does not contain all of any line or any circle.

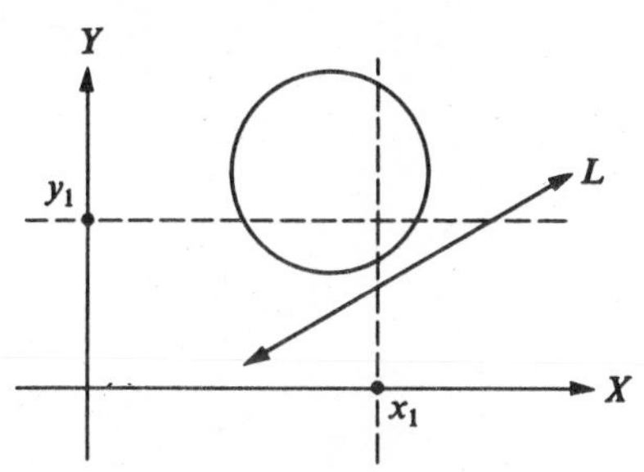

FIGURE 19.20

In fact, if x_1 and y_1 are nonsurds, then the lines $x = x_1$ and $y = y_1$ contain no points of $\overline{E}$ at all. The lines of this type cut the plane (in a manner of speaking) into pieces only one point wide.

On the other hand, if you were to investigate the surd plane by experimenting with ruler and compass, you could never tell that any points were missing. The only lines and circles you could draw would be S-lines and S-circles. If two S-lines, L_1 and L_2, intersect in a point, this point is in $\overline{E}$. (See Theorem 7, Section 19.6.) Therefore $\overline{L}_1$ intersects $\overline{L}_2$. If an S-line L intersects an S-circle C the intersection points are in $\overline{E}$ (same theorem). Therefore

$$\overline{L} \cap \overline{C} = L \cap C.$$

If two S-circles C_1, C_2 intersect, then so also do the surd circles $\overline{C}_1$, $\overline{C}_2$, in the same points.

Thus, although surd lines and surd circles are "full of holes," they "never pass through each other where the holes are"; they intersect each other everywhere we expect them to. We sum this up in the following theorem.

Theorem 7-1. Every ruler-and-compass construction which is possible in the plane is also possible in the surd plane.

In the light of this theorem, the questions of trisecting angles and duplicating cubes become questions of fact, as follows.

(1) Is it true that, in the surd plane, every angle has a trisector?

(2) Is $\sqrt[3]{2}$ a surd?

If the answer to (1) is "No," then there is no general method of trisecting angles with ruler and compass; the only points that we can construct are the surd points, and therefore we cannot construct a line that contains fewer than two surd points.

Similarly, if $\sqrt[3]{2}$ is not a surd, we cannot construct a segment whose length is $\sqrt[3]{2}$.

To answer questions (1) and (2), we need to do some algebra.

19.8 QUADRATIC EXTENSIONS OF FIELDS. CONJUGATES IN A QUADRATIC EXTENSION FIELD

Let F be a subfield of the real number system. Let k be a positive number belonging to F, and suppose that $\sqrt{k}$ does not belong to F. Let

$$F(k) = \{x + y\sqrt{k} | x, y \in F\}.$$

The set $F(k)$ is called a *quadratic extension* of F.

For example, if F is the field Q of rational numbers, then $2 \in F$ and $\sqrt{2} \notin F$. Therefore we can form the quadratic extension

$$F(k) = \mathrm{Q}(2) = \{x + y\sqrt{2} | x, y \in \mathrm{Q}\}.$$

We found, in Problem 6 of Problem Set 1.3, that these numbers form a field; and in fact this is what always happens, as we shall soon see.

Theorem 1. Let $F(k)$ be a quadratic extension of F. If $a, b \in F$, and

$$a + b\sqrt{k} = 0,$$

then $a = b = 0$.

Proof. Suppose that $b \neq 0$. Then $\sqrt{k} = -a/b$, and $\sqrt{k} \in F$, which is false. Therefore $b = 0$, and $a + b\sqrt{k} = a = 0$.

Theorem 2. Let $a, b, c, d \in F$. If $a + b\sqrt{k} = c + d\sqrt{k}$, then $a = c$ and $b = d$.

Proof by Theorem 1:

$$a + b\sqrt{k} = c + d\sqrt{k} \Rightarrow (a - c) + (b - d)\sqrt{k} = 0$$
$$\Rightarrow a = c \text{ and } b = d.$$

Thus every element of $F(k)$ can be expressed in only one way as a linear combination $a + b\sqrt{k}$.

Theorem 3. Every quadratic extension of a field forms a field.

Proof. Let F be a subfield of the real numbers, and let $F(k)$ be a quadratic extension of F. The associative, commutative, and distributive laws hold automatically in $F(k)$, because they hold for all real numbers. It is also easy to see that the numbers of the form $x + y\sqrt{k}$ $(x, y \in F)$ are closed under addition and multiplication, that 0 is among them $(= 0 + 0\sqrt{k})$, and that $-(x + y\sqrt{k})$ is always a number of the same form. It remains only to verify that if

$$x + y\sqrt{k} \neq 0$$

then

$$\frac{1}{x + y\sqrt{k}} \in F(k).$$

Of course we know that the reciprocal *exists*, because the real numbers form a field; the question is whether the reciprocal belongs to $F(k)$.

Lemma. If $x + y\sqrt{k} \neq 0$, then $x - y\sqrt{k} \neq 0$.

Proof. Here x and y cannot both be 0, because $x + y\sqrt{k}$ would then be 0. Therefore neither x nor y is 0. Therefore $x - y\sqrt{k} \neq 0$, by Theorem 1.

By the lemma, we can write

$$\begin{aligned}\frac{1}{x + y\sqrt{k}} &= \frac{1}{x + y\sqrt{k}} \cdot \frac{x - y\sqrt{k}}{x - y\sqrt{k}} \\ &= \frac{x - y\sqrt{k}}{x^2 - ky^2} \\ &= \frac{x}{x^2 - ky^2} + \frac{-y}{x^2 - ky^2}\sqrt{k},\end{aligned}$$

which belongs to $F(k)$.

If $z = x + y\sqrt{k} \in F(k)$, then the *conjugate* $\bar{z}$ of z is defined by the equation

$$\bar{z} = x - y\sqrt{k}.$$

(Note that $\bar{z}$ is determined if z is known: by Theorem 2, z determines x and y, and therefore z determines $\bar{z} = x - y\sqrt{k}$.) The operation of conjugation in a quadratic extension field is closely analogous to the corresponding operation for the complex numbers. It may be worthwhile to review this, to bring out the analogy.

Given a complex number

$$z = x + yi,$$

we define

$$\bar{z} = x - yi.$$

The basic properties of the operation $z \mapsto \bar{z}$ are given in the following theorems.

Theorem A. The conjugate of the sum is the sum of the conjugates. That is, if

$$z_1 = x_1 + y_1 i, \qquad z_2 = x_2 + y_2 i,$$

then

$$\overline{z_1 + z_2} = \bar{z}_1 + \bar{z}_2.$$

This is trivial to check.

Theorem B. The conjugate of the product is the product of the conjugates. That is,

$$\overline{z_1 z_2} = \bar{z}_1 \bar{z}_2.$$

Verification.

$$\begin{aligned}z_1 z_2 &= (x_1 + y_1 i)(x_2 + y_2 i) = x_1 x_2 - y_1 y_2 + (x_1 y_2 + x_2 y_1)i, \\ \bar{z}_1 \bar{z}_2 &= (x_1 - y_1 i)(x_2 - y_2 i) = x_1 x_2 - y_1 y_2 - (x_1 y_2 + x_2 y_1)i.\end{aligned}$$

Obviously $\overline{z_1 z_2} = \bar{z}_1 \bar{z}_2$.

By induction we get the following theorem.

Theorem C. $\overline{z^n} = \bar{z}^n$.

Theorem D. If a is a real number, then $\bar{a} = a$.

By a *polynomial* of degree $n > 0$, we mean (as usual) a function f defined by an equation

$$f(z) = a_n z^n + a_{n-1} z^{n-1} + \cdots + a_1 z + a_0,$$

where $a_n \neq 0$. We allow also the "zero polynomial" which is 0 for every z.

Theorem E. If f is a polynomial with all coefficients real, then $f(\bar{z}) = \overline{f(z)}$ for every z.

Proof. Let $f(z)$ be as in the definition of a polynomial. Then

$$\begin{aligned} f(\bar{z}) &= a_n \bar{z}^n + a_{n-1} \bar{z}^{n-1} + \cdots + a_1 \bar{z} + a_0 \\ &= \bar{a}_n \bar{z}^n + \bar{a}_{n-1} \bar{z}^{n-1} + \cdots + \bar{a}_1 \bar{z} + \bar{a}_0, \end{aligned}$$

by Theorem D. This is

$$= \bar{a}_n \overline{z^n} + \bar{a}_{n-1} \overline{z^{n-1}} + \cdots + \bar{a}_1 \bar{z} + \bar{a}_0,$$

by Theorem C. This is

$$= \overline{a_n z^n} + \overline{a_{n-1} z^{n-1}} + \cdots + \overline{a_1 z} + \bar{a}_0,$$

by Theorem B. And this is $= \overline{f(z)}$, by repeated applications of Theorem A.

Theorem F. Let f be a polynomial with all coefficients real. If $f(z_0) = 0$, then

$$f(\bar{z}_0) = 0.$$

That is, if the coefficients are real, then the roots of the equation occur in conjugate pairs z_0, $\bar{z}_0$.

The proof is trivial, because if $f(z_0) = 0$ we have

$$f(\bar{z}_0) = \overline{f(z_0)} = \bar{0} = 0.$$

For quadratic extension fields $F(k)$, we have a similar sequence of theorems in which F acts like the field of real numbers, $F(k)$ acts like the field of complex numbers, and $\sqrt{k}$ acts like i. We shall merely restate the theorems and leave to the reader the easy task of verifying that the same proofs work in the same way. (The equations in the proof that $\overline{z_1 z_2} = \bar{z}_1 \bar{z}_2$ take a slightly different form.) Throughout these theorems it should be understood that $F(k)$ is a quadratic extension of F, and that conjugates are defined by

$$\overline{x + y\sqrt{k}} = x - y\sqrt{k}.$$

Theorem 4. In $F(k)$, the conjugate of the sum is the sum of the conjugates.

Theorem 5. The conjugate of the product is the product of the conjugates.

Theorem 6. $\overline{z^n} = \bar{z}^n$, for every $z \in F(k)$.

Theorem 7. If $a \in F$, then $\bar{a} = a$.

Theorem 8. If f is a polynomial with all coefficients in F, then $f(\bar{z}) = \overline{f(z)}$ for every z in $F(k)$.

Theorem 9. Let f be a polynomial with all coefficients in F. If $z_0 \in F(k)$, and

$$f(z_0) = 0,$$

then

$$f(\bar{z}_0) = 0.$$

Thus, for polynomial equations with coefficients in F, the roots in $F(k)$ occur in conjugate pairs.

Theorems F and 9 describe a phenomenon which is familiar from elementary work with quadratic equations. The roots of the equation

$$ax^2 + bx + c = 0$$

are given by the formula

$$x = \frac{-b \pm \sqrt{b^2 - 4ac}}{2a}.$$

If $b^2 - 4ac = -d < 0$, then the roots are the complex numbers

$$\frac{-b}{2a} \pm \frac{\sqrt{d}}{2a} i$$

which are conjugate, as predicted by Theorem F.

Suppose now that a, b, and c are rational, and $b^2 - 4ac = e > 0$. The roots are then the real numbers

$$\frac{-b}{2a} \pm \frac{1}{2a}\sqrt{e}.$$

These are conjugate elements of the quadratic extension field $F(k) = \mathbf{Q}(e)$, as predicted by Theorem 9.

19.9 SURD FIELDS OF ORDER n; SURDS OF ORDER n

Suppose that we have an ascending finite sequence of fields, starting with the rationals and proceeding by a quadratic extension at every step. Thus our fields are

$$F_0, F_1, \ldots, F_n,$$

where

$$F_0 = \mathbf{Q}$$

and

$$F_{i+1} = F_i(k_{i+1}).$$

Here, for each i, k_{i+1} is in F_i, but $\sqrt{k_{i+1}}$ is not. In this case we say that F_n is a *surd field of order n*.

Theorem 1. All elements of F_n are surds.

Because they are obtainable from rational numbers by a finite number of operation $+, \cdot, -, \div, \sqrt{\ }$. (For a formal proof, we would use induction: all elements of F_0 are surds; and if F_i has this property, so does F_{i+1}.) The converse is also true:

Theorem 2. Every surd belongs to a surd field of some order.

Proof. Every surd x can be built from rational numbers by a finite number of operations $+, \cdot, -, \div, \sqrt{\ }$. If n root-extractions are needed in this process, we say that x is a *surd of order n*. Thus every rational number is a surd of order 0; $\sqrt{2}$ is a surd of order 1, and so on.

Given that x is a surd of order n, let

$$k_1, k_2, \ldots, k_n$$

be the numbers whose roots we extracted in forming x, in the order in which these roots were extracted. Let

$$F_0 = \mathbf{Q}, \qquad F_1 = F_0(k_1),$$

and in general

$$F_{i+1} = F_i(k_{i+1}).$$

Between the ith root-extraction and the $(i + 1)$st, we may have used $+$, $\cdot$, $-$, and $\div$, but these operations can all be performed in F_i. Therefore all the numbers formed in the intermediate stages are in the fields $F_0, \ldots, F_n$; and in particular, $x \in F_n$, which was to be proved.

Note that all of the indicated quadratic extensions are genuine; we have $\sqrt{k_{i+1}} \notin F_i$, because if $\sqrt{k_{i+1}} \in F_i$, we could reduce the number of root extractions used in forming x, and x would not be a surd of order n after all.

Note also what Theorem 2 does *not* say: it does *not* say that there is one particular F_n which contains all surds of order n. In fact, the latter statement is not true; in general, F_n depends not merely on the order n of x, but also on x. For example, $\sqrt{2}$ and $\sqrt{3}$ are surds of order 1, but no one field F_1 contains both of them, because $\sqrt{2} + \sqrt{3}$ is a surd of order 2.

Problem Set 19.9

1. Show that if nm is even and n is odd, then m is even.
2. Show that $\sqrt{\frac{3}{2}}$ is irrational.
3. Now prove the statement made in the last sentence of this section. That is, prove that no quadratic extension $\mathbf{Q}(k)$ of the rational numbers contains both $\sqrt{2}$ and $\sqrt{3}$.

19.10 APPLICATIONS TO CUBIC EQUATIONS WITH RATIONAL COEFFICIENTS

Given a cubic equation

$$f(z) = z^3 + a_2 z^2 + a_1 z + a_0 = 0,$$

where the coefficients are real numbers. We recall from the theory of equations that the polynomial on the left always has a factorization of the form

$$g(z) = (z - z_1)(z - z_2)(z - z_3) = 0,$$

where z_1, z_2, and z_3 are the roots. They may not all be different, and two of them may be complex. Why is it impossible for exactly one of them to be complex?

(For a full review of the algebraic background of this section, including proofs of all theorems cited here without proof, see Chapter 30 at the end of the book.)

This gives

$$g(z) = z^3 - (z_1 + z_2 + z_3)z^2 + (z_1z_2 + z_1z_3 + z_2z_3)z - z_1z_2z_3 = 0.$$

Now $f(z) = g(z)$ for every z. The only way this can happen, for polynomials, is for the corresponding coefficients to be equal. Therefore, in particular, we have

$$-(z_1 + z_2 + z_3) = a_2.$$

Suppose that a, b, and c are in a subfield F of the real numbers and that the equation

$$z^3 + az^2 + bz + c = 0$$

has a root z_1, in a quadratic extension $F(k)$ of F. Then $\bar{z}_1$ is also a root, by Theorem 9 of the preceding section. If the third root is z_3, then we have

$$-(z_1 + \bar{z}_1 + z_3) = a,$$

or

$$z_3 = -(z_1 + \bar{z}_1 + a).$$

Now $z_1 + \bar{z}_1 \in F$. (Why?) Therefore z_3 is in F. Thus we have the following theorem.

Theorem 1. Given an equation,

$$z^3 + az^2 + bz + c = 0,$$

with coefficients in a field F. If the equation has a root lying in a quadratic extension of F, then the equation has a root lying in F.

If the coefficients are rational, we can draw a stronger conclusion.

Theorem 2. Given a cubic equation,

$$z^3 + az^2 + bz + c = 0,$$

where the coefficients are rational. If the equation has a root in the surd field, then it also has a rational root.

Proof. Suppose that a surd z_1, of order n, is a root. Then we have a sequence,

$$F_0, F_1, F_2, \ldots, F_n,$$

of quadratic extensions, as in the preceding section, with

$$F_0 = \mathbf{Q},$$

$$F_{i+1} = F_i(k_{i+1}),$$

for every i, and $z_1 \in F_n$.

Now $a, b, c \in \mathbf{Q}$, so that our coefficients are in F_{n-1}. By Theorem 1, our equation has a root in F_{n-1}. Repeating the same reasoning another $n - 1$ times, we conclude that our equation has a root in F_0. Since $F_0 = \mathbf{Q}$, this proves the theorem.

19.11 THE TRISECTION OF THE ANGLE

Some angles can be trisected with ruler and compass. For example, a right angle can be trisected:

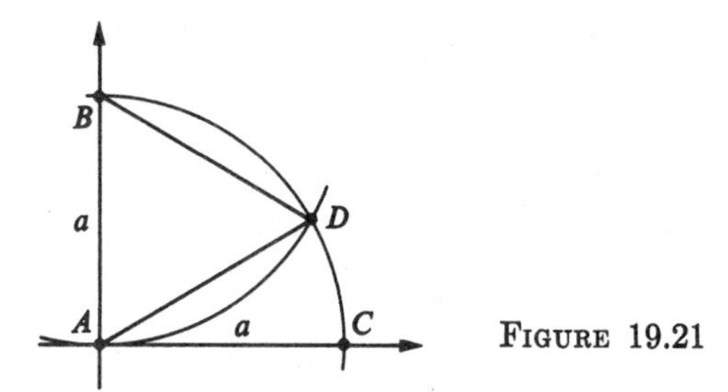

FIGURE 19.21

Given a right angle $\angle A$. Draw a circle with center at A, with any radius a, intersecting the sides of $\angle A$ in points B and C. Draw the circle with center at B, containing A. The two circles will intersect in two points, one of which will be a point D in the interior of $\angle A$. (Both points of intersection will lie on the same side of $\overleftrightarrow{AC}$ as B; and exactly one of them will lie on the same side of $\overleftrightarrow{AB}$ as C.) Now $\triangle ABD$ is equilateral, and hence equiangular. Therefore $m\angle BAD = 60$. Therefore $m\angle DAC = 90 - 60 = 30 = \frac{1}{3} \cdot 90$, and $\angle DAC$ is a trisector of $\angle BAC$. To get the other trisector, we draw the circle with center at C, containing A, and join A to the point where this circle intersects our first circle in the interior of $\angle BAC$.

For some angles—in particular, for angles of 60°—no such construction is possible. The proof is as follows.

The surd plane contains an angle of 60° because the surd plane contains an equilateral triangle. Now any angle can be "copied" with ruler and compass. (See Construction 7 of Section 17.1.) It follows that there is a 60° angle $\angle BAC$, with $\overrightarrow{AC}$ as the positive x-axis and B in the upper half plane, as in the figure below.

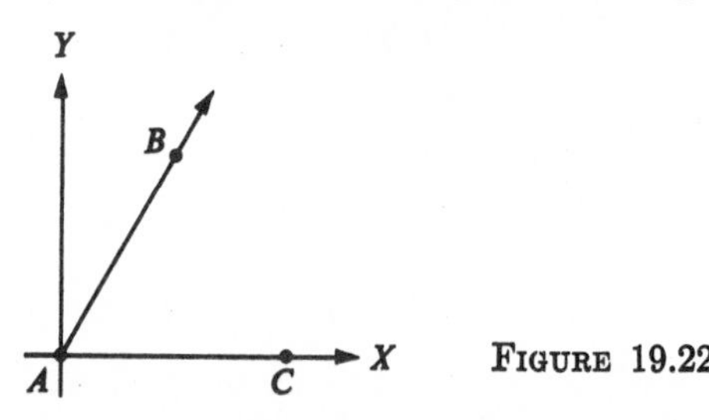

FIGURE 19.22

We have proved that the surd plane is a ruler-and-compass geometry. Surd lines and surd circles always intersect each other in the same way as the corresponding lines and circles in the complete plane. If there is a method of trisecting every angle with ruler and compass, then this general method must apply to $\angle BAC$; and the construction can be carried out in the surd plane. Thus we arrive at the following conclusions.

(1) (?) *The surd plane contains an angle* $\angle DAC$, *with degree measure* $= 20$. (?)

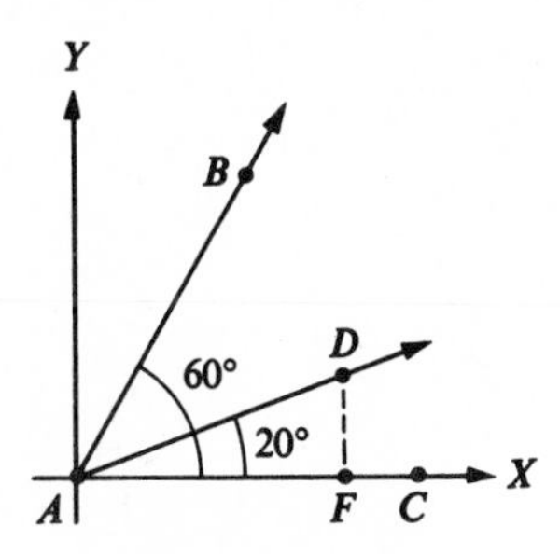

FIGURE 19.23

We shall show that this is impossible. It will follow that there is no general method for trisecting angles with ruler and compass. We have given that D is a point of the surd plane. Let F be the foot of the perpendicular from D to the x-axis. Then F is a surd point; the y-coordinate of F is 0, and the x-coordinate is the same as that of D. The distance between any two surd points is a surd. Therefore the number

$$y = \frac{AF}{AD}$$

is a surd. But $y = \cos 20°$. Thus we conclude that

(2) (?) $\cos 20°$ *is a surd.* (?)

We now need to do a little trigonometry, as follows:

$$\begin{aligned}\cos 3\theta &= \cos (2\theta + \theta)\\ &= \cos 2\theta \cdot \cos \theta - \sin 2\theta \cdot \sin \theta\\ &= (\cos^2 \theta - \sin^2 \theta) \cos \theta - 2 \sin \theta \cos \theta \sin \theta\\ &= (2 \cos^2 \theta - 1) \cos \theta - 2(1 - \cos^2 \theta) \cos \theta\\ &= 4 \cos^3 \theta - 3 \cos \theta.\end{aligned}$$

Setting $\theta = 20$, and recalling that $\cos 60° = \frac{1}{2}$, we get

$$\tfrac{1}{2} = 4 \cos^3 20° - 3 \cos 20°, \qquad \text{or} \qquad 8 \cos^3 20° - 6 \cos 20° - 1 = 0.$$

Thus we have

(3) $\cos 20°$ *is a root of the equation*

$$8y^3 - 6y - 1 = 0.$$

Here we omit the question marks, because (3), unlike (1) and (2), is actually correct; it is not merely a statement whose consequences we propose to investigate.

Setting $x = 2y$, we conclude that

(4) *the number* 2 cos 20° *is a root of the equation*

$$x^3 - 3x - 1 = 0.$$

We shall prove that no surd is a root of this cubic equation. This will mean that 2 cos 20° is not a surd, so that cos 20° is not a surd. This will mean that (2) is impossible, so that a ruler-and-compass trisection of a 60° angle is impossible.

It is easy to see that our cubic has no rational root. If p/q is a rational root, expressed in lowest terms, then each of the integers p and q is a divisor of 1 and hence is $=1$ or $= -1$. Therefore 1 and -1 are the only possible rational roots, and neither of them works. Therefore our cubic has no rational roots. By Theorem 2, Section 19.10, it follows that no root of the equation is a surd.

The impossibility of this classical problem is surprising to most people. In fact, many people refuse to believe it and go on trying to devise a method. It is easy to see why this happens.

In the first place, the proof is much too hard to be capable of popularization. For this reason, amateurs are in no position to understand why their enterprise is impossible. In the second place, an informal statement of the problem is misleading. It sounds as if we are saying that something or other "cannot be done"; and many times in the past, defeatist answers to questions like this have turned out to be false. For example, people said that flying machines could not be built and that matter could not be created or destroyed. People went on saying these things until the supposedly impossible projects were carried out. If you think of the trisection problem in these terms, then you may believe that you can do it, if you are more ingenious and persistent than the people who have tried and failed.

But once the problem is given an exact formulation, the negative answer seems natural. In the surd plane, all ruler-and-compass constructions are possible. Therefore, if a figure can be constructed with ruler and compass, this means that in the surd plane, the figure *exists*. Thus, if angle trisection were always possible, we would have the following theorem.

(?) **Theorem.** In the surd plane, every angle has a trisector.

If you think of the problem in these terms, then the negative answer is not surprising. The proposed theorem describes a completeness property of the surd plane; it says that when you look for rays in certain places, you will find them. But since the surd plane is "all full of holes," no kind of completeness condition has any plausibility until it is proved. Only the most incorrigible optimist would be surprised if he looked for a ray, in such a "geometry" as this, and failed to find one.

19.12 THE DUPLICATION OF THE CUBE

Another of the impossible construction problems of antiquity is the duplication of the cube. Given a segment $\overline{AB}$, we want to construct a segment $\overline{CD}$ such that a cube with $\overline{CD}$ as an edge has exactly twice the volume of a cube with $\overline{AB}$ as an edge.

As before, if this construction is possible, it is possible in the surd plane. Suppose then that A and B are surd points. Granted that $\overline{CD}$ can be constructed, C and D must be surd points. Thus the distances AB and CD are surds; and under the conditions of the problem we must have

$$CD^3 = 2AB^3,$$

or

$$\left(\frac{CD}{AB}\right)^3 = 2.$$

It follows that

(1) (?) *The equation* $x^3 - 2 = 0$ *has at least one surd as a root.* (?)

But this is impossible. The only possible rational roots of this cubic are 1, -1, 2, and -2. None of these work. Therefore no rational number is a root. Therefore no surd is a root.

Chapter 20
From Eudoxus to Dedekind

20.1 PROPORTIONALITIES WITHOUT NUMBERS

In the early chapters of this book we have distinguished between two different approaches to the concept of congruence for segments. In the metric approach, a distance function

$$d: S \times S \to \mathbf{R}$$

is given. Congruence for segments is defined in terms of distance; the definition states that $\overline{AB} \cong \overline{CD}$ if $AB = CD$. The properties of congruence now become theorems, proved on the basis of the metric definition.

The synthetic approach takes congruence for segments as a basic idea, left undefined, and governed by certain postulates. In this treatment, the idea of distance does not appear at all; in fact, the only numbers that appear are the natural numbers 1, 2, ... We may think of congruence for segments as the idea of "same distance."

The treatment of similarity, in this book, has been strictly metric. We recall that a correspondence

$$ABC \leftrightarrow DEF$$

is called a *similarity* if corresponding angles are congruent and the lengths of corresponding sides are proportional.

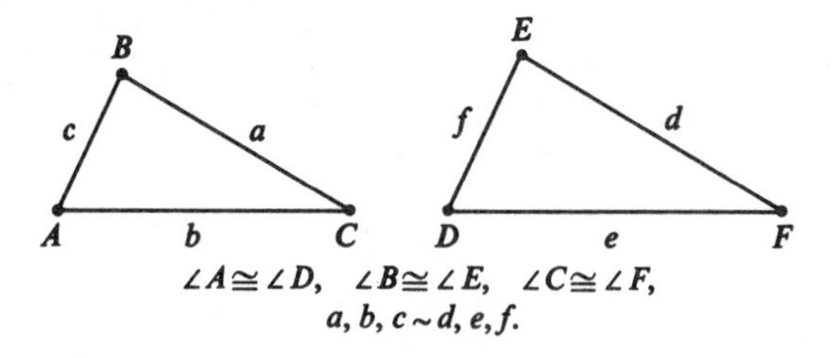

FIGURE 20.1

It is easy to treat angle congruence in a purely synthetic way; we have discussed this in Chapter 8. But proportionality seems to be another matter. We defined the expression $a, b, c \sim d, e, f$ to mean that

$$\frac{d}{a} = \frac{e}{b} = \frac{f}{c}.$$

Here the indicated divisions make sense, because a, b, c, d, e, and f are positive real numbers. We say that the *sides* of $\triangle ABC$ and $\triangle DEF$ are proportional if the *lengths* of these sides are proportional in the sense that we have just defined.

This treatment of similarity is by now very nearly universal, even in books which use a synthetic approach insofar as practicality permits. In nearly every elementary book on "synthetic" geometry there is a page on which the idea of distance is introduced, in order to permit a metric treatment of proportionality.

Indeed, if you are not allowed to measure distances and perform divisions, it is not easy to see how you could even explain what is *meant* by proportionality for segments, let alone prove anything about how proportionalities work. Nevertheless, this can be done, using no numbers at all except the positive integers. The method is used in Euclid's *Elements*; and the mathematical ideas that make it work are attributed to Eudoxus. This suggests two questions.

(1) What were the purely synthetic ideas that Eudoxus used as a substitute for algebra?

(2) Given a synthetic geometry, how can we define a distance function satisfying the ruler postulate?

The first of these questions may seem to be of purely historical interest. But this is hardly true of the second: we need to answer it, to introduce coordinate systems. Moreover, it turns out that the two questions are so closely related that if you answer one of them, the other becomes easy.

And the ideas of Eudoxus took on a new importance in the nineteenth century, when Richard Dedekind found that they were what was needed in setting up the foundations of the real number system. For these reasons, the purposes of the present chapter are only incidentally historical.

20.2 EUDOXUS' SYNTHETIC DEFINITION OF PROPORTIONALITY

We shall work our way gradually toward Eudoxus' idea, starting with the metric concept which we know, and gradually stripping it of its algebraic apparatus. To start, we define the expression

$$\overline{AB} : \overline{CD} :: \overline{EF} : \overline{GH} \tag{1}$$

to mean that

$$AB, CD \sim EF, GH; \tag{2}$$

that is,

$$\frac{EF}{AB} = \frac{GH}{CD}. \tag{3}$$

The first of these expressions is pronounced "$\overline{AB}$ is to $\overline{CD}$ as $\overline{EF}$ is to $\overline{GH}$." We shall find a way to define this expression without mentioning any numbers except positive integers. Of course, it can easily happen that the segments $\overline{AB}$ and $\overline{EF}$ are incommensurable. In this case, the proportionality constant

$$x = \frac{EF}{AB} = \frac{GH}{CD}$$

will be irrational. Our first step toward Euclid will be to express Condition (3) in the form of a statement about *rational* numbers. In the light of the comparison theorem, this can be done as follows.

(4) If p/q is rational and

$$\frac{p}{q} < \frac{EF}{AB},$$

then

$$\frac{p}{q} < \frac{GH}{CD}.$$

Conversely, if the second of these inequalities holds, then so does the first. This in turn can be expressed without mentioning division.

(5) If p and q are positive integers, and

$$p \cdot AB < q \cdot EF,$$

then

$$p \cdot CD < q \cdot GH.$$

And conversely, if the second of these inequalities holds, then so does the first.

We are now almost done, because (5) is very close to being a statement about segments. We recall that addition can be defined for *congruence classes* of segments. For each segment $\overline{XY}$, we let $[\overline{XY}]$ be the set of all segments that are congruent to $\overline{XY}$. Given any two segments $\overline{AB}$ and $\overline{CD}$, we take points X, Y, Z such that

$$X\text{-}Y\text{-}Z, \qquad \overline{AB} \cong \overline{XY}, \qquad \overline{CD} \cong \overline{YZ};$$

and we then define the sum $[\overline{AB}] + [\overline{CD}]$ by the formula

$$[\overline{AB}] + [\overline{CD}] = [\overline{XZ}].$$

We showed, in Chapter 8, that the sum of the two congruence classes depends only on the congruence classes $[\overline{AB}]$ and $[\overline{CD}]$, and is independent of the choice of A, B, C, D, X, Y, and Z. When $\overline{AB} \cong \overline{CD}$, we write

$$[\overline{AB}] + [\overline{AB}] = 2 \cdot [\overline{AB}].$$

And for any positive integer n, we use the shorthand

$$n[\overline{AB}] = [\overline{AB}] + [\overline{AB}] + \cdots + [\overline{AB}] \quad \text{(to } n \text{ terms)}.$$

Thus, if we write

$$n[\overline{AB}] = [\overline{XZ}],$$

this means that if you take n congruent copies of $\overline{AB}$ and lay them end to end, you get a segment congruent to $\overline{XZ}$.

In Chapter 7, we defined the expression

$$\overline{AB} < \overline{CD}$$

to mean that there is a point B', between C and D, such that $\overline{AB} \cong \overline{CB'}$. This, of course, is a synthetic way of conveying the idea that $AB < CD$.

We are now ready to give Eudoxus' formulation of (5). It reads as follows.

(6) Let p and q be any positive integers. If

$$p[\overline{AB}] < q[\overline{EF}],$$

then

$$p[\overline{CD}] < q[\overline{GH}].$$

And conversely, if the second of these inequalities holds, then so does the first.

This was Eudoxus' working definition of the statement

$$\overline{AB} : \overline{CD} :: \overline{EF} : \overline{GH}.$$

Euclid used it, throughout the *Elements*, whenever he dealt with proportionality (except in the commensurable case). This was an extraordinary *tour de force*, because even the simplest theorems under this scheme become formidable even to state. Every proportionality that we write down becomes a complicated statement about what happens when congruent segments are laid end to end.

For example, Proposition 4 of Book V of the *Elements* reads as follows.

Propositon 4. "If a first magnitude has to a second the same ratio as a third to a fourth, any equimultiples whatever of the first and third will also have the same ratio to any equimultiple whatever of the second and fourth respectively, taken in corresponding order."

We can rewrite this in the following form.

Proposition 4′. If $\overline{AB} : \overline{CD} :: \overline{EF} : \overline{GH}$, and m and n are any positive integers, then

$$m\overline{AB} : m\overline{EF} :: n\overline{CD} : n\overline{GH}.$$

Here $m\overline{AB}$ denotes any segment in the class $m[\overline{AB}]$, and so on.

The corresponding algebraic theorem is simple. It says that if

$$AB, CD \sim EF, GH,$$

and m and n are any positive integers, then

$$mAB, mEF \sim nCD, nGH.$$

Writing these proportionalities as equations between fractions, we get the following proposition.

Proposition 4″. If

$$\frac{EF}{AB} = \frac{GH}{CD},$$

and m and n are positive integers, then

$$\frac{nCD}{mAB} = \frac{nGH}{mEF}.$$

The first of the equations means that

$$EF \cdot CD = AB \cdot GH;$$

and the second means that

$$mnEF \cdot CD = mnAB \cdot GH.$$

Thus the "proof" doesn't amount to much. All the theorems in Book V of the *Elements* evaporate in the same way, as soon as they are interpreted algebraically. This, however, does not mean that Euclid was being silly. The point is that in Euclid's time the algebra of the real numbers had not yet been discovered. Getting along without it in the study of geometry was a formidable achievement.

We observed, when we were regarding congruence synthetically, that we could not talk about distance; congruence involves only the idea of same distance. Much the same thing happens if we use a purely synthetic treatment of proportionality. We cannot speak of the ratio $\overline{AB} : \overline{CD}$. All we can say is that $\overline{AB} : \overline{CD}$ and $\overline{EF} : \overline{GH}$ are the *same ratio*. The easiest way to see this is to observe that the ratio $\overline{AB} : \overline{CD}$ would have to be a real number; and in synthetic geometry the only numbers are the positive integers.

Finally, we should confess that in giving what are supposed to be Euclid's formulations of certain ideas, we have made no attempt to copy his literary style. In the Heath translation, his definition of proportionality is as follows.

"Magnitudes are said to be *in the same ratio*, the first to the second and the third to the fourth, when, if any equimultiples whatever be taken of the first and third, and any equimultiples whatever of the second and fourth, the former equimultiples alike exceed, are alike equal to, or alike fall short of, the latter equimultiples respectively taken in corresponding order."

This is the statement which we gave above, in a rewritten form, as Condition (6).

Our account of Book V has also been simplified in other ways. For one thing, Euclid also gave a synthetic formulation of the statement that the ratio $\overline{AB} : \overline{CD}$ is less than the ratio $\overline{EF} : \overline{GH}$. For another thing, he did not use, even tacitly, the Archimedean postulate. He merely provided that segments had to behave in an Archimedean fashion for one to be able to talk about proportionalities between them. That is, to write

$$\overline{AB} : \overline{CD} :: \overline{EF} : \overline{GH},$$

you must first know that

(1) $p[\overline{AB}] > [\overline{CD}]$ for some p,
(2) $q[\overline{CD}] > [\overline{AB}]$ for some q,
(3) $r[\overline{EF}] > [\overline{GH}]$ for some r, and
(4) $s[\overline{GH}] > [\overline{EF}]$ for some s.

At no point did Euclid commit himself on the question whether these conditions held for any two given segments; he merely announced that he was not going to talk about ratios except in the cases where they did hold. The resulting treatment was delicate in the extreme. The fact that such a program was carried out, with only occasional slips in matters of detail, reminds us that the Greeks were no more primitive in mathematics than they were in the arts.

PROBLEM SET 20.2

The following is a selection of propositions from Book V of the *Elements*. Interpret each of these propositions algebraically, and prove the resulting theorem. This problem set is designed to convince you that the language of mathematics has progressed in important ways.

Proposition 5. "If a magnitude be the same multiple of a magnitude that a part subtracted is of a part subtracted, the remainder will also be the same multiple of the remainder that the whole is of the whole."

Proposition 6. "If two magnitudes be equimultiples of two magnitudes, and any magnitudes subtracted from them be equimultiples of the same, the remainders also are either equal to the same or equimultiples of them."

Proposition 9. "Magnitudes which have the same ratio to the same are equal to one another; and magnitudes to which the same has the same ratio are equal."

Proposition 15. "Parts have the same ratio as the same multiples of them taken in corresponding order."

Proposition 19. "If, as a whole is to a whole, so is a part subtracted to a part subtracted, the remainder will also be to the remainder as whole to whole."

Proposition 24. "If a first magnitude has to a second the same ratio as a third has to a fourth, and also a fifth has to the second the same ratio as a sixth to the fourth, the first and fifth added together will have to the second the same ratio as the third and sixth have to the fourth."

Proposition 25. "If four magnitudes be proportional, the greatest and the least are greater than the remaining two."

(This last one is very ambiguous. You should try to find an interpretation that makes it true.)

20.3 THE ALGEBRA OF SEGMENT ADDITION

We shall not give a full development of the Euclidean theory of proportion, but we shall need to know some of the simplest facts about what happens when we lay segments end to end.

Theorem 1. *The Commutative Law.*

$$[\overline{AB}] + [\overline{CD}] = [\overline{CD}] + [\overline{AB}].$$

This follows from the definition of segment addition. The order in which the segments were named never even seemed to matter; the same is true of the next theorem.

Theorem 2. *The Associative Law.*

$$([\overline{AB}] + [\overline{CD}]) + [\overline{EF}] = [\overline{AB}] + ([\overline{CD}] + [\overline{EF}]).$$

It follows, as in Section 1.10, that there is an n-fold addition for congruence classes $[\overline{AB}]$, and that the n-fold operation satisfies the general associative law. (The results of Section 1.10 were stated for addition and multiplication of real numbers, but special properties of real numbers were not used. For any associative operation, defined on any set, the same results hold, and the proofs are exactly the same.)

Theorem 3. *The Distributive Law.* For every positive integer n,

$$n([\overline{AB}] + [\overline{CD}]) = n[\overline{AB}] + n[\overline{CD}].$$

Proof. For each n, let P_n be the proposition given by the formula. Then P_1 is true. To prove the theorem by induction, we need to show that if P_n is true, then so also is P_{n+1}. Now

$$\begin{aligned}(n + 1)([\overline{AB}] + [\overline{CD}]) &= n([\overline{AB}] + [\overline{CD}]) + ([\overline{AB}] + [\overline{CD}]) \\ &= (n[\overline{AB}] + n[\overline{CD}]) + ([\overline{AB}] + [\overline{CD}]) \\ &= (n[\overline{AB}] + n[\overline{CD}]) + ([\overline{CD}] + [\overline{AB}]) \\ &= n[\overline{AB}] + ((n[\overline{CD}] + [\overline{CD}]) + [\overline{AB}]) \\ &= n[\overline{AB}] + ((n + 1)[\overline{CD}] + [\overline{AB}]) \\ &= n[\overline{AB}] + ([\overline{AB}] + (n + 1)[\overline{CD}]) \\ &= (n[\overline{AB}] + [\overline{AB}]) + (n + 1)[\overline{CD}] \\ &= (n + 1)[\overline{AB}] + (n + 1)[\overline{CD}].\end{aligned}$$

What is the reason for each step?

Theorem 4. *Preservation of Order.* If

$$[\overline{AB}] > [\overline{CD}],$$

then

$$n[\overline{AB}] > n[\overline{CD}]$$

for every positive integer n.

Proof. If $[\overline{AB}] > [\overline{CD}]$,
then

$$[\overline{AB}] = [\overline{CD}] + [\overline{EF}]$$

for some segment $\overline{EF}$. (Recall the definition of $>$ for congruence classes.) By the preceding theorem,

$$n[\overline{AB}] = n[\overline{CD}] + n[\overline{EF}],$$

which is $> n[\overline{CD}]$, as desired.

Theorem 5. If

$$n[\overline{AB}] > n[\overline{CD}],$$

for some n, then

$$[\overline{AB}] > [\overline{CD}].$$

(If not, we would have a contradiction of the preceding theorem.) It is also easy to see the following theorem.

Theorem 6. If A-B-C, then

$$n[\overline{AC}] > n[\overline{AB}]$$

for every n.

The reason is that if A-B-C, it follows that $[\overline{AC}] > [\overline{AB}]$.

Theorem 7. If $[\overline{AB}] < [\overline{CD}]$, then $[\overline{AB}] + [\overline{EF}] < [\overline{CD}] + [\overline{EF}]$ for every $[\overline{EF}]$.

Proof. Let W, X, and Z be collinear points such that W-X-Z, $\overline{WX} \cong \overline{EF}$, and $\overline{XZ} \cong \overline{CD}$. Since $[\overline{AB}] < [\overline{CD}]$, there is a point Y such that X-Y-Z and $\overline{XY} \cong \overline{AB}$. Then

$$[\overline{AB}] + [\overline{EF}] = [\overline{WY}], \qquad \text{and} \qquad [\overline{CD}] + [\overline{EF}] = [\overline{WZ}].$$

Since W-X-Z and X-Y-Z, we have W-Y-Z. Therefore $[\overline{WY}] < [\overline{WZ}]$, and the theorem follows.

20.4 HOW TO DEFINE RATIOS: THE SUPREMUM

We have observed that in the Euclidean theory of proportionality, we cannot talk about ratios; we can only talk about the relation of *same ratio*. The relation

$$\overline{AB} : \overline{CD} :: \overline{EF} : \overline{GH}$$

says that $\overline{AB}$ and $\overline{CD}$ are in the same ratio as $\overline{EF}$ and $\overline{GH}$; but the "ratios" $\overline{AB} : \overline{CD}$ and $\overline{EF} : \overline{GH}$ have no meaning at all when they stand alone. If the real number system is available, however, we can assign a meaning to $\overline{AB} : \overline{CD}$ without using metric geometry. As a guide in framing the definition, let us recall that the ratio should turn out to be the number AB/CD. Thus, by the comparison theorem, we can say

$$\frac{p}{q} < \overline{AB} : \overline{CD} \tag{1}$$

if and only if

$$\frac{p}{q} < \frac{AB}{CD}. \tag{2}$$

This is equivalent to

$$pCD < qAB, \tag{3}$$

which in turn is equivalent to

$$p[\overline{CD}] < q[\overline{AB}]. \tag{4}$$

Thus $p/q < \overline{AB} : \overline{CD}$ if and only if p copies of $[\overline{CD}]$, laid end to end, form a segment shorter than a segment formed by q copies of $[\overline{AB}]$, laid end to end. Let

$$K = \left\{ \frac{p}{q} \;\middle|\; p[\overline{CD}] < q[\overline{AB}] \right\}.$$

Officially, this set of rational numbers has been defined synthetically, without reference to distance. Unofficially, we observe that in the metric scheme it must be true that

$$K = \left\{ \frac{p}{q} \;\middle|\; \frac{p}{q} < \frac{AB}{CD} \right\}.$$

The relation between the set K and the number AB/CD is simple; it can be conveyed by the two conditions that follow.

(1) AB/CD is an *upper bound* of K. That is, every element of K is $\leqq AB/CD$. (In fact, every element of K is strictly less than AB/CD. But this stronger condition is not required in the general definition of an upper bound of a set of numbers.)

(2) AB/CD is the *least* of all the upper bounds of K. That is, every other upper bound of K is greater than AB/CD.

If a set K of numbers and a number s are related in this way, we write

$$s = \sup K,$$

and we say that s is the *supremum*, or the *least upper bound* of K. To repeat: $s = \sup K$ if (1) s is an upper bound of K, and (2) no number less than s has this property.

More examples of this follow. Let

$$K_1 = \left\{ \frac{1}{n} \;\middle|\; n \text{ a positive integer} \right\}.$$

Here

$$\sup K_1 = 1;$$

and $\sup K_1$ belongs to K_1. On the other hand, if

$$K_2 = \left\{ 2 - \frac{1}{n} \;\middle|\; n \text{ a positive integer} \right\},$$

then

$$\sup K_2 = 2;$$

and $\sup K_2$ does not belong to K_2. If K_3 is the set $\mathbf{N}$ of all positive integers, then there is no such thing as $\sup K_3$, because K_3 has no upper bounds at all.

To define ratios of segments, in terms of Euclid's scheme, we need the following two basic postulates, one dealing with geometry and the other dealing with the real number system.

THE ARCHIMEDEAN POSTULATE. Given any two segments $\overline{AB}$ and $\overline{CD}$, there is a positive integer n such that $n[\overline{AB}] > [\overline{CD}]$.

(We recall that Euclid's theory of proportionality was restricted to pairs of segments that behaved in this way.)

THE DEDEKIND POSTULATE. Given a nonempty set K of real numbers. If K has an upper bound, then K has a supremum sup K.

This is our final and crucial postulate for the real number system. To indicate what it means, as a completeness condition, we shall explain why it fails to hold in the rational number system. Consider, for example, the set

$$K = \left\{\frac{p}{q} \;\middle|\; p, q > 0 \quad \text{and} \quad \frac{p}{q} < \sqrt{2}\right\}.$$

This can be described purely in terms of rational numbers:

$$K = \left\{\frac{p}{q} \;\middle|\; p, q > 0 \quad \text{and} \quad \frac{p^2}{q^2} < 2\right\}.$$

In the *real* number system, K has a least upper bound sup $K = \sqrt{2}$. But K has no sup in the rational number system. The reason is that if r/s is an upper bound of K, and

$$\sqrt{2} < \frac{t}{u} < \frac{r}{s},$$

then t/u is also an upper bound of K. Thus no rational upper bound of K is smaller than all other rational upper bounds of K.

In the same way, if we delete from the real number system any one number x, the resulting set $\mathbf{R}'$ of numbers does not satisfy the Dedekind postulate. The reason is that in the reduced system $\mathbf{R}'$, the set

$$K = \{y | y < x\}$$

has many upper bounds, but none of them is a least upper bound. (If z is an upper bound, then any number between x and z is an upper bound.)

Given two segments $\overline{AB}$, $\overline{CD}$. Let

$$K = \left\{\frac{p}{q} \;\middle|\; p[\overline{CD}] < q[\overline{AB}]\right\}.$$

On the basis of the Archimedean postulate, we shall prove the following theorems.

Theorem 1. K contains at least one positive number p/q.

Theorem 2. K has an upper bound.

From these two theorems it will follow, by the Dedekind postulate, that:

Theorem 3. K has a least upper bound sup K, and sup $K > 0$.

These theorems will justify the following definition:

DEFINITION.

$$\overline{AB} : \overline{CD} = \sup K.$$

And we shall then know by Theorem 3 that:

Theorem 4. For every two segments $\overline{AB}, \overline{CD}$,

$$\overline{AB} : \overline{CD} > 0.$$

Let us now prove Theorems 1 and 2.

Proof of Theorem 1. Given $\overline{AB}$ and $\overline{CD}$. By the Archimedean postulate, there is an n, such that

$$n[\overline{AB}] > [\overline{CD}].$$

Let $p = 1$ and let $q = n$. Then p/q is positive, and belongs to K.

Proof of Theorem 2. Given $\overline{AB}$ and $\overline{CD}$. By the Archimedean postulate, there is a number n, such that

$$n[\overline{CD}] > [\overline{AB}]. \tag{1}$$

We assert that n is an upper bound of K. We shall prove this by showing that if $p/q > n$, then p/q does not belong to K.

If $p/q > n$, then $p > nq$. Therefore

$$p[\overline{CD}] > nq[\overline{CD}].$$

It follows from (1), by Theorem 4, Section 20.3, that

$$nq[\overline{CD}] > q[\overline{AB}].$$

Therefore

$$p[\overline{CD}] > q[\overline{AB}],$$

so that p/q does not belong to K.

To see where the following theorem comes from, let us first state its metric form, which is trivial.

Theorem. If $A \neq B$, and C-D-E, then

$$\frac{CE}{AB} = \frac{CD}{AB} + \frac{DE}{AB}.$$

The reason is that $CD + DE = CE$.

The corresponding synthetic theorem is not trivial.

Theorem 5. *The Addition Theorem.* If $A \neq B$, and C-D-E, then

$$\overline{CE} : \overline{AB} = \overline{CD} : \overline{AB} + \overline{DE} : \overline{AB}.$$

The proof is long and tricky. It will be given in the next section.

20.5 PROOF OF THE ADDITION THEOREM

Under our definition of ratios, the addition theorem is a statement about suprema. To prove the theorem, we must first interpret it in terms of the definition.

Given $A \neq B$ and C-D-E. Let

$$K_1 = \left\{ \frac{p}{q} \,\middle|\, p[\overline{AB}] < q[\overline{CD}] \right\},$$

so that

$$\overline{CD} : \overline{AB} = \sup K_1 = k_1.$$

Let

$$K_2 = \left\{ \frac{r}{s} \,\middle|\, r[\overline{AB}] < s[\overline{DE}] \right\},$$

so that

$$\overline{DE} : \overline{AB} = \sup K_2 = k_2.$$

Finally, let

$$K = \left\{ \frac{t}{m} \,\middle|\, t[\overline{AB}] < m[\overline{EC}] \right\},$$

so that

$$\overline{CE} : \overline{AB} = \sup K = k.$$

The addition theorem states that $k_1 + k_2 = k$.

The proof is long. We have tried to make it as easy as possible by splitting it up into a series of subsidiary theorems, each of which has a fairly short proof. Unless you have encountered this sort of proof before, it may be confusing. On the other hand, the proof will repay study, because the ideas that come up in it are fundamental in analysis.

Theorem 1. If p/q belongs to K, and $r/s < p/q$, then r/s belongs to K.

Proof. Given

$$p[\overline{AB}] < q[\overline{CE}]$$

and

$$qr < ps,$$

we need to show that

$$r[\overline{AB}] < s[\overline{CE}].$$

By Theorem 4, Section 20.3, we have

$$pr[\overline{AB}] < qr[\overline{CE}],$$

and

$$qr[\overline{CE}] < ps[\overline{CE}],$$

because $qr < ps$. Therefore

$$pr[\overline{AB}] < ps[\overline{CE}].$$

Now if it were true that $r[\overline{AB}] = s[\overline{CE}]$, then we would have $pr[\overline{AB}] = ps[\overline{CE}]$; and if $r[\overline{AB}] > s[\overline{CE}]$, then $pr[\overline{AB}] > ps[\overline{CE}]$. In each case, the consequence is false. Therefore $r[\overline{AB}] < s[\overline{CE}]$, which was to be proved.

The following theorem is not logically necessary for the proof of the Addition Theorem, but it helps clear the air, and establishes a connection between the present discussion and the standard apparatus of analysis.

Theorem 2. K has no greatest element.

(Obviously this applies to the K defined by any pair of segments.)

Proof. Suppose that $p/q \in K$. Then $p[\overline{AB}] < q[\overline{CE}]$. Therefore there is a segment $\overline{FG}$ such that $q[\overline{CE}] = p[\overline{AB}] + [\overline{FG}]$. Therefore $nq[\overline{CE}] = np[\overline{AB}] + n[\overline{FG}]$ for each n. By the Archimedean Postulate there is an n such that

$$n[\overline{FG}] < [\overline{AB}].$$

It follows by Theorem 7 of Section 20.3 that

$$np[\overline{AB}] + n[\overline{FG}] > np[\overline{AB}] + [\overline{AB}].$$

Therefore

$$(np + 1)[\overline{AB}] < np[\overline{AB}] + n[\overline{FG}] = nq[\overline{CE}],$$

and

$$\frac{np + 1}{nq} \in K.$$

Since

$$\frac{p}{q} < \frac{np + 1}{nq},$$

it follows that p/q is not largest in K. Therefore K has no greatest element.

Thus the set K is of a special type, called a *cut in the rational numbers*. To be exact, a set Z is called a cut in the rational numbers if

(1) Z is a set of positive rational numbers,
(2) Z is not empty,
(3) Z has an upper bound,
(4) if

$$0 < \frac{p}{q} < \frac{r}{s};$$

and r/s belongs to Z, then p/q belongs to Z, and

(5) K has no greatest element.

In this language, we can sum up:

Theorem 3. If $\overline{AB}$ and $\overline{CE}$ are any two segments, and

$$K = \left\{\frac{p}{q} \;\middle|\; p[\overline{AB}] < q[\overline{CE}]\right\}$$

then K is a cut in the rational numbers.

Theorem 1 tells us that K satisfies (4), Theorem 2 gives (5), and we knew already that K satisfies the other three conditions of the definition.

The following theorem is logically trivial, but very useful.

Theorem 4. If $z = \sup Z$, and e is any positive number, then some element x of Z is greater than $z - e$.

Proof. If there were no such x, then every element x of Z would be $\leqq z - e$. Therefore $z - e$ would be an upper bound of Z. This is impossible, because $\sup Z$ is the smallest of all of the upper bounds of Z.

Combining Theorems 3 and 4, we get a simple description of the set K.

Theorem 5. Let $\overline{AB}$ and $\overline{DE}$ be segments. Let

$$K = \left\{\frac{p}{q} \;\middle|\; p[\overline{AB}] < q[\overline{DE}]\right\}.$$

Let

$$k = \sup K.$$

Then

$$K = \left\{\frac{r}{s} \;\middle|\; 0 < \frac{r}{s} < k\right\}.$$

Proof. Let r/s be a rational number between 0 and k. Let

$$e = k - \frac{r}{s}.$$

Then $e > 0$. By Theorem 4 there is an element p/q of K such that

$$\frac{p}{q} > k - e.$$

This means that

$$\frac{r}{s} < \frac{p}{q}.$$

By Theorem 1, $r/s \in K$. Thus

$$\left\{\frac{r}{s} \;\middle|\; 0 < \frac{r}{s} < k\right\} \subset K.$$

If $k \in K$, then k is the largest element of K, which is impossible. Therefore every element of K is strictly between 0 and k, and

$$K \subset \left\{\frac{r}{s} \;\middle|\; 0 < \frac{r}{s} < k\right\}.$$

The theorem follows.

Thus we can write

$$K_1 = \left\{\frac{p}{q} \;\middle|\; 0 < \frac{p}{q} < k_1\right\},$$

$$K_2 = \left\{\frac{r}{s} \;\middle|\; 0 < \frac{r}{s} < k_2\right\},$$

$$K = \left\{\frac{t}{u} \;\middle|\; 0 < \frac{t}{u} < k\right\}.$$

We now return to our geometry.

Theorem 6. If p/q belongs to K_1 and r/s belongs to K_2, then $p/q + r/s$ belongs to K.

Proof. We have given that $p[\overline{AB}] < q[\overline{CD}]$ and $r[\overline{AB}] < s[\overline{DE}]$. Obviously

$$\frac{p}{q} + \frac{r}{s} = \frac{qr + ps}{qs}.$$

Thus we need to prove that

$$(qr + ps)[\overline{AB}] < qs[\overline{CE}].$$

Now

$$qr[\overline{AB}] < qs[\overline{DE}]$$

and

$$ps[\overline{AB}] < qs[\overline{CD}],$$

by Theorem 4, Section 20.3. Therefore

$$(qr + ps)[\overline{AB}] < qs([\overline{CD}] + [\overline{DE}]) = qs[\overline{CE}],$$

which was to be proved.

Theorem 7. If p/q does not belong to K_1, and r/s does not belong to K_2, then $p/q + r/s$ does not belong to K.

Proof. Given $p[\overline{AB}] \geqq q[\overline{CD}]$ and $r[\overline{AB}] \geqq s[\overline{DE}]$, we need to show that

$$(qr + ps)[\overline{AB}] \geqq qs[\overline{CE}].$$

The proof is very much like that of the preceding theorem. First,

$$qr[\overline{AB}] \geqq qs[\overline{DE}]$$

and

$$ps[\overline{AB}] \geqq qs[\overline{CD}],$$

by Theorem 4, Section 20.3. Therefore

$$\begin{aligned}(qr + ps)[\overline{AB}] &\geqq qs([\overline{CD}] + [\overline{DE}]) \\ &= qs[\overline{CE}],\end{aligned}$$

which was to be proved.

We are now ready to finish the proof of the addition theorem. In the notation of this section, the theorem says that

$$k_1 + k_2 = k.$$

If this is false, then either

$$k_1 + k_2 > k \tag{1}$$

or

$$k_1 + k_2 < k. \tag{2}$$

We shall show that both of these are impossible.

If (1) holds, then $k_1 + k_2 - k > 0$. Let

$$e = k_1 + k_2 - k,$$

so that

$$k = k_1 + k_2 - e.$$

By Theorem 4, there is a p/q in K_1 such that

$$\frac{p}{q} > k_1 - \frac{e}{2},$$

and there is an r/s in K_2 such that

$$\frac{r}{s} > k_2 - \frac{e}{2}.$$

It follows that

$$\frac{p}{q} + \frac{r}{s} > k_1 + k_2 - e.$$

Therefore $p/q + r/s$ does not lie in K. But by Theorem 6, $p/q + r/s$ must lie in K. This gives a contradiction, showing that (1) is impossible.

If (2) holds, then $k - k_1 - k_2 > 0$. Let

$$e = k - k_1 - k_2,$$

so that

$$k = k_1 + k_2 + e.$$

Let p/q and r/s be rational numbers such that

$$k_1 < \frac{p}{q} < k_1 + \frac{e}{2}$$

and

$$k_2 < \frac{r}{s} < k_2 + \frac{e}{2}.$$

Then p/q does not belong to K_1, and r/s does not belong to K_2. By Theorem 7, $p/q + r/s$ does not belong to K. Since

$$\left\{\frac{t}{u} \,\middle|\, 0 < \frac{t}{u} < k\right\} \subset K,$$

it follows that

$$\frac{p}{q} + \frac{r}{s} \geqq k.$$

But this is impossible. Adding our previous inequalities for p/q and r/s, we get

$$\frac{p}{q} + \frac{r}{s} < k_1 + \frac{e}{2} + k_2 + \frac{e}{2} = k_1 + k_2 + e = k.$$

This gives a contradiction, showing that (2) is impossible. Therefore the addition theorem is true.

20.6 THE METRIZATION THEOREM

In this book, we have considered Euclidean geometry from two viewpoints, the metric and the purely synthetic. In the metric approach, the postulates tell us a great deal about the relation between our geometry and the real number system: the structure is

$$[S, \mathcal{L}, \mathcal{P}, d, m].$$

where d is the distance function

$$d : S \times S \to \mathbf{R}.$$

The distance function obeys the ruler postulate; and congruence and betweenness are defined in terms of distance.

In the purely synthetic approach, which Euclid used, the real number system is nowhere mentioned. The structure is

$$[S, \mathcal{L}, \mathcal{P}, \cong, \mathcal{B}].$$

Here $\cong$ and $\mathcal{B}$ are undefined relations of congruence and betweenness, subject to the postulates stated in Chapter 8; and the only numbers that get used are the positive integers, which are used to count things.

The first really big step in geometry after the Greeks was the invention of coordinate systems, by René Descartes. Obviously, to set up a coordinate system in the plane (or in space), you must have a distance function. In fact, to label the points of the x-axis with numbers, you must have gotten, from somewhere, a coordinate system for the x-axis, satisfying the conditions of the ruler postulate.

We shall now show that in a purely synthetic geometric system we can *define* a distance function satisfying the metric postulates. To be exact, we shall prove the following theorem.

Theorem 1. Given a geometric structure

$$[S, \mathcal{L}, \mathcal{P}, \cong, \mathcal{B}],$$

satisfying the postulates of Chapter 8, and satisfying the Archimedean postulate. Let A and B be any two points. Then there is a function

$$d : S \times S \to \mathbf{R},$$

such that

(1) d satisfies the ruler postulate,
(2) $\overline{CD} \cong \overline{EF}$ if and only if the distances CD and EF are the same;
(3) C-D-E if and only if $CD + DE = CE$, and
(4) $AB = 1$.

That is, we can always define a distance function which gives us back the same congruence relation for segments and the same betweenness relation for points that we had before; and we can always set up our distance function in such a way that any given segment $\overline{AB}$ is the "unit of length."

When we say that d "satisfies the ruler postulate," we mean that for any line L there is a coordinate function

$$f: L \to \mathbf{R}$$

of L *into* the real numbers, such that some point has coordinate 0, and such that if $x = f(P)$ and $y = f(Q)$, then

$$PQ = |x - y|.$$

It is not claimed that every real number x is the coordinate $f(P)$ of some point P. In fact the latter statement cannot be proved on the basis of the synthetic postulates that we have stated. (The easiest way to see this is to observe that the surd plane satisfies the synthetic postulates; and in the surd plane, only surds are needed as coordinates.) If the coordinate functions are one-to-one correspondences $f: L \leftrightarrow R$, then the geometric structure $[S, \mathcal{L}, \mathcal{P}, \cong, \mathcal{B}]$ is *called complete in the sense of Dedekind.*

We now proceed to the proof. Fortunately, the hard part of the proof, which is to set up our distance function in such a way that any given segment $\overline{AB}$ is the "unit of length," is already over with. For any two points P, Q, let

$$PQ = d(P, Q) = \overline{PQ} : \overline{AB},$$

where $\overline{PQ} : \overline{AB}$ is the ratio defined and studied earlier in this chapter. Since the ratio was defined, in the first place, in terms of congruence classes $[\overline{PQ}]$, we know immediately that (2) is satisfied.

To prove (3), we first observe that if C-D-E, it follows by the addition theorem that

$$\overline{CE} : \overline{AB} = \overline{CD} : \overline{AB} + \overline{DE} : \overline{AB}.$$

Therefore

$$CE = CD + DE,$$

by our definition of distance. (This is the first of the two things that the addition theorem is good for.) Suppose, conversely, that

$$CE = CD + DE, \tag{1}$$

where C, D, and E are all different. If it were true that D-E-C, then it would follow that

$$DC = DE + EC,$$

so that

$$CE = CD - DE; \tag{2}$$

and from (1) and (2), we get

$$CD + DE = CD - DE,$$

or $2DE = 0$, or $DE = 0$. Therefore $D = E$, which is a contradiction. If we suppose that D-C-E, this leads to a contradiction in the same way. But one of the three points must be between the other two. Since D-E-C and D-C-E are both false, it follows that C-D-E must be true.

Also, (4) is satisfied. Obviously $p[\overline{AB}] < q[\overline{AB}]$ precisely if $p < q$, which means that $p/q < 1$. Therefore

$$\overline{AB}:\overline{AB} = \sup\left\{\frac{p}{q} \;\middle|\; 0 < \frac{p}{q} < 1\right\} = 1,$$

and $d(A, B) = 1$.

It remains only to check the ruler postulate. Given a line L, and three points C, D, E of L such that C-D-E (Fig. 20.2).

C P? D P? E

FIGURE 20.2

We shall set up a function

$$f: L \to \mathbf{R},$$

and show that f is a coordinate system satisfying the ruler postulate.

If P belongs to the ray $\overrightarrow{DE}$, let $f(P) = DP$.

We shall check that if P and Q are points of $\overrightarrow{DE}$, with coordinates x, y, then

$$PQ = |x - y|,$$

as required in the ruler postulate. If either P or Q is $= D$, this is easy to see. For example, if $Q = D$, then

$$PQ = DP = f(P) = x = |x| = |x - 0|.$$

If $P = Q$, it is also trivial, because then $PQ = 0 = |x - x|$. Suppose then, finally, that P, Q, and D are all different. Then either D-P-Q or D-Q-P. If D-P-Q, then

$$\overline{DQ}:\overline{AB} = \overline{DP}:\overline{AB} + \overline{PQ}:\overline{AB},$$

by the addition theorem. (This is the second of the two things for which the addition theorem is useful.) In terms of distance, this tells us that

$$DQ = DP + PQ,$$

or

$$PQ = DQ - DP = y - x.$$

Here $y - x > 0$, because $y - x = PQ$. Therefore

$$PQ = |x - y|,$$

which was to be proved.

We now define the coordinate function f, for points P of the opposite ray $\overrightarrow{DC}$, by the condition

$$f(P) = -DP.$$

We now have

$$PQ = |x - y|,$$

if P and Q both belong to $\overrightarrow{DC}$. (The proof is almost exactly the same as for the ray $\overrightarrow{DE}$: the point is that $|x - y|$ is unchanged when the signs of x and y are reversed.) If P belongs to $\overrightarrow{DC}$ and Q belongs to $\overrightarrow{DE}$, then

$$PQ = PD + DQ, \qquad x = -PD, \qquad y = DQ.$$

Therefore

$$\begin{aligned} PQ &= y - x \\ &= |x - y|, \end{aligned}$$

which was to be proved.

The reader will observe that the proof of the metrization theorem requires all of the apparatus set up in the present chapter. It is plain that such proofs as this do not form a part of elementary mathematics. Nevertheless it is usual, in elementary geometry courses, to give some sort of "indication of proof" for what amounts to a metrization theorem. Usually this is done just before the treatment of similarity, so as to permit an algebraic treatment of proportionality. Usually the same discussion is repeated, when coordinate systems are introduced. The reader should now be able to judge the adequacy of such "proofs."

In Theorem 1, the coordinate systems $f: L \to \mathbf{R}$ did not necessarily use all the real numbers. That is, $f(L) = F_L \subset \mathbf{R}$, but F_L was not necessarily all of $\mathbf{R}$.

Theorem 2. In Theorem 1, the sets F_L are the same for all lines, $= F$ for every L. If the geometric structure satisfies the Two-Circle Postulate of Section 16.5, and satisfies the conditions of the Line-Circle Theorem of Section 16.2, then F forms a Euclidean ordered field.

Proof. (1) Let L and L' be two lines, and let f and f' be coordinate systems for them, so that $f(P) = 0$ for some P and $f'(P') = 0$ for some P'. If $f(Q) = x$ for some Q in L, then it follows by the segment construction postulate that $f'(Q') = x$ for some Q' in L'. (There are two cases: $x > 0$ or $x < 0$.)

(2) If the Two-Circle Postulate and the Line-Circle Theorem hold, then all ruler-and-compass constructions can be carried out. We found in Section 19.2 that for positive numbers, addition, subtraction, multiplication, division, and square root extraction can be carried out with ruler and compass. The segment construction postulate implies that F contains the negative of each of its elements. It follows that F forms a Euclidean ordered field.

If our "plane" is the surd plane, then F is precisely the surd field.

20.7 THE DEDEKIND CUT

If the ring $\mathbf{Z}$ of integers is given, it is not hard to set up the field $\mathbf{Q}$ of rational numbers. To pass from the rational numbers to the real numbers is another matter. This, however, is what we need to do,-to show that there is a number system which satisfies the field postulates of Chapter 1 and also the Dedekind postulate.

One of the most elegant approaches to this problem was devised by Dedekind, leaning heavily on the ideas of Eudoxus. For the sake of convenience, we shall restrict ourselves to positive numbers. (Once we have them, it is not hard to set up their negatives.)

We recall that in Section 20.4 we defined

$$\overline{AB}:\overline{CD} = \sup K,$$

where

$$K = \left\{\frac{p}{q} \;\middle|\; p[\overline{CD}] < q[\overline{AB}]\right\}.$$

It appeared in Section 20.5 that K was always a *cut* in the rational numbers. That is:

(1) K is a set of rational numbers.
(2) K is not empty.
(3) K has an upper bound.
(4) If $0 < p/q < r/s$, and $r/s \in K$, then $p/q \in K$.
(5) K has no greatest element.

If the real number system $\mathbf{R}$ is regarded as given, then to every cut K there corresponds a unique positive real number $\sup K$. On the other hand, cuts are defined purely in terms of the rational number system $\mathbf{Q}$. We can use this fact to define a set of objects which can be regarded as the positive reals.

DEFINITION. *A positive real number* is a cut in the positive rationals.

Let $\mathbf{R}^+$ be the set of all cuts. In $\mathbf{R}^+$ we need to define addition, multiplication, and order.

(1) If K and L are cuts, then $K + L = \{x + y | x \in K, y \in L\}$.
(2) $K \cdot L = \{xy | x \in K, y \in L\}$.

We need to show, of course, that the sum and product of any two cuts are also cuts. It is then easy to check that $\mathbf{R}^+$ satisfies the usual algebraic identities given in the field postulates. For example,

$$K + L = L + K,$$

because

$$\{x + y | x \in K, y \in L\} = \{y + x | y \in L, x \in K\}.$$

(3) We define $K < L$ to mean that K is a proper subset of L, that is, $K \subset L$ and $K \neq L$.

Under this definition, when we say that K is an upper bound of Z, this means simply that every cut belonging to Z is a *subset* of K. It is now easy to see that every bounded set of cuts has a supremum: $\sup Z$ is simply the *union* of all of the sets belonging to Z. This forms a cut; it is an upper bound of Z; and no smaller cut is an upper bound of Z.

The above discussion is, of course, merely a sketch. For details, see the latter portion of E. Landau's *Foundations of Analysis*.

A good case can be made out for speaking not of the Dedekind cut but of the Eudoxian cut. The crux of Dedekind's procedure was to use cuts as a working definition of real numbers; and this is what Eudoxus had done, over two thousand years before.

Chapter 21
Length and Plane Area

21.1 THE DEFINITION OF ARC LENGTH

Given an arc $\widehat{AB}$ of a circle C:

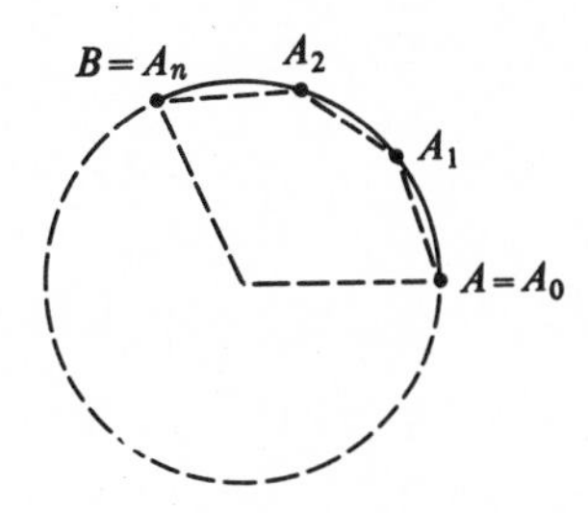

FIGURE 21.1

We take a sequence of points

$$A = A_0, A_1, A_2, \ldots, A_n = B,$$

in the order from A to B on the arc; and for each pair of successive points A_{i-1}, A_i we draw the segment $\overline{A_{i-1}A_i}$, as indicated in the figure. The union of these segments is called an *inscribed broken line*; and the sum of their lengths is denoted by p_n. Thus

$$\begin{aligned} p_n &= A_0A_1 + A_1A_2 + \cdots + A_{n-1}A_n \\ &= \sum_{i=1}^{n} A_{i-1}A_i. \end{aligned}$$

There are now various ways that we might define the length of the arc $\widehat{AB}$. If we merely want to *state* a definition, as a matter of form, without intending to use it, then our problem is simple. We agree to use equally spaced points $A_0, A_1, \ldots, A_n$. The length p_n of our broken line is now determined by n, and we can define the length to be

$$p = \lim_{n\to\infty} p_n.$$

To justify this, we would need to explain what is meant by $\lim_{n\to\infty}$, and we would have to show that the indicated limit exists, for every circular arc.

The following definition, however, is more manageable. Let P be the set of all numbers p_n which are lengths of broken lines inscribed in $\widehat{AB}$. Thus

$$P = \left\{p_n \;\middle|\; p_n = \sum A_{i-1}A_i\right\}.$$

Let

$$p = \sup P.$$

To justify this, we need to prove the following theorem.

Theorem 1. P has an upper bound.

It will then follow that P has a *least* upper bound sup P. The proof is easy, on the basis of the following preliminary result. Let $\triangle PQR$ be an isosceles triangle, with $PQ = PR$.

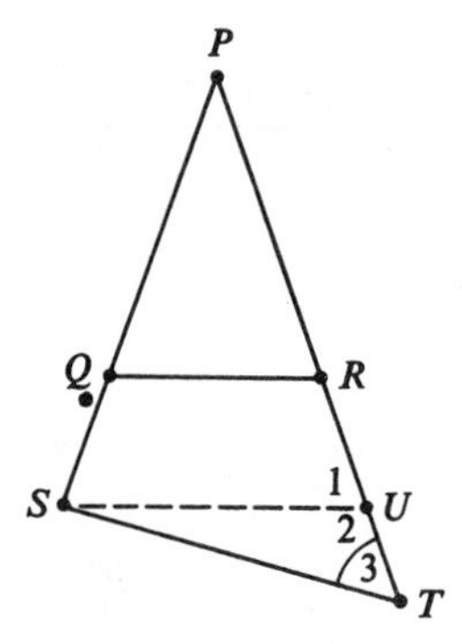

FIGURE 21.2

We assert that if P-Q-S and P-R-T, then

$$ST > QR.$$

Suppose that $PS < PT$, as in the figure, and take U between R and T so that $\overline{SU} \parallel \overline{QR}$. Then

$$\frac{SU}{QR} = \frac{PS}{PQ} > 1.$$

Therefore

$$SU > QR.$$

We shall now show that $ST > SU$. Evidently $\angle 1$ is acute, because $\angle 1$ is a base angle of an isosceles triangle. Therefore $\angle 2$ is obtuse. Therefore $\angle 3$ is acute. Therefore $m\angle 3 < m\angle 2$. Therefore $ST > SU$. (Why?)

We now return to our circular arc. Draw any square that contains the whole circle in its interior (Fig. 21.3). We project each point A_i onto the square, as indicated in the figure. That is, A'_i is the point where $\overrightarrow{DA_i}$ intersects the square. Then $A_{i-1}A_i < A'_{i-1}A'_i$. Therefore p_n is always less than $\sum_{i=1}^{n} A'_{i-1}A'_i$. Therefore p_n is always less than the perimeter of the square. Thus the perimeter of the square is the upper bound that we are looking for.

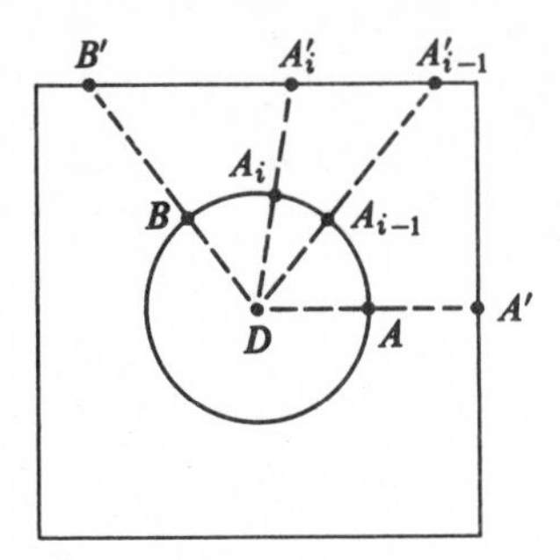

FIGURE 21.3

This justifies our definition

$$p = \sup P.$$

Of course a circle is not an arc. But we can define the circumference of a circle in an analogous way, by setting up an inscribed polygon with vertices

$$A_0, A_1, \ldots, A_{n-1}, A_n = A_0.$$

We then let p_n be the perimeter

$$\sum_{i=1}^{n} A_{i-1}A_i,$$

we let P be the set of all such perimeters p_n, and we define the circumference as

$$p = \sup P.$$

21.2 THE EXISTENCE OF π

We now want to prove the existence of the number π. To do this, we need to show that the ratio of the circumference to the diameter is the same for all circles. This is a theorem about suprema, and to prove it, we need a preliminary result.

Let P be any bounded set of positive numbers, and let k be a positive number. Then kP denotes the set of all numbers of the form kp, where p belongs to P. For example, if

$$P = [0, 1] = \{x | 0 \leqq x \leqq 1\},$$

and $k = 3$, then

$$kP = 3P = [0, 3].$$

If $P = [1, 2]$ and $k = \frac{2}{3}$, then $kP = [\frac{2}{3}, \frac{4}{3}]$; and so on.

This "multiplication" is associative. That is,

$$j(kP) = (jk)P,$$

because

$$\{jx | x \in kP\} = \{j(kp) | p \in P\} = \{(jk)p | p \in P\}.$$

Thus, for example, we always have

$$\frac{1}{k}(kP) = P.$$

Lemma 1. If b is an upper bound of P, then kb is an upper bound of kP.

Reason. If $p \leqq b$, then $kp \leqq kb$.

Lemma 2. If c is an upper bound of kP, then c/k is an upper bound of P.

Proof. Since $P = (1/k)(kP)$, this follows from Lemma 1.

These lemmas give us the following theorem.

Theorem 1. $\sup(kP) = k \sup P$.

Proof. Let $b = \sup P$. Then b is an upper bound of P. By Lemma 1, kb is an upper bound of kP.

Suppose that kP has an upper bound

$$c < kb.$$

Then c/k is an upper bound of P, by Lemma 2. This is impossible, because $c/k < b$, and b was the *least* upper bound of P.

We can now prove the theorem which establishes the existence of π. What is needed is the following theorem.

Theorem 2. Let C and C' be circles with radii r, r' and circumferences p, p'. Then

$$\frac{p}{2r} = \frac{p'}{2r'}.$$

That is, the ratio of the circumference to the diameter is the same for all circles. This common ratio is denoted by π.

Proof. Suppose that the circles have the same center. (This involves no loss of generality.) In the figure, we indicate the ith side $\overline{A_{i-1}A_i}$ of a polygon inscribed

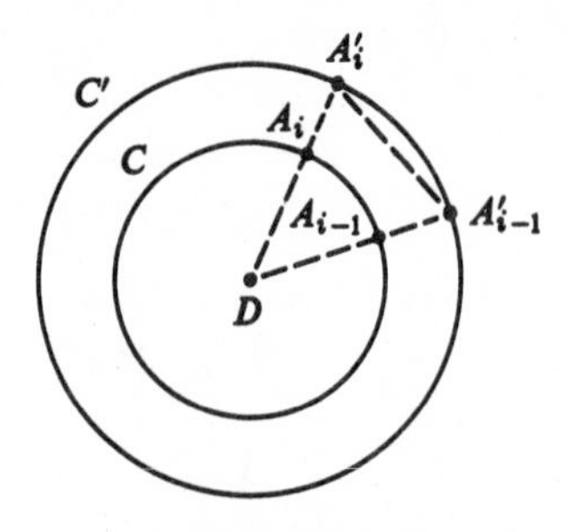

FIGURE 21.4

in C. To each such polygon there corresponds a polygon incribed in C', obtained by projection outward (or perhaps inward) from the common center D. We then have

$$\triangle DA_{i-1}A_i \sim \triangle DA'_{i-1}A'_i.$$

Therefore

$$\frac{A'_{i-1}A'_i}{A_{i-1}A_i} = \frac{DA'_i}{DA_i} = \frac{r'}{r}.$$

If the perimeters of our polygons are p_n and p'_n, then we have

$$p'_n = \frac{r'}{r} \cdot p_n.$$

Let

$$p = \sup P, \qquad \text{and} \qquad p' = \sup P',$$

where $P = \{p_n\}$ and $P' = \{p'_n\}$, as usual. Then

$$P' = \frac{r'}{r} \cdot P.$$

Therefore, by the preceding theorem, using $k = r'/r$, we have

$$\sup P' = \frac{r'}{r} \sup P, \qquad \text{or} \qquad p' = \frac{r'}{r} p, \qquad \text{or} \qquad \frac{p'}{r'} = \frac{p}{r}.$$

Dividing both sides by 2, we get the equation called for in the theorem.

An analogous theorem holds for circular arcs.

Theorem 3. Let $\widehat{AB}$ and $\widehat{A'B'}$ be arcs of the same degree measure, in circles of radius r and r', respectively. Let the lengths of $\widehat{AB}$ and $\widehat{A'B'}$ be p and p'. Then

$$\frac{p}{r} = \frac{p'}{r'}.$$

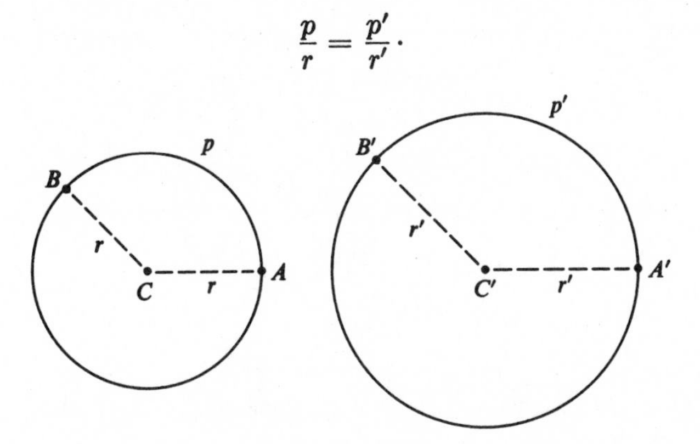

FIGURE 21.5

This ratio is called the radian measure of the arc $\widehat{AB}$. If $\widehat{AB}$ is a minor arc, then p/r is the *radian measure* of the angle $\angle BCA$. The theorem tells us that the radian measure really depends only on the angle, or on the degree measure of the arc, and is independent of the radius of the circle.

The proof is virtually identical with the proof of the preceding theorem.

21.3 LIMITS AS THE MESH APPROACHES ZERO

Given

$$p = \sup P,$$

we know that we can find numbers p_n in the set P, as close to $\sup P$ as we please. More precisely, if e is any positive number, there is a p_n in P such that

$$p_n > p - e.$$

This was Theorem 4, Section 20.5; and it is true because otherwise $p - e$ would be an upper bound of P.

We ought, however, to be able to make a stronger statement than this about the numbers p_n and their supremum. To get p_n close to p, we should not have to choose the inscribed broken line in any special way. It ought to be true that p_n is close to p whenever the sides of the inscribed broken line are sufficiently short. We shall make this idea precise in the next theorem.

Let

$$A = A_0, A_1, \ldots, A_n = B$$

be the vertices of a broken line B_n inscribed in the arc $\widehat{AB}$. By the *mesh* of the broken line B_n, we mean the largest of the numbers $A_{i-1}A_i$. Thus the mesh of a broken line B_n is the length of its longest segment. In this language, the statement that we want to prove can be stated roughly as follows.

p_n is as close to p as we please, if the mesh of B_n is small.

This suggests the idea, but it is not exact enough to form the basis of a proof, because it involves nonmathematical terms, notably the terms *please* and *enough*. Statements like this are like bowling balls without finger holes: they are easy to look at, but awkward to handle. The corresponding mathematical statement follows.

Theorem 1. Let $\widehat{AB}$ be an arc of length

$$p = \sup P,$$

where P is the set of lengths p_n of inscribed broken lines B_n. For every positive number e there is a positive number d such that if the mesh of B_n is less than d, then $p_n > p - e$.

Since we always have $p_n \leqq p$, the inequality $p_n > p - e$ means that $|p_n - p| < e$, that is, p_n is within a distance e from p.

The proof of this theorem is not as difficult as its statement. It is as follows.

(1) Let B'_n be an inscribed broken line, with length p'_n, such that

$$p'_n > p - \frac{e}{2}.$$

(We know that there is such a broken line, because $p = \sup P$.)

(2) Now let B_m be *any* inscribed broken line, of length p_m. Let B''_r be the broken line obtained by using *all* of the vertices of B_m and *all* of the vertices of B'_n. Let p''_r be the length of B''_r. Here the points marked with crosses are vertices of B'_n. The points marked with little circles are vertices of B_m. The end points A and B are marked both ways because they must be vertices of both. And all the indicated points are vertices of B''_r.

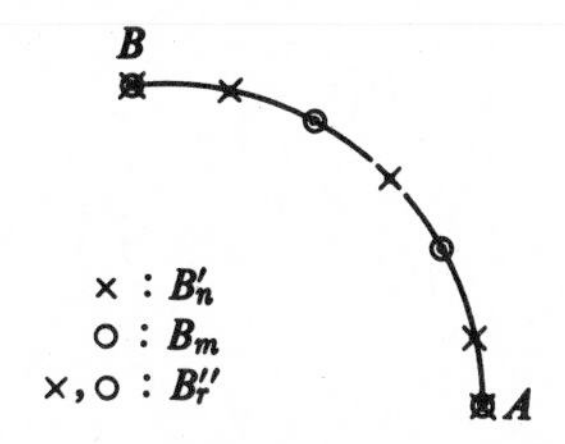

FIGURE 21.6

By repeated applications of the triangular inequality, we have

$$p''_r \geqq p'_n.$$

Therefore

$$p''_r \geqq p - \frac{e}{2}.$$

This merely says that when you insert new vertices, the length of a broken line increases.

Of course we cannot claim that $p_m \geqq p''_r$. It can easily happen that B_m is shorter than B''_r, because B_m may take "shortcuts" past vertices of B''_r (see Fig. 21.7). Here $PR < PQ + QR$. Thus the length of $\overline{PR}$ is less than the sum of the lengths of the corresponding sides $\overline{PQ}$, $\overline{QR}$, of B''_r.

The question is how many times you can gain by these shortcuts, and how much you can gain each time. The number of times the shortcuts can appear is surely no more than $n - 1$, because there are no more than $n - 1$ possibilities for Q. The saving at each shortcut is

$$PQ + QR - PR.$$

This is no bigger than PR, because $\overline{PR}$ is the longest side of $\triangle PQR$. If k is the mesh of B_m, then $PR \leqq k$. Thus there are at most $n - 1$ shortcuts, and the distance saved at each of them is $\leqq k$. Therefore the total saving is

$$p''_r - p_m \leqq (n - 1)k.$$

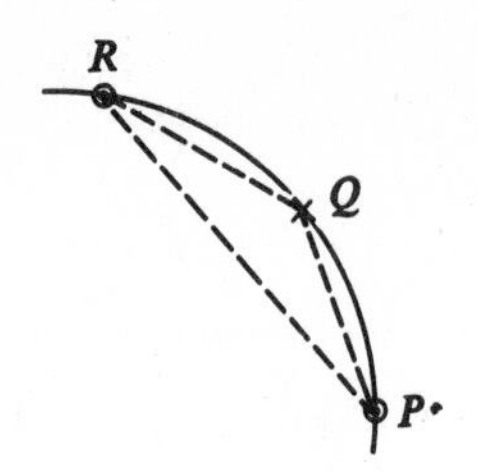

FIGURE 21.7

Now we know that

$$p - p_r'' \leqq \frac{e}{2}.$$

Therefore

$$p - p_m \leqq \frac{e}{2} + (n - 1)k.$$

What we want to get is

$$p - p_m < e.$$

This will hold if

$$(n - 1)k < \frac{e}{2},$$

or

$$k < \frac{e}{2(n - 1)}.$$

Our problem is now solved: let

$$d = \frac{e}{2(n - 1)}.$$

If the mesh k of B_m is less than d, then $p - p_m$ is less than e, as we wanted it to be.

We state Theorem 1 briefly by saying that p_n approaches p as the mesh of B_n approaches zero. This is, of course, the same sort of limiting process that is used in the theory of definite integrals.

21.4 THE ADDITION THEOREM FOR ARC LENGTH

One of our postulates for angular measure was the addition postulate. This said that if C lies in the interior of $\angle DAB$, then $m\angle DAB = m\angle DAC + m\angle CAB$.

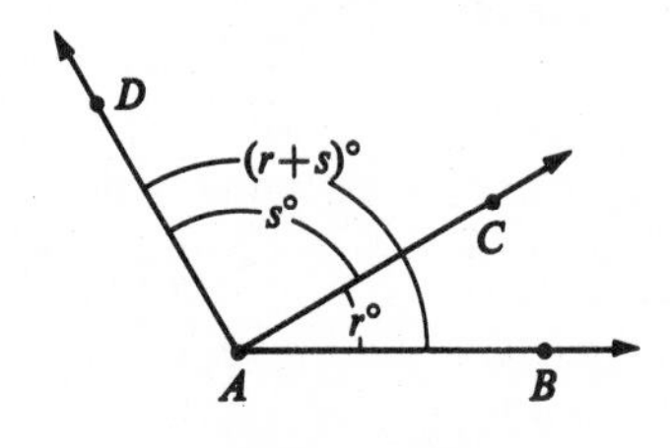

FIGURE 21.8

The corresponding statement about degree measure of circular arcs was a theorem. The theorem stated that $m\widehat{ABC} = m\widehat{AB} + m\widehat{BC}$. A corresponding theorem holds for arc length.

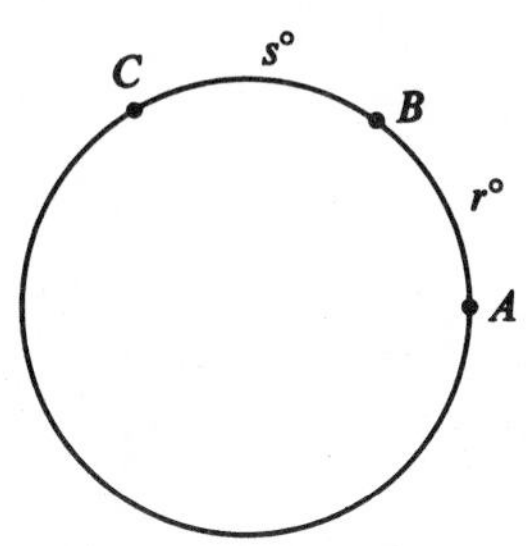

FIGURE 21.9

Theorem 1. Let $\widehat{AB}$ and $\widehat{BC}$ be arcs of the same circle, with only the point B in common. Let s_1 and s_2 be the lengths of $\widehat{AB}$ and $\widehat{BC}$; and let s be the length of $\widehat{ABC}$. Then

$$s_1 + s_2 = s.$$

The proof is an easy exercise in the use of least upper bounds and limits as the mesh approaches 0.

(1) Suppose that $s_1 + s_2 > s$, so that $s_1 + s_2 - s > 0$. Let

$$e = s_1 + s_2 - s.$$

Let B_n be a broken line inscribed in $\widehat{AB}$, with length p_n, such that

$$p_n > s_1 - \frac{e}{2}.$$

Let B'_m be a broken line inscribed in $\widehat{BC}$, with length p'_m, such that

$$p'_m > s_2 - \frac{e}{2}.$$

Fitting these broken lines end to end, we get a broken line B''_{m+n}, of length p''_{m+n}, such that

$$p''_{m+n} = p_n + p'_m.$$

Since B''_{m+1} is inscribed in $\widehat{ABC}$, it follows that $p''_{m+n} \leqq s$. On the other hand, by addition, we get

$$p''_{m+n} = p_n + p'_m > s_1 + s_2 - e = s.$$

Thus $p''_{m+n} \leqq s$ and $p''_{m+n} > s$, which is a contradiction.

(2) Suppose that $s_1 + s_2 < s$, so that $s - s_1 - s_2 > 0$. Let

$$e = s - s_1 - s_2,$$

so that

$$s = s_1 + s_2 + e.$$

Let d be a positive number such that if B_n is inscribed in $\widehat{ABC}$ and has length p_n and mesh less than d, then

$$p_n > s - e.$$

(By Theorem 1, Section 21.3, there is such a d. That is, p_n is as close to s as we please if only the mesh of B_n is small enough.)

Now let B_n be a broken line inscribed in $\widehat{ABC}$, of mesh less than d, *such that* B *is a vertex of* B_n. Let p_n be the length of B_n. Then B_n can be broken up into two broken lines B'_m, B''_r $(m + r = n)$, one inscribed in $\widehat{AB}$ and the other inscribed in $\widehat{BC}$. If the lengths of these broken lines are p'_m and p''_r, then we have

$$p'_m + p''_r = p_n,$$

$$p'_m \leqq s_1,$$

$$p''_r \leqq s_2.$$

Therefore

$$p_n \leqq s_1 + s_2.$$

But $s_1 + s_2 = s - e$. Therefore

$$p_n \leqq s - e,$$

which gives a contradiction.

If you review the proof of the addition theorem for ratios $\overline{AB}:\overline{CD}$, in Section 20.5, you will find that this proof is very similar to it. The technique involved here is quite important enough to be worth going through twice, or more.

We remember that when we defined arc length by means of broken lines, we allowed broken lines whose sides were not necessarily of the same length. This possibility was important in the second part of the proof of the preceding theorem.

21.5 APPROACHING THE AREA OF A CIRCULAR SECTOR FROM BELOW

So far in this chapter, the mathematics has been exact and complete. We have given a definition of the length of a circular arc, and we have proved theorems on the basis of the definition. We shall approach the problem of plane area, for the corresponding circular sectors, in a more backhanded fashion. First we shall calculate the areas of such figures numerically, by a method which will no doubt be familiar to the reader in one form or another. We shall then reexamine the situation, and see just what we needed to assume, to justify our calculation. Then, in the following chapter, we shall develop a theory of plane area, sufficiently general to apply to the familiar elementary figures, and show that in this theory, the area formulas of this chapter become genuine theorems. This backhanded approach will involve very little lost motion. And the theory given in the next chapter will be easier to understand if we first survey the situation in a particular case, and get a rough notion of how the theory ought to work.

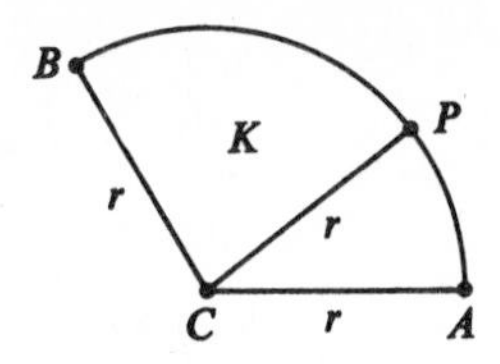

FIGURE 21.10

By a *circular sector* we mean a figure like the one above. To be exact, if $\widehat{AB}$ is an arc of a circle with center at C and radius r, and K is the union of all radii $\overline{CP}$, where P is in $\widehat{AB}$, then K is a *sector;* r is called its *radius*, and $\widehat{AB}$ is its *boundary arc*.

If we use *all* of the circle (instead of an arc $\widehat{AB}$) then the union of the radii $\overline{CP}$ is the circle plus its interior. Such a figure is called a *disk*. We shall begin our investigation by proving the following theorem.

Theorem 1. Let K be a circular sector with radius r and boundary arc of length s. Then there is a sequence of polygonal regions

$$K_1, K_2, \ldots,$$

all contained in K, such that

$$\lim_{n\to\infty} \alpha K_n = \tfrac{1}{2}rs.$$

Here αK_n denotes the area of K_n.

Proof. We begin by inscribing in $\widehat{AB}$ a broken line B_n in which all the sides are congruent, of length b_n:

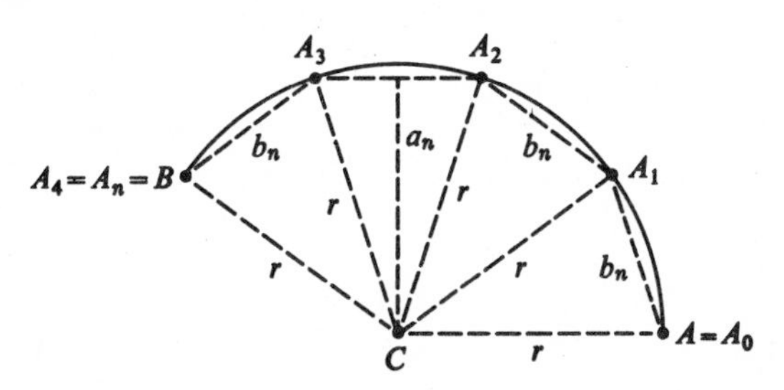

FIGURE 21.11

In the figure, r is the radius of the circle. Obviously all the triangles $\triangle A_{i-1}CA_i$ are congruent. Therefore they all have the same altitude (measured from C to the opposite side $\overline{A_{i-1}A_i}$). This common altitude is called the *apothem*, and is denoted by a_n. Thus the area of each of our triangles is $\frac{1}{2}a_nb_n$. Let K_n be the polygonal region which is their union. Then

$$\alpha K_n = \tfrac{1}{2}na_nb_n.$$

Let s be the length of $\widehat{AB}$.

We now want to see what happens to b_n, a_n and αK_n, as $n \to \infty$.

(1) The number nb_n is the length of our inscribed broken line. Therefore

$$nb_n \leqq s.$$

Therefore

$$b_n \leqq \frac{s}{n}.$$

Since

$$\lim_{n\to\infty} \frac{s}{n} = s \lim_{n\to\infty} \frac{1}{n} = s \cdot 0 = 0,$$

it follows by the squeeze principle (Theorem 6, Chapter 31) that

$$\lim b_n = 0.$$

(2) Since the mesh of the inscribed broken line is b_n, and $\lim_{n\to\infty} b_n = 0$, it follows that

$$\lim_{n\to\infty} nb_n = s.$$

Here we are using the fact that the length of the inscribed broken line approaches s as the mesh approaches 0.

(3) Examining a typical triangle $\triangle A_{i-1}CA_i$, we see that

$$r < a_n + \frac{b_n}{2}.$$

Therefore

$$r - \frac{b_n}{2} < a_n < r.$$

Since $\lim_{n\to\infty} b_n = 0$, it follows that

$$\lim_{n\to\infty} \left(r - \frac{b_n}{2}\right) = r.$$

By the squeeze principle, this means that

$$\lim_{n\to\infty} a_n = r.$$

(4) Fitting together (2) and (3), we get

$$\lim_{n\to\infty} (\alpha K_n) = \lim_{n\to\infty} [\tfrac{1}{2} \cdot a_n \cdot nb_n] = \tfrac{1}{2}rs.$$

This proves our theorem.

21.6 APPROACHING THE AREA OF A CIRCULAR SECTOR FROM ABOVE

It would now seem natural to show that there is a sequence $L_1, L_2, \ldots$ of polygonal regions, each of them containing the sector K, such that

$$\lim_{n\to\infty} \alpha L_n = \tfrac{1}{2}rs.$$

The following theorem, however, is easier to prove, and will be sufficient for our purposes.

Theorem 1. Let K be a circular sector with radius r and boundary arc of length s, and let e be any positive number. Then there is a polygonal region L, containing K, such that

$$\alpha L < \tfrac{1}{2}rs + e.$$

To find such a polygonal region, we draw a circular sector K' with the same center, and radius $r' > r$:

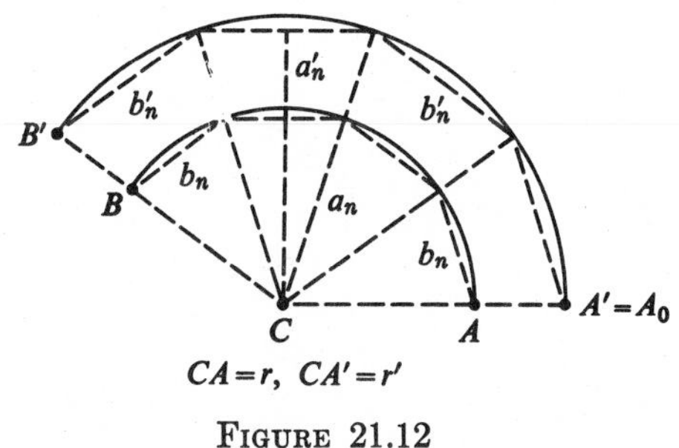

FIGURE 21.12

In the figure, the broken line inscribed in the smaller circle is the same as before; and we have also indicated the corresponding broken line inscribed in the larger circle. For the second broken line, the mesh is b'_n and the apothem is a'_n. Let L be the polygonal region inscribed in the larger sector. Then

$$\alpha L = \frac{n}{2}\, a'_n b'_n.$$

Let s' be the length of $\widehat{A'B'}$. By Theorem 3, Section 21.2, we have

$$\frac{s'}{r'} = \frac{s}{r}.$$

Therefore

$$s' = \frac{r's}{r}.$$

Therefore

$$nb'_n \leqq \frac{r's}{r}.$$

Also, by similarity,

$$\frac{a'_n}{a_n} = \frac{r'}{r},$$

so that

$$a'_n = \frac{a_n r'}{r}, \qquad \text{and} \qquad \alpha L \leqq \frac{1}{2}\,\frac{a_n r'}{r}\cdot\frac{r's}{r} \leqq \frac{1}{2}\,\frac{r'^2 s}{r}.$$

So far, all of this holds true for every $r' > r$, and for every n. What we need is to choose r' so that $\alpha L < \frac{1}{2}rs + e$, and then to choose n so that $K \subset L$.

(1) We want

$$\frac{r'^2 s}{r} < rs + 2e, \qquad \text{or} \qquad r'^2 < r^2 + \frac{2er}{s}$$

and this will be true whenever

$$r' < \sqrt{r^2 + 2er/s}.$$

We take an $r' > r$, satisfying this condition.

(2) Since $\lim_{n\to\infty} a'_n = r'$, it follows that $a'_n > r$ for some n. Such an n gives us an L that contains K.

21.7 THE AREA OF A SECTOR

We can now show that if area theory works in a reasonable fashion, then the area of a circular sector must be given by the formula

$$\alpha K = \tfrac{1}{2}rs.$$

Our conception of reasonableness is conveyed by the following assumptions.

Assumption 1. There is an area function α, defined for a class $\mathfrak{M}$ of figures. The class $\mathfrak{M}$ contains, at least, all polygonal regions and all circular sectors and disks.

Assumption 2. If K is a polygonal region, then αK is the area of K in the elementary sense.

Assumption 3. (Monotonicity.) If K and L belong to $\mathfrak{M}$ and $K \subset L$, then $\alpha K \leqq \alpha L$.

Under these three assumptions, we can prove that our formula holds. Let K_1, $K_2, \ldots$ be as in Theorem 5, Section 21.5. Then $\alpha K_n \leqq \alpha K$ for every n. Therefore

$$\tfrac{1}{2}rs = \lim_{n\to\infty} \alpha K_n \leqq \alpha K. \qquad \text{(Why?)}$$

It cannot be true that $\frac{1}{2}rs < \alpha K$. If so, let

$$e = \alpha K - \tfrac{1}{2}rs,$$

and let L be as in Theorem 1, Section 21.6. Then

$$\alpha K \leqq \alpha L < \tfrac{1}{2}rs + e = \alpha K.$$

Thus $\alpha K < \alpha K$, which is impossible. (You have a good deal of choice, in deciding how to express the contradiction in proofs like this.)

Thus there is no reasonable area theory in which the formula $\alpha k = \frac{1}{2}rs$ fails to hold. In the following chapter we shall replace this negative statement by a positive one: we shall show that there is an area function α for which our three assumptions are valid.

Chapter 22
Jordan Measure in the Plane

22.1 THE BASIC DEFINITION

It is fairly easy to define an area function for which the three assumptions that we made in the preceding section are valid. The definition goes like this. First, let α be the usual area function for polygonal regions. Now given a set K of points of the plane, let P_I be the set of all polygonal regions P that lie in K, and let N_I be the set of all numbers αP which are areas of elements of P_I. Let

$$m_I K = \sup N_I.$$

The number $m_I K$ is called the *inner measure* of K. If it happens that K contains no polygonal regions at all, then we agree that $m_I K = 0$. Thus the inner measure of a point or a segment is always $= 0$.

Suppose now that K is contained in at least one polygonal region. Let P_0 be the set of all polygonal regions P that *contain* K. Let N_0 be the set of all numbers αP which are areas of elements of P_0. Let

$$m_0 K = \inf N_0.$$

Here $\inf N_0$ is the greatest lower bound of N_0. The number $\inf N_0$ is called the *outer measure* of K.

If $P \in P_I$ and $P' \in P_0$, then

$$P \subset K \subset P'.$$

Therefore

$$P \subset P', \qquad \text{and} \qquad \alpha P \leqq \alpha P'.$$

Thus every element of N_I is less than or equal to every element of N_0. The two sets of numbers must therefore look like the figures below. The figures suggest that we must have

$$m_I K \leqq m_0 K.$$

FIGURE 22.1

And in fact this always holds. If it were true that

$$m_0K < m_IK,$$

then there would be a polygonal region P, lying in K, such that

$$m_0K < \alpha P \leqq m_IK.$$

And αP is a lower bound of N_0, because

$$\alpha P \leqq \alpha P'$$

for every P' in P_0. Therefore m_0K is not the greatest lower bound of N_0; and this is impossible, because m_0K was defined to be inf N_0.

Thus

$$m_IK \leqq m_0K.$$

If the equality holds, then we say that K is *measurable in the sense of Jordan,* and we define the *measure* of K to be

$$mK = m_IK = m_0K.$$

Since we shall be talking about only one kind of measure theory in this book, we shall say for short that K is *measurable* if $m_IK = m_0K$. We now have the following theorems.

Theorem 1. Every polygonal region P is measurable; and $mP = \alpha P$.

Proof. P belongs to P_I, because $P \subset P$. And if $P' \subset P$, then $\alpha P' \leqq \alpha P$. Therefore $m_IP = \alpha P$.

Similarily, P belongs to P_0, because $P \supset P$. And if $P' \supset P$, then $\alpha P' \geqq \alpha P$. Therefore $m_0P = \alpha P$. Therefore

$$mP = m_0P = m_1P = \alpha P,$$

which was to be proved.

Theorem 2. Outer measure is monotonic. That is, if $A \subset B$, then $m_0A \leqq m_0B$.

Theorem 3. Measure is monotonic. That is, if A and B are measurable, and $A \subset B$, then $mA \leqq mB$.

The proofs are left as exercises.

Theorem 4. Every circular sector K is measurable. If the radius is r, and the boundary arc has length s, then

$$mK = \tfrac{1}{2}rs.$$

Proof. The results of the preceding section tell us that

$$m_I K \geqq \tfrac{1}{2}rs, \qquad \text{and} \qquad m_O K \leqq \tfrac{1}{2}rs.$$

Since $m_I K \leqq m_O K$, it follows that

$$m_I K = m_O K = \tfrac{1}{2}rs,$$

which was to be proved.

If you review our investigation of circular arcs in the preceding chapter, you will see that the definition of Jordan measure is modeled on it. One way of putting it is to say that in the preceding chapter we gave a proof, and that in the present section our task has been to find the theorem that our proof proves.

By a discussion exactly analogous to the discussion in Section 21.7, we can show that if K is a disk of radius r and hence of circumference $s = 2\pi r$, then

$$m_I K = \tfrac{1}{2}rs = m_O K.$$

Thus we have the following theorem.

Theorem 5. Every disk K is measurable, and

$$mK = \pi r^2,$$

where r is the radius.

PROBLEM SET 22.1

1. Show that every point P forms a measurable set, and that $mP = 0$.
2. Show that every segment $\overline{AB}$ is measurable, and that $m\overline{AB} = 0$.
3. Prove Theorem 2.
4. Prove Theorem 3.
5. Consider a plane with a coordinate system. Let K be the unit square:

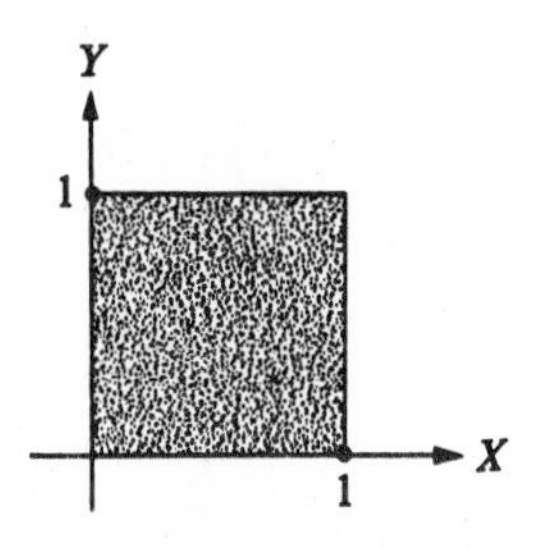

FIGURE 22.2

Let L be the set of all rational points of K. That is, (x, y) belongs to L if x and y are both rational, $0 \leqq x \leqq 1$ and $0 \leqq y \leqq 1$. Is L a measurable set? Why or why not?

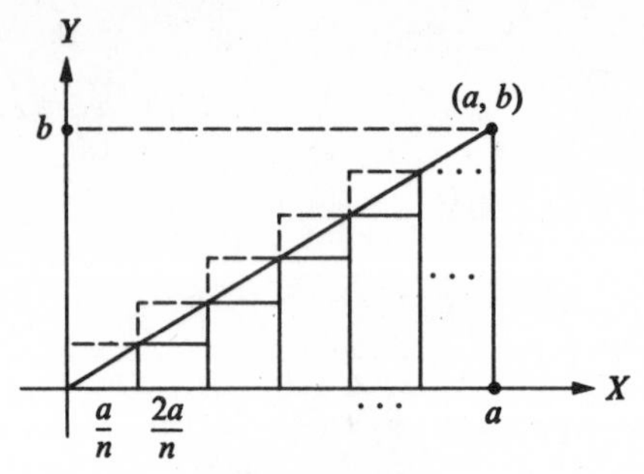

FIGURE 22.3

6. In a coordinate plane, consider a right triangle in the figure above. We divide the interval [0, a] into n congruent segments, each of length a/n, and construct an inscribed polygonal region P_n and a circumscribed polygonal region P'_n. (In the figure, the boundary of P_n is drawn solid, and that of P'_n is dashed.) Calculate αP_n and $\alpha P'_n$. Verify algebraically that

$$\sup\{\alpha P_n\} = \inf\{\alpha P'_n\} = \tfrac{1}{2}ab.$$

This problem throws some light on the theory. It means that if we know only about areas of rectangles, we can calculate the Jordan measures of right triangles in the indicated position.

22.2 THE CLASS OF MEASURABLE SETS

Let $\mathfrak{M}$ be the class of all measurable sets in the plane. (Here, as usual, we use the word *class* as a synonym for the word set.) We shall show that this class has the following simple properties.

Theorem 1. If M_1 and M_2 belong to $\mathfrak{M}$, then so also does $M_1 \cup M_2$.

It will follow by induction, that:

Theorem 2. *Finite Additivity.* If each of the sets $M_1, M_2, \ldots, M_n$ belongs to $\mathfrak{M}$, then so also does their union

$$M = \bigcup_{i=1}^{n} M_i.$$

Theorem 3. If M_1 and M_2 belong to $\mathfrak{M}$, then so also does $M_1 - M_2$.

Theorem 4. If M_1 and M_2 belong to $\mathfrak{M}$, then so also does $M_1 \cap M_2$.

Theorem 5. If M_1 and M_2 belong to $\mathfrak{M}$, and

$$M_1 \cap M_2 = 0,$$

then

$$m(M_1 \cup M_2) = mM_1 + mM_2.$$

Theorem 6. If M_1 and M_2 belong to $\mathfrak{M}$, and $M_1 \subset M_2$, then

$$m(M_2 - M_1) = mM_2 - mM_1.$$

Theorem 7. If M_1 and M_2 belong to $\mathfrak{M}$, and

$$m(M_1 \cap M_2) = 0,$$

then

$$m(M_1 \cup M_2) = mM_1 + mM_2.$$

The proofs begin with the following lemma.

Lemma 1. Let M be a measurable set in the plane. Then for every positive number e there are polygonal regions P and P' such that

$$P \subset M \subset P' \tag{1}$$

and

$$\alpha P' - \alpha P < e. \tag{2}$$

Proof. Take P so that $P \subset M$ and

$$\alpha P > m_I M - \frac{e}{2}.$$

Take P' so that $M \subset P'$ and

$$\alpha P' < m_0 M + \frac{e}{2}.$$

Then

$$-\alpha P < -m_I M + \frac{e}{2};$$

and by addition, we get

$$\alpha P' - \alpha P < e.$$

(We are using, of course, the fact that $m_0 M - m_I M = 0$.)

The converse is also true.

Lemma 2. Let M be a set of points in the plane. Suppose that for every positive number e there are polygonal regions P, P' such that

$$P \subset M \subset P' \tag{1}$$

and

$$\alpha P' - \alpha P < e. \tag{2}$$

Then M is measurable.

Proof. Given such regions P, P', we have

$$m_0 M \leqq \alpha P', \qquad \alpha P \leqq m_I M.$$

Therefore

$$-m_I M \leqq -\alpha P,$$

and so

$$m_0 M - m_I M \leqq \alpha P' - \alpha P < e.$$

Since $m_0M - m_IM < e$ for every positive number e, and $m_0M - m_IM \geqq 0$, it follows that $m_0M - m_IM$ must be $= 0$. Therefore M is measurable, which was to be proved.

We shall now prove Theorem 1. Consider two measurable sets M_1, M_2. Let e be any positive number:

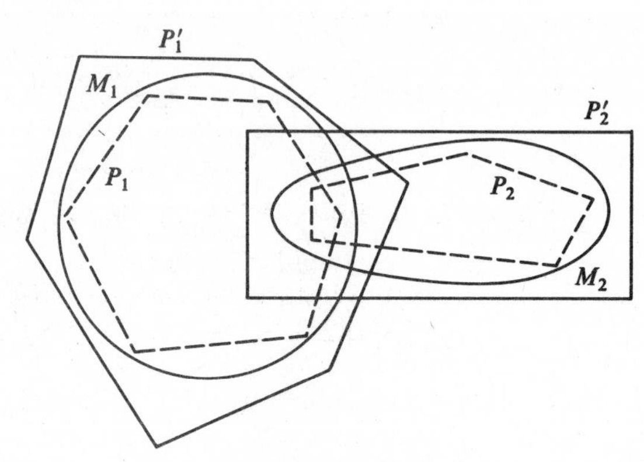

FIGURE 22.4

Take P_1, P'_1 for M_1, as in Lemma 1, so that

$$P_1 \subset M_1 \subset P'_1$$

and

$$\alpha P'_1 - \alpha P_1 < \frac{e}{2}.$$

Take P_2, P'_2 for M_2 so that

$$P_2 \subset M_2 \subset P'_2$$

and

$$\alpha P'_2 - \alpha P_2 < \frac{e}{2}.$$

Let

$$P = P_1 \cup P_2,$$

and let

$$P' = P'_1 \cup P'_2.$$

Then

$$\alpha P' - \alpha P \leqq (\alpha P'_1 - \alpha P_1) + (\alpha P'_2 - \alpha P_2).$$

To see how this works, consider a simpler figure:

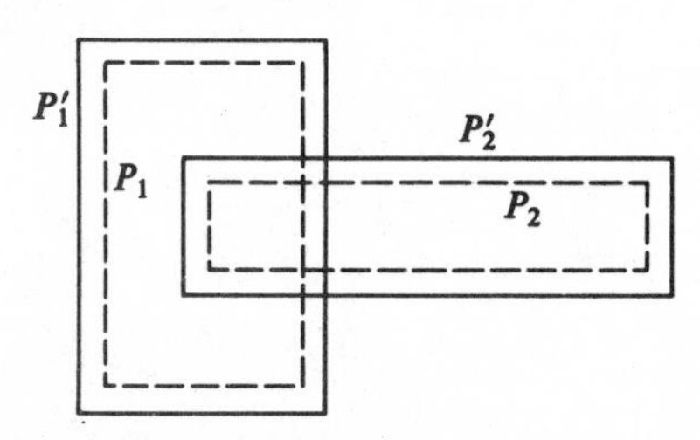

FIGURE 22.5

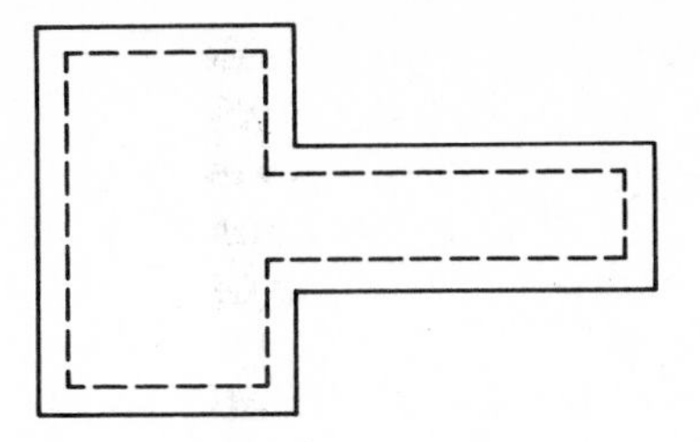

FIGURE 22.6

Here $\alpha P' - \alpha P$ is the area of the thin strip in the figure above. This is $\leqq$ the sum of the differences

$$\alpha P_1' - \alpha P_1, \qquad \alpha P_2' - \alpha P_2.$$

The same principle applies in the general case. If we form a suitable triangulation of $P_1' \cup P_2'$, then each of the three differences

$$\alpha P' - \alpha P, \qquad \alpha P_1' - \alpha P_1, \qquad \alpha P_2' - \alpha P_2$$

is the sum of the areas of a collection of triangular regions, and every triangle that contributes to $\alpha P - \alpha P'$ must contribute at least once, and perhaps twice, to $(\alpha P_1' - \alpha P_1) + (\alpha P_2' - \alpha P_2)$.

Therefore

$$\alpha P' - \alpha P < \frac{e}{2} + \frac{e}{2} = e,$$

and so $M_1 \cup M_2$ is measurable, which was to be proved.

Theorem 2 follows, as we pointed out, from Theorem 1.

To prove Theorem 3, we use the same figures. To prove that $M_1 - M_2$ is measurable, we form a triangulation of $P_1' \cup P_2'$ in which each of the regions P_1, P_1', P_2, P_2' is a union of triangular regions, intersecting only in edges and vertices. Let P', this time, be the union of all the triangular regions that lie in P_1', but not in P_2. Let P be the union of all of the triangular regions that lie in P_1, but not in P_2'. Then P and P' look like this:

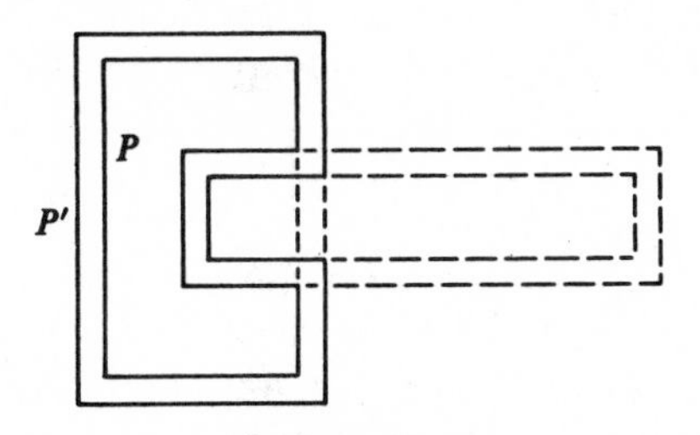

FIGURE 22.7

Here the boundaries of P and P' are drawn solid, and the rest of the figure is dashed merely to remind us of how P and P' were defined. We recall that e was any positive number, and $\alpha P_1' - \alpha P_1 < e/2$; $\alpha P_2' - \alpha P_2 < e/2$.

Evidently,

$$\alpha P' - \alpha P \leqq (\alpha P_1' - \alpha P_1) + (\alpha P_2' - \alpha P_2).$$

Therefore

$$\alpha P' - \alpha P < \frac{e}{2} + \frac{e}{2} = e;$$

and so $M_1 - M_2$ is measurable, which was to be proved.

To prove that $M_1 \cap M_2$ is measurable, we merely need to manipulate sets, without using either geometry or algebra. A figure will make it easier to keep track of what we are doing.

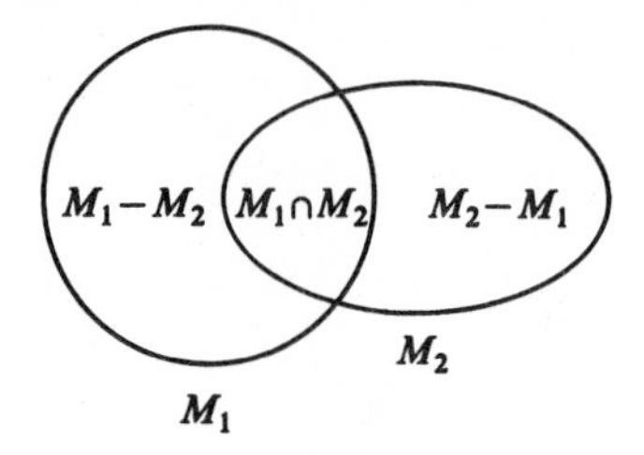

FIGURE 22.8

Given $M_1 \in \mathfrak{M}$, $M_2 \in \mathfrak{M}$. By Theorem 1,

$$M_1 \cup M_2 \in \mathfrak{M}.$$

By Theorem 3,

$$M_2 - M_1 \in \mathfrak{M}, \qquad M_1 - M_2 \in \mathfrak{M}.$$

Therefore, by Theorem 1,

$$(M_1 - M_2) \cup (M_2 - M_1) \in \mathfrak{M}.$$

Hence, by Theorem 3.

$$(M_1 \cup M_2) - [(M_1 - M_2) \cup (M_2 - M_1)] \in \mathfrak{M}.$$

This proves Theorem 4, because the set we have described is precisely $M_1 \cap M_2$.

Now for Theorem 5, which says that if M_1 and M_2 do not intersect, then

$$m(M_1 \cup M_2) = mM_1 + mM_2.$$

First we observe that if P is a polygonal region lying in $M_1 \cup M_2$, then $P = P_1 \cup P_2$, where $P_1 \subset M_1$ and $P_2 \subset M_2$:

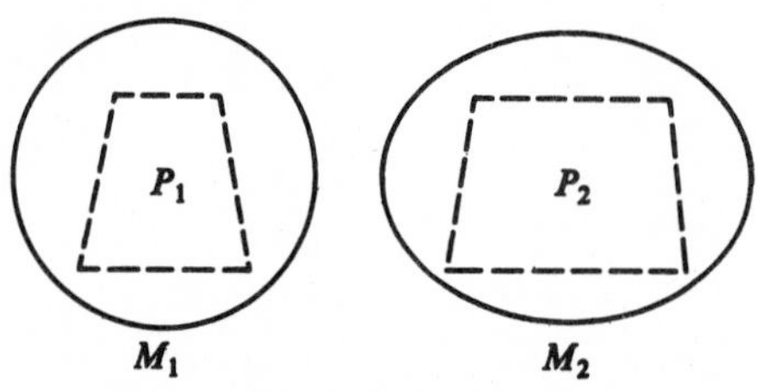

FIGURE 22.9

Therefore

$$\alpha P = \alpha P_1 + \alpha P_2 \leqq m_I M_1 + m_I M_2.$$

Therefore

$$m_I(M_1 \cup M_2) \leqq m_I M_1 + m_I M_2,$$

because $m_I(M_1 \cup M_2)$ is the *least* of the upper bounds of the set of numbers αP.

On the other hand, given any $e > 0$, we can find $P_1 \subset M_1$ and $P_2 \subset M_2$ so that

$$\alpha P_1 > m_I M_1 - \frac{e}{2}, \qquad \alpha P_2 > m_I M_2 - \frac{e}{2}.$$

We then have

$$\alpha P > m_I M_1 + m_I M_2 - e.$$

Therefore

$$m_I(M_1 \cup M_2) \geqq m_I M_1 + m_I M_2 - e$$

for every $e > 0$. Hence

$$m_I(M_1 \cup M_2) \geqq m_I M_1 + m_I M_2.$$

Since we already have the reverse inequality, it follows that the equality holds. But all three of our sets are measurable. Therefore

$$m(M_1 \cup M_2) = mM_1 + mM_2,$$

which was to be proved.

This is the last algebraic proof in this section; we get the rest of our theorems from the earlier ones.

Proof of Theorem 6. Given $M_1 \subset M_2$, we have

$$M_2 = (M_2 - M_1) \cup M_1,$$

and the two sets on the right do not intersect. Therefore

$$mM_2 = m(M_2 - M_1) + mM_1,$$

so that

$$m(M_2 - M_1) = mM_2 - mM_1,$$

as desired.

Proof of Theorem 7. Given $m(M_1 \cap M_2) = 0$, we want to show that

$$m(M_1 \cup M_2) = mM_1 + mM_2.$$

First we observe (Fig. 22.10) that

$$M_1 - M_2 = M_1 - (M_1 \cap M_2)$$

and

$$M_2 - M_1 = M_2 - (M_1 \cap M_2).$$

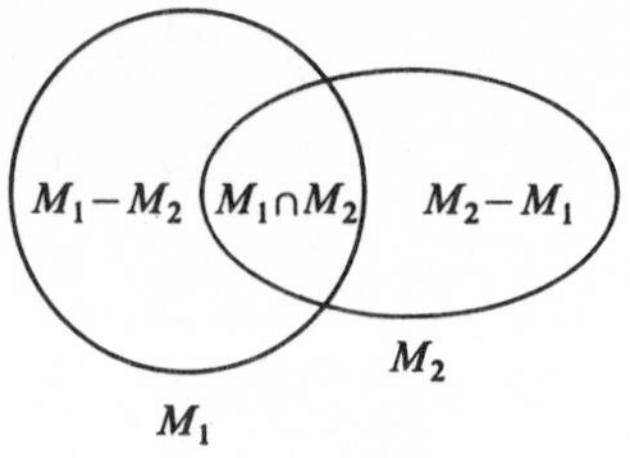

FIGURE 22.10

Therefore

$$m(M_1 - M_2) = mM_1 - 0, \qquad \text{and} \qquad m(M_2 - M_1) = mM_2 - 0.$$

Now

$$M_1 \cup M_2 = (M_1 - M_2) \cup (M_1 \cap M_2) \cup (M_2 - M_1);$$

therefore

$$m(M_1 \cup M_2) = mM_1 + mM_2 + 0,$$

which was to be proved.

PROBLEM SET 22.2

All of the following problems are to be solved on the basis of the theorems proved in this chapter, plus, of course, our old theorems on areas of polygonal regions.

1. Let C be a circle of radius r, let L be the interior of C, and let K be the disk $C \cup L$. What is $m_I L$? What is $m_O L$? Why, in each case?

2. Show that every circle is measurable, and that its measure is $= 0$.

3. Show the same, for arcs or circles.

4. Show that if $m_O M = 0$, then every subset of M is measurable, and has measure $= 0$.

5. Show that the interior of a triangle is always measurable. What is the measure of such a set?

6. A *segment* of a circle is a figure like this:

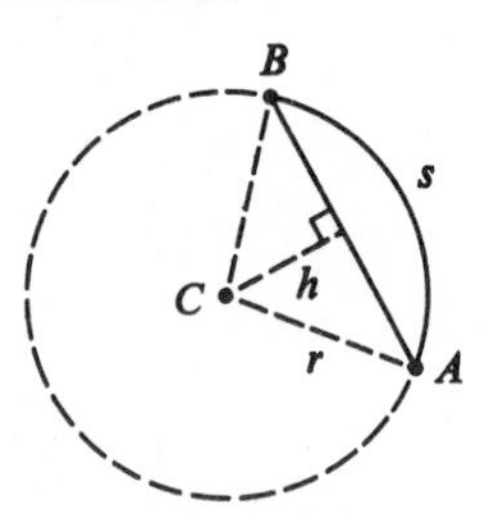

FIGURE 22.11

Prove that a segment of a circle is always measurable, and find its measure. [*Warning:* If K is the sector with boundary arc $\widehat{AB}$, and L is the segment, then it is *not* true that $L = K - \triangle ABC$. The first step in solving the problem is to get a correct expression for L.]

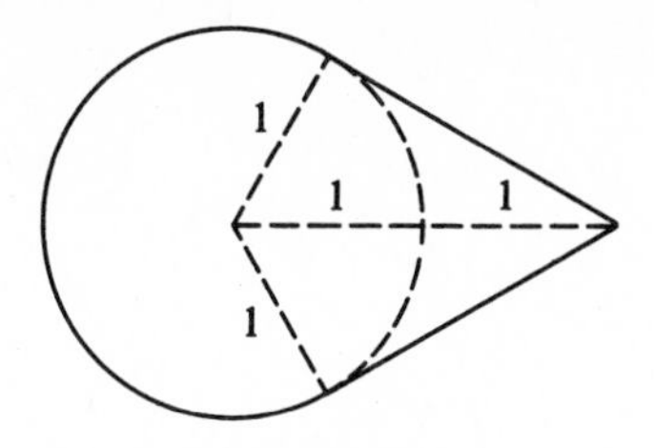

FIGURE 22.12

7. Show that the region indicated in the figure above is measurable, and find its measure.

22.3 AREAS UNDER THE GRAPHS OF CONTINUOUS FUNCTIONS

Jordan measure is the theory that is needed to fill a certain gap in elementary calculus. Suppose that we have a region R, bounded by the graphs of two continuous functions.

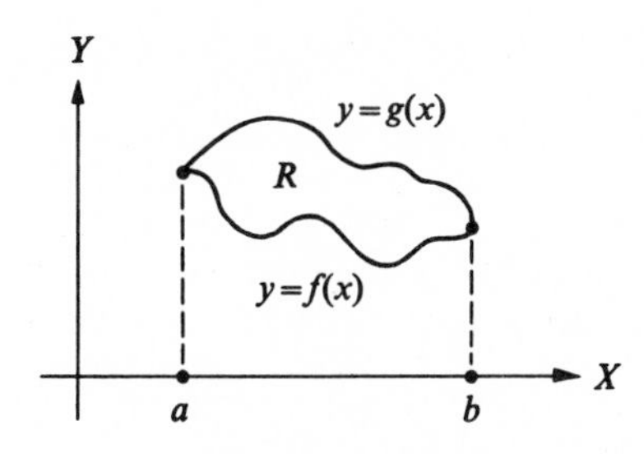

FIGURE 22.13

Here

$$f(x) \leqq g(x), \qquad a \leqq x \leqq b,$$

and

$$R = \{(x, y) | a \leqq x \leqq b \text{ and } f(x) \leqq y \leqq g(x)\}.$$

We compute the area of R by the formula

$$mR = \int_a^b [g(x) - f(x)]\, dx.$$

The usual derivations of this area formula are not proofs in any strict sense, because they do not appeal to any valid definition of area. These derivations are persuasive. And now that we know about Jordan measure, it is easy to see that they show that the definite integral gives the Jordan measure of the region. The situation here is much the same as for circles. The elementary discussion becomes adequate, as soon as we supply the definition to which it tacitly appeals.

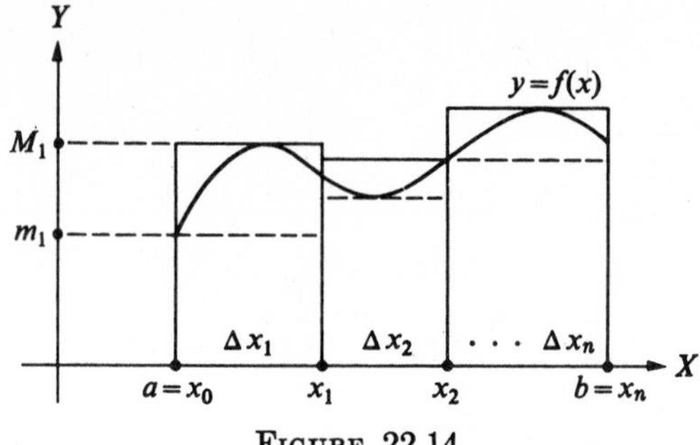

FIGURE 22.14

To see this, consider first the case where R is the region under a single positive continuous function $f(x)$ as shown in the figure above. Here

$$f(x) \geqq 0, \qquad a \leqq x \leqq b,$$

and

$$R = \{(x, y) | a \leqq x \leqq b \text{ and } 0 \leqq y \leqq f(x)\}.$$

As in the definition of the definite integral, we take an ascending sequence of points,

$$a = x_0 < x_1 < \cdots < x_n = b,$$

on the interval $[a, b]$. Let Δx_i be the length of the ith subinterval. Let m_i be the minimum value of $f(x)$ on the ith subinterval, and let M_i be the maximum value of $f(x)$ on the ith subinterval. (In the figure, we have indicated m_1 and M_1.) Then

$$\sum_{i=1}^{n} m_i \Delta x_i$$

is the sum of the areas of the *inscribed* rectangles (with dashed upper bases in the figure); and

$$\sum_{i=1}^{n} M_i \Delta x_i$$

is the sum of the areas of the circumscribed rectangles (drawn solid in the figure).

There are various ways of setting up the definite integral; we shall not review the theory here. But in any case it turns out that

$$\Sigma m_i \Delta x_i \leqq \int_a^b f(x)\, dx \leqq \Sigma M_i \Delta x_i.$$

And, for continuous functions $f(x)$, we can make the difference,

$$\Sigma M_i \Delta x_i - \Sigma m_i \Delta x_i,$$

as small as we please, merely by taking all of the numbers Δx_i sufficiently small. Thus, for any $e > 0$ there is a sequence,

$$a = x_0 < x_1 < \cdots < x_n = b,$$

for which

$$\Sigma M_i \Delta x_i - \Sigma m_i \Delta x_i < e.$$

Here the first sum is the area αP of a polygonal region lying in R, and the second is the area of a polygonal region P' containing R. By Lemma 2 of the preceding section, this means that R is measurable: $m_I R = m_O R$. Since the integral is *an* upper bound of the numbers αP, we have

$$m_I R \leqq \int_a^b f(x)\, dx.$$

Similarly, the integral is a lower bound of the numbers $\alpha P'$, and so

$$\int_a^b f(x)\, dx \leqq m_O R.$$

Since $m_I R = m_O R = mR$, it follows that

$$mR = \int_a^b f(x)\, dx,$$

which was to be proved.

The extension to the case where R is bounded by the graphs of two continuous functions is not hard. We shall not go into it here.

Many calculus books, including some otherwise excellent ones, try to *define* the area of such a region as the definite integral. This will not work. The area of a region ought to depend merely on its size and shape, and not on the way it is placed relative to the axes. That is, the area of a region ought to be unchanged under rigid motions. This is not clear if we use the definite integral to *define* the area. The trouble is that the same region may be described in infinitely many ways as the region between the graphs of two continuous functions. Here R_1

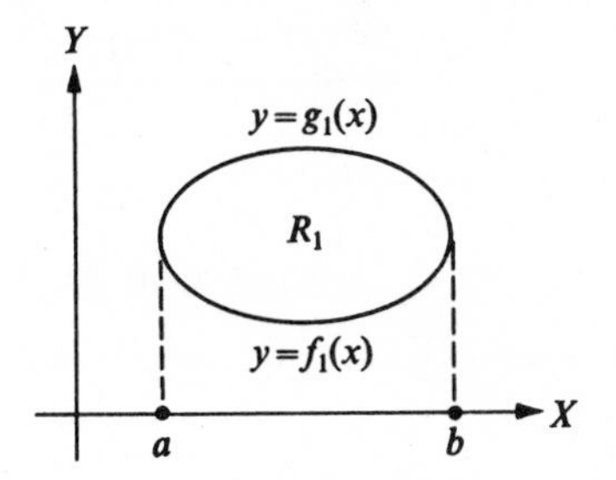

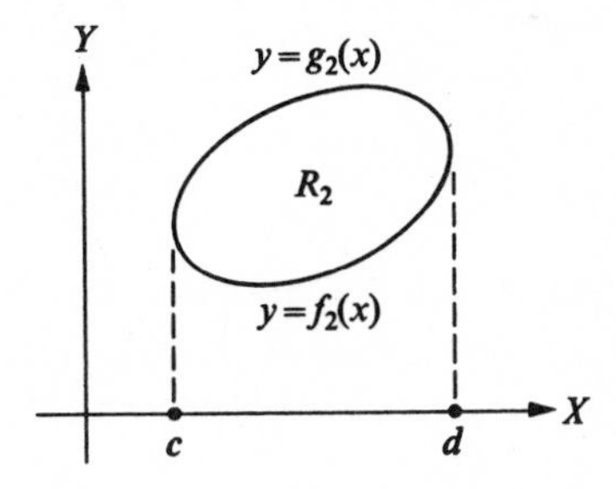

FIGURE 22.15

and R_2 are isometric. In this chapter of this book we can infer that

$$\int_a^b [g_1(x) - f_1(x)]\, dx = \int_c^d [g_2(x) - f_2(x)]\, dx.$$

The reason is that both of the integrals give the right answer to the same question. But if your definition of measure is stated in terms of a coordinate system, you are left with the problem of proving *by calculus* that the integrals have the same value; and this is not a practical enterprise.

If you review the theorems given in this chapter so far, you will see that Jordan measure theory has, at this point, been brought down to earth. Given a figure whose measure you would normally expect to calculate, the chances are that you can prove that the figure is measurable in the sense of Jordan, by using the theorems of this section and the preceding one. (In fact, a certain amount of effort is required to think of a figure which is *not* measurable in the sense of Jordan.) Moreover, the theory at this point is adequate to justify the elementary methods of calculating plane areas. The theory of measure has been generalized, by Henri Lebesgue, in such a way as to assign areas to an even larger class of figures. But Jordan measure is adequate for the purposes of elementary mathematics.

Problem Set 22.3

1. Given that $f(x)$ is continuous and $\geqq 0$ for $a \leqq x \leqq b$. Let

$$R' = \{(x, y) | a \leqq x \leqq b \text{ and } 0 \leqq y < f(x)\}.$$

(Here $y < f(x)$ is not a misprint; the graph itself is not in R'.) Show that R' is measurable, and that $mR' = mR$.

2. Show that if F is the graph of a continuous function, for $a \leqq x \leqq b$, then $mF = 0$.

3. Show that if R is a region of the sort described at the beginning of this section, then mR is the integral of the difference of the two functions.

Chapter 23
Solid Mensuration: The Elementary Theory

23.1 BASIC ASSUMPTIONS FOR THE THEORY OF VOLUME

The theory of volume, carried out in a spirit like that of the preceding chapter, is technically difficult. Moreover, the work required is not worthwhile, because anyone pursuing the theory at such length should study not the theory of Jordan but that of Lebesgue, which has superseded it for the purposes of advanced mathematics. For this reason, we shall treat the theory of volume only in a style analogous to that of Chapter 21, basing our derivations on postulates, and not attempting to describe a volume function which satisfies our postulates.

Suppose, then, that we have given a class $\mathcal{V}$ of sets of points in space, called *measurable* sets. Here $\mathcal{V}$ stands for *volume.* We suppose also that we have a function

$$v : \mathcal{V} \to \mathbf{R}$$

of $\mathcal{V}$ into the nonnegative real numbers. If $M \in \mathcal{V}$, then vM will be called the *volume* of M.

Our first two postulates are designed to ensure that the elementary solids whose volumes we propose to discuss really do have volumes.

V-1. Every convex set is in $\mathcal{V}$.

V-2. If M and N belong to $\mathcal{V}$, then $M \cup N$, $M \cap N$, and $M - N$ also belong to $\mathcal{V}$.

The rest of our assumptions deal with the volume function v.

V-3. v is monotonic. That is, if $M, N \in \mathcal{V}$, and $M \subset N$, then $vM \leqq vN$.

V-4. If $M, N \in \mathcal{V}$, and $v(M \cap N) = 0$, then

$$v(M \cup N) = vM + vN.$$

Obviously, if v is a volume function satisfying these conditions, then we can get more such functions by multiplying v by any positive constant. To "determine the unit of volume," we state the following.

V-5. The volume of a parallelepiped is the product of its altitude and the area of its base.

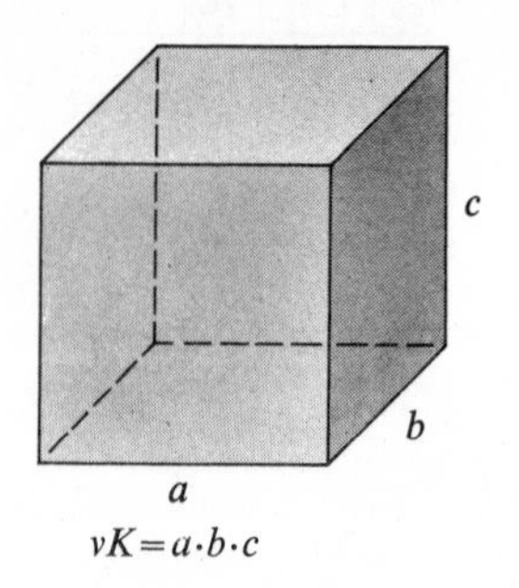

FIGURE 23.1

Here we are assuming that you know what a parallelepiped is. If not, see Section 23.3 below. The figure and the formula remind us that any face of a parallelepiped can be regarded as the base.

We recall that triangular regions with the same base and altitude always have the same area. Our final assumption for volume is analogous to this:

V-6. *Cavalieri's Principle.* Let K and K' be figures in space. Let E_0 be a plane. If $E \parallel E_0$, then $K \cap E$ and $K' \cap E$ will be called corresponding horizontal cross sections of K and K', respectively. Suppose that

(1) $K \in \mathcal{V}$;
(2) $K' \in \mathcal{V}$;
(3) every horizontal cross section of K is measurable in its own plane;
(4) every horizontal cross section of K' is measurable in its own plane, and
(5) corresponding horizontal cross sections D, D' of K and K' have the same measure in their plane. Then

$$vK = vK'.$$

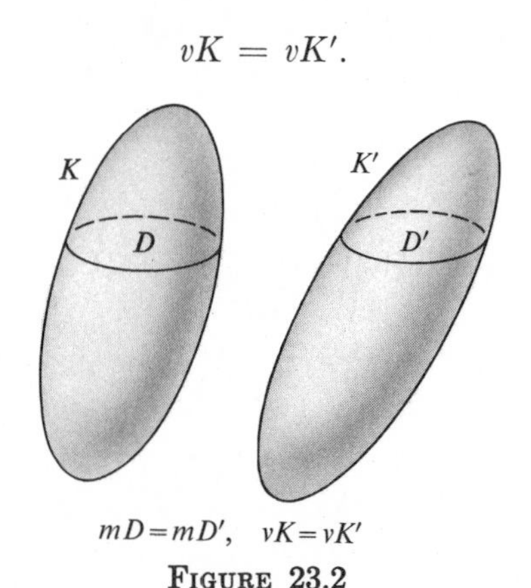

FIGURE 23.2

This principle is the key to the study of volume; it deserves a careful reading. Obviously hypothesis (5) is crucial; without it, Cavalieri's principle couldn't possibly be right. But the other four hypotheses are necessary too. In fact, if $K \in \mathcal{V}$, it does not follow that all horizontal cross sections of K are measurable in their own planes. Also, if all horizontal cross sections of K are measurable in their own planes, it does not follow that K is measurable, and similarly, of course, for K'.

Before we can proceed to calculate volumes, we need to do a little preliminary geometry.

23.2 CROSS SECTIONS OF CONES AND PYRAMIDS

Let D be a (circular) disk in a plane E, and let V be a point not in E. The *cone with base D and vertex V* is the union of all segments $\overline{VP}$, where P belongs to D. The *altitude* of the cone is the perpendicular distance from V to E.

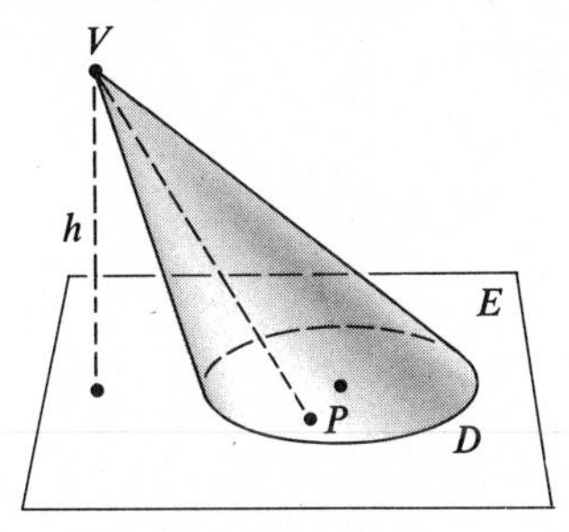

FIGURE 23.3

Now let E_k be a plane parallel to E, on the same side of E as V, lying at a perpendicular distance $k < h$ from E. Let C be the center of D, and let r be the radius. For each point B of D, let B' be the point in which $\overline{VB}$ intersects E_k. Let D_k be the set of all such points B'. We call D_k the *cross section of the cone at altitude k*.

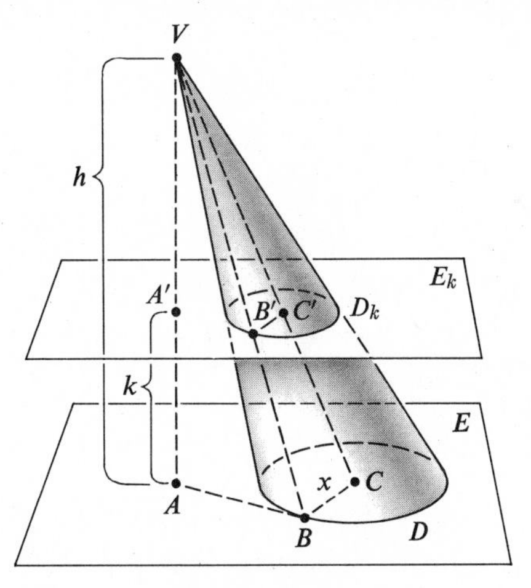

FIGURE 23.4

Theorem 1. D_k is a disk of radius

$$r' = \frac{h-k}{h} r.$$

Proof. Let A, A', C and C' be as in the figure above. Let B be any point of D, and let B' be the corresponding point of D'. Since $E_k \parallel E$, any plane that intersects E_k and E intersects them in a pair of parallel lines. Therefore

$$\overline{A'B'} \parallel \overline{AB} \quad \text{and} \quad \overline{B'C'} \parallel \overline{BC}.$$

Hence $\qquad \Delta VA'B' \sim \Delta VAB \qquad$ and $\qquad \Delta VB'C' \sim \Delta VBC.$

Thus
$$\frac{VA'}{VA} = \frac{h-k}{h} = \frac{A'B'}{AB} = \frac{C'B'}{CB}.$$
Then
$$C'B' = \frac{h-k}{h} CB.$$
If $CB \leqq r$, it follows that
$$C'B' \leqq \frac{h-k}{h} r,$$
and conversely. Therefore D_k is a disk of radius
$$r' = \frac{h-k}{h} r,$$
which was to be proved.

Using the formula $mD = \pi r^2$, we get the following.

Theorem 2. Let K be a cone with base area a and altitude h. Let D_k be the cross section of K at altitude k. Then
$$mD_k = \left(\frac{h-k}{h}\right)^2 a.$$

Proof. We know that
$$a = \pi r^2 \qquad \text{and} \qquad mD_k = \pi r'^2,$$
where r is the radius of the base, and r' is the radius of D_k. Therefore
$$mD_k = \pi \left(\frac{h-k}{h}\right)^2 r^2 = \left(\frac{h-k}{h}\right)^2 a,$$
which was to be proved.

If we use a triangular region as base, instead of a disk, we have a similar discussion leading to a similar theorem.

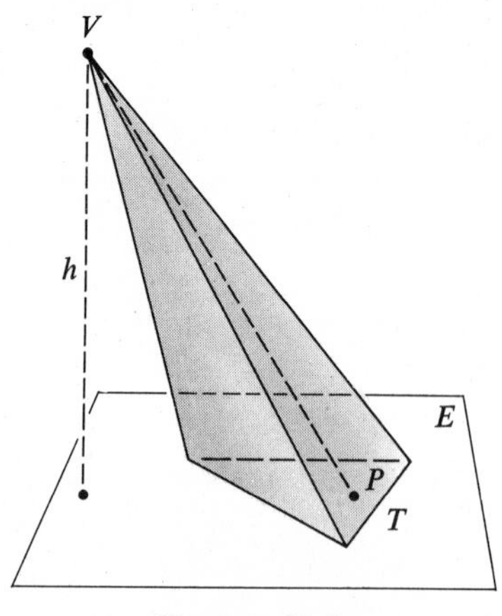

FIGURE 23.5

Let T be a triangular region lying in a plane E, and let V be a point not in E. The *pyramid with base T and vertex V* is the union of all segments $\overline{VP}$, where P belongs to T. The *altitude* of the pyramid is the perpendicular distance h from V to E.

Let E_k be the plane parallel to E, on the side of E that contains V, such that the perpendicular distance between E and E_k is $k < h$:

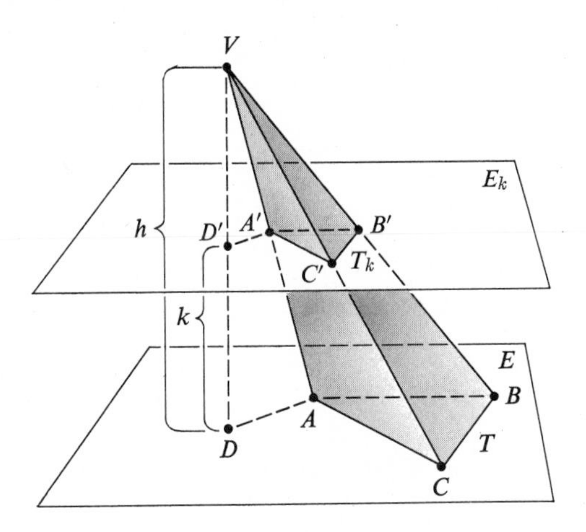

FIGURE 23.6

Let the boundary of T be $\triangle ABC$, and let A', B', and C' be the points where E_k intersects $\overline{VA}$, $\overline{VB}$, and $\overline{VC}$. Let T_k be the triangular region corresponding to $\triangle A'B'C'$. We then have

$$\triangle VAD \sim \triangle VA'D', \qquad \triangle VBC \sim \triangle VB'C',$$
$$\triangle VAB \sim \triangle VA'B', \qquad \triangle VAC \sim \triangle VA'C'.$$

The reason is that if a plane intersects each of two parallel planes, it intersects them in two parallel lines. Therefore

$$\overline{D'A} \parallel \overline{DA}, \qquad \overline{A'B'} \parallel \overline{AB},$$
$$\overline{B'C'} \parallel \overline{BC}, \qquad \overline{A'C'} \parallel \overline{AC}.$$

From these parallelisms, our similarities follow. Therefore

$$\frac{VA'}{VA} = \frac{VD'}{VD} = \frac{h-k}{h}, \qquad \frac{B'C'}{BC} = \frac{h-k}{h},$$

$$\frac{A'C'}{AC} = \frac{VA'}{VA} = \frac{h-k}{h}, \qquad \frac{A'B'}{AB} = \frac{h-k}{h}.$$

Hence

$$\triangle A'B'C' \sim \triangle ABC,$$

by the SSS similarity theorem (Theorem 3, Section 12.2). And the areas of the corresponding regions T, T_k are related by the formula

$$mT_k = \left(\frac{h-k}{h}\right)^2 mT.$$

(This follows from Theorem 8, Section 13.2.) Thus we have the following theorem.

Theorem 3. Let P be a pyramid with base area a and altitude h. Let T_k be its cross section at altitude k. Then

$$mT_k = \left(\frac{h-k}{h}\right)^2 a.$$

23.3 PRISMS AND CYLINDERS

Given a pair of parallel planes E_1, E_2, a set B_1 lying in E_1, and a line L which intersects E_1 in exactly one point:

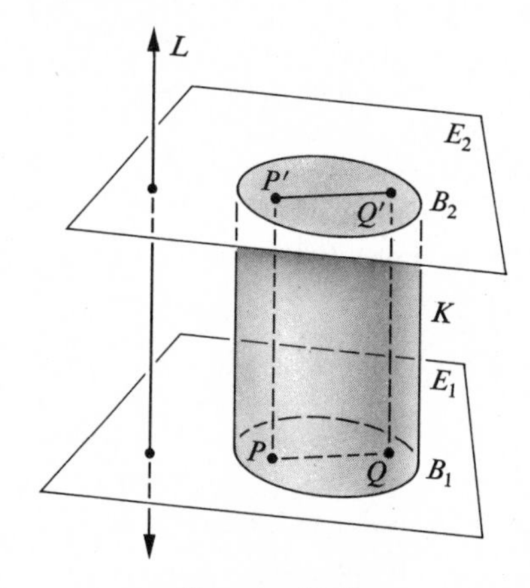

FIGURE 23.7

Through each point P of B_1 we take a segment $\overline{PP'}$, parallel to L, with P' in E_2. The union K of all such segments $\overline{PP'}$ is called a *cylinder;* B_1 is called its lower *base,* and L is called its *directrix.* The intersection B_2 of the cylinder with E_2 is called its upper base. (Thus B_2 is the set of all points P'.) The perpendicular distance between E_1 and E_2 is called the *altitude* of the cylinder. And the segments $\overline{PP'}$ are called *elements* of the cylinder.

If B_1 is a disk, then K is called a *circular cylinder.* If B_1 is a polygonal region, then K is called a *prism.* If L is perpendicular to E_1, then K is called a *right cylinder.* If a cylinder has a parallelogram region as its base, then it is called a *parallelepiped.* If P and Q are points of B_1, and P' and Q' are the corresponding points of B_2, then it is not hard to see that $\square PP'Q'Q$ is a parallelogram.

Proof. The lines $\overleftrightarrow{PP'}$, $\overleftrightarrow{QQ'}$ are parallel, because they are both parallel to L (unless one of them is $=L$, in which case the same conclusion follows). Therefore $\overleftrightarrow{PP'}$ and $\overleftrightarrow{QQ'}$ are automatically coplanar, lying in a plane E; E intersects each of the planes E_1, E_2, and E therefore intersects them in a pair of parallel lines. Therefore $\overleftrightarrow{P'Q'} \parallel \overleftrightarrow{PQ}$. Therefore $\square PP'Q'Q$ is a parallelogram. Thus we have the following theorem.

Theorem 1. In any cylinder, the lower base and the upper base are isometric.

Proof. In defining the cylinder, we used a correspondence $P \leftrightarrow P'$ between the upper base and the lower base. Since opposite sides of a parallelogram are congruent, we always have

$$PQ = P'Q'.$$

Therefore our correspondence is an isometry.

In particular, if the cylinder is circular, the two bases are always disks with the same radius. If the cylinder is a prism with a triangular base, then the two bases are always congruent. If the cylinder is a prism, with any polygonal region whatever as base, then the two bases, being isometric, always have the same area. All of this is conveyed by Theorem 1; the basic idea here has nothing to do with the shape of the base.

By a *horizontal cross section* of a cylinder, we mean the intersection of the cylinder with a plane parallel to the base.

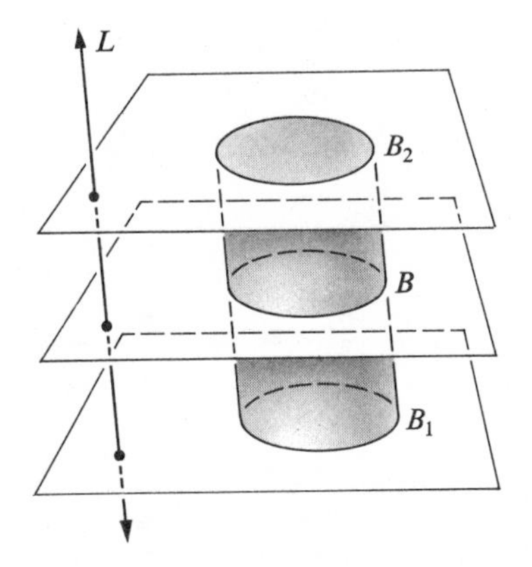

FIGURE 23.8

Theorem 2. All horizontal cross sections of a cylinder are isometric.

This follows immediately from Theorem 1. To see this, we merely need to observe that every horizontal cross section B is the upper base of a cylinder with base B_1.

From Theorem 2 we get the same sort of special conclusions that we got from Theorem 1, i.e., if the base is a disk then all the horizontal cross sections are disks with the same radius; if the base is triangular, then all of the cross sections are congruent; and if the base is a polygonal region, then all of the horizontal cross sections are polygonal regions with the same area.

23.4 VOLUMES OF PRISMS AND CYLINDERS

To make use of the additivity postulate V-4, we need the following theorem.

Theorem 1. If $M \in \mathcal{V}$, and M is a bounded set lying in a plane, then $vM = 0$.

Proof. M, being bounded, lies in a rectangular region R in its own plane. Let $\alpha R = ab$. Then for every positive number h, M lies in a parallelepiped K, with

$$vK = abh.$$

Obviously vK is as small as we please if h is small enough. More precisely, given any $e > 0$, we have

$$vK = abh < e,$$

whenever

$$h < \frac{e}{ab}.$$

Therefore

$$vM \leqq vK \leqq e$$

for every $e > 0$. Therefore $vM = 0$.

Theorem 2. Let K be a right prism, of altitude h, with a triangular region T as base.

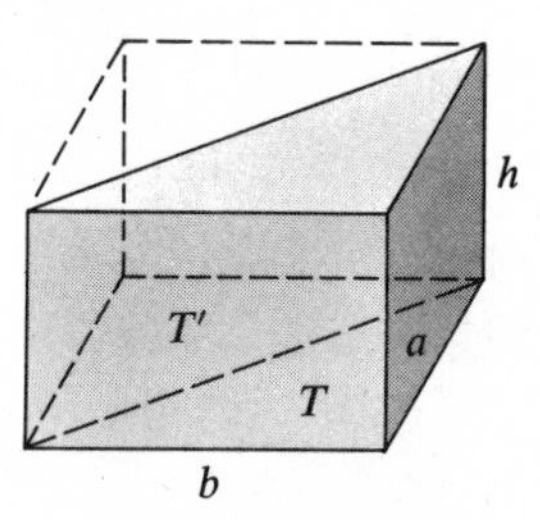

FIGURE 23.9

Then

$$vK = h \cdot \alpha T.$$

Proof. Construct another right prism K' with base T', in such a way that $T \cup T'$ is a parallelogram region, and $K \cup K'$ is a parallelepiped. If a and b are as in Fig. 23.9, then

$$v(K \cup K') = abh.$$

By Cavalieri's principle

$$vK = vK'.$$

By Theorem 1,

$$v(K \cap K') = 0.$$

Therefore

$$v(K \cup K') = vK + vK' = 2vK.$$

Hence

$$vK = \tfrac{1}{2}v(K \cup K') = \tfrac{1}{2}abh = h \cdot \alpha T,$$

which was to be proved.

More generally, we have the following theorem.

Theorem 3. Let K be a right prism, with lower base B and altitude h. Then

$$vK = h \cdot \alpha B.$$

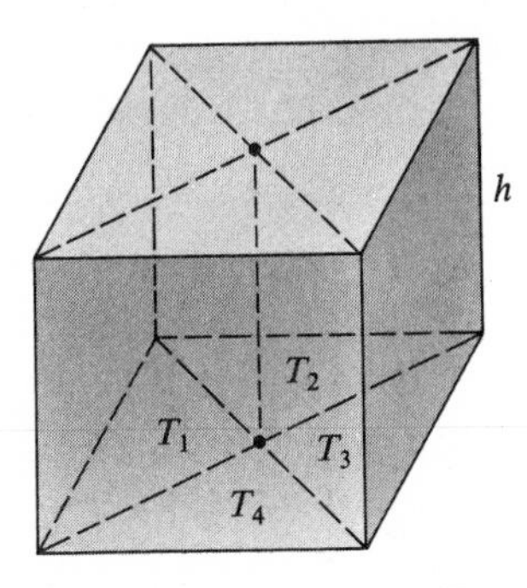

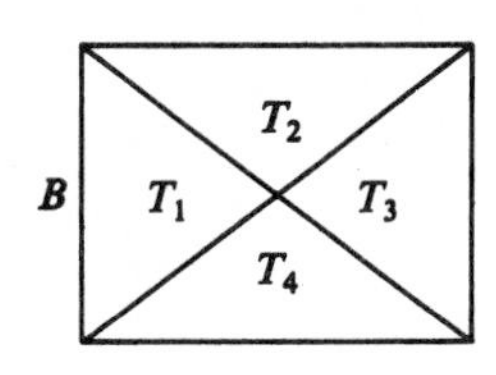

FIGURE 23.10

Proof. B is the union of a finite collection of triangular regions T_i. Thus K is the union of a finite collection of prisms K_i with the T_i's as bases. We know by Theorem 2 that

$$vK_i = h \cdot \alpha T_i$$

for each i. And

$$vK = \sum_{i=1}^{n} vK_i,$$

because each K_i intersects the others in a set of volume $= 0$. Therefore

$$\begin{aligned} vK &= \sum_{i=1}^{n} h \cdot \alpha T_i \\ &= h \cdot \sum_{i=1}^{n} \alpha T_i \\ &= h \cdot \alpha B. \end{aligned}$$

Theorem 4. Let K be any prism (right or not). Then vK is the product of the altitude and the base area.

Proof. Let K' be a *right* prism, with the same base area as K and the same altitude.

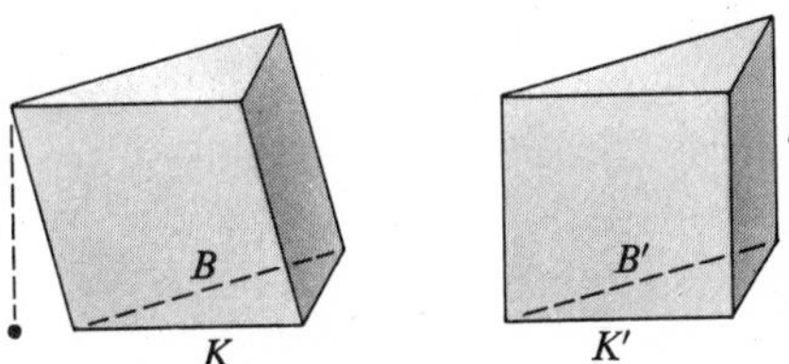

FIGURE 23.11

By Theorem 3,

$$vK' = h \cdot \alpha B' = h \cdot \alpha B.$$

By Cavalieri's principle

$$vK' = vK.$$

Therefore

$$vK = h \cdot \alpha B,$$

which was to be proved.

For cylinders with circular bases, the volume formula and its derivation are the same.

Theorem 5. Let K be a cylinder whose altitude is h and whose base is a disk of radius r. Then

$$vK = \pi r^2 h.$$

To prove this, we take a right prism L, with base area πr^2 and altitude h. Then

$$vL = \pi r^2 h.$$

By Theorem 2, Section 23.3, all horizontal cross sections of K and L have the same area. Therefore, by Cavalieri's principle, we have

$$vK = vL = \pi r^2 h,$$

which was to be proved.

23.5 VOLUMES OF PYRAMIDS AND CONES

Theorem 1. Given two pyramids with their bases in the same plane and their vertices on the same side of this plane. If they have the same base area and the same altitude, then they have the same volume.

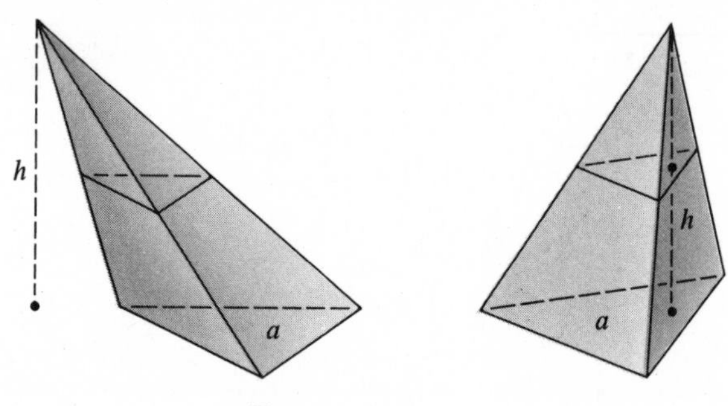

FIGURE 23.12

Proof. Let the common base area be a, and let the common altitude be h. By Theorem 3, Section 23.2, the horizontal cross sections at height k have the same area, namely

$$a_k = \left(\frac{h - k}{h}\right)^2 a.$$

By Cavalieri's principle, the two pyramids have the same volume.

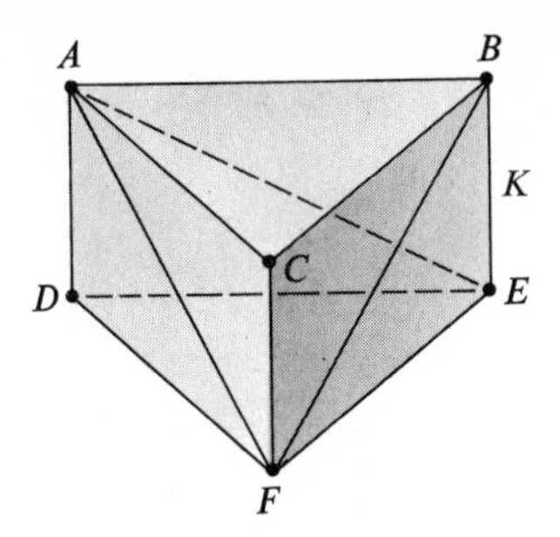

FIGURE 23.13

We can now compute the volume of a pyramid, by the following ingenious device. Consider first a right prism K with a triangular base, as in the figure above. The figure below indicates how the prism can be expressed as the union of three

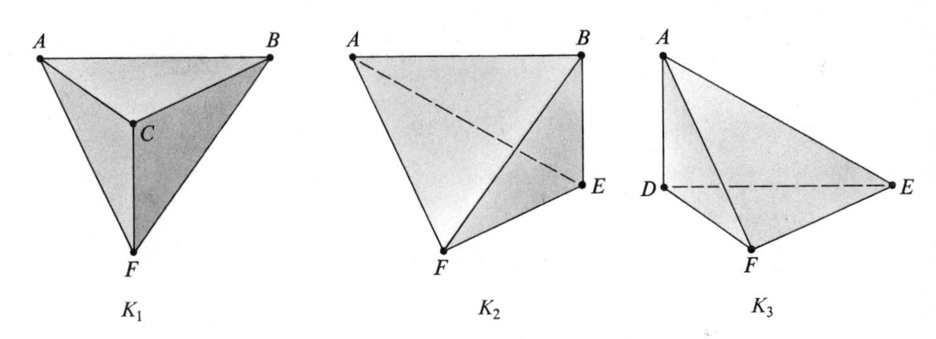

FIGURE 23.14

Now $\triangle BCF \cong \triangle BEF$. If we regard K_1 and K_2 as pyramids with A as "top vertex," then K_1 and K_2 have the same base area; and they have the same altitude, because the altitude of each of them is the perpendicular distance from A to the plane that contains B, C, E, and F. By the preceding theorem, we have

$$vK_1 = vK_2.$$

Now let us regard K_2 and K_3 as pyramids with F as "top vertex." Then K_2 and K_3 have the same base area, because $\triangle ABE \cong \triangle ADE$, and they have the same altitude, because the altitude of each of them is the perpendicular distance from F to the plane that contains A, B, D, and E. By the preceding theorem, we have

$$vK_2 = vK_3.$$

Since K_1, K_2, and K_3 intersect each other only in planar sets, which have volume $= 0$, we have

$$\begin{aligned} vK &= vK_1 + vK_2 + vK_3 \\ &= 3vK_2. \end{aligned}$$

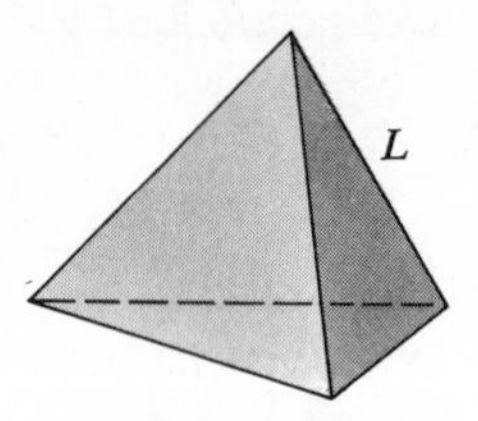

FIGURE 23.15

Suppose now that we start with a pyramid L with a triangular base. Let K_3 be a pyramid with the same base and altitude, with a vertical edge $\overline{AD}$, as in Fig. 23.14. And let K be a right prism with the same base and the same altitude as K_3. Then

$$\begin{aligned} vK &= 3vK_3 \\ &= 3vL. \end{aligned}$$

Therefore

$$\begin{aligned} vL &= \tfrac{1}{3}vK \\ &= \tfrac{1}{3}ah, \end{aligned}$$

where a and h are the base area and altitude of L. Thus we have the following.

Theorem 2. The volume of a pyramid is one-third the product of its base area and its altitude.

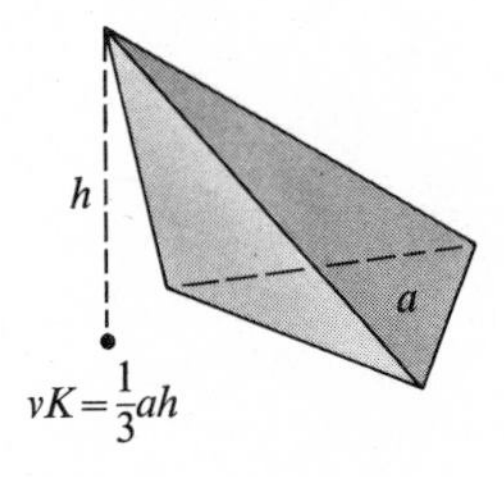

FIGURE 23.16

If we use the full force of Cavalieri's principle, then the corresponding formula for the volume of a cone becomes almost trivial. Given a cone with base area $a = \pi r^2$ and altitude h (Fig. 23.17). Take *any* pyramid with its base in the same plane, with the same base area a and the same altitude h. For the two figures illustrated in Fig. 23.17, cross sections at the same height have the same area; by the theorems of Section 23.2, the cross sections at height k have area

$$a_k = \left(\frac{h-k}{h}\right)^2 a.$$

By Cavalieri's principle, the cone and the pyramid have the same volume. Since we know that the volume of the pyramid is $\frac{1}{3}ah$, we have the following.

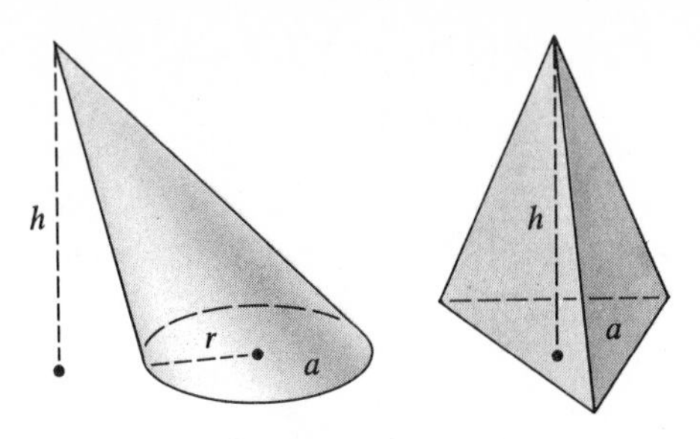

FIGURE 23.17

Theorem 3. The volume of a cone is one-third the product of its base area and its altitude.

Your only "problem" for this section is the following project. Close the book; draw a figure showing the decomposition of the right triangular prism K into the three pyramids K_1, K_2 and K_3; and then draw figures showing the three pyramids separately. If you can do this, the chances are that you understand this whole section, and conversely.

23.6 THE VOLUME OF A SPHERE

To find the volume of a sphere, we first circumscribe a cylinder around it:

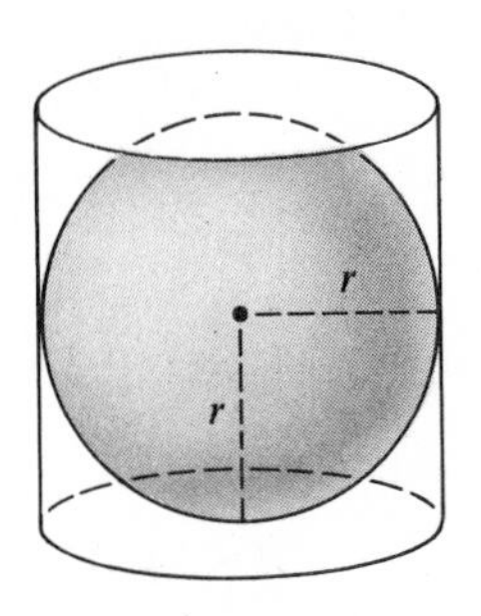

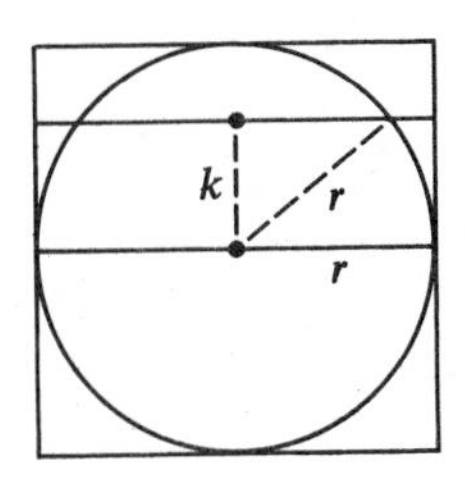

FIGURE 23.18

Let K be the sphere plus its interior; let C be the circumscribed cylinder, and let

$$L = C - K,$$

so that

$$C = K \cup L.$$

We know how to find vC:

$$\begin{aligned} \iota C &= \pi r^2 \cdot 2r \\ &= 2\pi r^3. \end{aligned}$$

Therefore, if we can find vL, we can calculate vK by subtraction as

$$vK = vC - vL.$$

In calculating vL, we shall use Cavalieri's principle in the same way as in the last two sections. That is, we shall find a figure whose volume we know, which has the same cross-sectional areas as L.

The areas of the cross sections of L are easy to calculate. Figure 23.18 shows a *vertical* cross section of our sphere and its circumscribed cylinder. Thus the cylinder appears as a rectangle and the sphere appears as a circle. The cross section of L at height k is a figure like this:

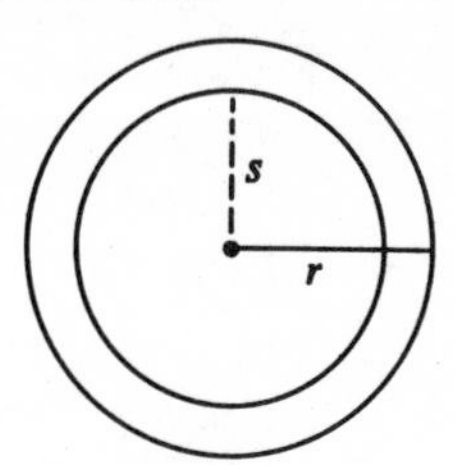

FIGURE 23.19

The outer radius is r, and the inner radius is

$$s = \sqrt{r^2 - k^2},$$

by the Pythagorean theorem. Therefore the cross-sectional area of L at height k is

$$\begin{aligned} a_k &= \pi r^2 - \pi s^2 \\ &= \pi(r^2 - s^2) \\ &= \pi[r^2 - (r^2 - k^2)] \\ &= \pi k^2. \end{aligned}$$

Now consider the figure L' shown below. On the right we show a vertical cross

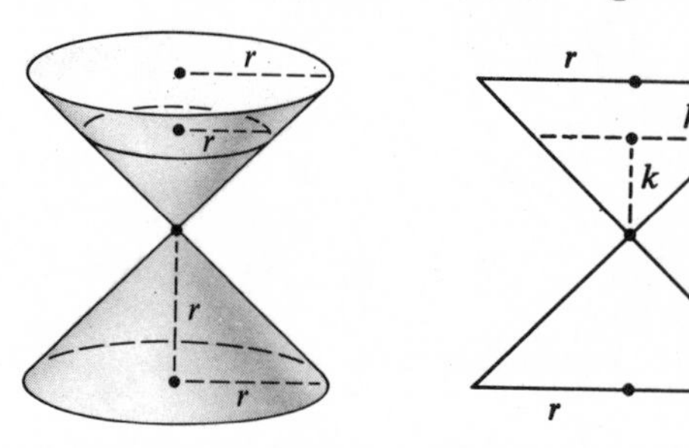

FIGURE 23.20

section. The cross section of L at height k is a disk of radius k. Therefore the cross-sectional area at height k is

$$a_k = \pi k^2.$$

Thus

$$vL = vL'.$$

But

$$vL' = 2 \cdot \tfrac{1}{3}\pi r^2 \cdot r$$
$$= \tfrac{2}{3}\pi r^3.$$

Therefore

$$vL = \tfrac{2}{3}\pi r^3.$$

Hence

$$vK = vC - vL$$
$$= 2\pi r^3 - \tfrac{2}{3}\pi r^3$$
$$= \tfrac{4}{3}\pi r^3.$$

For reference, we record this as a theorem.

Theorem 1. The volume of a sphere of radius r is $\frac{4}{3}\pi r^3$.

Problem Set 23.6

1. If we pass a secant plane through a sphere, it cuts the sphere into two parts, one of which is lens-shaped like this:

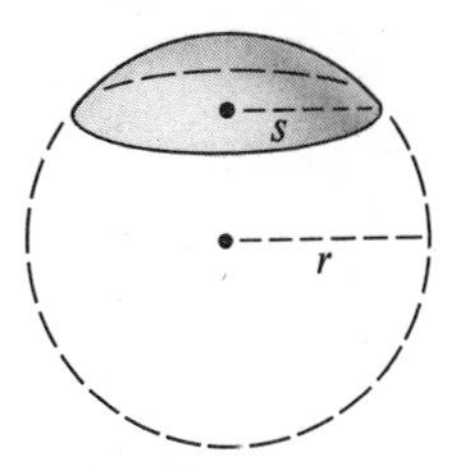

Figure 23.21

Such a figure is called a *spherical segment*. Find the volume, in terms of r and s.

2. By a *spherical sector* we mean a figure like this:

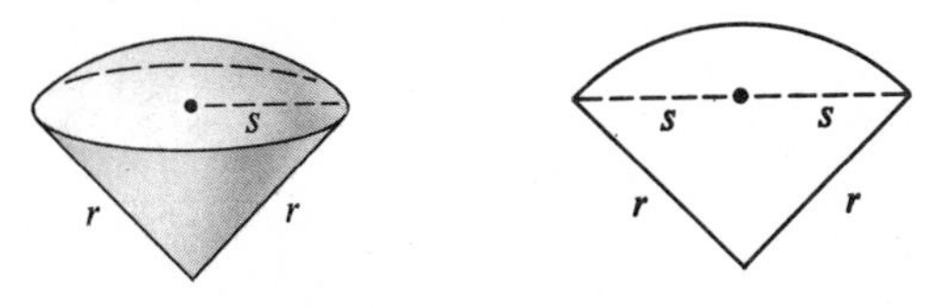

Figure 23.22

On the right we show a vertical cross section, which forms a sector of a disk. Find the volume, in terms of r and s.

Chapter 24
Hyperbolic Geometry

24.1 ABSOLUTE GEOMETRY, CONTINUED: THE CRITICAL FUNCTION

In this chapter, we shall make heavy use of the incidence and separation theorems of Chapter 4. For convenience, we briefly restate two of them:

The Postulate of Pasch. Given $\triangle ABC$ and a line L (in the same plane). If L intersects $\overline{AB}$ at a point between A and B, then L also intersects either $\overline{AC}$ or $\overline{BC}$.

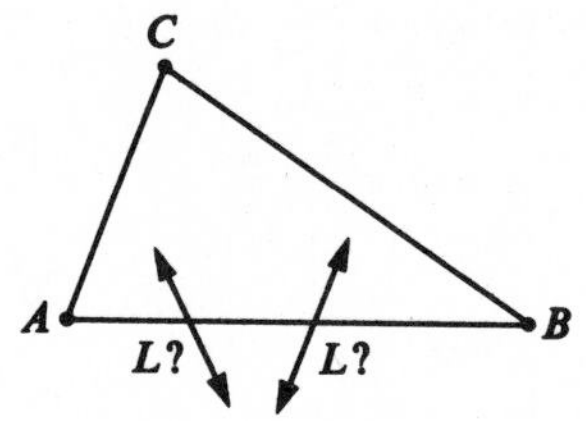

FIGURE 24.1

(This was Theorem 1, Section 4.1.)

The Crossbar Theorem. If D is in the interior of $\angle BAC$, then $\overrightarrow{AD}$ intersects $\overline{BC}$.

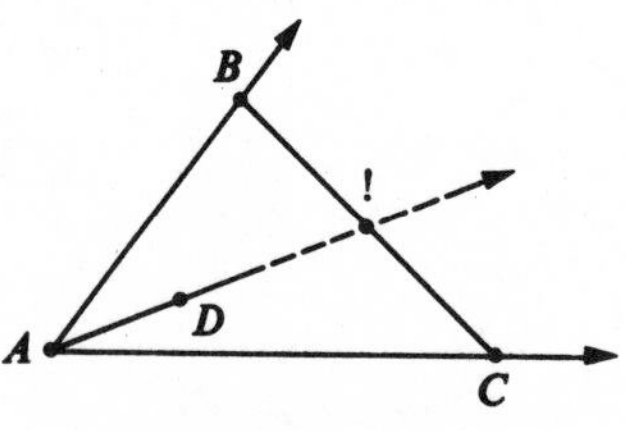

FIGURE 24.2

(This was Theorem 3, Section 4.3.)

Given a line L and an external point P. Let A be the foot of the perpendicular from P to L, and let B be any other point of L (Fig. 24.3). For each number r between 0 and 180 there is exactly one ray $\overrightarrow{PD}$, with D on the same side of $\overleftrightarrow{AP}$ as B, such that

$$m\angle APD = r.$$

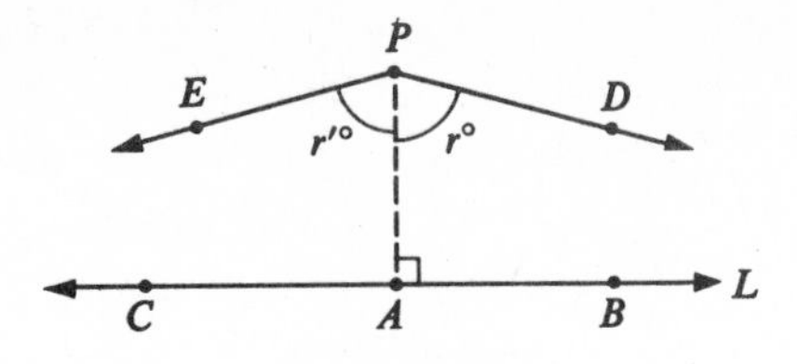

FIGURE 24.3

Obviously, for some numbers r, $\overrightarrow{PD}$ will intersect $\overrightarrow{AB}$. (For example, take $r = m\angle APB$.) For $r \geqq 90$, $\overrightarrow{PD}$ will not intersect $\overrightarrow{AB}$. Let

$$K = \{r | \overrightarrow{PD} \text{ intersects } \overrightarrow{AB}\}.$$

Then K is nonempty, and has an upper bound. Therefore K has a supremum. Let

$$r_0 = \sup K.$$

The number r_0 is called the *critical number* for P and $\overrightarrow{AB}$. The angle $\angle APD$ with measure $= r_0$ is called the *angle of parallelism* of $\overrightarrow{AB}$ and P.

Theorem 1. If $m\angle APD = r_0$, then $\overrightarrow{PD}$ does not intersect $\overrightarrow{AB}$.

Proof. Suppose that $\overrightarrow{PD}$ intersects $\overrightarrow{AB}$ at Q:

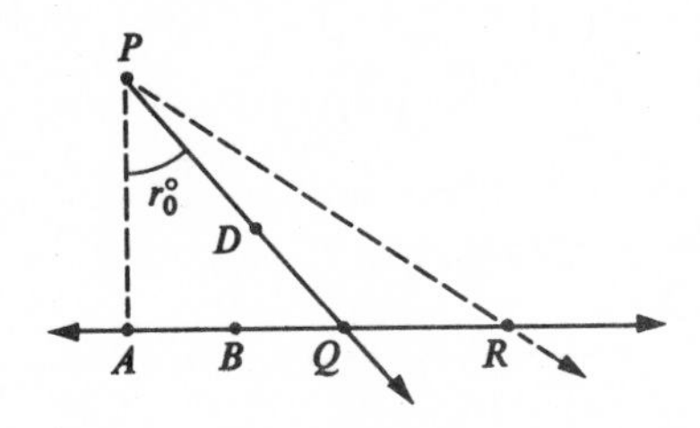

FIGURE 24.4

If R is any point such that A-Q-R, then $m\angle APR > r_0$, so that r_0 is not an upper bound of K.

Theorem 2. If $m\angle APD < r_0$, then $\overrightarrow{PD}$ intersects $\overrightarrow{AB}$.

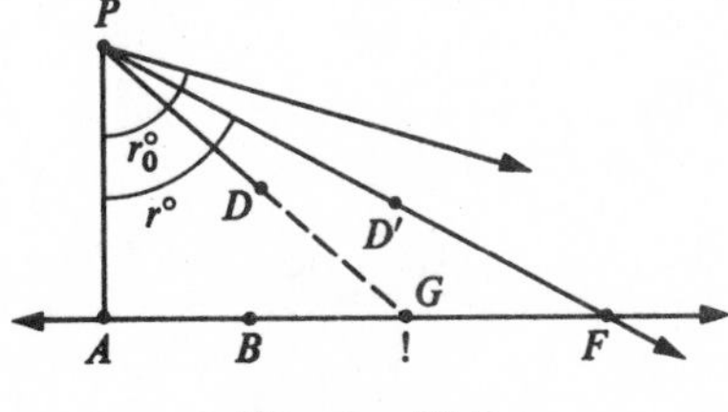

FIGURE 24.5

Proof. Since $r_0 = \sup K$, and $m\angle APD < r_0$, it follows that $m\angle APD$ is not an upper bound of K. Therefore some r in K is $> m\angle APD$. Let D' be such that $m\angle APD' = r$. Then $\overrightarrow{PD'}$ intersects $\overrightarrow{AB}$ in a point F. But $\overrightarrow{PD}$ is in the interior of $\angle APD'$. Therefore, by the crossbar theorem, $\overrightarrow{PD}$ intersects $\overline{AF}$. Therefore $\overrightarrow{PD}$ intersects $\overrightarrow{AB}$. Thus there is a certain "critical ray" $\overrightarrow{PD}$, with $m\angle APD = r_0$; $\overrightarrow{PD}$ does not intersect $\overrightarrow{AB}$, but if F is in the interior of $\angle APD$, then $\overrightarrow{PF}$ *does* intersect $\overrightarrow{AB}$.

[Hereafter, if F is in the interior of $\angle APD$, we shall say that $\overrightarrow{AF}$ is an *interior ray* of $\angle APD$.]

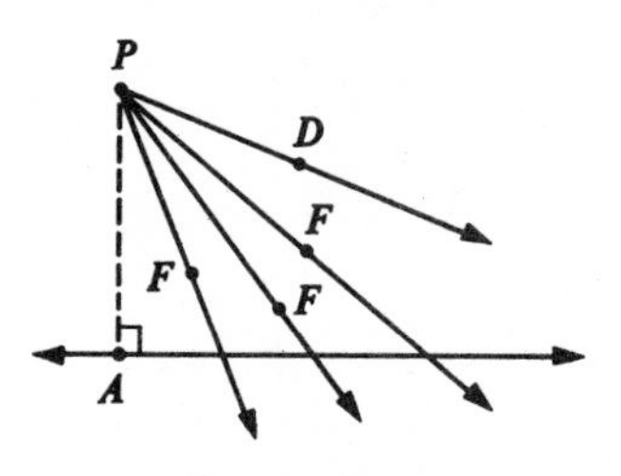

FIGURE 24.6

Note that r_0 was defined in terms of P, A, and B. It turns out, however, that r_0 depends only on the distance AP.

Theorem 3. Let P, A, B and also P', A', B' be as in the definition of the critical number. If $AP = A'P'$, then the critical numbers r_0, r_0' are the same.

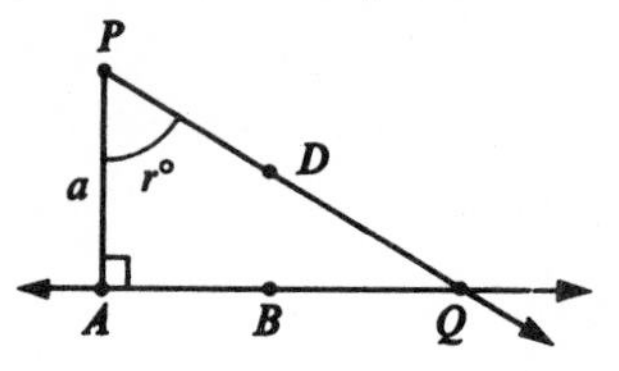

FIGURE 24.7

Proof. Let

$$K = \{r | \overrightarrow{PD} \text{ intersects } \overleftrightarrow{AB}\},$$

and let

$$K' = \{r | \overrightarrow{P'D'} \text{ intersects } \overrightarrow{A'B'}\},$$

as before. If $r \in K$, let Q be the point where $\overrightarrow{PD}$ intersects $\overrightarrow{AB}$, and let Q' be the point of $\overrightarrow{A'B'}$ for which $A'Q' = AQ$. Then $m\angle A'P'Q' = r$. (Why?) Therefore $r \in K'$. Thus $K \subset K'$; and similarly $K' \subset K$. Therefore

$$K' = K \quad \text{and} \quad \sup K' = \sup K.$$

We now have a function $AP \to r_0$. We shall denote this function by c, and call it the *critical function.* Thus, for every $a > 0$, $c(a)$ denotes the critical number

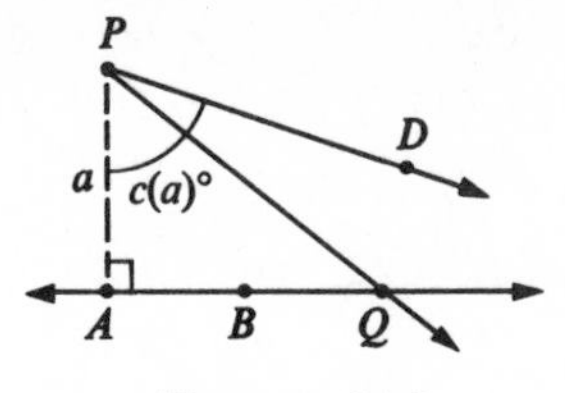

FIGURE 24.8

corresponding to $AP = a$. Thus $\overrightarrow{PD}$ intersects $\overrightarrow{AB}$ when $m\angle APD < c(a)$, but $\overrightarrow{PD}$ does not intersect $\overrightarrow{AB}$ when $m\angle APD \geqq c(a)$.

We shall now investigate the function c.

Theorem 4. c never increases as a increases. That is, if $a' > a$, then $c(a') \leqq c(a)$.

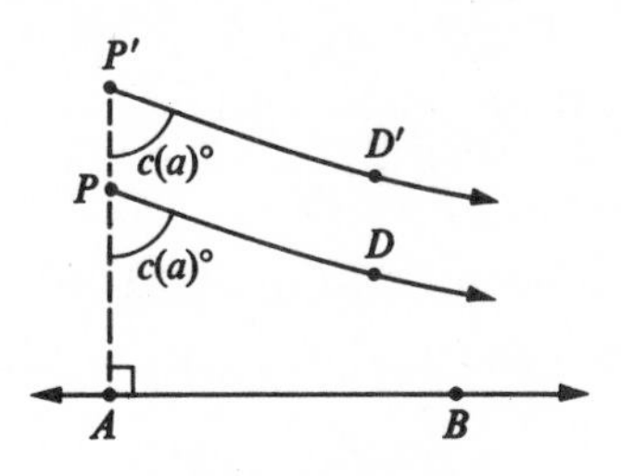

FIGURE 24.9

Proof. Given P, P', with $a = AP$, $a' = AP'$, as in Fig. 24.9. Take $\overrightarrow{PD}$ so that $m\angle APD = c(a)$, and take $\overrightarrow{P'D'}$ so that $m\angle AP'D' = c(a)$. Then $\overleftrightarrow{PD}$ and $\overleftrightarrow{P'D'}$ are parallel. Therefore all points of $\overrightarrow{P'D'}$ are on the side of $\overleftrightarrow{PD}$ that contains P'. And all points of $\overrightarrow{AB}$ are on the side of $\overleftrightarrow{PD}$ that contains A. Therefore $\overrightarrow{P'D'}$ does not intersect $\overrightarrow{AB}$.

Now let

$$K' = \{r \mid \overrightarrow{P'D''} \text{ intersects } \overrightarrow{AB}\},$$

as in the definition of the critical angle, so that

$$c(a') = \sup K'.$$

Then $c(a)$ is an upper bound of K', because $\overrightarrow{P'D'}$ does not intersect $\overrightarrow{AB}$. And $c(a')$ is the *least* upper bound of K'. Therefore $c(a') \leqq c(a)$, which was to be proved.

In the Euclidean case, this theorem cannot be strengthened to give the strict inequality $c(a') < c(a)$ for $a' > a$. (The reason, obviously, is that in Euclidean geometry we have $c(a) = 90$ for every a.) In hyperbolic geometry, however, not only do we have the strict inequality but we actually have $c(a) \to 0$ as $a \to \infty$. (It would be worthwhile to figure out how this works in the Poincaré model described in Chapter 9.)

Theorem 4 allows the possibility that $c(a) < 90$ when a is large, but $c(a) = 90$ when a is sufficiently small. But in fact this cannot happen, as the following two theorems show.

Theorem 5. If $c(a) < 90$, then $c(a/2) < 90$.

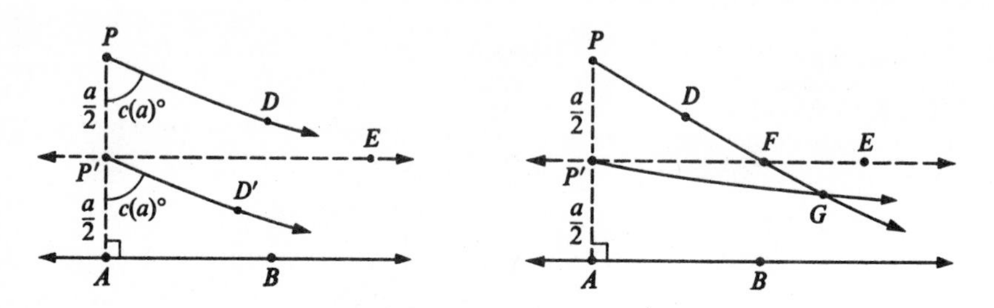

FIGURE 24.10

Proof. Given P, P' as in the figures, with $AP = a$, $AP' = a/2$. Take $\overrightarrow{PD}$ so that $m\angle APD = c(a) < 90$, and take $\overleftrightarrow{P'E} \perp \overleftrightarrow{AP}$ at P'. If $\overrightarrow{PD}$ fails to intersect $\overrightarrow{P'E}$, as on the left, then obviously $c(a/2) < 90$.

Suppose, then, that $\overrightarrow{PD}$ does intersect $\overrightarrow{P'E}$, at a point F. (It will turn out, later in the theory, that this is what always happens.) Let G be any point such that P-F-G. Then $\angle AP'G$ is acute.

Now (1) $\overrightarrow{AB}$ cannot intersect $\overrightarrow{P'G}$ except perhaps in a point of $\overline{P'G}$; the reason is that all other points of $\overrightarrow{P'G}$ lie on the "wrong side" of $\overleftrightarrow{PG}$. And (2) $\overleftrightarrow{AB}$ does not contain P' or G. Finally (3) $\overleftrightarrow{AB}$ does not contain a point between P' and G; if so, it would follow from the postulate of Pasch that $\overleftrightarrow{AB}$ intersects $\overline{P'F}$ or $\overline{FG}$, which is false.

Therefore $\overrightarrow{P'G}$ does not intersect $\overrightarrow{AB}$, and $c(a/2) < 90$, which was to be proved.

Theorem 6. If $c(a_0) < 90$ for some a_0, then $c(a) < 90$ for every a.

Proof. For each n, let

$$a_n = \frac{a_0}{2^n}.$$

By induction based on Theorem 5, we have

$$c(a_n) < 90 \quad \text{for every } n.$$

Suppose now that $c(b) = 90$ for some b. Since

$$\lim_{n\to\infty} a_n = 0,$$

we have

$$a_k < b \quad \text{for some } k.$$

Thus $a_k < b$ but $c(a_k) < c(b)$, and this contradicts Theorem 4.

This theorem clarifies the meaning of the parallel postulate; it tells us that the situation described in the postulate holds either always or never.

Theorem 7. *The All-or-None Theorem.* If parallels are unique for one line and one external point, then parallels are unique for all lines and all external points.

Proof. Given P, L, P', and L', with $AP = a$ and $A'P' = a'$, as in the figure.

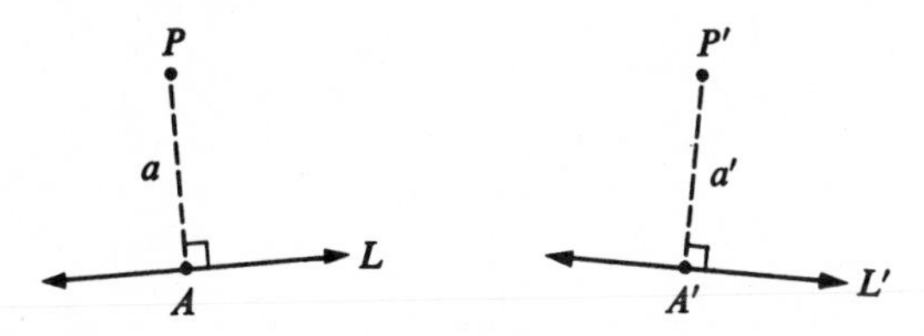

FIGURE 24.11

It is now easy to see that each of the statements below is equivalent to the next.

(1) There is only one parallel to L through P.
(2) $c(a) = 90$.
(3) $c(a') = 90$.
(4) There is only one parallel to L' through P'.

Therefore (1) and (4) are equivalent, which was to be proved.

Thus we can state our two possible parallel postulates in seemingly weak but actually quite adequate forms.

(I) *Euclidean.* For some line and some external point, parallels are unique.

(II) *Lobachevskian.* For some line and some external point, parallels are not unique.

We have already remarked, in Chapter 9, that the same all-or-none principle applies in other connections. If the formula $m\angle A + m\angle B + m\angle C = 180$ holds for even one triangle, then it holds for all triangles; if even one pair of triangles are similar without being congruent, then the geometry is Euclidean; and so on.

24.2 ABSOLUTE GEOMETRY: OPEN TRIANGLES AND CRITICALLY PARALLEL RAYS

Given rays $\overrightarrow{AB}$, $\overrightarrow{PD}$ and the segment $\overline{AP}$, no two of these figures being collinear. Suppose that B and D are on the same side of $\overleftrightarrow{AP}$, and that $\overleftrightarrow{AB} \parallel \overleftrightarrow{PD}$. Then $\overrightarrow{PD} \cup \overline{PA} \cup \overrightarrow{AB}$ is called an *open triangle*, and is denoted by $\triangle DPAB$.

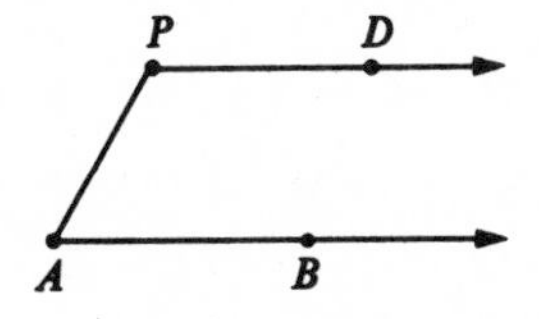

FIGURE 24.12

Here, when we write $\overleftrightarrow{AB} \parallel \overleftrightarrow{PD}$, we mean that the lines are parallel in the usual sense of not intersecting one another.

Suppose now that $\triangle DPAB$ is an open triangle, and every interior ray of $\angle APD$ intersects $\overrightarrow{AB}$:

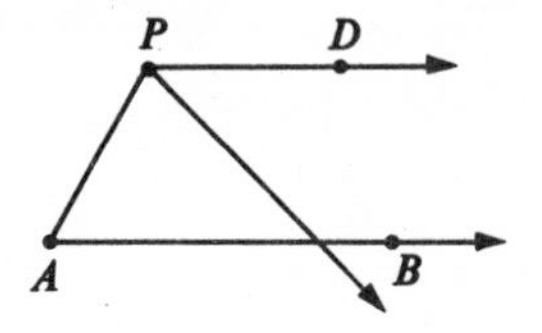

FIGURE 24.13

We then say that $\overrightarrow{PD}$ is *critically parallel* to $\overrightarrow{AB}$, and we write $\overrightarrow{PD}|\overrightarrow{AB}$. Here the single vertical stroke is supposed to suggest that $\overrightarrow{PD}$ is parallel to $\overrightarrow{AB}$ with no room to spare.

Note that $\overrightarrow{PD}$ and $\overrightarrow{AB}$ do not appear symmetrically in this definition. Thus if $\overrightarrow{PD}|\overrightarrow{AB}$, it does not immediately follow that $\overrightarrow{AB}|\overrightarrow{PD}$. Note also that the relation $\overrightarrow{PD}|\overrightarrow{AB}$ (as we have defined it) depends not only on the "directions" of the two rays, but also on the initial points:

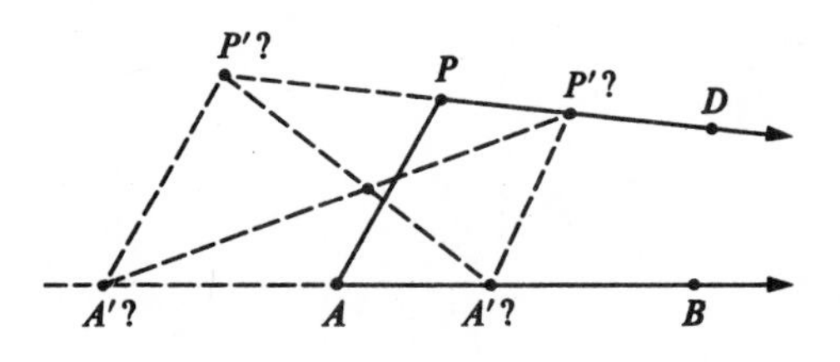

FIGURE 24.14

Thus, if $\overrightarrow{PD}|\overrightarrow{AB}$, we cannot conclude immediately that $\overrightarrow{P'D}|\overrightarrow{A'B}$. We shall see, however, in the next few theorems, that the conclusion is true.

Theorem 1. If $\overrightarrow{PD}|\overrightarrow{AB}$, and C-P-D, then $\overrightarrow{CD}|\overrightarrow{AB}$.

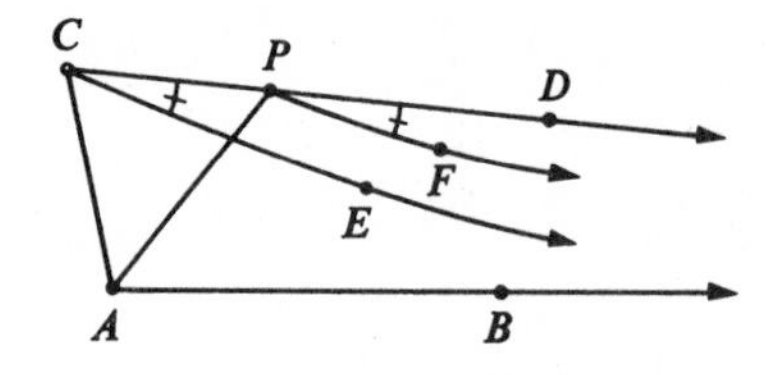

FIGURE 24.15

Proof. Let $\overrightarrow{CE}$ be an interior ray of $\angle ACD$, and suppose that $\overrightarrow{CE}$ does not intersect $\overrightarrow{AB}$. By the exterior angle theorem (which was, fortunately, proved without the use of the parallel postulate), we know that $\angle APD > \angle ACD$. Therefore there is an interior ray $\overrightarrow{PF}$ of $\angle APD$ such that $\angle DPF \cong \angle DCE$. Therefore $\overleftrightarrow{PF} \parallel \overleftrightarrow{CE}$. Therefore $\overrightarrow{PF}$ does not intersect $\overrightarrow{AB}$, because these rays lie on opposite sides of $\overleftrightarrow{CE}$. This contradicts the hypothesis $\overrightarrow{PD}|\overrightarrow{AB}$.

Theorem 2. If $\overrightarrow{PD}|\overrightarrow{AB}$, and P-C-D, then $\overrightarrow{CD}|\overrightarrow{AB}$.

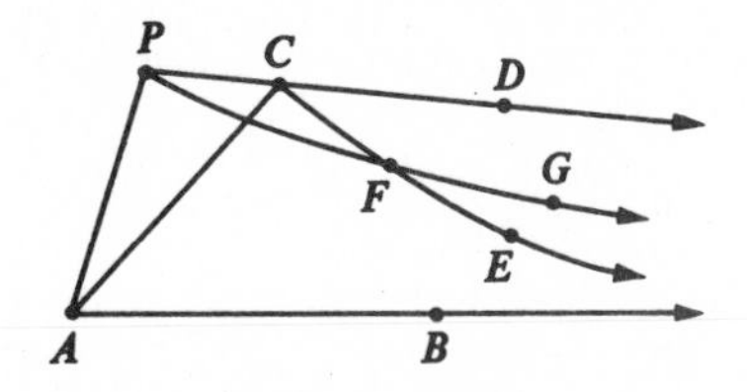

FIGURE 24.16

We give the proof briefly. Suppose that there is an interior ray $\overrightarrow{CE}$ of $\angle ACD$ such that $\overrightarrow{CE}$ does not intersect $\overrightarrow{AB}$. Let F be any point of $\overrightarrow{CE} - C$, and take G so that P-F-G. Then

(1) F is in the interior of $\angle APC$;
(2) $\overline{PF}$ does not intersect $\overrightarrow{AB}$;
(3) $\overrightarrow{FG}$ does not intersect $\overrightarrow{AB}$;
(4) $\overrightarrow{PF}$ does not intersect $\overrightarrow{AB}$.

Statements (1) and (4) contradict the hypothesis $\overrightarrow{PD}|\overrightarrow{AB}$.

Two rays R and R' are called *equivalent* if one of them contains the other. We then write $R \sim R'$. Obviously the symbol $\sim$ represents an equivalence relation. Fitting together the preceding two theorems, we get:

Theorem 3. If $R|\overrightarrow{AB}$, and R and R' are equivalent, then $R'|\overrightarrow{AB}$.

Somewhat easier proofs show that the relation $\overrightarrow{PD}|\overrightarrow{AB}$ depends only on the equivalence class of $\overrightarrow{AB}$. We leave these proofs to you.

Theorem 4. If $R_1|R_2$, $R_1' \sim R_1$ and $R_2' \sim R_2$, then $R_1'|R_2'$.

Given $\overrightarrow{PD}|\overrightarrow{AB}$, let C be the foot of the perpendicular from P to $\overleftrightarrow{AB}$, and let $PC = a$.

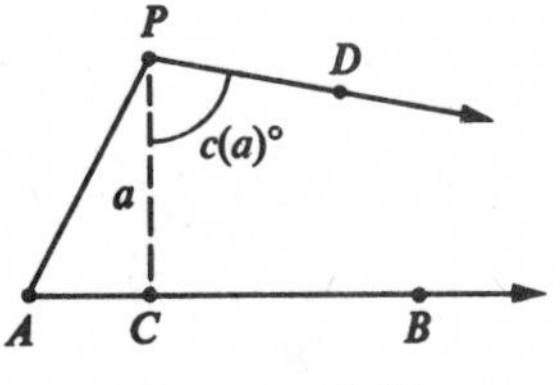

FIGURE 24.17

Then $\overrightarrow{PD}|\overrightarrow{CB}$ (providing, of course, that B is chosen so that A-C-B, as in the figure). Therefore $m\angle CPD = c(a)$. Now on the side of $\overleftrightarrow{PC}$ that contains B there is only one ray $\overrightarrow{PD}$ for which $m\angle CPD = c(a)$. Thus we have:

Theorem 5. The critical parallel to a given ray, through a given external point, is unique.

Two open triangles are called *equivalent* if the rays that form their sides are equivalent. An open triangle $\Delta DPAB$ is called *isosceles* if $\angle P \cong \angle A$.

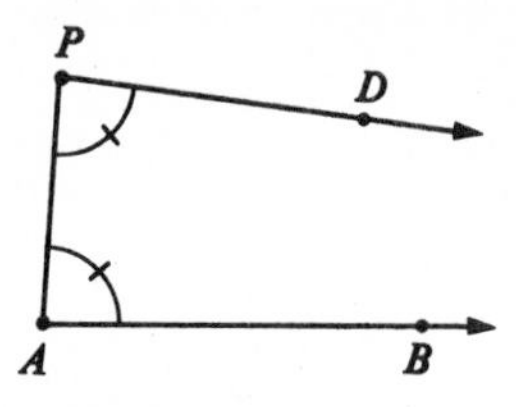

FIGURE 24.18

Theorem 6. If $\overrightarrow{PD}|\overrightarrow{AB}$, then $\Delta DPAB$ is equivalent to an isosceles open triangle which has P as a vertex.

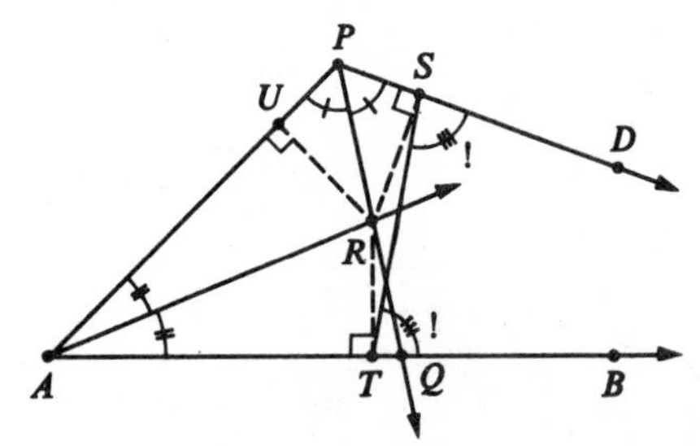

FIGURE 24.19

Proof. Since $\overrightarrow{PD}|\overrightarrow{AB}$, the bisecting ray of $\angle APD$ intersects $\overrightarrow{AB}$ in a point Q. By the crossbar theorem, the bisecting ray of $\angle PAB$ intersects $\overline{PQ}$ at a point R. Let S, T, and U be the feet of the perpendiculars from R to $\overleftrightarrow{PD}$, $\overleftrightarrow{AB}$ and $\overleftrightarrow{AP}$. Then $RU = RT$ and $RU = RS$. Therefore $RS = RT$, and $\angle RST \cong \angle RTS$. Hence (by addition or subtraction) $\angle DST \cong \angle BTS$; and $\Delta DSTB$ is isosceles.

To make P a vertex, we take V on the ray opposite to $\overrightarrow{TB}$, such that $TV = SP$.

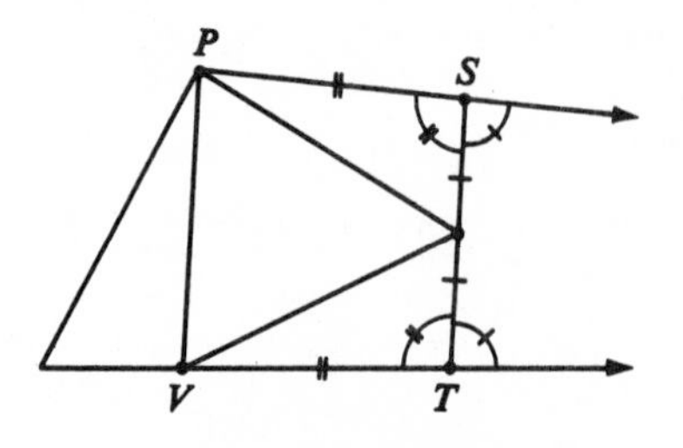

FIGURE 24.20

Theorem 7. Critical parallelism is a symmetric relation. That is, if $\overrightarrow{PD}|\overrightarrow{AB}$, then $\overrightarrow{AB}|\overrightarrow{PD}$.

Proof. By Theorems 4 and 6, we may suppose that $\triangle DPAB$ is an isosceles open triangle:

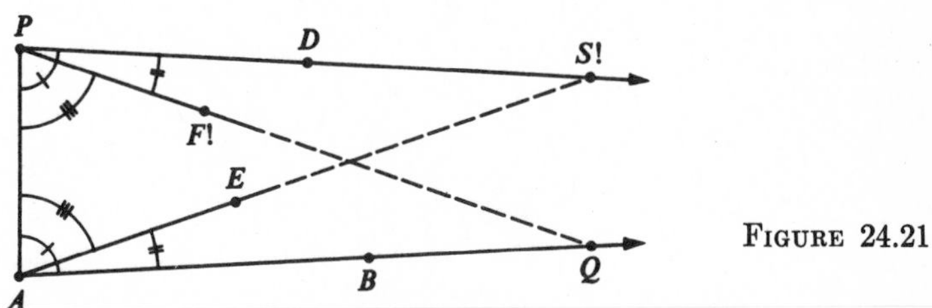

FIGURE 24.21

Let $\overrightarrow{AE}$ be any interior ray of $\angle PAB$. Let $\overrightarrow{PF}$ be an interior ray of $\angle APD$, such that $\angle DPF \cong \angle BAE$. Then $\overrightarrow{PF}$ intersects $\overrightarrow{AB}$ at a point Q. It follows that $\overrightarrow{AE}$ intersects $\overrightarrow{PD}$ at the point S where $PS = AQ$. (Proof? The whole figure is symmetric, from top to bottom.)

Theorem 8. If two nonequivalent rays are critically parallel to a third ray, then they are critically parallel to each other.

RESTATEMENT. If $\overrightarrow{AB}|\overrightarrow{CD}$, $\overrightarrow{CD}|\overrightarrow{EF}$, and $\overrightarrow{AB}$ and $\overrightarrow{EF}$ are not equivalent, then $\overrightarrow{AB}|\overrightarrow{EF}$.

Proof. (1) Suppose that $\overrightarrow{AB}$ and $\overrightarrow{EF}$ lie on opposite sides of $\overrightarrow{CD}$. Then $\overline{AE}$ intersects $\overleftrightarrow{CD}$, and by Theorem 4 we can assume that the point of intersection is C.

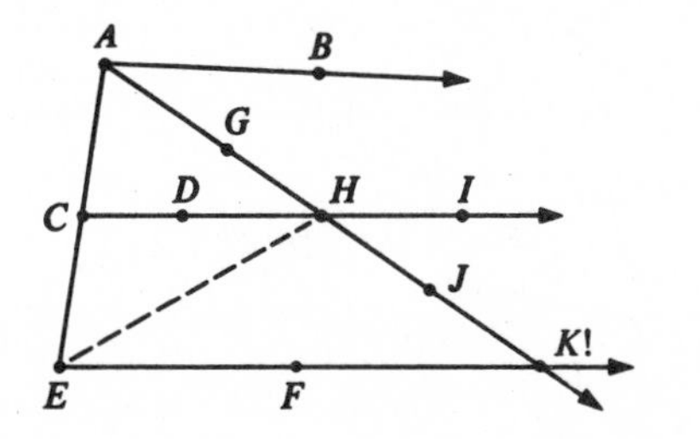

FIGURE 24.22

Let $\overrightarrow{AG}$ be any interior ray of $\angle EAB$. Then $\overrightarrow{AG}$ intersects $\overrightarrow{CD}$ at a point H. Take I so that C-H-I and take J so that A-H-J. Then $\overrightarrow{HI}|\overrightarrow{EF}$, by Theorem 4; and $\overrightarrow{HJ}$ is an interior ray of $\angle EHI$. Therefore $\overrightarrow{HJ}$ intersects $\overrightarrow{EF}$ at a point K. Therefore $\overrightarrow{AG}$ intersects $\overrightarrow{EF}$, which was to be proved.

(2) If $\overrightarrow{CD}$ and $\overrightarrow{EF}$ are on opposite sides of $\overleftrightarrow{AB}$, then the same conclusion follows.

Here we may suppose that $\overleftrightarrow{AB} \cap \overline{EC} = A$, for the same reasons as in the first case. Through E there is exactly one ray $\overrightarrow{EF'}$ critically parallel to $\overrightarrow{AB}$. By the result in Case (1), $\overrightarrow{EF'}|\overrightarrow{CD}$. Since critical parallels are unique, $\overrightarrow{EF'} = \overrightarrow{EF}$ and $\overrightarrow{EF}|\overrightarrow{AB}$, which was to be proved.

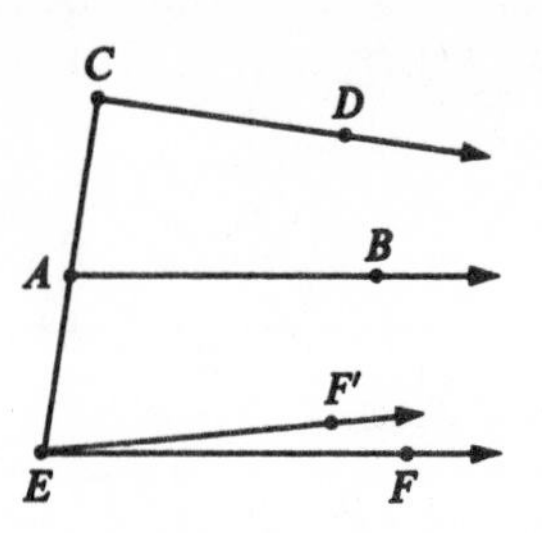

FIGURE 24.23

There remains a sticky point which some authors have overlooked. How do we know that some two of our three rays lie on opposite sides of the line containing the third? In the Euclidean case, this is easy to see, because any three parallel lines have a common transversal; in fact, any line which crosses one of them must cross the other two. But this is far from true in hyperbolic geometry, as an examination of the Poincaré model will easily show. Given three nonintersecting lines, it can easily happen that every two of them are on the same side of the third. Therefore the conditions $\overleftrightarrow{AB} \parallel \overleftrightarrow{CD}$, $\overleftrightarrow{CD} \parallel \overleftrightarrow{EF}$ are not enough for our purpose; to get a valid proof, we need to use the full force of the hypothesis $\overrightarrow{AB}|\overrightarrow{CD}$, $\overrightarrow{CD}|\overrightarrow{EF}$. We shall show, under these conditions, that (3) *some line intersects all three of the rays* $\overrightarrow{AB}$, $\overrightarrow{CD}$, $\overrightarrow{EF}$. (Surely this will be enough.)

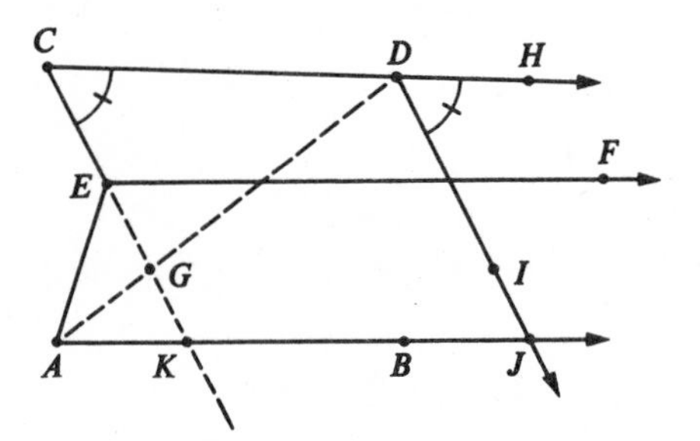

FIGURE 24.24

If A and E are on opposite sides of $\overleftrightarrow{CD}$, then $\overline{AE}$ intersects $\overleftrightarrow{CD}$, and (3) follows. Suppose, then, that (a) *A and E are on the same side of* $\overleftrightarrow{CD}$. If A and D are on the same side of $\overleftrightarrow{EC}$, then $\overrightarrow{CA}$ is an interior ray of $\angle C$, so that $\overrightarrow{CA}$ intersects $\overrightarrow{EF}$, and (3) follows. If A lies on $\overleftrightarrow{CE}$, then (3) holds. We may therefore suppose that (b) *A and D are on opposite sides of* $\overleftrightarrow{CE}$. Therefore $\overline{AD}$ intersects $\overleftrightarrow{CE}$ at a point G.

Take H so that C-D-H. Then $\overrightarrow{DH}|\overrightarrow{AB}$. By the exterior angle theorem, $\angle HDA > \angle C$. Therefore there is an interior ray $\overrightarrow{DI}$ of $\angle HDA$ such that $\angle HDI \cong \angle C$. Then $\overleftrightarrow{DI} \parallel \overleftrightarrow{CE}$, but $\overrightarrow{DI}$ intersects $\overrightarrow{AB}$ at a point J.

Now $\overleftrightarrow{CE}$ intersects $\overline{AD}$ at G. Therefore $\overleftrightarrow{CE}$ intersects another side of $\triangle ADJ$. Since $\overleftrightarrow{CE}$ does not intersect $\overline{DJ}$, $\overleftrightarrow{CE}$ intersects $\overline{AJ}$ at a point K. Now (3) follows; the line that we wanted is $\overleftrightarrow{CE}$.

The oversight leading to the incomplete proof of this theorem is illustrious. It is due originally to Gauss, and has been faithfully reproduced by good authors ever since.

Problem Set 24.2

1. By the *interior* of an open triangle $\triangle DPAB$, we mean the intersection of the interiors of $\angle P$ and $\angle A$. If a line intersects the interior of an open triangle, does it follow that the line intersects one of the sides? Why or why not?

2. Same question, for the case where $\overrightarrow{PD}|\overrightarrow{AB}$.

3. In a Euclidean plane, if a line intersects the interior of an angle, does it follow that the line intersects the angle?

4. Same question, in a hyperbolic plane.

5. Given $\angle ABC$, we define the *crossbar interior* of $\angle ABC$ to the set of all points P such that B'-P-C', for some B' in $\overrightarrow{AB} - A$ and some C' in $\overrightarrow{AC} - A$. In a Euclidean plane, is the crossbar interior the same as the interior?

6. Same question, in a hyperbolic plane.

24.3 HYPERBOLIC GEOMETRY: CLOSED TRIANGLES AND ANGLE SUMS

So far in this chapter, we have been doing absolute geometry. To mention the hyperbolic parallel postulate in our proofs would have been misleading, because in the Euclidean case, our theorems, so far, are not false but merely trivial, and the difference between falsity and triviality is important.

In this section we deal specifically with the hyperbolic case. To avoid confusion, throughout this chapter, we shall mention the hyperbolic parallel postulate in every theorem whose proof requires it. We shall abbreviate the name of the postulate as HPP.

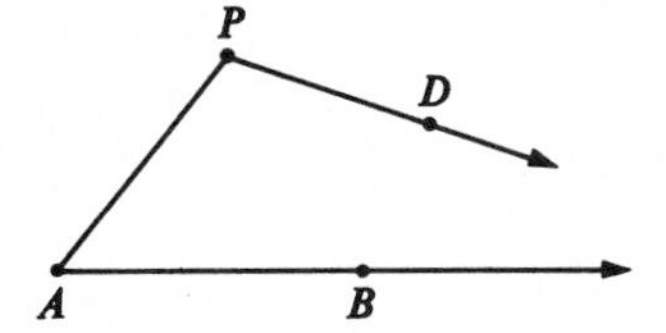

Figure 24.25

If $\overrightarrow{PD}|\overrightarrow{AB}$, then $\triangle DPAB$ is called a *closed triangle*.

Note that every closed triangle is an open triangle, but under HPP the converse is false, because through P there is more than one line parallel to $\overleftrightarrow{AB}$. Closed triangles have important properties in common with genuine triangles.

Theorem 1. *The Exterior Angle Theorem.* Under HPP, in every closed triangle, each exterior angle is greater than its remote interior angle.

Restatement. If $\overrightarrow{PD}|\overrightarrow{AB}$ and Q-A-B, then $\angle QAP > \angle P$.

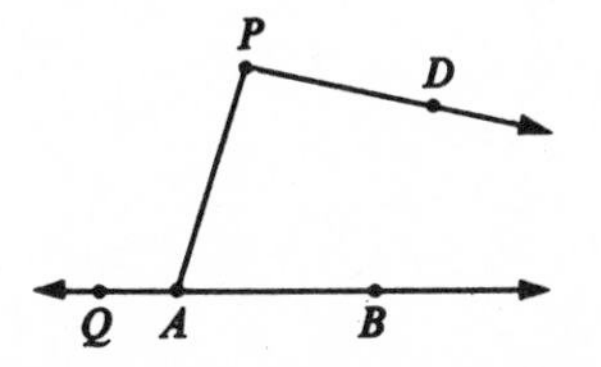

Figure 24.26

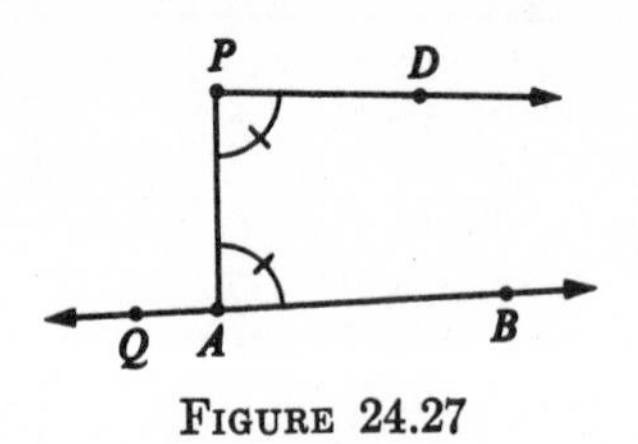

FIGURE 24.27

Proof. If $\triangle DPAB$ is isosceles, this is obvious. Here, if HPP holds, then $\angle P$ and $\angle PAB$ are acute (because $c(a) < 90$ for every a), and therefore $\angle QAP$ is obtuse.

Suppose then that $\triangle DPAB$ is not isosceles. By Theorem 6, Section 24.2, $\triangle DPAB$ is equivalent to an isosceles open triangle $\triangle DPCB$, and this open triangle is also closed:

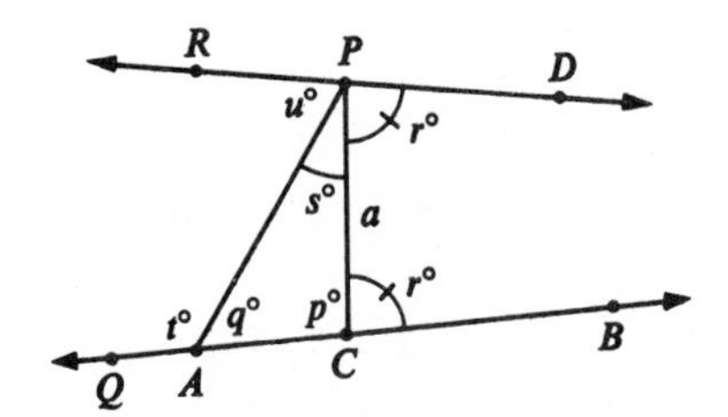

FIGURE 24.28

If $C = A$, there is nothing to prove. For the case A-C-B, let the degree measures of the various angles be as in the figure. Then

$$p > r,$$

because $c(a) < 90$. And

$$p + q + s \leqq 180,$$

by Theorem 6, Section 10.4. Therefore

$$t = 180 - q \geqq p + s > r + s,$$

and

$$t > r + s,$$

which proves half of our theorem.

To prove the other half, we need to show that $u > q$. This follows from

$$t = 180 - q > 180 - u = r + s.$$

We found, in Theorem 3, Section 24.1, that the critical function c was nonincreasing. That is, if $a' > a$, then $c(a') \leqq c(a)$. Using the exterior angle theorem, we can sharpen this result.

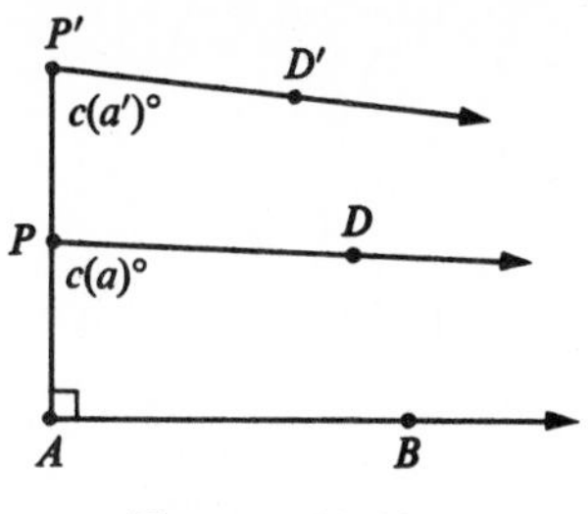

FIGURE 24.29

Theorem 2. Under HPP, the critical function is strictly decreasing. That is, if $a' > a$, then $c(a') < c(a)$.

Proof. In the figure, $AP = a$ and $AP' = a'$, $\overrightarrow{PD}|\overrightarrow{AB}$ and $\overrightarrow{P'D'}|\overrightarrow{AB}$, so that $\overrightarrow{PD}|\overrightarrow{P'D'}$. Therefore $\Delta D'P'PD$ is a closed triangle. Therefore $c(a) > c(a')$, which was to be proved.

Theorem 3. Under HPP, the upper base angles of a Saccheri quadrilateral are always acute.

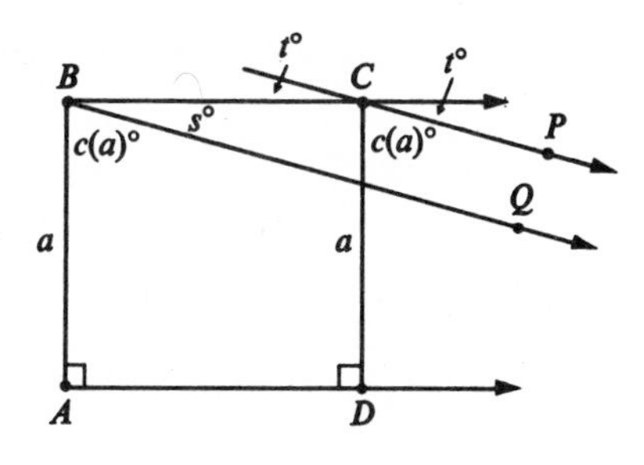

FIGURE 24.30

(We already know, from Chapter 10, that they are congruent, and cannot be obtuse.)

In the figure, $\overrightarrow{BQ}$ and $\overrightarrow{CP}$ are the critical parallels to $\overrightarrow{AD}$, through B and C. Therefore

$$m\angle ABQ = c(a) = m\angle DCP,$$

as indicated. Applying the exterior angle theorem to the closed triangle $\Delta PCBQ$, we see that

$$t > s.$$

Therefore

$$t + c(a) > s + c(a).$$

Therefore

$$s + c(a) < 90,$$

which proves our theorem.

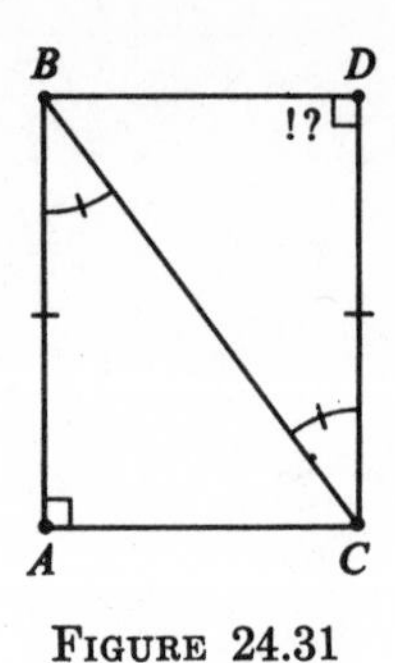

FIGURE 24.31

Theorem 4. Under HPP, in every right triangle $\triangle ABC$, we have

$$m\angle A + m\angle B + m\angle C < 180.$$

Proof. Suppose not. Then, if $\angle A$ is the right angle, $\angle B$ and $\angle C$ must be complementary. Take D on the opposite side of $\overleftrightarrow{BC}$ from A, so that $\angle BCD \cong \angle ABC$ and $CD = AB$. Then $\triangle ABC \cong \triangle DCB$, by SAS; and $\square ABDC$ is a Saccheri quadrilateral. This is impossible, because $\angle D$ is a right angle.

Theorem 5. Under HPP, for every triangle $\triangle ABC$, we have

$$m\angle A + m\angle B + m\angle C < 180.$$

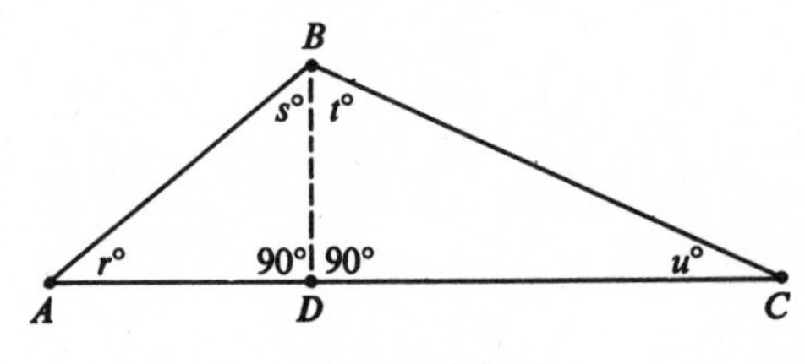

FIGURE 24.32

Proof. Let $\overline{AC}$ be a longest side of $\triangle ABC$, and let $\overline{BD}$ be the altitude from B to $\overleftrightarrow{AC}$. Then

$$r + s + 90 < 180,$$

and

$$t + u + 90 < 180.$$

Therefore

$$r + (s + t) + u < 180,$$

which proves the theorem.

Soon we shall see that under HPP this theorem has a true converse: for every number $x < 180$ there is a triangle for which the angle sum is x. Thus 180 is not merely an upper bound for the angle sums of triangles, but is precisely their supremum.

24.4 HYPERBOLIC GEOMETRY: THE DEFECT OF A TRIANGLE AND THE COLLAPSE OF SIMILARITY THEORY

The *defect* of $\triangle ABC$ is defined to be

$$180 - m\angle A - m\angle B - m\angle C.$$

The defect of $\triangle ABC$ is denoted by $\delta\triangle ABC$. Under HPP we know that the defect of any triangle is positive, and obviously it is less than 180. (Later we shall see that the converse holds: every number between 0 and 180 is the defect of some triangle.)

The following theorem is easy to check, regardless of HPP.

Theorem 1. Given $\triangle ABC$, with B-D-C. Then

$$\delta\triangle ABC = \delta\triangle ABD + \delta\triangle ADC.$$

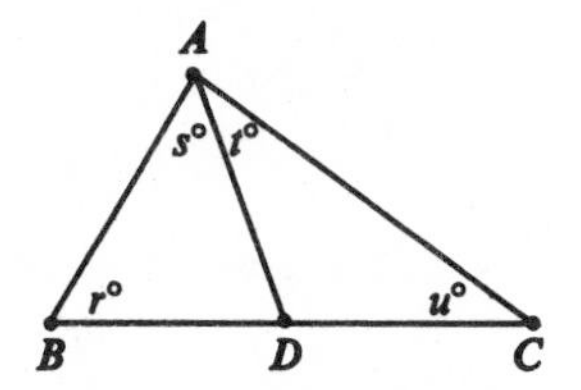

FIGURE 24.33

It has, however, an important consequence.

Theorem 2. Under HPP, every similarity is a congruence. That is, if $\triangle ABC \sim \triangle DEF$, then $\triangle ABC \cong \triangle DEF$.

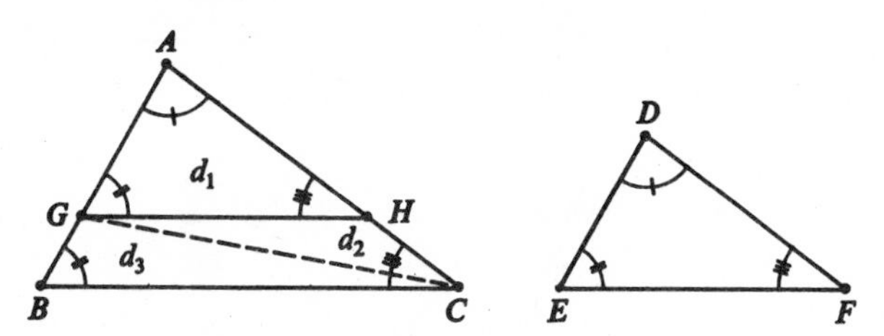

FIGURE 24.34

First we take G on $\overrightarrow{AB}$ so that $AG = DE$; and we take H on $\overrightarrow{AC}$ so that $AH = DF$. We then have $\triangle AGH \cong \triangle EDF$, by SAS; therefore

$$\triangle AGH \sim \triangle ABC.$$

If $G = B$, then $H = C$, and the theorem follows. We shall show that the contrary assumption $G \neq B$, $H \neq C$ (as shown in the figure) leads to a contradiction.

Let the defects of $\triangle AGH$, $\triangle GHC$, and $\triangle GBC$ be d_1, d_2, and d_3, as indicated in the figure; let d be the defect of $\triangle ABC$. By two applications of the preceding theorem, we have $d = d_1 + d_2 + d_3$. This is impossible, because the angle congruences given by the similarity $\triangle ABC \sim \triangle AGH$ tell us that $d = d_1$.

The additivity of the defect, described in Theorem 1, gives us more information about the critical function c. What we know so far is that (1) $0 < c(a) < 90$ for every $a > 0$, and (2) c decreases as a increases. There remains the question of how small the numbers $c(a)$ eventually become when a is very large. We might have either of the following situations:

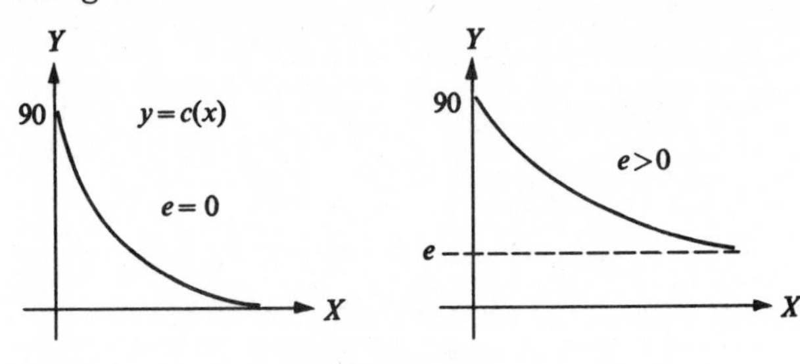

FIGURE 24.35

In each case, $e = \inf \{c(a)\}$, that is, the greatest lower bound of the numbers $c(a)$. In each case, it follows from (2) that $\lim_{a\to\infty} c(a) = e$. To prove the following theorem, therefore, we need merely show that $e > 0$ is impossible.

Theorem 3. $\lim_{a\to\infty} c(a) = 0$.

Proof. Suppose that $c(a) > e > 0$ for every a.

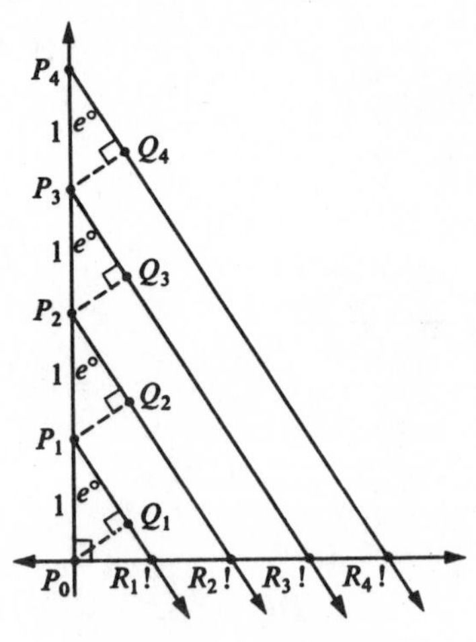

FIGURE 24.36

The markings in the figure should be self-explanatory. For each n, $\overrightarrow{P_nQ_n}$ intersects $\overrightarrow{P_0R_1}$, because $e < c(n)$. The right triangles $\triangle P_nP_{n+1}Q_{n+1}$ are all congruent,

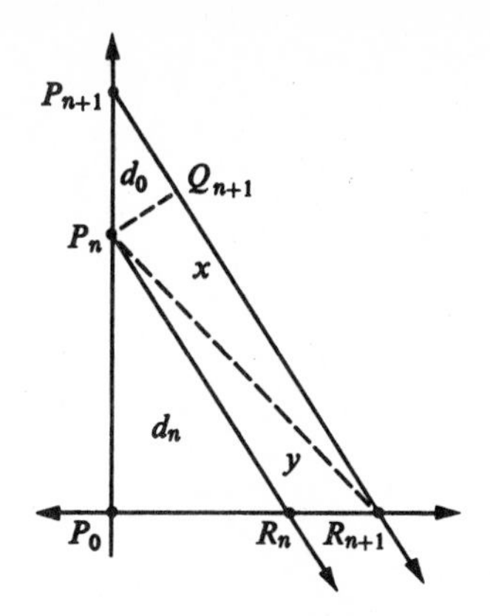

FIGURE 24.37

and therefore have the same defect d_0. Consider now what happens to the defect d_n of $\triangle P_0P_nR_n$ when n is increased by 1. The letters in the interiors of the triangles denote their defects. We have

$$\delta\triangle P_0P_nR_{n+1} = d_n + y,$$
$$\delta\triangle P_{n+1}P_nR_{n+1} = d_0 + x,$$
$$d_{n+1} = (d_n + y) + (d_0 + x),$$

by Theorem 1 in each case. Therefore

$$d_{n+1} > d_n + d_0.$$

Thus

$$d_2 > d_1 + d_0, \qquad d_3 > d_2 + d_0 > d_1 + 2d_0;$$

and by induction, we have

$$d_n > d_1 + (n - 1)d_0.$$

When n is sufficiently large, we have $d_n > 180$, by the Archimedean postulate. This is impossible, because the defect of a triangle is 180 minus the angle sum. Therefore $c(a) > e > 0$ is impossible, which was to be proved.

Consider now what happens to the measure $r(a)$ of the base angles of an isosceles right triangle, as the length a of the legs becomes large.

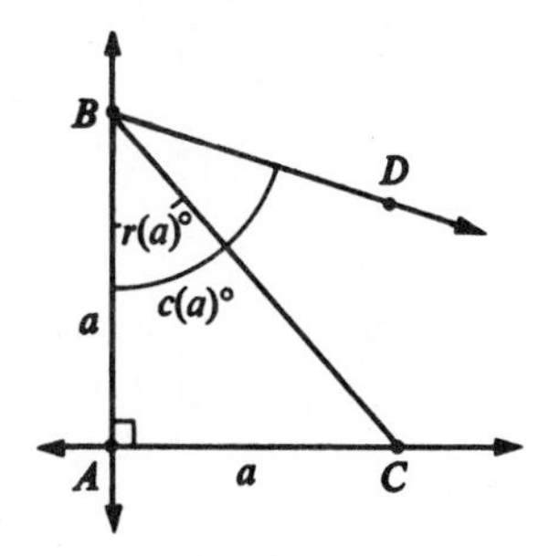

FIGURE 24.38

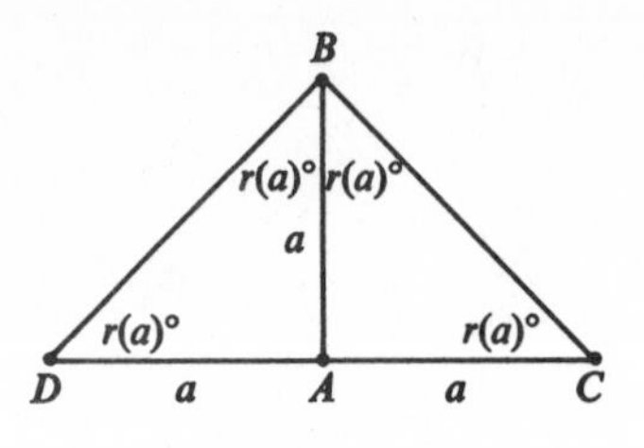

FIGURE 24.39

Here $\overrightarrow{BD}|\overrightarrow{AC}$. Therefore we always have $r(a) < c(a)$. Therefore $\lim_{a\to\infty} r(a) = 0$. Let us now make the figure symmetrical by copying $\triangle ABC$ on the other side of $\overleftrightarrow{AB}$. For $\triangle DBC$, the angle sum is $4r(a)$. Therefore the defect $180 - 4r(a)$ can be made as close to 180 as we please; we merely need to take a sufficiently large. Thus 180 is not merely *an* upper bound of the numbers which are the defects of triangles; 180 is precisely their supremum.

Theorem 4. For every number $x < 180$ there is a triangle whose defect is greater than x.

24.5 ABSOLUTE GEOMETRY: TRIANGULATIONS AND SUBDIVISIONS

Let R be a polygonal region. As in Chapter 14, by a *triangulation* of R we mean a finite collection,

$$K = \{T_1, T_2, \ldots, T_n\},$$

of triangular regions T_i, such that (1) the T_i's intersect only at edges and vertices, and (2) their union is R. A collection K which satisfies (1) is called a *complex*. Evidently every complex K forms a triangulation of the union of its elements.

Given two complexes

$$K = \{T_1, T_2, \ldots, T_n\},$$
$$K' = \{T'_1, T'_2, \ldots, T'_m\}.$$

If every T'_i lies in some one of the sets T_j, then K' is called a *subdivision* of K. Triangulations of certain types will be especially useful.

Given a polygon with vertices $P_1, P_2, \ldots, P_n$. Suppose that for each pair of successive vertices P_i, P_{i+1} all other vertices of the polygon lie on the same side

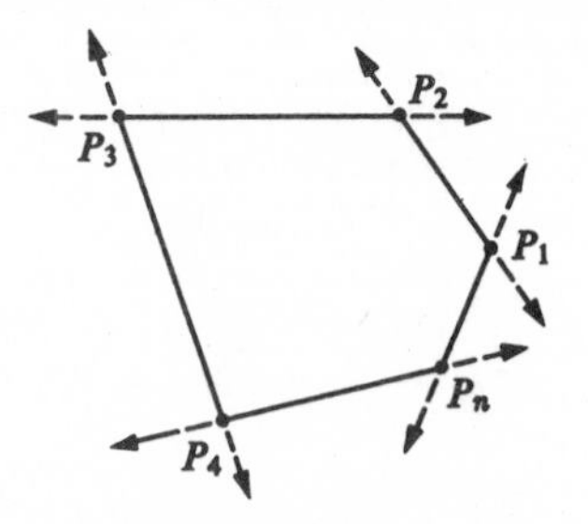

FIGURE 24.40

of $\overleftrightarrow{P_iP_{i+1}}$. Then the polygon is *convex*. From this it follows that if P_i, P_{i+1}, P_{i+2} are successive, then the other vertices (if any) all lie in the interior of $\angle P_iP_{i+1}P_{i+2}$. (To get this, we merely apply the definition of the interior of an angle.) By the *interior* of a convex polygon, we mean the intersection of the interiors of its angles. By a *convex polygonal region*, we mean the union of a convex polygon and its interior. By a *star triangulation* of such a region, we mean a complex like this:

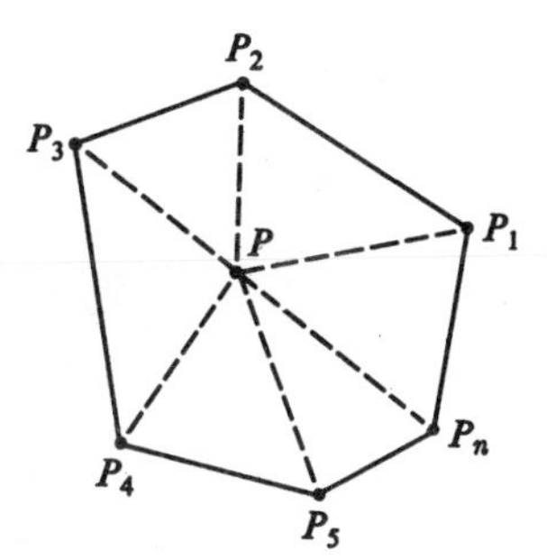

FIGURE 24.41

(More precise definition?) Obviously every convex polygonal region has a star triangulation; and any point P of the interior can be used as the central vertex.

Theorem 1. Let R be a convex polygonal region, and let L be any line which intersects the interior of R. Then L decomposes R into two convex polygonal regions.

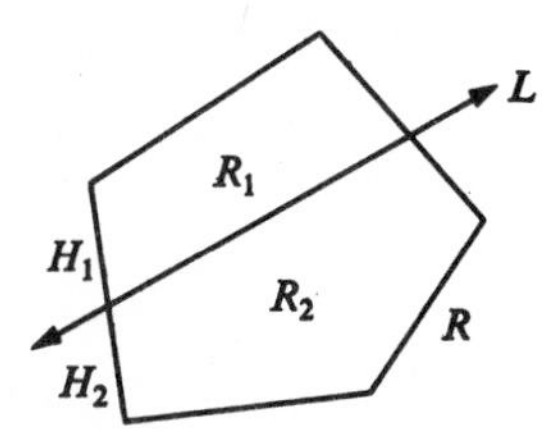

FIGURE 24.42

Proof. Let H_1 and H_2 be the half planes with L as edge; let $\overline{H_1} = H_1 \cup L$ and let $\overline{H_2} = H_2 \cup L$. Let

$$R_1 = R \cap \overline{H_1}, \qquad R_2 = R \cap \overline{H_2}.$$

Then R_1 and R_2 are convex sets, because each of them is the intersection of two convex sets; and it is easy to check that they are convex polygonal regions in the sense that we have just defined.

We shall use this as a lemma in proving the following theorem.

Theorem 2. Every two triangulations of the same polygonal region have a common subdivision.

That is, if K_1 and K_2 are triangulations of R, then there is a triangulation K of R which is a subdivision both of K_1 and K_2.

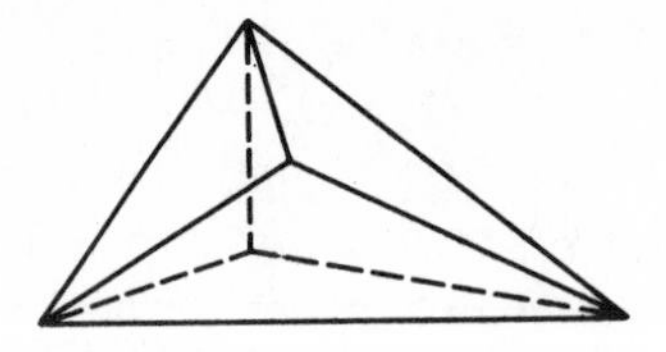

FIGURE 24.43

In the figure, the edges of K_1 are drawn solid, and the edges of K_2 are dashed. (As in Chapter 14, an adequate figure is not easy to draw or to look at.)

Proof. Let

$$L_1, L_2, \ldots, L_n$$

be the lines which contain either an edge of K_1 or an edge of K_2. (For Fig. 24.43, we would have $n = 9$.)

Now L_1 decomposes each T_i in K_1 (and each T'_j in K_2) into two convex regions, if L_1 intersects the interiors of these sets at all:

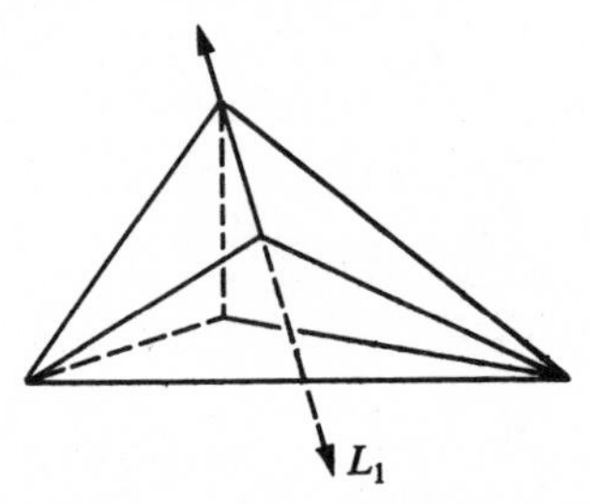

FIGURE 24.44

(In the figure, we show a possibility for L_1.) By induction it now follows that the union of all of the L_i's decomposes R into a finite collection,

$$C = \{C_1, C_2, \ldots, C_n\},$$

of convex polygonal regions, like this:

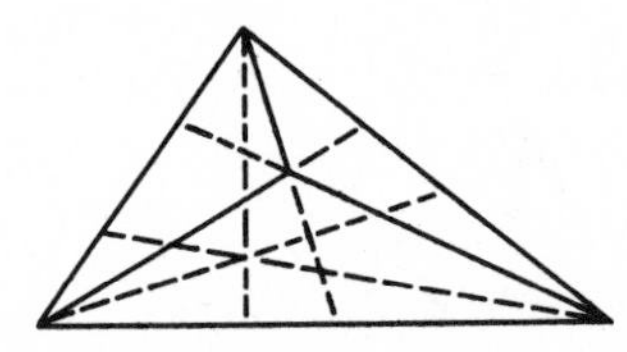

FIGURE 24.45

Evidently every C_i lies in some *one* $T_j \in K_1$ and in some *one* $T'_k \in K_2$. For each C_i we take a star triangulation. These fit together to give our common subdivision K.

Let R_1 and R_2 be polygonal regions. Suppose that they have triangulations

$$K_1 = \{T_1, T_2, \ldots, T_n\},$$
$$K_2 = \{T'_1, T'_2, \ldots, T'_n\},$$

such that for each i we have

$$T_i \cong T'_i.$$

Then we say that R_1 and R_2 are *equivalent by finite decomposition*, and we write

$$R_1 \equiv R_2.$$

Here by $T_i \cong T_i'$ we mean that the corresponding triangles are congruent in the elementary sense. In fact, this is equivalent to saying that T_i and T_i' are isometric in the sense of Chapter 18.

Equivalence by finite decomposition is a familiar idea in simple cases; it means, intuitively, that you can cut R_1 into little triangular pieces, with scissors, and then put the pieces back together again, usually in a different way, to get R_2.

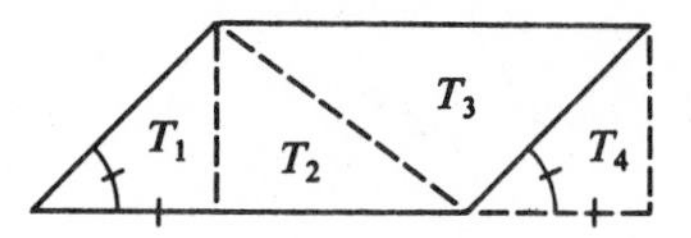

FIGURE 24.46

For example, in the figure above, we have $T_1 \cong T_4$. Therefore

$$T_1 \cup T_2 \cup T_3 \equiv T_4 \cup T_2 \cup T_3.$$

In fact, this is the observation that people usually make to infer that the parallelogram and the rectangle have the same area.

Theorem 3. Equivalence by finite decomposition is an equivalence relation.

Proof. Trivially, $\equiv$ is reflexive and symmetric. We must now show that if $R_1 \equiv R_2$ and $R_2 \equiv R_3$, then $R_1 \equiv R_3$.

Let K_1 and K_2 be the triangulations used in exhibiting that $R_1 \equiv R_2$. That is,

$$K_1 = \{T_1, T_2, \ldots, T_n\},$$
$$K_2 = \{T_1', T_2', \ldots, T_n'\},$$

with

$$T_i \cong T_i'.$$

Let K_2' and K_3 be the triangulations exhibiting that $R_2 \equiv R_3$. Let K be a common subdivision of K_2 and K_2' (Fig. 24.47). Given any $T_i \in K_1$, we observe that the corresponding $T_i' \in K_2$ has been subdivided in a certain way. We copy this subdivision scheme in T_i, following the congruence $T_i \cong T_i'$ backwards. This gives a subdivision K_1' of K_1, shown by the dotted lines in the figure. Similarly, we get a subdivision K_3' of K_3. We can now match up the elements of K_1' and K_3' in such a way as to show that $R_1 \equiv R_3$. (We omit the details, on the ground that a careful inspection of the figures is likely to convey the idea adequately and more easily.)

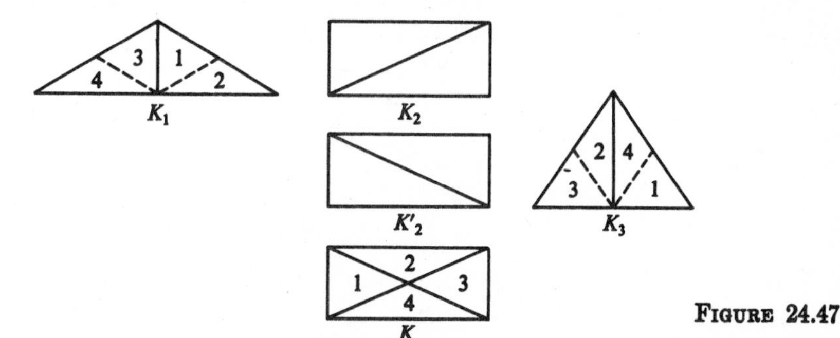

FIGURE 24.47

24.6 EUCLIDEAN GEOMETRY: BOLYAI'S THEOREM

If two triangular regions are equivalent by finite decomposition, then surely they have the same area. Thus if

$$R_1 = T_1 \cup T_2 \cup \ldots \cup T_n,$$
$$R_2 = T'_1 \cup T'_2 \cup \ldots \cup T'_n,$$

with

$$T_i \cong T'_i$$

for each i, then we have $\alpha T_i = \alpha T'_i$, from which we get $\alpha R_1 = \alpha R_2$ by addition.

It was discovered by Bolyai that the *converse is also true;* if $\alpha R_1 = \alpha R_2$, it follows that $R_1 \equiv R_2$. This section will be devoted to the proof of Bolyai's theorem.

Given $\triangle ABC$, with $\overline{BC}$ considered to be the base.

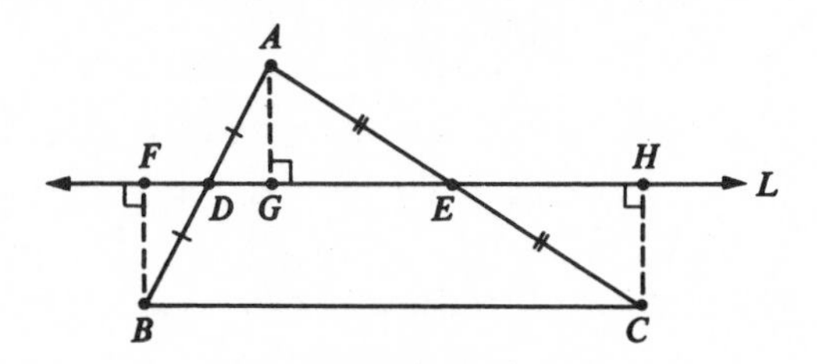

FIGURE 24.48

Let D and E be the midpoints of $\overline{AB}$ and $\overline{AC}$; let $L = \overleftrightarrow{DE}$; and let $\overline{BF}$, $\overline{AG}$, and $\overline{CH}$ be perpendicular to L. Then $\square BFHC$ is a rectangle. (Remember the *SAA* Theorem.) We shall call $\square BFHC$ the rectangle *associated with* $\triangle ABC$. (Of course, $\square BFHC$ depends on the choice of the base, but it will always be clear which base is meant.)

Theorem 1. Every triangular region is equivalent by finite decomposition to its associated rectangular region.

Proof. The preceding figure shows the easiest case. Here G is the foot of the perpendicular from A to $\overleftrightarrow{DE}$, and D-G-E. In this case the relation $\triangle ABC \equiv \square BFHC$ follows from two triangle congruences.

The proof in the general case requires a different method, as follows. As before, let D and E be the midpoints of $\overline{AB}$ and $\overline{AC}$. Let G be the point such that D-E-G and $DE = EG$.

By SAS, we have $\triangle ADE \cong \triangle CGE$, so that

$$\triangle ABC \equiv \square BDGC.$$

As indicated in the figure, $DB = GC$, and $\angle GDB \cong \angle IGC$. $\square BDGC$ will be called *the parallelogram associated with* $\triangle ABC$.

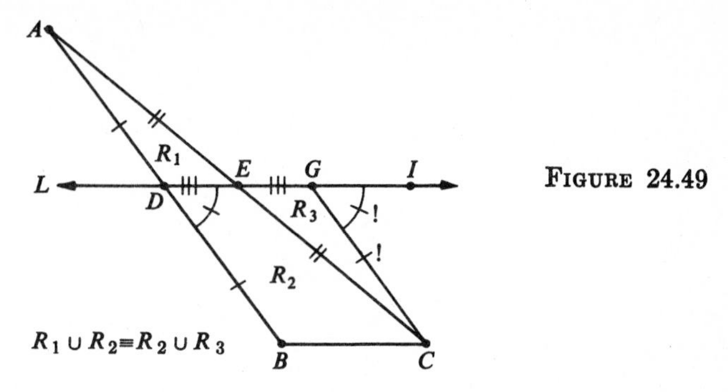

FIGURE 24.49

If $D = F$, then it follows easily that $G = H$, so that $\square BDGC = \square BFHC$, and the theorem follows.

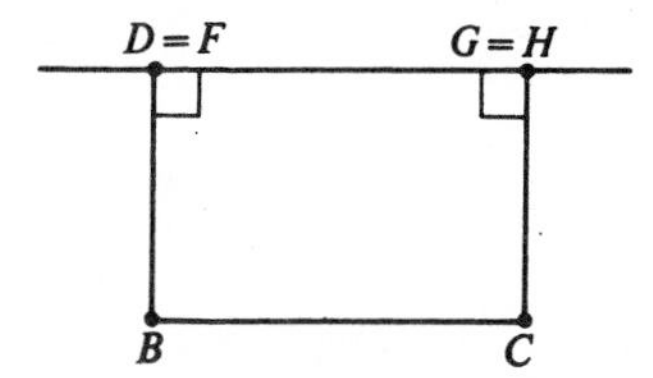

FIGURE 24.50

Hereafter we assume that $D \neq F$. We may then suppose that D-F-H, so that D "lies to the left of $\overline{FH}$."

CASE 1. Suppose that F lies on $\overline{DG}$ (with $D \neq F$), as in the figure below.

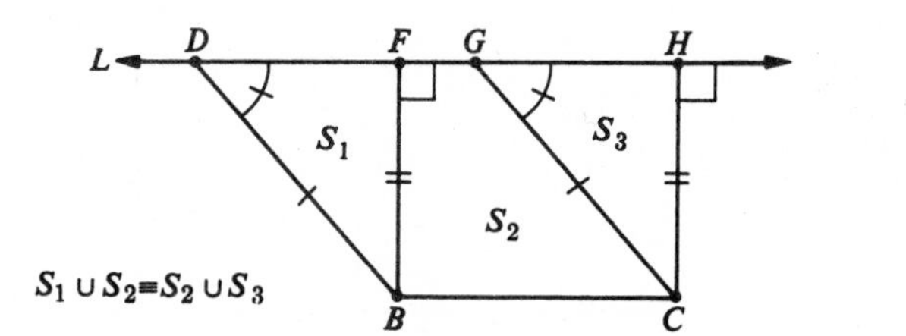

FIGURE 24.51

Here $\triangle BDF \cong \triangle CGH$, so that $\square BDGC \equiv \square BFHC$. We shall show that the general case can always be reduced to this special case.

CASE 2. Suppose that D-G-F. (The case of interest is the case in which D and G are "far to the left of F.") Consider the geometric operation conveyed by the following figure.

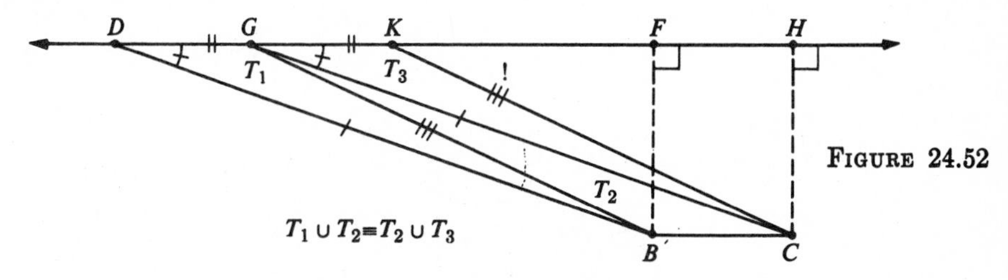

FIGURE 24.52

Here D-G-K and $DG = GK$. Since $T_1 \cong T_3$, it follows that

$$\square BDGC \equiv \square BGKC.$$

In replacing the first of these quadrilaterals by the second, we have "moved $\overline{DG}$ one step to the right." Since the real number system is Archimedean, we know that $nDG > GF$ for some positive integer n. Thus, in a finite number of steps we can move $\overline{DG}$ far enough to the right to get Case 1, in which F lies on $\overline{DG}$. The theorem follows.

Theorem 2. If two triangular regions have the same base and the same area, then they are equivalent under finite decomposition.

Proof. Let the triangular regions be T and T', and let the associated rectangular regions be R and R'. Then R and R' have the same base b. Since $\alpha R = \alpha T = \alpha T' = \alpha R'$, we have $\alpha R = \alpha R'$, and R and R' have the same altitude. Therefore $R \equiv R'$. (Proof?) Thus

$$T \equiv R \equiv R' \equiv T',$$

and $T \equiv T'$, which was to be proved.

Theorem 3. *Bolyai's Theorem.* If two triangular regions have the same area, they are equivalent under finite decomposition.

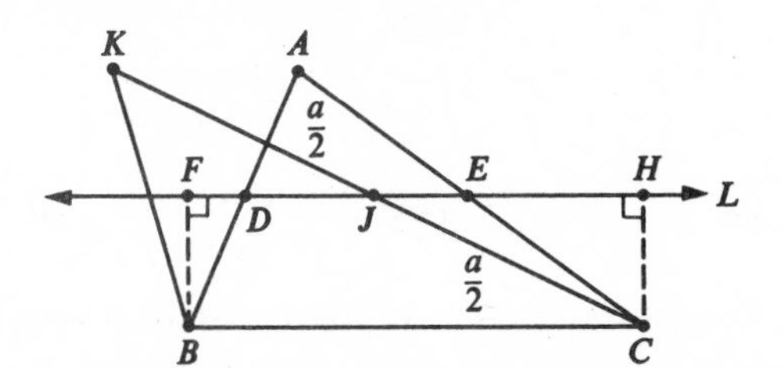

FIGURE 24.53

Proof. Given T_1, T_2, with $\alpha T_1 = \alpha T_2$. Suppose that T_1 is the union of $\triangle ABC$ and its interior. If a side of T_1 is congruent to a side of T_2, then $T_1 \equiv T_2$ by Theorem 2. Suppose, then, that T_2 has a side of length $a > AC$. If we prove the theorem for this case, it will follow in general merely by a change of notation. As before, let L be the line through the bisectors D and E of $\overline{AB}$ and $\overline{AC}$. Let J be a point of L such that $CJ = a/2$. [*Query:* How do we know that there is such a point?] Now take K so that C-J-K and $JK = a/2$. Let T_3 be the union of $\triangle KBC$ and its interior. By Theorem 2, we have

$$T_3 \equiv R \quad \text{and} \quad T_2 \equiv T_3.$$

Therefore $T_2 \equiv R$. Since we know already that $R \equiv T_1$, we have $T_1 \equiv T_2$, which was to be proved.

Note that the use of the transitivity of the relation $\equiv$ has spared us some almost impossibly complicated figures, exhibiting the equivalence of T_1 and T_2.

We shall see that Bolyai's Theorem can be extended to polygonal regions in general. To show this, we use the following theorem.

Theorem 4. In a Euclidean plane, every polygonal region is equivalent by finite decomposition to a triangular region.

Proof. Given a polygonal region R, with a triangulation

$$K = \{T_1, T_2, \ldots, T_n\}.$$

For each i, let

$$a_i = \alpha T_i.$$

There is now a complex which looks like this:

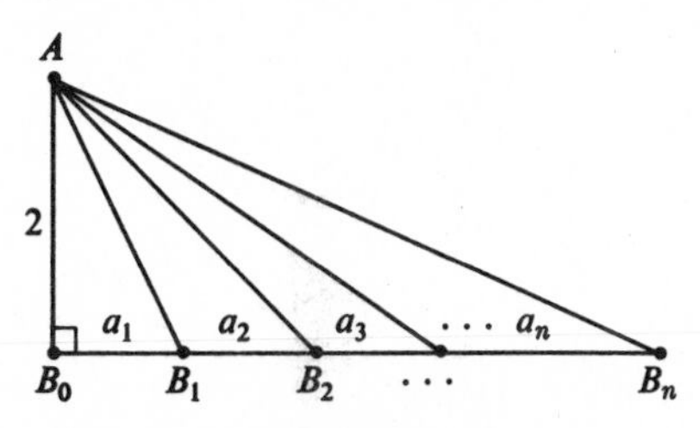

FIGURE 24.54

If T'_i is the ith triangular region in this figure, then

$$\alpha T'_i = \alpha T_i.$$

Therefore

$$T'_i \equiv T_i$$

for every i, by Theorem 3. Therefore R is equivalent by finite decomposition to the union of the regions T'_i.

This is *not* a theorem of absolute geometry; it will turn out that under HPP the triangular regions are—in a certain sense—of bounded area, and the polygonal regions in general are not. In the following section we shall elucidate this idea by showing the form that area theory takes in hyperbolic geometry. Meanwhile we generalize Bolyai's theorem.

Theorem 5. Let R and R' be polygonal regions in a Euclidean plane. If $\alpha R = \alpha R'$, then $R' \equiv R$.

Proof. Let T and T' be triangular regions such that

$$T \equiv R, \qquad T' \equiv R'.$$

Then $\alpha T = \alpha R$ and $\alpha T' = \alpha R'$. Therefore $\alpha T = \alpha T'$ and $T \equiv T'$. Therefore $R \equiv R'$, which was to be proved.

24.7 HYPERBOLIC AREA-THEORY: THE DEFECT OF A POLYGONAL REGION

We have already found, in Theorem 1, Section 24.4, that the defect of a triangle is additive in the same way that area is:

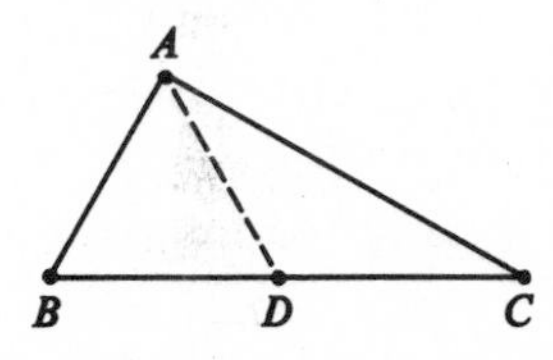

FIGURE 24.55

That is, if B-D-C, then the defect of $\triangle ABC$ is the sum of the defects of $\triangle ABD$ and $\triangle ADC$. This simple fact is the key to the development of an area theory in hyperbolic geometry. We begin by defining the *area* of a triangular region T to be the defect of the corresponding triangle. We denote the hyperbolic area by δT, where δ stands for *defect*, just as α stood for *area*. Under HPP, we know that $\delta T > 0$ for every T. And we know that the additivity postulate holds, insofar as it applies at all.

We would like to define our "area function" δ more generally so as to make it apply to all polygonal regions. We shall do this in several stages.

First, given a complex

$$K = \{T_1, T_2, \ldots, T_n\},$$

we define

$$\delta K = \delta T_1 + \delta T_2 + \cdots + \delta T_n.$$

We are now in the same situation as in Chapter 14. Every polygonal region R has infinitely many triangulations K; we would like to define δR as δK; but to do this we must first show that δK depends only on R, and is independent of the choice of the triangulation K. This is easy to see for star triangulations of a convex polygonal region.

Theorem 1. If K_1 and K_2 are star triangulations of the same polygonal region R, then $\delta K_1 = \delta K_2$.

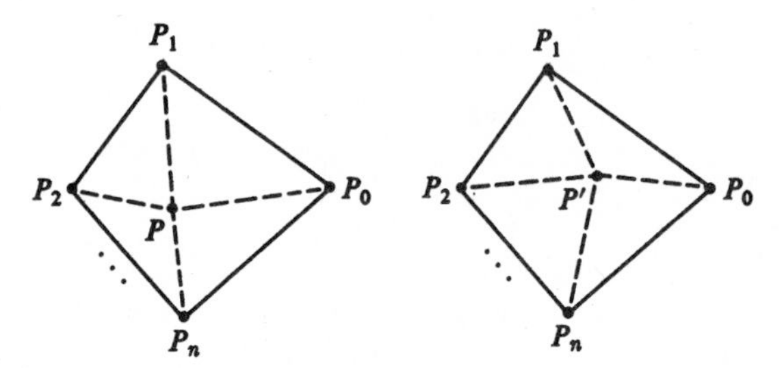

FIGURE 24.56

The reason is that the total defect in each star triangulation is

$$\begin{aligned}(n+1)180 - (m\angle P_0 + m\angle P_1 + \cdots + m\angle P_n + 360)\\ = (n-1)180 - (m\angle P_0 + m\angle P_1 + \cdots + m\angle P_n).\end{aligned}$$

This theorem justifies the following definition.

DEFINITION. The defect δR of a convex polygonal region R is the number which is the defect of every star triangulation of R.

A *border triangulation* of a convex polygonal region is one which looks like Fig. 24.57. (Exact definition?)

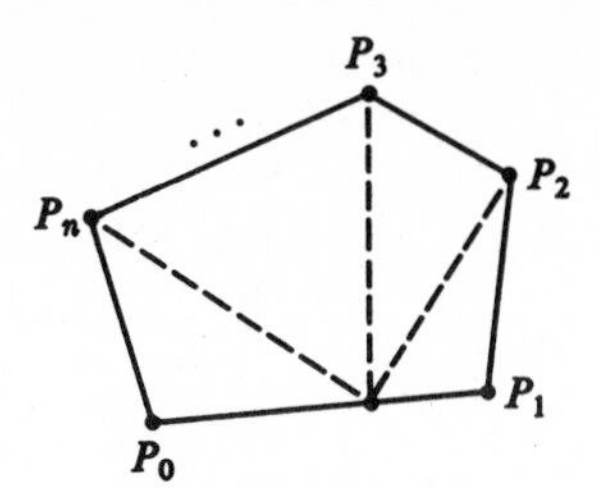

FIGURE 24.57

Theorem 2. The defect of a border triangulation is the same as the defect of the region.

The reason is that the defect of the border triangulation is

$$n(180) - (m\angle P_0 + m\angle P_1 + \cdots + m\angle P_n + 180),$$

which gives the same answer as for star triangulations.

Theorem 3. If a convex polygonal region is decomposed by a line into two such regions, then the defect of the union is the sum of the defects.

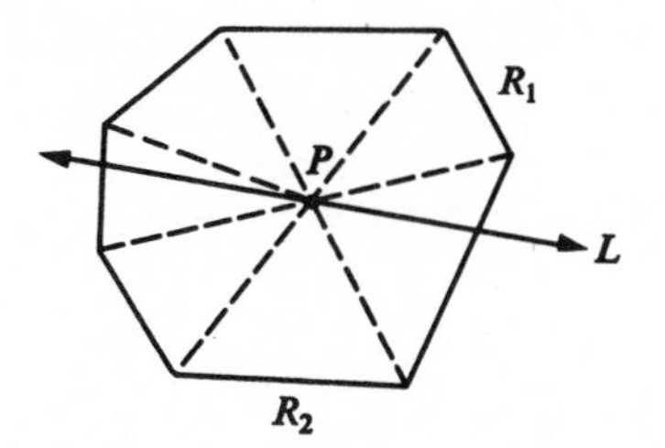

FIGURE 24.58

Proof. Given $R_1 \cup R_2 = R$, $R_1 \cap R_2 \subset L$, as in the figure. Let P be any point of L in the interior of R; let K_1 and K_2 be the border triangulations in which P is the extra vertex. Let $K = K_1 \cup K_2$. Then, trivially, we have

$$\delta K = \delta K_1 + \delta K_2.$$

Since $\delta K = \delta R$, $\delta K_1 = \delta R_1$ and $\delta K_2 = \delta R_2$, this proves the theorem.

Theorem 4. If K_1 and K_2 are triangulations of the same polygonal region R, then $\delta K_1 = \delta K_2$.

The proof is very similar indeed to the proof of Theorem 2, Section 24.5. Exactly as in that proof, we let

$$L_1, L_2, \ldots, L_n$$

be the lines that contain either an edge of K_1 or an edge of K_2. As before, we use the lines L_i, one at a time, to cut up the triangular regions in K_1 and K_2 into smaller convex polygonal regions. At each stage, we know by Theorem 3 that the total defect is unchanged. When all the lines have been used, we form a star triangulation of each of the resulting convex polygonal regions; this final step also leaves the total defect unchanged. (In fact, the defect of a convex polygonal region C was *defined* to be the total defect in any star triangulation.) We now have a common subdivision K of K_1 and K_2, with

$$\delta K = \delta K_1$$

and

$$\delta K = \delta K_2.$$

Therefore $\delta K_1 = \delta K_2$, which was to be proved.

DEFINITION. The area δR of a polygonal region R is the number which is the defect δK of every triangulation K of R.

It is not hard to see that our area function δ satisfies the postulates A-1 through A-5 of Section 13.1. (You may wonder, at first, about A-5, which says that the area of a square region is the square of the length of its edges. But under HPP, even this last postulate holds, for the odd reason that there are no square regions to which it can be applied.)

24.8 BOLYAI'S THEOREM FOR TRIANGLES IN THE HYPERBOLIC CASE

We shall see that Bolyai's theorem holds under HPP. That is, if $\delta R_1 = \delta R_2$, then it follows that $R_1 \equiv R_2$. Although a part of the proof follows the lines of Section 24.6, the technique is more complicated, and so we shall need some preliminaries.

Theorem 1. If two Saccheri quadrilaterals have the same upper base and the same defect, then their upper base angles are congruent.

RESTATEMENT. Let $\square ABCD$ and $\square A'B'C'D'$ be Saccheri quadrilaterals (with right angles at A, D, A', and D'). If $BC = B'C'$ and $\delta\square ABCD = \delta\square A'B'C'D'$, then $\angle B \cong \angle B'$ and $\angle C \cong \angle C'$.

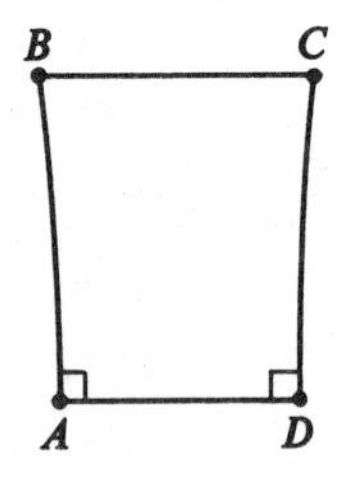

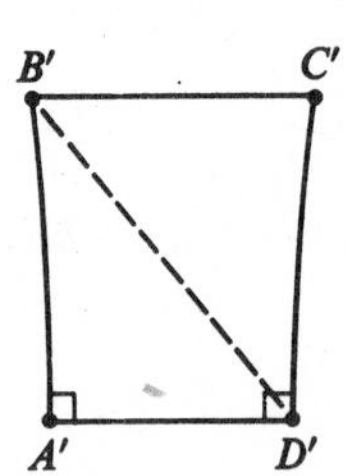

FIGURE 24.59

Proof. It is not hard to calculate that

$$\delta\square ABCD = 180 - (m\angle B + m\angle C) = 180 - 2m\angle B.$$

Therefore $m\angle B$ is determined by the defect; and from this the theorem follows.

Theorem 2. Under HPP, if $\square ABCD$ and $\square A'B'C'D'$ are as in Theorem 1, then

$$\square ABCD \cong \square A'B'C'D'.$$

(Here the indicated congruence means that the correspondence $ABCD \leftrightarrow A'B'C'D'$ preserves lengths of sides and measures of angles.)

Proof. Let E and F be points of $\overrightarrow{BA}$ and $\overrightarrow{CD}$ such that

$$BE = CF = B'A' = C'D'.$$

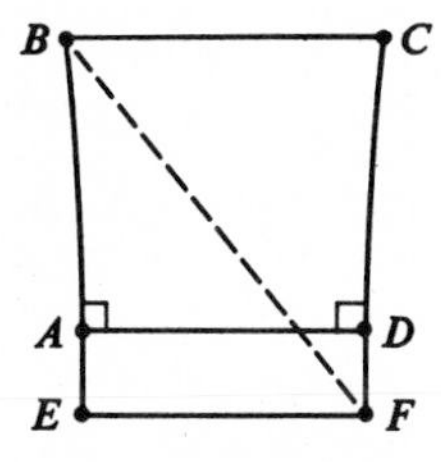

FIGURE 24.60

By SAS, we have $\triangle BCF \cong \triangle B'CD$. Therefore $BF = B'D'$. By angle subtraction, we have $\angle EBF \cong \angle A'B'D'$. By SAS, we have $\triangle BEF \cong \triangle B'A'D'$. Therefore $\angle E$ is a right angle. In the same way, we conclude that $\angle CFE$ is a right angle. Thus we have

$$\square EBCF \cong \square A'B'C'D'.$$

If $E = A$ and $F = D$, this proves the theorem. If not, $\square EADF$ is a rectangle, which is absurd; there is no such thing as a rectangle.

The reader is warned that, hereafter in this section, to draw our triangles right side up, we are going to draw our Saccheri quadrilaterals upside down.

Given $\triangle ABC$, with $\overline{BC}$ considered to be the base:

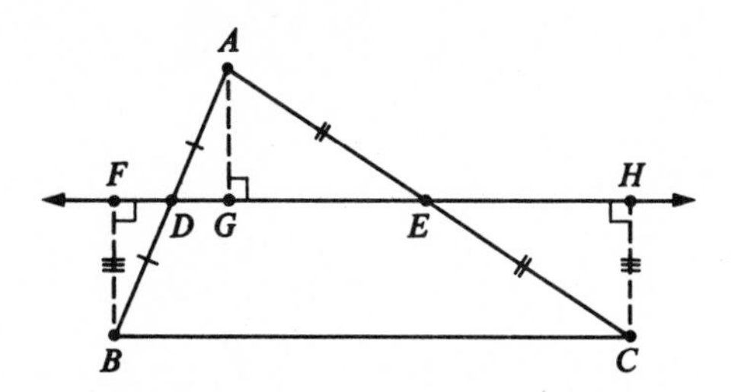

FIGURE 24.61

As before, let D and E be the bisectors of $\overline{AB}$ and $\overline{AC}$; let F, G, and H be the feet of the perpendiculars from B, A, and C to L. As in the Euclidean case, we have

$$\triangle FBD \cong \triangle GAD,$$
$$\triangle GAE \cong \triangle HCE,$$
$$FB = GA = HC.$$

(The elementary theory of congruence is a part of absolute geometry.) Therefore $\square HCBF$ is a Saccheri quadrilateral. We shall call it the quadrilateral *associated with* $\triangle ABC$. It depends on the choice of the base, but it will always be clear which base we mean.

Theorem 3. Every triangular region is equivalent by finite decomposition to its associated quadrilateral region.

The proof is exactly like the proof of Theorem 1, Section 24.6; this proof depended only on congruences and the SAA theorem.

Theorem 4. Every triangular region has the same defect as its associated quadrilateral region.

Because the two are equivalent by finite decomposition.

Theorem 5. If $\triangle ABC$ and $\triangle DEF$ have the same defect and a pair of congruent sides, then the two triangular regions are equivalent by finite decomposition.

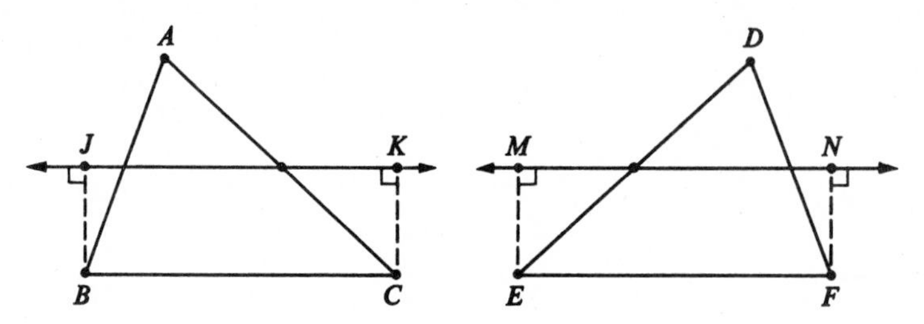

FIGURE 24.62

Proof. It does no harm to suppose that the congruent sides are the "bases" $\overline{BC}$ and $\overline{EF}$. If the associated Saccheri quadrilaterals are as indicated, then by Theorem 4 they have the same defect. By Theorem 2, $\square KCBJ \cong \square NFEM$. Let T_1, T_2, R_1, R_2 be the regions determined by our triangles and quadrilaterals. Then $R_1 \equiv R_2$. (This requires a proof based on our congruence, but the proof is immediate.) Since $T_1 \equiv R_1$ and $R_2 \equiv T_2$, by Theorem 3, we have $T_1 \equiv T_2$, which was to be proved.

Theorem 6. Given $\triangle ABC$, D, E, and $L = \overleftrightarrow{DE}$, as in the definition of the associated quadrilateral. Let A' be a point on the same side of $\overleftrightarrow{BC}$ as A. If L contains the midpoint E' of $\overline{A'C}$, then L also contains the midpoint D' of $\overline{A'B}$, and $\triangle ABC$ and $\triangle A'BC$ have the same associated quadrilateral.

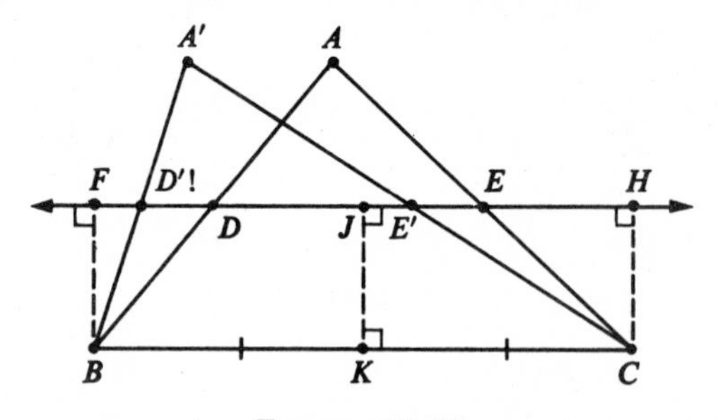

FIGURE 24.63

Proof. Let J and K be the midpoints of $\overline{FH}$ and $\overline{BC}$. Then $\overline{JK} \perp \overleftrightarrow{BC}$ and $\overline{JK} \perp \overleftrightarrow{DE}$, as indicated in the figure. Therefore

(1) $\overleftrightarrow{DE}$ is the perpendicular, through E, to the perpendicular bisector of $\overline{BC}$.

Applying precisely the same reasoning to $\triangle A'BC$, we get

(2) $\overleftrightarrow{D'E'}$ is the perpendicular, through E', to the perpendicular bisector of $\overline{BC}$.

Since we know that E' lies on $\overleftrightarrow{DE}$, it follows that $\overleftrightarrow{DE} = \overleftrightarrow{D'E'}$, and D' lies on L. Therefore $\triangle ABC$ and $\triangle A'BC$ have the same associated quadrilateral $\square HCBF$, which was to be proved.

Theorem 7. If T_1 and T_2 are triangular regions, and $\delta T_1 = \delta T_2$, then $T_1 \equiv T_2$.

Proof. Let the associated triangles be $\triangle ABC$ and $\triangle A'B'C'$. If a side of one is congruent to a side of the other, then $T_1 \equiv T_2$, by Theorem 5. If not, we may suppose that

$$a = A'C' > AC.$$

(This proof is going to be very similar to that of Theorem 5, Section 24.6.)

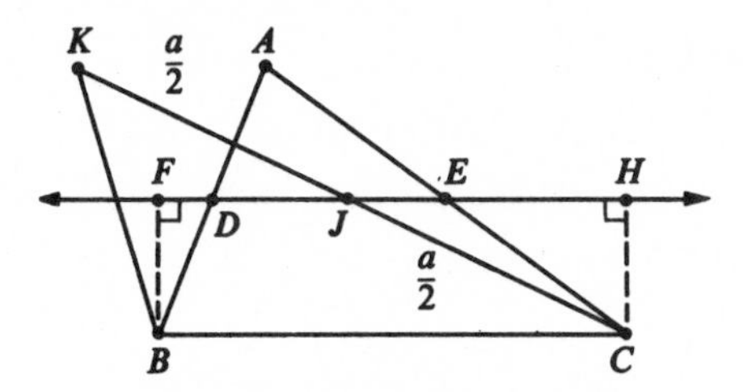

FIGURE 24.64

As indicated in the figure, let $\square HCBF$ be the quadrilateral associated with $\triangle ABC$. Let J be a point of L such that $CJ = a/2$. Then take K so that C-J-K and $JK = a/2$. By Theorem 6, $\square HCBF$ is the quadrilateral associated with $\triangle KBC$. Thus

$$\delta\triangle KBC = \delta\square HCBF = \delta\triangle ABC = \delta\triangle A'B'C'.$$

Now $\triangle KBC$ and $\triangle A'B'C'$ have a pair of congruent sides and the same defect. Therefore the corresponding regions T_3, T_2 are equivalent by finite decomposition. Let R be the region corresponding to $\square HCBF$. Then we have

$$T_1 \equiv R \equiv T_3 \equiv T_2.$$

Therefore $T_1 \equiv T_2$, which was to be proved.

24.9 DEFECTS OF SMALL TRIANGLES

We shall show, in this section, that the defect of a triangle is as small as we please, if the triangle itself is sufficiently small. More precisely:

Theorem 1. Let e be any positive number. Then there is a positive number d such that if all the sides of $\triangle ABC$ have length less than d, then $\delta\triangle ABC < e$.

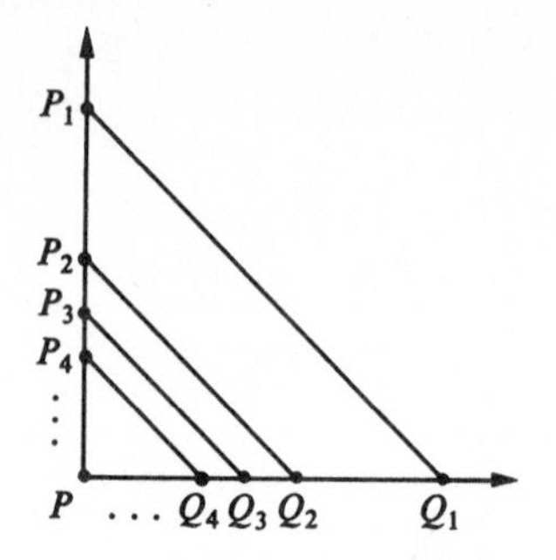

FIGURE 24.65

We proceed to the proof. We have given a positive number e. Consider a right angle $\angle P_1PQ_1$, with $P_1P = PQ_1 = 1$ (Fig. 24.65). For each n, take P_n and Q_n, as indicated, so that $PP_n = PQ_n = 1/n$. Thus we get a sequence of convex quadrilaterals $\square P_1P_2Q_2Q_1$, $\square P_2P_3Q_3Q_2, \ldots$. Let d_n be the defect of the nth quadrilateral $\square P_nP_{n+1}Q_{n+1}Q_n$, and let $d_0 = \delta\triangle P_1PQ_1$. Then

$$d_1 + d_2 + \cdots + d_n < d_0$$

for every n. Therefore the infinite series,

$$d_1 + d_2 + \cdots + d_n + \cdots,$$

is convergent. Therefore $\lim_{n\to\infty} d_n = 0$. Therefore $d_n < e$ for some n.

The sole purpose of the above discussion was to demonstrate the following statement.

There is a convex quadrilateral $\square PQRS$, *of defect less than* e.

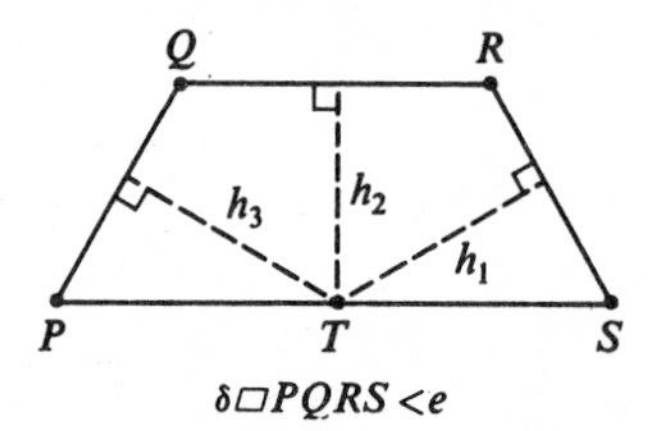

FIGURE 24.66

Let T be any point between P and S. Let h_1, h_2, h_3 be the perpendicular distances from T to the other three sides of $\square PQRS$, and let d be the smallest of the numbers h_1, h_2, h_3.

It is now easy to check that d is a number of the sort that we wanted. Given any triangle $\triangle ABC$, with sides of length less than d:

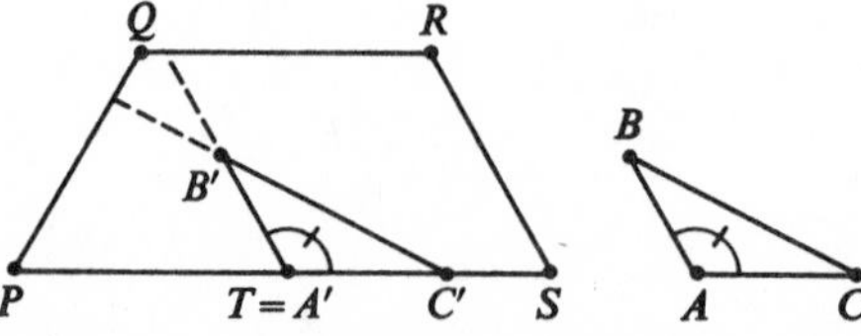

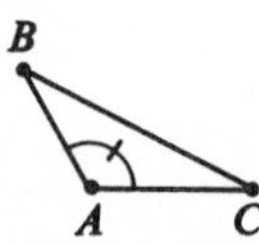

FIGURE 24.67

By SAS, we can construct a congruent copy $\triangle A'B'C'$ of $\triangle ABC$ in the quadrilateral region, with $T = A'$. The copy really will lie inside the quadrilateral, because its sides are too short to reach the other three sides of the quadrilateral. But now we are done, because

$$\delta\triangle ABC = \delta\triangle A'B'C' < \delta\square PQRS < e.$$

This theorem tells us that it may be very hard to tell the difference between a hyperbolic plane and a Euclidean plane, if you are allowed to inspect only a small portion of it. One of the possibilities for physical space is that planes are hyperbolic, but that the portions of them that we can examine from the earth are "small," so that the deviation from the Euclidean angle-sum formula $m\angle A + m\angle B + m\angle C = 180$ is too small to be detected, for every triangle small enough for us to examine. C. F. Gauss made a test of this sort, using the peaks of three neighboring mountains as the vertices of his triangle. He was unable to observe a deviation from the Euclidean formula, but obviously it is possible that his mountains were too neighboring.

24.10 THE CONTINUITY OF THE DEFECT

Given $\triangle ABC$ and $\triangle DEF$, with $\delta\triangle ABC > \delta\triangle DEF$. (Note that in the light of the results in the previous section, a plausible figure should make the second triangle look smaller.)

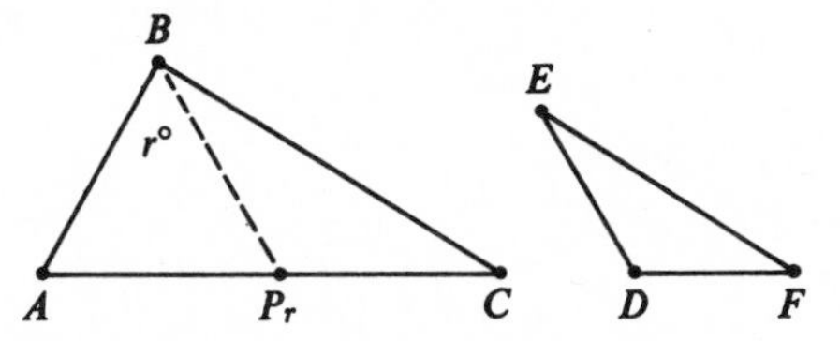

FIGURE 24.68

For $0 < r \leqq m\angle B$, let P_r be a point of $\overline{AC}$ such that $m\angle ABP_r = r$. (By the cross-bar theorem, for each such r there is a point P_r.) We know that

$$\delta\triangle ABP_r + \delta\triangle P_rBC = \delta\triangle ABC.$$

For $0 < r \leqq m\angle B$, let

$$f(r) = \delta\triangle ABP_r.$$

As a *definition* of $f(0)$, we provide further that

$$f(0) = 0.$$

(Of course, the definition is reasonable: in effect, we are defining the defect of a segment to be 0.)

It is easy to see that f is a strictly increasing function: if $r < s$, then we have

$$f(s) = f(r) + \delta\triangle P_rBP_s,$$

so that

$$f(r) < f(s).$$

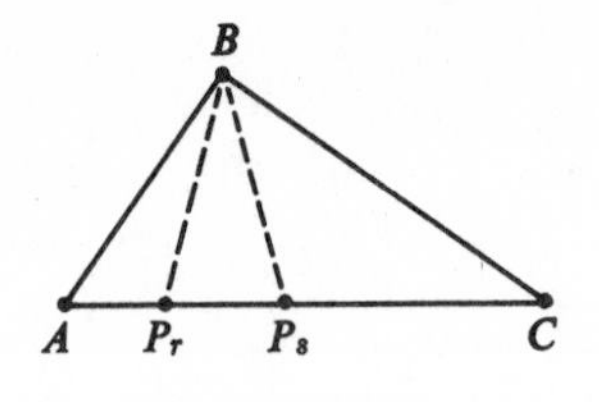

FIGURE 24.69

It is reasonable to suppose that f is continuous. If this is not merely reasonable but also true, then we will have the following theorem.

Theorem 1. If $\delta\triangle ABC > \delta\triangle DEF$, then there is a point P between A and C such that $\delta\triangle ABP = \delta\triangle DEF$.

The reason is this. The graph of $y = f(r)$ looks like this:

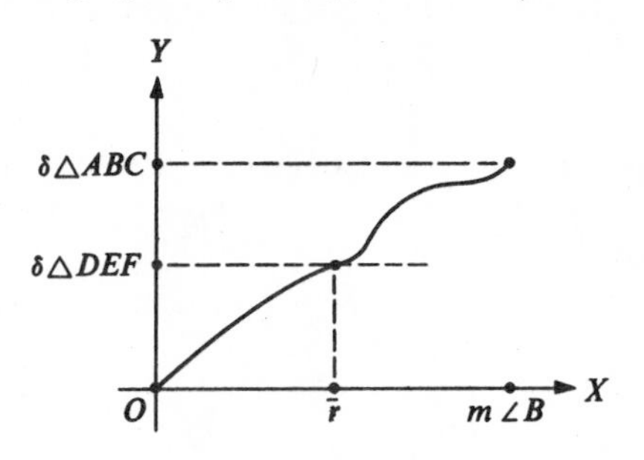

FIGURE 24.70

Being continuous, f takes on every value between its initial value 0 and its final value $\delta\triangle ABC$. Therefore $f(\bar{r}) = \delta\triangle DEF$ for some $\bar{r}$. Let $P = P_{\bar{r}}$.

We shall need this result to generalize Bolyai's theorem to arbitrary polygonal regions in the hyperbolic case. We therefore complete its proof.

Lemma. f is continuous.

Given $0 \leqq k \leqq m\angle B$, we need to show that

$$\lim_{r\to k} f(r) = f(k).$$

By definition, this means that the following condition holds.

(1) Given $0 \leqq k \leqq m\angle B$. For every $e > 0$ there is a $d > 0$ such that if

$$|r - k| < d,$$

then

$$|f(r) - f(k)| < e.$$

Let us interpret this geometrically in terms of our definition of f.

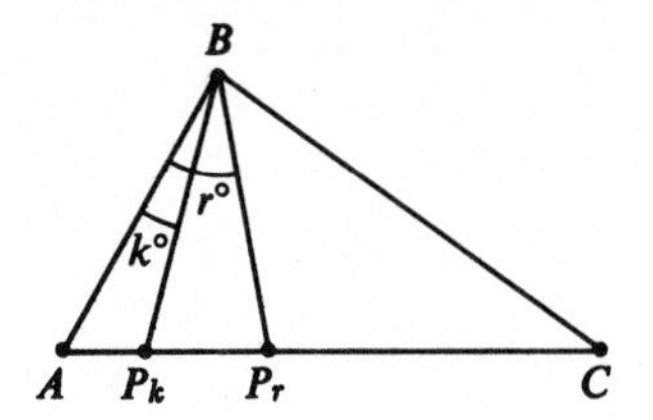

FIGURE 24.71

We have

$$|r - k| = m\angle P_kBP_r.$$

(Note that if $r < k$, which it may be, we need the absolute value signs to make this formula correct.) Also

$$\begin{aligned} |f(r) - f(k)| &= |\delta\triangle ABP_r - \delta\triangle ABP_k| \\ &= \delta\triangle P_kBP_r. \end{aligned}$$

(Here again the absolute value signs are needed to take care of the possibility $r < k$.)

In these terms, Condition (1) takes the following form.

(2) Given $0 \leqq k \leqq m\angle B$. For every $e > 0$ there is a $d > 0$ such that if

$$m\angle P_kBP_r < d,$$

then

$$\delta\triangle P_kBP_r < e.$$

We shall prove (2). First let us suppose that $BC > BA$, as the figures suggest.

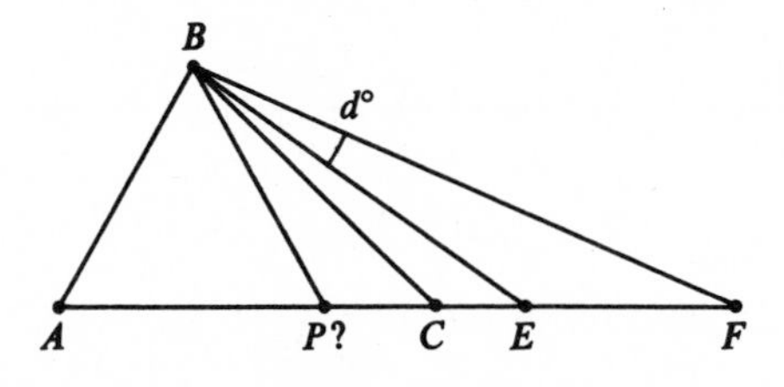

FIGURE 24.72

If A-C-E, then $BP < BE$ for every point P of $\overline{AC}$. (Proof?) We assert that there are points E and F such that A-C-E, A-C-F and

$$\delta\triangle EBF < e.$$

The reader should be able to produce a proof, following a scheme suggested by the proof of an analogous result in the preceding section. Given such an E, F, we let

$$d = m\angle EBF.$$

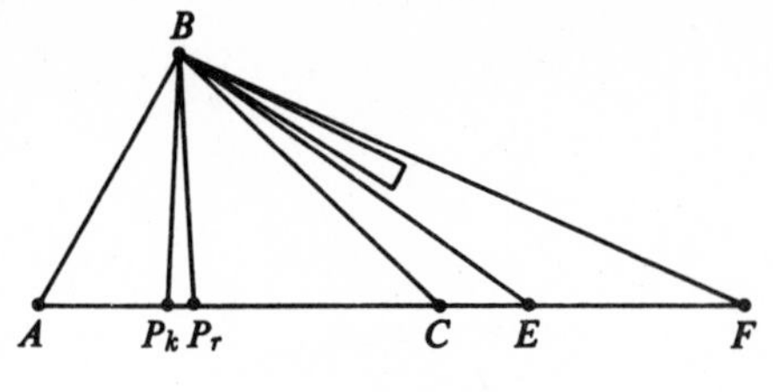

FIGURE 24.73

This is a number d of the sort that we wanted. The point is that if $m\angle P_kBP_r < m\angle EBF$, then $\triangle EBF$ contains a congruent copy $\triangle P_k'BP_r'$ of $\triangle P_kBP_r$. Evidently,

$$\delta\triangle P_kBP_r = \delta\triangle P_k'BP_r' < \delta\triangle EBF.$$

Thus for $m\angle P_kBP_r < d$, we have $\delta\triangle P_kBP_r < e$, which was to be proved.

24.11 BOLYAI'S THEOREM FOR POLYGONAL REGIONS IN THE HYPERBOLIC CASE

Theorem 1. Under HPP, if two polygonal regions have the same area, then they are equivalent by finite decomposition.

That is, if $\delta R_1 = \delta R_2$, then $R_1 \equiv R_2$.

To prove this, we take any triangulations K_1, K_2 of R_1 and R_2:

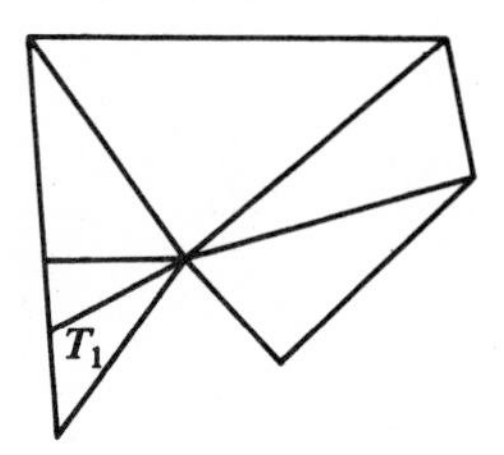

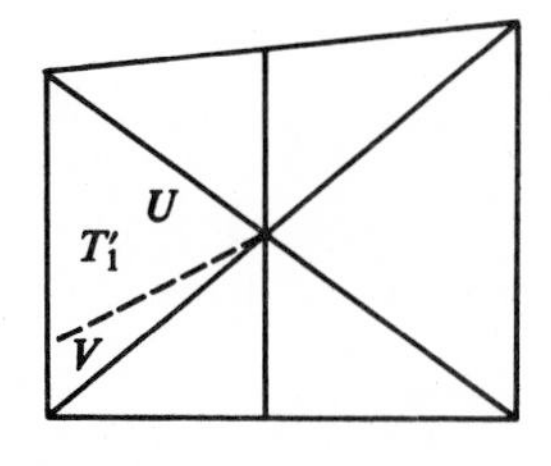

FIGURE 24.74

Some one triangular region in one of these complexes must have minimum defect. That is, some T in either K_1 or K_2 must have the property that $\delta T \leqq \delta T'$ for every T' in K_1 or K_2. Suppose that this is $T_1 \in K_1$. Let $T_1' \in K_2$. If it happens that $\delta T_1 = \delta T_1'$, we delete T_1 from K_1 and delete T_1' from K_2. This gives new complexes K_1', K_2', with fewer elements than K_1 and K_2. Let R_1', R_2' be the corresponding regions. If $R_1' \equiv R_2'$, then $R_1 \equiv R_2$.

If $\delta T_1 < \delta T_1'$, then we know by Theorem 1, Section 24.10, that T_1' can be subdivided into two triangular regions U, V, such that

$$\delta V = \delta T_1.$$

We now delete T_1 from K_1, and we replace T_1' by U in K_2. If the resulting regions are equivalent by finite decomposition, then so also are R_1 and R_2.

Thus, in either case we can reduce our theorem to a case in which the total number of triangular regions is smaller than it was to start with. In a finite number of such steps, we can reduce the theorem to the case of two triangular regions, for which the theorem is known to be true.

24.12 THE IMPOSSIBILITY OF EUCLIDEAN AREA-THEORY IN HYPERBOLIC GEOMETRY

To define the area of a polygonal region as the number which is the total defect of each of its triangulations may seem to be a peculiar proceeding. We shall show, however, that this peculiarity is inevitable. Under HPP, it is impossible to define an area function which has even a minimal resemblance to the Euclidean area function. In the following theorem, it should be understood as usual that $\mathcal{R}$ is the set of all polygonal regions and that R, R_i, and so on, denote polygonal regions.

Theorem 1. Under HPP there does *not* exist a function

$$\alpha : \mathcal{R} \to \mathbf{R}$$

such that

(1) $\alpha R > 0$ for every R;
(2) if R_1 and R_2 intersect only in edges and vertices, then

$$\alpha(R_1 \cup R_2) = \alpha R_1 + \alpha R_2;$$

(3) if T_1 and T_2 are triangular regions with the same base and altitude, then $\alpha T_1 = \alpha T_2$.

(Surely these are minimum requirements for an area function of the Euclidean type.)

Suppose that there is such a function α. Then by (2) and (3), we have:

(4) if $R_1 \equiv R_2$, then $\alpha R_1 = \alpha R_2$, because congruent triangles have the same bases and altitudes.

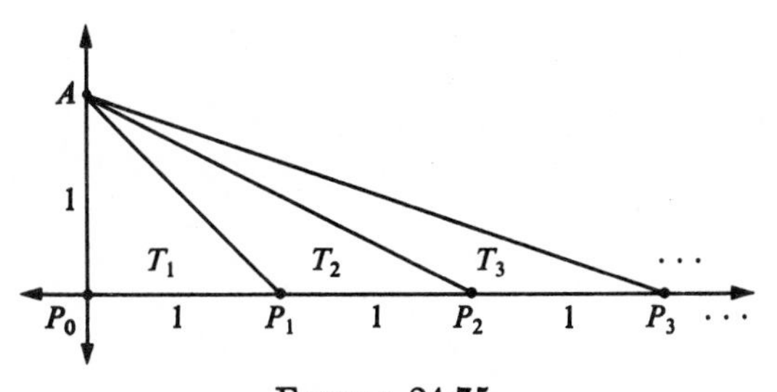

FIGURE 24.75

Consider now a right angle $\angle AP_0P_1$, with $AP_0 = P_0P_1 = 1$. For each n, let P_n be the point of $\overrightarrow{P_0P_1}$ such that $P_0P_n = n$. This gives a sequence of triangles

$$\triangle AP_0P_1, \triangle AP_1P_2, \ldots,$$

and a corresponding sequence of triangular regions

$$T_1, T_2, \ldots.$$

By (3), all the regions T_i have the same "area" $\alpha T_i = A$.
Now consider the corresponding defects

$$d_i = \delta T_i.$$

For each n,

$$d_1 + d_2 + \cdots + d_n = \delta\triangle AP_0P_n < 180.$$

Since the finite sums $d_1 + d_2 + \cdots + d_n$ are bounded, it follows that the infinite series,

$$d_1 + d_2 + \cdots + d_n + \cdots,$$

is convergent. Therefore

$$\lim_{n\to\infty} d_n = 0.$$

Hence

$$d_n < d_1 \text{ for some } n.$$

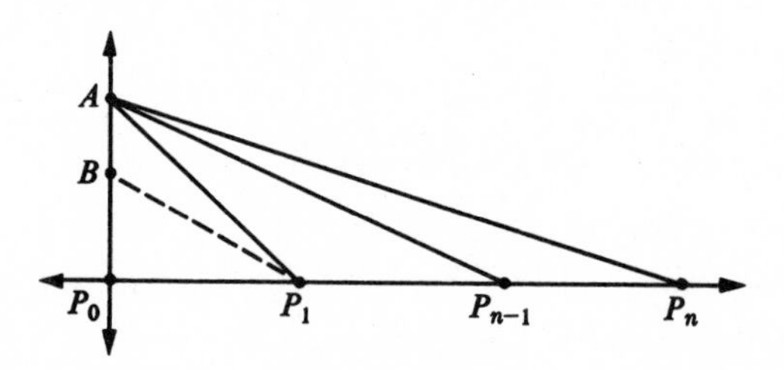

FIGURE 24.76

By Theorem 1, Section 24.10, there is a point B, between A and P_0 such that

$$\delta\triangle BP_0P_1 = \delta\triangle AP_{n-1}P_n = d_n.$$

Therefore, by Bolyai's theorem (Theorem 7, Section 24.8), the regions T, T_n determined by these triangles are equivalent by finite decomposition. By (4), this means that

$$\alpha T = \alpha T_n.$$

But

$$\alpha T_n = \alpha T_1 = A.$$

Therefore

$$\alpha T = \alpha T_1.$$

But this is impossible, because $\alpha\triangle ABP_1 > 0$ and

$$\alpha T_1 = \alpha T + \alpha\triangle ABP_1.$$

24.13 THE UNIQUENESS OF HYPERBOLIC AREA THEORY

Any reasonable area function α should have the following properties:

(1) $\alpha R > 0$ for every R;
(2) if R_1 and R_2 intersect only in edges and vertices, then

$$\alpha(R_1 \cup R_2) = \alpha R_1 + \alpha R_2;$$

(3) if $R_1 \equiv R_2$, then $\alpha R_1 = \alpha R_2$.

We know of one such function, namely δ, and it is plain that there are lots of others. If k is any positive real number, and

$$\alpha R = k\delta R$$

for every polygonal region R, then α satisfies (1), (2), and (3). On the other hand, this trivial way of getting an area function different from δ is in fact the only way.

Theorem 1. Let

$$\alpha : R \to \mathbf{R}$$

be an area function satisfying (1), (2), and (3). Then there is a $k > 0$ such that

$$\alpha R = k\delta R$$

for every R.

In the proof, it will surely be sufficient to find a $k > 0$ such that $\alpha R = k\delta R$ whenever R is a triangular region; the general formula will then follow by addition. In fact, it will be sufficient to prove the following lemma:

Lemma. For every two triangles $\triangle ABC$ and $\triangle DEF$, we have

$$\frac{\alpha\triangle DEF}{\alpha\triangle ABC} = \frac{\delta\triangle DEF}{\delta\triangle ABC}.$$

If this holds, then we have

$$\frac{\alpha\triangle ABC}{\delta\triangle ABC} = \frac{\alpha\triangle DEF}{\delta\triangle DEF},$$

and the desired k is the fraction on the left; for every $\triangle DEF$, we have

$$\alpha\triangle DEF = k\delta\triangle DEF.$$

Proof. For the case when $\delta\triangle ABC = \delta\triangle DEF$, we have $\triangle ABC \equiv \triangle DEF$ by Bolyai's theorem; by (3) it follows that $\alpha\triangle ABC = \alpha\triangle DEF$. Therefore the lemma holds.

We may therefore assume that

$$\delta\triangle ABC > \delta\triangle DEF.$$

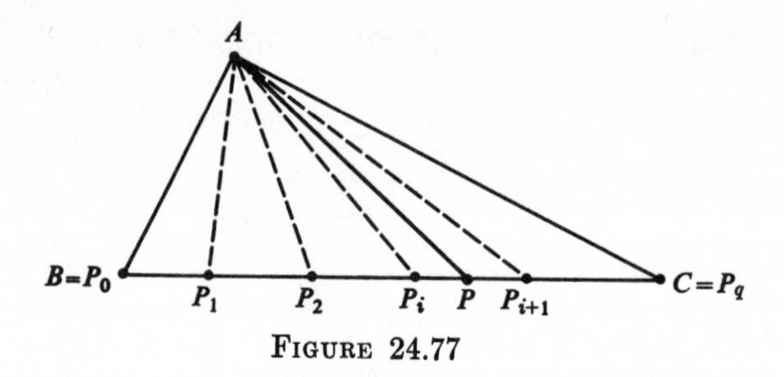

FIGURE 24.77

By Theorem 1, Section 24.10, there is a point P between A and C such that

$$\delta\triangle ABP = \delta\triangle DEF.$$

Since the defect is continuous, we can take a sequence

$$B = P_0, P_1, \ldots, P_i, \ldots, P_q = C,$$

of points, in the stated order on $\overline{BC}$ (in the figure above) such that

$$\delta\triangle AP_iP_{i+1} = \frac{1}{q}\,\delta\triangle ABC$$

for every i. Thus the segments $\overline{AP_i}$ cut $\triangle ABC$ into q triangles with the *same defect*. If we remember Bolyai's theorem, and Conditions (2) and (3), we can easily see that each of the following conditions is equivalent to the next:

(a) $\dfrac{i}{q} < \dfrac{\alpha\triangle DEF}{\alpha\triangle ABC}$,

(b) $\dfrac{i}{q} < \dfrac{\alpha\triangle ABP}{\alpha\triangle ABC}$,

(c) $\dfrac{i}{q}\,\alpha\triangle ABC < \alpha\triangle ABP$,

(d) $B\text{-}P_i\text{-}P$,

(e) $\dfrac{i}{q}\,\delta\triangle ABC < \delta\triangle ABP$,

(f) $\dfrac{i}{q} < \dfrac{\delta\triangle ABP}{\delta\triangle ABC}$,

(g) $\dfrac{i}{q} < \dfrac{\delta\triangle DEF}{\delta\triangle ABC}$.

By the comparison theorem, the lemma follows.

24.14 ALTERNATIVE FORMS OF THE PARALLEL POSTULATE

We observed in Section 24.1, as a consequence of the all-or-none theorem, that the following statement could be used as a substitute for the Euclidean parallel postulate, EPP.

(1) *For some line and some point, parallels are unique.*

Many of the other theorems of this chapter give us such alternative forms of EPP.

(2) *The plane contains at least one rectangle.*

See Theorem 3, Section 24.3, which tells us that under HPP the plane contains no rectangles at all.

(3) *The plane contains at least one triangle for which the angle sum is 180.*

See Theorem 5, Section 24.3, which tells us that under HPP the angle-sum equality never holds.

(4) *The plane contains at least two triangles which are similar without being congruent.*

See Theorem 2, Section 24.4, which tells us that under HPP, similarity without congruence cannot occur even once.

(5) *There is an area-function*

$$\alpha : \mathcal{R} \to \mathbf{R},$$

such that αR is always positive, α is additive for regions intersecting only in edges and vertices, and $\alpha \triangle ABC$ depends only on the base and altitude of $\triangle ABC$.

See Theorem 1, Section 24.12, which says that under HPP there is no such function.

The persuasiveness of these statements may make it easier to understand the state of mind of Saccheri and others.

Chapter 25
The Consistency of the Hyperbolic Postulates

25.1 INTRODUCTION

In this chapter, we shall show that the Poincaré model, described in Chapter 9, satisfies all the postulates of hyperbolic geometry. Our analysis of the model will depend, of course, on Euclidean geometry, and so our consistency proof will be conditional. At the end of the chapter we shall know not that the hyperbolic postulates are consistent, but merely that they are *as consistent as* the Euclidean postulates. In the following chapter, we shall investigate the consistency of the Euclidean postulates. (See the discussion at the end of Chapter 9.)

25.2 INVERSIONS OF A PUNCTURED PLANE

Given a point A of a Euclidean plane E and a circle C with center at A and radius a. The set $E - A$ is called a *punctured plane*. The *inversion of* $E - A$ *about* C is a function,

$$f: E - A \leftrightarrow E - A,$$

defined in the following way. For each point P of $E - A$, let $P' = f(P)$ be the point of $\overrightarrow{AP}$ for which

$$AP' = \frac{a^2}{AP}.$$

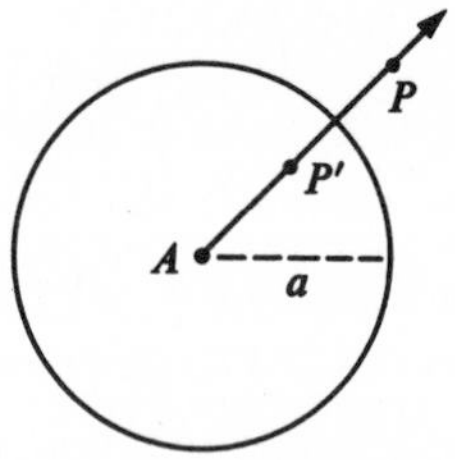

FIGURE 25.1

(Thus, for $a = 1$, we have $AP' = 1/AP$.) Since $a^2/a = a$, we have the following theorems.

Theorem 1. If $P \in C$, then $f(P) = P$.

Theorem 2. If P is in the interior of C, then $f(P)$ is in the exterior of C, and conversely.

Theorem 3. For every P, $f(f(P)) = P$.

That is, when we apply an inversion twice, this gets us back to wherever we started.

Proof. $f(P)$ is the point of $\overrightarrow{AP}$ for which $Af(P) = a^2/AP$, and $f(f(P))$ is the point of the same ray for which

$$Af(f(P)) = \frac{a^2}{Af(P)} = \frac{a^2}{a^2/AP} = AP.$$

Therefore $f(f(P)) = P$.

Theorem 4. If L is a line through A, then $f(L - A) = L - A$.

Here by $f(L - A)$ we mean the set of all image points $f(P)$, where $P \in L - A$. In general, if

$$K \subset E - A,$$

then

$$f(K) = \{P' = f(P) | P \in K\}.$$

It is also easy to see that "if P is close to A, then P' is far from A," and conversely; the reason is that "a^2/AP is large when AP is small." In studying less obvious properties of inversions, it will be convenient to use both rectangular and polar coordinates, taking the origin of each coordinate system at A. The advantage of polar coordinates is that they allow us to describe the inversion in the simple form

$$f: E - A \leftrightarrow E - A,$$
$$: (r, \theta) \leftrightarrow (s, \theta),$$

where

$$s = \frac{a^2}{r}$$

and

$$r = \frac{a^2}{s}.$$

In rectangular coordinates, we have

$$P = (x, y) = (r \cos \theta, r \sin \theta),$$
$$f(P) = (u, v) = (s \cos \theta, s \sin \theta),$$

where r and s are related by the same equations as before. Evidently

$$s^2 = u^2 + v^2$$

just as

$$r^2 = x^2 + y^2.$$

These equations will enable us to tell what happens to lines and circles under inversions. We allow the cases in which the lines and circles contain the origin A, so that they appear in $E - A$ as "punctured lines" and "punctured circles." Thus we shall be dealing with four types of figures, namely, lines and circles, punctured and unpunctured. For short, we shall refer to such figures as *k-sets*. The rest of this section will be devoted to the proof that if K is a k-set, then so also is $f(K)$. Let us look first, however, at a special case.

Let K be the line $x = a$.

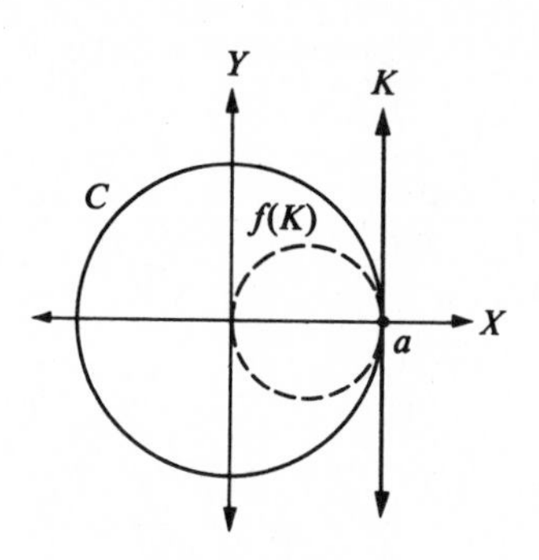

FIGURE 25.2

Then K is the graph of the polar equation

$$r \cos \theta = a.$$

Since $r = a^2/s$, where $f(r, \theta) = (s, \theta)$, it follows that $f(K)$ is the graph of the condition

$$\frac{a^2}{s} \cos \theta = a, \qquad s \neq 0$$

or

$$s = a \cos \theta, \qquad s \neq 0$$

or

$$s^2 = as \cos \theta, \qquad s \neq 0.$$

In rectangular form, this is

$$u^2 + v^2 = au, \qquad u^2 + v^2 \neq 0.$$

Replacing u and v by x and y (to match the labels on the axes), we see that $f(K)$ is the graph of

$$x^2 - ax + y^2 = 0, \qquad x^2 + y^2 \neq 0,$$

and is hence the punctured circle with center at $(a/2, 0)$ and radius $a/2$. Thus f has pulled the upper half of the line K onto the upper semicircle, and the lower half onto the lower semicircle. It is easy to see that points far from the x-axis (either above or below) go onto points near the origin.

More generally, we have the following theorem.

Theorem 5. If K is a line in $E - A$, then $f(K)$ is a punctured circle.

Proof. Since we can choose the axes any way we want, we are free to assume that K is the graph of a rectangular equation

$$x = b > 0,$$

and hence of a polar equation

$$r \cos \theta = b > 0.$$

As before, setting $r = a^2/s$, we conclude that $f(K)$ is the graph of

$$\frac{a^2}{s} \cos \theta = b, \qquad s \neq 0,$$

or

$$s^2 = \frac{a^2}{b} s \cos \theta, \qquad s \neq 0,$$

or

$$u^2 - \frac{a^2}{b} u + v^2 = 0, \qquad u^2 + v^2 \neq 0,$$

or

$$x^2 - \frac{a^2}{b} x + y^2 = 0, \qquad x^2 + y^2 \neq 0.$$

Therefore $f(K)$ is a punctured circle, with center at $(a^2/2b, 0)$ and radius $a^2/2b$.

It is easy to see that (1) every punctured circle is described by the above formula for some choice of b and some choice of the axes. Therefore (2) every punctured circle L is $= f(K)$ for some line K. But Theorem 3 tells us that $f(f(P)) = P$ for every P. Therefore

$$f(L) = f(f(K)) = K.$$

Thus we have the following theorem.

Theorem 6. If L is a punctured circle, then $f(L)$ is a line in $E - A$.

We now know, from Theorem 4, that under f, punctured lines go onto punctured lines; and we know, by Theorems 5 and 6, that lines go onto punctured circles and vice-versa. Now we must see what happens to circles.

Theorem 7. If M is a circle in $E - A$, then $f(M)$ is a circle in $E - A$.

Proof. M is the graph of a rectangular equation

$$x^2 + y^2 + Ax + By + C = 0,$$

where $C \neq 0$ because the circle is not punctured. In polar form, this is

$$r^2 + Ar \cos \theta + Br \sin \theta + C = 0.$$

Since $r = a^2/s$, this tells us that $f(M)$ is the graph of the equation

$$\frac{a^4}{s^2} + A \cdot \frac{a^2}{s} \cos\theta + B \cdot \frac{a^2}{s} \sin\theta + C = 0,$$

or

$$a^4 + Aa^2 s \cos\theta + Ba^2 s \sin\theta + Cs^2 = 0,$$

or

$$a^4 + Aa^2 u + Ba^2 v + C(u^2 + v^2) = 0.$$

Replacing u and v by x and y, to match the labels on the axes, we get an equation for $f(M)$ in the form

$$x^2 + y^2 + \frac{Aa^2}{C} x + \frac{Ba^2}{C} y + \frac{a^2}{C} = 0.$$

The graph $f(M)$ is a circle; this circle is not punctured, because $a^2/C \neq 0$.

Theorem 8. If K is a k-set, then so also is $f(K)$.

25.3 PRESERVATION OF THE CROSS RATIO UNDER INVERSIONS

We recall, from Section 9.2, the definition of distance in the Poincaré model.

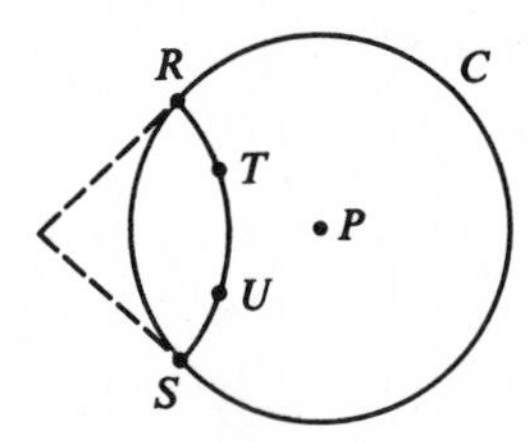

FIGURE 25.3

If T and U are points of the L-line with end points R, S on the boundary circle C, then the non-Euclidean distance is defined by the formula

$$d(T, U) = \left| \log_e \frac{TR/TS}{UR/US} \right| .$$

The fraction whose logarithm gets taken in this formula is called the *cross ratio* of the quadruplet R, S, T, U, and is commonly denoted by (R, S, T, U). Thus

$$(R, S, T, U) = \frac{TR \cdot US}{UR \cdot TS},$$

and changing the notation slightly, we have

$$(P_1, P_2, P_3, P_4) = \frac{P_1P_3 \cdot P_2P_4}{P_1P_4 \cdot P_2P_3}.$$

We shall show that inversions preserve the cross ratio. In the following theorem, f is an inversion of a punctured plane $E - A$ about a circle with center at A and radius a, as in the preceding section.

Theorem 1. If $P'_i = f(P_i)$ $(i = 1, 2, 3, 4)$, then

$$(P_1, P_2, P_3, P_4) = (P'_1, P'_2, P'_3, P'_4).$$

Proof. For each i from 1 to 4, let the polar coordinates of P_i be (r_i, θ_i). By the usual polar distance formula, we have

$$P_iP_j^2 = r_i^2 + r_j^2 - 2r_ir_j \cos(\theta_i - \theta_j).$$

Now

$$P'_i = (s_i, \theta_i) = \left(\frac{a^2}{r_i}, \theta_i\right).$$

Therefore

$$(P_1, P_2, P_3, P_4)^2 = \frac{[r_1^2 + r_3^2 - 2r_1r_3 \cos(\theta_1 - \theta_3)][r_2^2 + r_4^2 - 2r_2r_4 \cos(\theta_2 - \theta_4)]}{[r_1^2 + r_4^2 - 2r_1r_4 \cos(\theta_1 - \theta_4)][r_2^2 + r_3^2 - 2r_2r_3 \cos(\theta_2 - \theta_3)]},$$

and

$$(P'_1, P'_2, P'_3, P'_4)^2 = \frac{\left[\frac{a^4}{r_1^2} + \frac{a^4}{r_3^2} - 2\frac{a^4}{r_1r_3}\cos(\theta_1 - \theta_3)\right]\left[\frac{a^4}{r_2^2} + \frac{a^4}{r_4^2} - 2\frac{a^4}{r_2r_4}\cos(\theta_2 - \theta_4)\right]}{\left[\frac{a^4}{r_1^2} + \frac{a^4}{r_4^2} - 2\frac{a^4}{r_1r_4}\cos(\theta_1 - \theta_4)\right]\left[\frac{a^4}{r_2^2} + \frac{a^4}{r_3^2} - 2\frac{a^4}{r_2r_3}\cos(\theta_2 - \theta_3)\right]}.$$

To reduce the second of these fractions to the first, we multiply in both the numerator and denominator by

$$\frac{r_1^2r_2^2r_3^2r_4^2}{a^8}.$$

This theorem will tell us, in due course, that inversions applied to the Poincaré model are *isometries*, relative to the non-Euclidean distance.

25.4 PRESERVATION OF ANGULAR MEASURE UNDER INVERSIONS

A reexamination of Section 25.2 will indicate that the image of an angle, under an inversion, is never an angle. The point is that every angle in $E - A$ has at least one side lying on a nonpunctured line, and the image of a nonpunctured line is always a punctured circle. Therefore the following theorem does not mean what it might seem to mean.

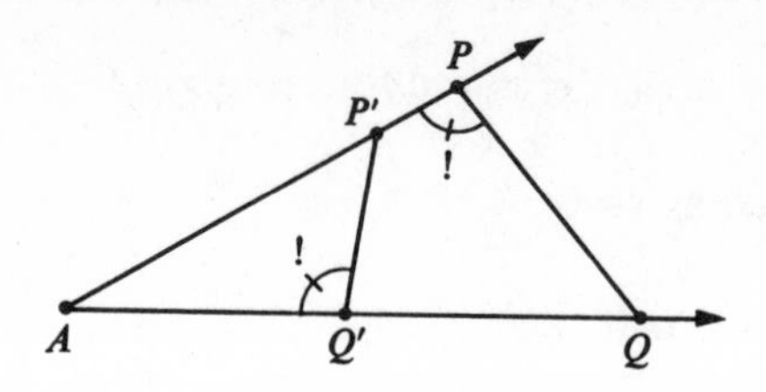

FIGURE 25.4

Theorem 1. If A, P, and Q are noncollinear, $P' = f(P)$ and $Q' = f(Q)$, then

$$m\angle APQ = m\angle AQ'P'.$$

Proof. Consider $\triangle PAQ$ and $\triangle Q'AP'$. They have the angle $\angle A$ in common. Since

$$AP' = \frac{a^2}{AP}, \qquad AQ' = \frac{a^2}{AQ},$$

we have

$$AP \cdot AP' = AQ \cdot AQ' = a^2,$$

so that

$$\frac{AP}{AQ'} = \frac{AQ}{AP'}.$$

By the SAS similarity theorem,

$$\triangle PAQ \sim \triangle Q'AP'.$$

(Note the reversal of order of vertices here.) Since $\angle APQ$ and $\angle AQ'P'$ are corresponding angles, they have the same measure.

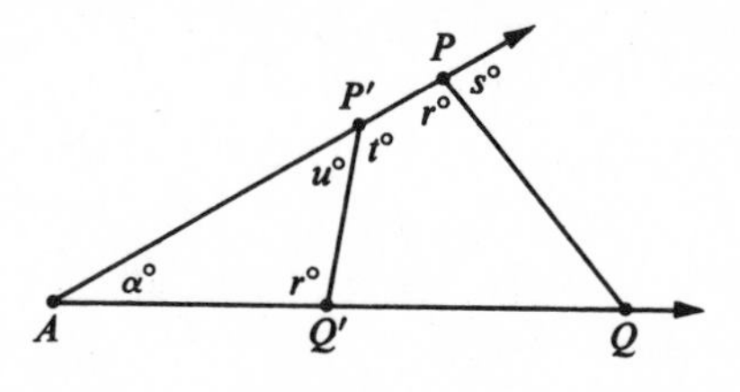

FIGURE 25.5

In the figure above, $P' = f(P)$ and $Q' = f(Q)$, as before. Here we have

$$\begin{aligned} u &= 180 - \alpha - r \\ &= (180 - r) - \alpha \\ &= s - \alpha. \end{aligned}$$

Therefore

$$s - u = \alpha.$$

The order of s and u depends on the order in which P and P' appear on the ray.

If P and P' are interchanged, we should interchange s and u, getting

$$u - s = \alpha.$$

Thus in general we have

$$|s - u| = \alpha.$$

Consider next the situation illustrated in the figure below:

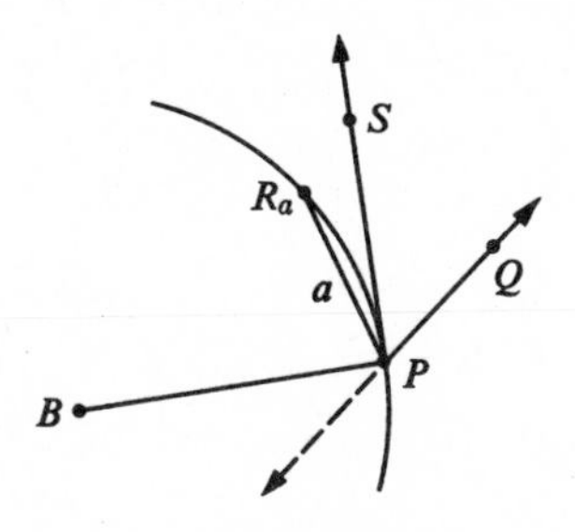

FIGURE 25.6

Here B is the center of a circular arc; $\overleftrightarrow{PQ}$ is a line intersecting the arc at P; $\overrightarrow{PS}$ is a tangent ray at P; and $R_aP = a$. We assert that

$$\lim_{a \to 0} m\angle R_aPQ = m\angle SPQ.$$

(Proof? The first step is to show that $\lim_{a\to 0} m\angle R_aPS = 0$.)

Consider now a circular arc $\widehat{QS}$ with end point at a point Q. For small positive

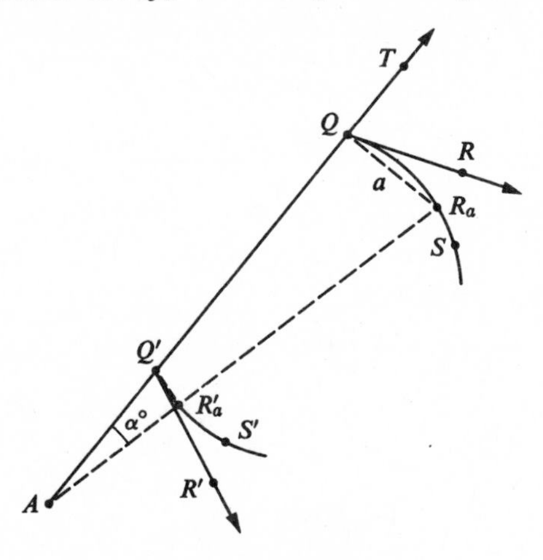

FIGURE 25.7

numbers a, let R_a be the point of the arc for which

$$QR_a = a.$$

Let $\widehat{Q'S'}$ be the image of $\widehat{QS}$; that is

$$\widehat{Q'S'} = f(\widehat{QS});$$

let $\overrightarrow{Q'R'}$ be the tangent ray at Q'.

We assert that

$$\angle AQ'R' \cong \angle TQR.$$

To see this, we observe that $m\angle TQR_a$ and $m\angle AQ'R'_a$ are the s and u that we discussed just after Theorem 1. Therefore

$$|m\angle TQR_a - m\angle AQ'R'_a| = \alpha.$$

Now

$$\lim_{a\to 0} m\angle TQR_a = m\angle TQR,$$

and

$$\lim_{a\to 0} m\angle AQ'R'_a = m\angle AQ'R'.$$

Therefore

$$\lim_{a\to 0} [m\angle TQR_a - m\angle AQ'R'_a] = m\angle TQR - m\angle AQ'R'.$$

But the absolute value of the quantity indicated in square brackets is $=\alpha$; and $\alpha \to 0$ as $a \to 0$. Therefore

$$m\angle TQR = m\angle AQ'R'.$$

Given two intersecting circles or lines, the tangent rays give us "tangent angles," like this:

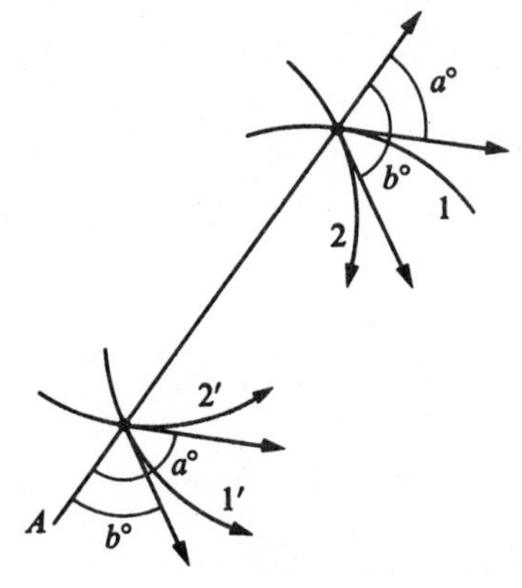

FIGURE 25.8

By the preceding result, we have the following theorem.

Theorem 2. Under inversions, corresponding tangent angles are congruent.

That is, if $\widehat{AB}$ and $\widehat{AC}$ are arcs with a tangent angle of measure r, then their images $f(\widehat{AB})$ and $f(\widehat{AC})$ have a tangent angle of measure r. Similarly for an arc and a segment or a segment and a segment.

25.5 REFLECTIONS ACROSS L-LINES IN THE POINCARÉ MODEL

We recall that the *points* in the Poincaré model are the points of the interior E of a circle C with center at P; the *L-lines* are (1) the intersection of E with lines through P and (2) the intersection of E with circles C' orthogonal to C.

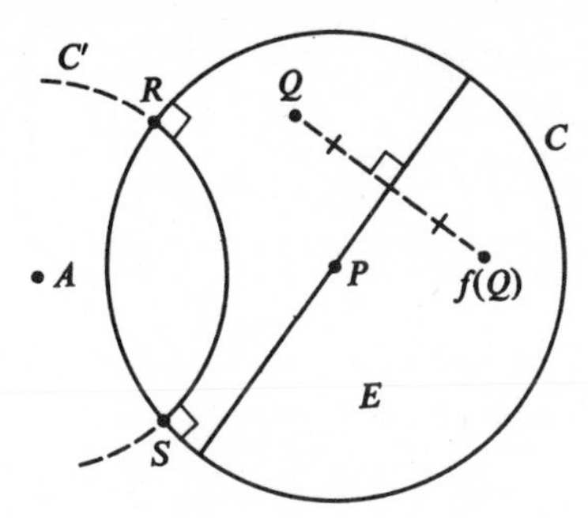

FIGURE 25.9

If L is an L-line of the first type, then the *reflection of E across L* is defined in the familiar fashion as a one-to-one correspondence,

$$f : E \leftrightarrow E,$$

such that for each point Q of E, Q and $f(Q)$ are symmetric across L.

If L is an L-line of the second type, then the *reflection of E across L* is the inversion of E about C'. To justify this definition, of course, we have to show that if f is an inversion about a circle C' orthogonal to C, then $f(E) = E$. But this is not hard to show. In the next few theorems, it should be understood that f is an inversion about C'; C' has center at A, and intersects C orthogonally at R and S; and $L = E \cap C'$.

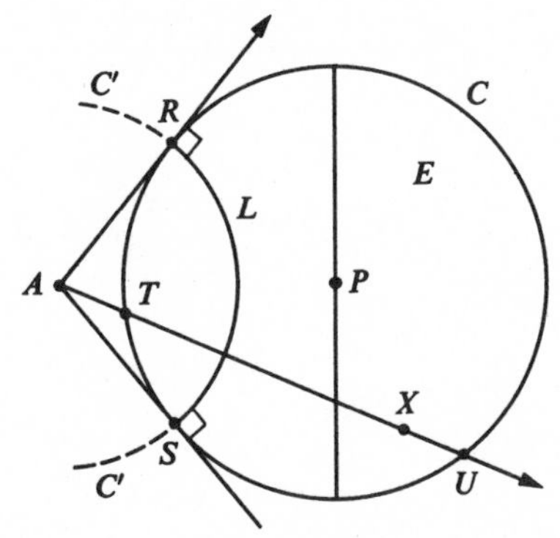

FIGURE 25.10

Theorem 1. $f(C) = C$.

Proof. $f(C)$ is a circle. This circle contains R and S, because $f(R) = R$ and $f(S) = S$. By Theorem 2 of the preceding section, $f(C)$ and C' are orthogonal. But there is only one circle C which crosses C' orthogonally at R and S. (Proof? Show that P must be the center of any such circle.) Therefore $f(C) = C$, which was to be proved.

Theorem 2. $f(E) = E$.

Proof. Let X be any point of E. Then $\overrightarrow{AX}$ intersects C at points T and U. Since $f(C) = C$, we have $U = f(T)$ and $T = f(U)$. But inversions preserve betweenness on rays starting at A. Therefore $f(\overline{TU}) = \overline{TU}$, and $f(X) \in E$. Thus $f(E) \subset E$.

We need to show, conversely, that $E \subset f(E)$. This is trivial: given that $f(E) \subset E$, we have $f(f(E)) \subset f(E)$. Since $f(f(E)) = E$, this gives $E \subset f(E)$.

Theorem 3. If M is an L-line, then so also is $f(M)$.

Proof. M is the intersection $E \cap D$, where D is either a circle orthogonal to C or a line orthogonal to C. Now $f(D)$ is orthogonal to C, and is a line or circle (punctured or unpunctured). Let D' be the corresponding complete line or circle. (Thus $D' = f(D)$ or $D' = f(D) \cup A$.) Then

$$\begin{aligned} f(M) &= f(D) \cap E \\ &= D' \cap E, \end{aligned}$$

which is an L-line.

We recall that an L-angle is the angle formed by two "rays" in the Poincaré model.

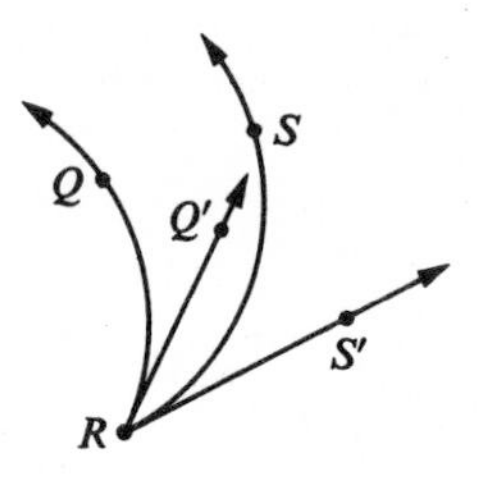

FIGURE 25.11

The measure of an L-angle is the measure of the angle formed by the tangent rays.

We can now sum up nearly all of the preceding discussion in the following theorem.

Theorem 4. Let f be a reflection of E across an L-line. Then,

(1) f is a one-to-one correspondence $E \leftrightarrow E$;
(2) f preserves the non-Euclidean distances between points;
(3) f preserves L-lines;
(4) f preserves measures of L-angles.

For L-lines of the first kind (passing through P) all this is trivial, because in this case f is an isometry in the Euclidean sense. It therefore preserves distances of both kinds, lines, circles, orthogonality, and angular measure. For L-lines of the second kind, Conditions (1) through (4) follow from the theorems of this section and the preceding two sections.

25.6 UNIQUENESS OF THE L-LINE THROUGH TWO POINTS

Given the center P of C, and some other point Q of E. We know that P and Q lie on only one (straight) line in the Euclidean plane. Therefore P and Q lie on only one L-line *of the first kind*. But P does not lie on any L-line of the second kind. (The reason is that on the right triangle $\triangle ARP$ in the figure, the hypotenuse, $\overline{AP}$, is the longest side.) It follows that the L-line through two points of E is unique, in the case where one of the points is P.

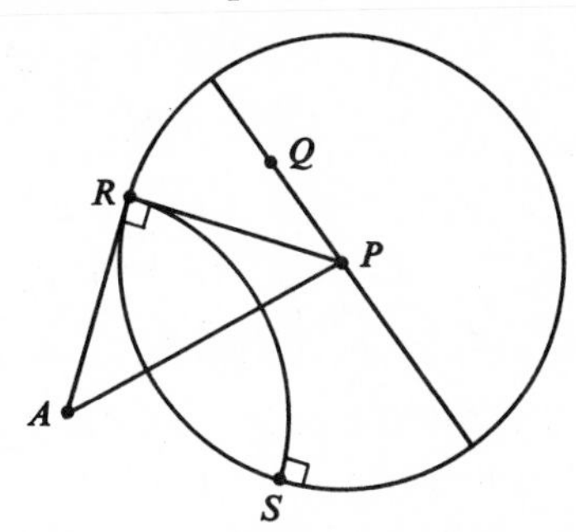

FIGURE 25.12

To prove that uniqueness always holds, we need the following theorem.

Theorem 1. For each point Q of E there is a reflection f such that $f(Q) = P$.

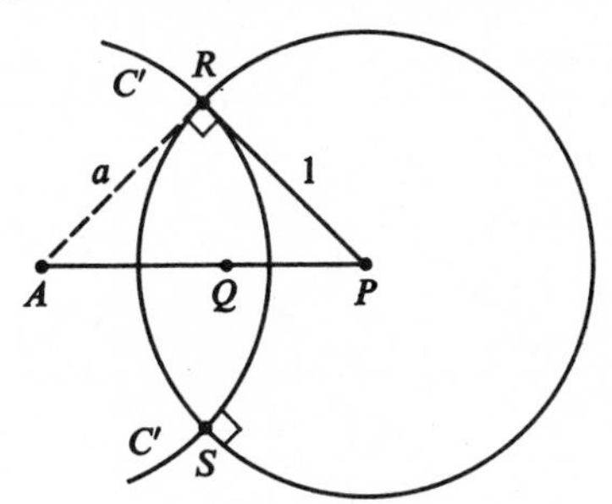

FIGURE 25.13

Proof. We start by the method of wishful thinking. If the inversion f about C' gives $f(Q) = P$, then

$$AP = \frac{a^2}{AQ}.$$

We recall that the radius $PR = 1$. Let $k = QP$, and let x be the unknown distance

AP (Fig. 25.13). Then the equation

$$AP \cdot AQ = a^2$$

takes the form

$$x(x - k) = x^2 - 1$$

or

$$kx = 1 \qquad \text{or} \qquad x = \frac{1}{k}.$$

Since Q is in E, we know that $k < 1$. Therefore $x > 1$, and A is outside C. If C' is the circle with center at A, orthogonal to C, then the reflection across $E \cap C'$ is the one that we wanted.

We can now prove the following theorem.

Theorem 2. In the Poincaré model, every two points lie on exactly one L-line.

Proof. Let Q and R be points of E. Let f be a reflection across an L-line such that $f(Q) = P$ and $f(R) = R'$. We know that P and R' lie on an L-line L. Therefore Q and R lie on the L-line $f(L)$. If there were two L-lines L_1, L_2, containing Q and R, then $f(L_1)$ and $f(L_2)$ would be different L-lines containing P and R', which is impossible.

By Theorem 1, we can speak of *the* L-line containing Q and R. We shall denote this by $\overleftrightarrow{QR}$; and to avoid confusion, we shall agree not to use this notation, in the rest of this chapter, to denote Euclidean lines.

25.7 THE RULER POSTULATE, BETWEENNESS AND PLANE SEPARATION

Our strategy in this chapter is to verify statements about L-lines, first for the easy case of L-lines through P, and then to use inversions to show that the "curved" L-lines behave in the same way as the "straight" ones. In this spirit, we first check the ruler postulate for L-lines through P.

Theorem 1. Every L-line through P has a coordinate system.

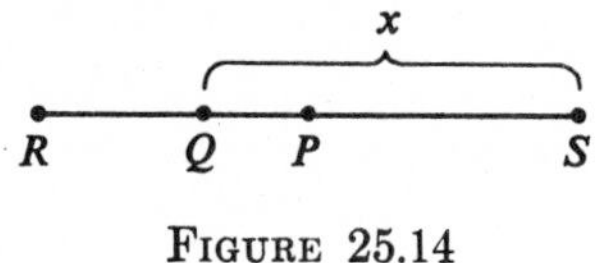

FIGURE 25.14

Proof. Suppose that L passes through P, and let its end points on C be R and S. For each point Q of L, let

$$f(Q) = \log_e \frac{QR/QS}{PR/PS}$$

$$= \log_e \frac{QR}{QS}.$$

(Because $PR = PS$.) Let $QS = x$. Then

$$QR = 2 - QS = 2 - x,$$

and we have

$$f(Q) = \log_e \frac{2 - x}{x}.$$

Obviously f is a function $L \to \mathbf{R}$ *into* the real numbers. We need to verify that f is a one-to-one correspondence $L \leftrightarrow \mathbf{R}$. Thus we need to show that every real number k is $=f(Q)$ for exactly one point Q. Thus we want

$$k = \log_e \frac{2 - x}{x}$$

or

$$e^k = \frac{2 - x}{x}$$

or

$$(e^k + 1)x = 2$$

or

$$x = \frac{2}{e^k + 1}.$$

For every k there is exactly one such x, and $0 < x < 2$, as it should be. Therefore every k is $=f(Q)$ for exactly one point Q of L.

We have already checked, in Chapter 9, that when the coordinate system f is defined in this way, the distance formula

$$d(T, U) = |f(T) - f(U)|$$

is always satisfied.

Before proceeding to generalize Theorem 1, we observe that the formulas above give us some more information:

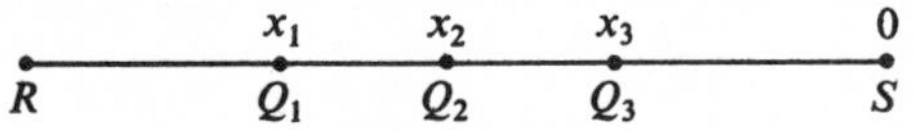

FIGURE 25.15

In the figure $x_i = Q_iS$ for $i = 1, 2, 3$. It is easy to check that $(2 - x)/x$ is a decreasing function. (Its derivative is $-2/x^2 < 0$.) And the logarithm is an increasing function. Therefore, if $x_1 < x_2 < x_3$, as in the figure, it follows that

$$f(Q_1) < f(Q_2) < f(Q_3),$$

and conversely. We recall that betweenness is defined in terms of distance, and that one point of a line is between two others if and only if its coordinate is between their coordinates. Thus we have:

Theorem 2. Let Q_1, Q_2, Q_3 be points of an L-line through P. Then Q_1-Q_2-Q_3 under the non-Euclidean distance if and only if Q_1-Q_2-Q_3 in the Euclidean plane.

Theorem 3. Every L-line has a coordinate system.

Proof. Given an L-line L. If L contains P, we use Theorem 1. If not, let Q be any point of L; let g be a reflection such that $g(Q) = P$; let $L' = g(L)$, and let

$$f : L' \leftrightarrow \mathbf{R}$$

be a coordinate system for L'. For each point T of L, let

$$f'(T) = f(g(T)).$$

That is, the coordinate of T is the coordinate of the corresponding point $g(T)$ of L'. Since f and g are one-to-one correspondences, so also is their composition $f(g)$. Given points T, U of L, we know that

$$d(T, U) = d(g(T), g(U)),$$

because inversions preserve the non-Euclidean distance. This in turn is

$$= |f(g(T)) - f(g(U))|,$$

because f is a coordinate system for L'. Therefore

$$d(T, U) = |f'(T) - f'(U)|,$$

which was to be proved.

Theorem 4. Every L-line through P separates E into two sets H_1 and H_2 such that (1) H_1 and H_2 are convex, and (2) if $Q \in H_1$ and $R \in H_2$, then $\overline{QR}$ intersects L.

Here $\overline{QR}$ means of course the *non-Euclidean* segment.

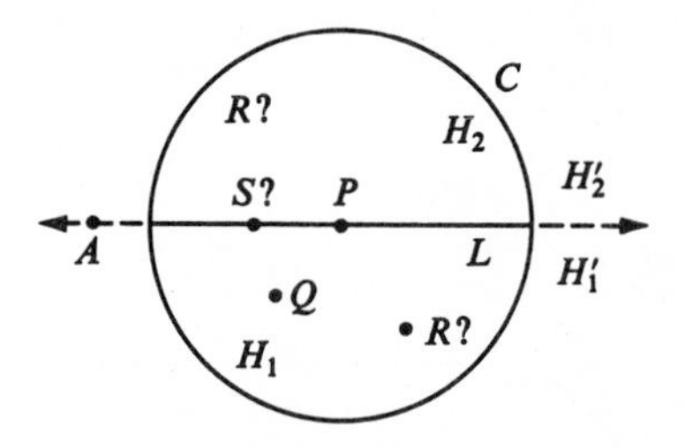

FIGURE 25.16

Proof. We know that the Euclidean line containing L separates the Euclidean plane into two half-planes H_1', H_2'. Let H_1 and H_2 be the intersections $H_1' \cap E$ and $H_2' \cap E$, as indicated in the figure.

Suppose that $Q, R \in H_1$, and suppose that $\overline{QR}$ intersects L in a point S. Let f be an inversion $E \leftrightarrow E$, about a circle with center A on the line containing L such that $f(S) = P$. Then $f(\overleftrightarrow{QR})$ is an L-line through P, and $f(Q)$ and $f(R)$ belong to H_1. Since Q-S-R, we have $f(Q)$-P-$f(R)$, in the non-Euclidean sense, because f preserves the non-Euclidean distance. Therefore $f(Q)$-P-$f(R)$ in the *Euclidean* sense, which is impossible, because $f(Q)$ and $f(R)$ are in the same Euclidean half plane.

It follows, in the same way, that H_2 is convex. Thus we have verified half of the plane separation postulate for the Poincaré model.

Suppose now that $Q \in H_1$ and $R \in H_2$. Let C' be the Euclidean circle that contains the L-line $\overleftrightarrow{QR}$:

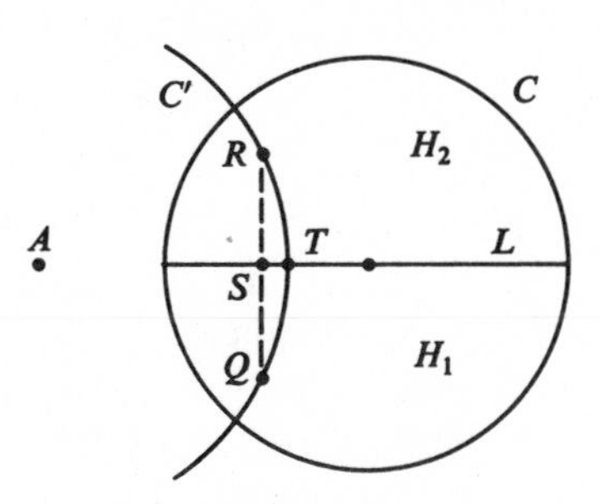

FIGURE 25.17

Then L contains a point S of the Euclidean segment from Q to R, and S is in the interior of C'. It follows that the Euclidean line containing L intersects C' in two points, one of which is a point T of L. Now we must verify that Q-T-R in the non-Euclidean sense. [*Hint:* Use an inversion $f: E \leftrightarrow E$, $H_1 \leftrightarrow H_1$, $H_2 \leftrightarrow H_2$, $T \leftrightarrow P$, and then apply Theorem 2.]

To extend this result to L-lines in general, we observe that:

Theorem 5. Reflections preserve betweenness.

Because they preserve lines and distance.

Theorem 6. Reflections preserve segments.

Because they preserve betweenness.

Theorem 7. Reflections preserve convexity.

Because they preserve segments.

Theorem 8. The plane separation postulate holds in the Poincaré model.

Proof. Let L be any L-line, and let Q be any point of L. Let f be a reflection such that $f(Q) = P$; let $L' = f(L)$; and let H_1' and H_2' be the half-planes in E determined by L'. Let

$$H_1 = f^{-1}(H_1') \quad \text{and} \quad H_2 = f^{-1}(H_2').$$

Since f^{-1} is also a reflection, and reflections preserve convexity, it follows that H_1 and H_2 are convex. This proves half of the plane separation postulate for L.

It remains to show that if $R \in H_1$ and $S \in H_2$, then $\overline{RS}$ intersects L. If $R' = f(R)$ and $S' = f(S)$, then $R' \in H_1'$ and $S' \in H_2'$, so that $\overline{R'S'}$ intersects L' at a point T'. Therefore $\overline{RS}$ intersects L at $T = f^{-1}(T')$.

Theorem 9. Reflections preserve half planes.

That is, if H_1 and H_2 are the half planes determined by L, then $f(H_1)$ and $f(H_2)$ are the half planes determined by $f(L)$. Proof?

Theorem 10. Reflections preserve interiors of angles.

Proof. The interior of $\angle ABC$ is the intersection of (1) the side of $\overleftrightarrow{AB}$ that contains C, and (2) the side of $\overrightarrow{BC}$ that contains A. Since reflections preserve half planes, they preserve intersections of half planes.

25.8 ANGULAR MEASURE IN THE POINCARÉ MODEL

We have defined the measure of an (non-Euclidean) angle as the measure of the (Euclidean) angle formed by the two tangent rays. We need to check whether this measure function satisfies the postulates of Section 5.1. For angles with vertex at P this is obvious. To verify it for angles with vertex at some other point Q, we throw Q onto P by a reflection f. Now f preserves angles, angular measure, lines, and interiors of angles. It is therefore trivial to check that if Postulates M-1 through M-5 hold at P, then they hold at Q.

25.9 THE SAS POSTULATE

We have now verified, for the Poincaré model, all the postulates of absolute plane geometry, with the sole exception of SAS. With heavy use of inversions, this turns out not to be difficult.

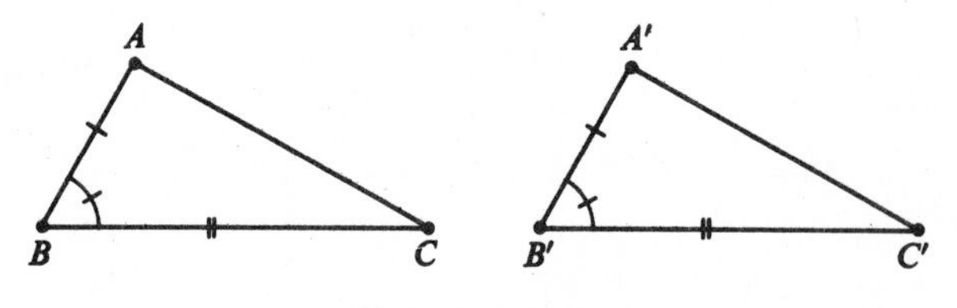

FIGURE 25.18

Given $\triangle ABC$, $\triangle A'B'C'$, and a correspondence

$$ABC \leftrightarrow A'B'C'$$

such that

$$\overline{AB} \cong \overline{A'B'}, \qquad \overline{BC} \cong \overline{B'C'}, \qquad \angle B \cong \angle B'.$$

(Here the segments and congruences are non-Euclidean.) We want to show that $\triangle ABC \cong \triangle A'B'C'$.

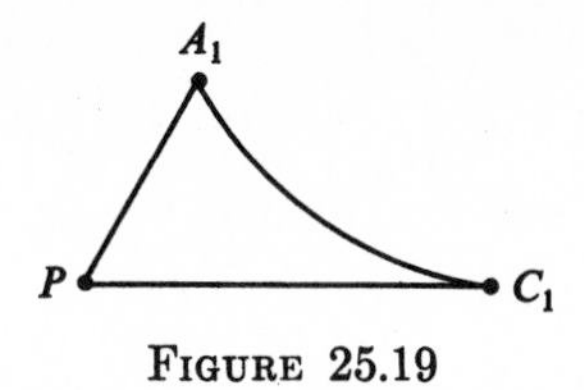

FIGURE 25.19

First let f_1 be an inversion such that $f_1(B) = P$, and let $\triangle A_1PC_1$ be $f_1(\triangle ABC)$. (Note that $\overline{PA}_1$ and $\overline{PC}_1$ look "straight," as they should.) Since f_1 preserves both distances and angular measure, we have

$$\triangle A_1PC_1 \cong \triangle ABC.$$

Next let f_2 be an inversion such that $f_2(B') = P$, and let $\triangle A_1'PC_1' = f_2(\triangle A'B'C')$.

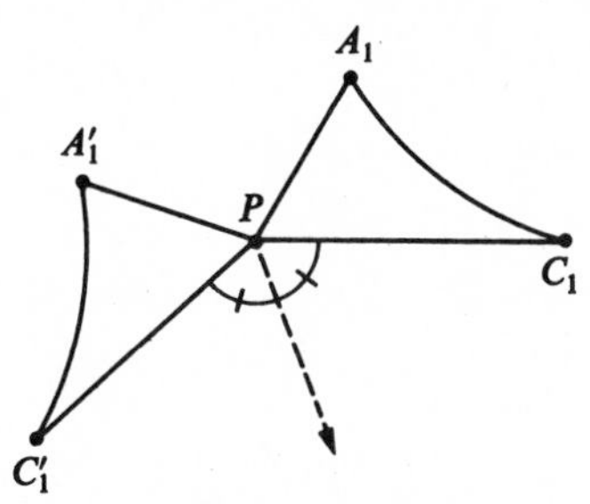

FIGURE 25.20

It is easy to see that there is a reflection f_3, across an L-line through P, such that $f_3(\overrightarrow{PC_1'}) = \overrightarrow{PC}_1$. (If P, C_1, and C_1' are not collinear, we reflect across the bisector of $\angle C_1'PC_1$, as indicated in the figure. If C_1'-P-C_1, we reflect across the perpendicular to $\overleftrightarrow{PC_1}$ at P. If we already have $\overrightarrow{PC_1} = \overrightarrow{PC_1'}$, we leave well enough alone.)

Let $\triangle A_2'PC_2' = f_3(\triangle A_1'PC_1')$. Since f_3 preserves distance and angular measure, we have

$$\triangle A_2'PC_2' \cong \triangle A_1'PC_1'.$$

And since $BC = PC_1$, and $B'C' = PC_1' = PC_2'$, we have $C_1 = C_2'$.

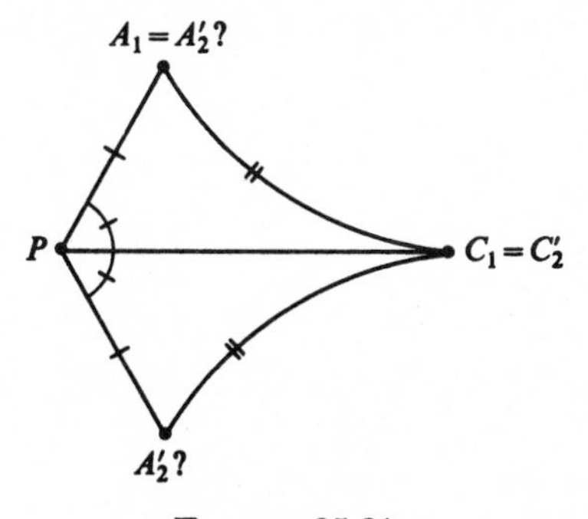

FIGURE 25.21

There are now only two possibilities:

(1) A'_2 is on the same side of $\overleftrightarrow{PC_1}$ as A_1. In this case, since $\angle A'_2PC'_2 \cong \angle A_1PC_1$, we have $\overrightarrow{PA'_2} = \overrightarrow{PA_1}$; and since $PA_1 = PA'_2$, we have $A_1 = A'_2$. Thus $P = P$, $A_1 = A'_2$ and $C_1 = C'_2$. Therefore

$$\triangle A_1PC_1 \cong \triangle A'_2PC'_2;$$

this fits together with our other congruences to give $\triangle ABC \cong \triangle A'B'C'$.

(2) A'_2 and A_1 are on opposite sides of $\overleftrightarrow{PC_1}$. In this case we reflect across $\overleftrightarrow{PC_1}$ and then proceed as in case (1).

Rather curiously, this proof has a great deal in common, intuitively, with Euclid's "proof" of SAS, by superposition, in Book I of the *Elements.* You really can prove things by superposition, if you carry out the process using a family of transformations (in this case, the reflections) which are known to preserve the properties that you are concerned with.

Chapter 26
The Consistency of Euclidean Geometry

26.1 INTRODUCTION

Our proof of the consistency of hyperbolic geometry, in the preceding chapter, was conditional. We showed that *if* there is a mathematical system satisfying the postulates for Euclidean geometry, *then* there is a system satisfying the postulates for hyperbolic geometry. We shall now investigate the *if*, by describing a model for the Euclidean postulates. Here again our consistency proof will be conditional. To set up our model, we shall need to assume that the real number system is given.

DEFINITION 1. $E = \mathbf{R} \times \mathbf{R}$.

That is, a point is defined to be an ordered pair of real numbers.

DEFINITION 2. A *line* is a set of the form

$$L = \{(x, y) | Ax + By + C = 0, \quad A^2 + B^2 > 0\}.$$

That is, a line is defined to be the graph of a linear equation in x and y.

DEFINITION 3. If $P = (x_1, y_1)$ and $Q = (x_2, y_2)$, then

$$d(P, Q) = \sqrt{(x_2 - x_1)^2 + (y_2 - y_1)^2}\,.$$

That is, distance is *defined* by the distance formula which appeared as a theorem in Chapter 17.

We define betweenness in terms of distance. (As usual, we abbreviate $d(P, Q)$ as PQ.) Segments and rays are defined in terms of betweenness; and angles are defined when rays are known.

It turns out that setting up an angular measure function is a formidable technical chore. We hope, therefore, that the reader will settle for a congruence relation $\cong$ for angles, satisfying the purely synthetic postulates C-6 through C-9 of Chapter 8. This relation is defined in the following way.

DEFINITION 4. An *isometry* is a one-to-one correspondence

$$f : E \leftrightarrow E,$$

preserving distance.

DEFINITION 5. Two angles $\angle ABC$ and $\angle DEF$ are *congruent* if there is an isometry $f : E \leftrightarrow E$ such that $f(\angle ABC) = \angle DEF$.

We have now given definitions, in the Cartesian model, for the terms used in the Euclidean postulates. Each of these postulates thus becomes a statement about a question of fact; and our task is to show that all of these statements are true.

26.2 THE RULER POSTULATE

By a *vertical* line we mean a line which is the graph of an equation $x = a$. The following are easy to check:

(1) Every nonvertical line is the graph of an equation $y = mx + b$.

(2) The graph of an equation $y = mx + b$ is never vertical.

(3) If $x = a$ and $x = b$ are equations of the same line, then $a = b$.

(4) If $y = m_1x + b_1$ and $y = m_2x + b_2$ are equations of the same line, then $m_1 = m_2$ and $b_1 = b_2$.

Theorem 1. Every vertical line L has a coordinate system.

Proof. For each point $P = (a, y)$ of L, let

$$f(P) = y.$$

Then f is a one-to-one correspondence $L \leftrightarrow \mathbf{R}$. If $P = (a, y_1)$ and $Q = (a, y_2)$, then

$$\begin{aligned} PQ = d(P, Q) &= \sqrt{(a - a)^2 + (y_2 - y_1)^2} \\ &= \sqrt{(y_2 - y_1)^2} \\ &= |y_2 - y_1| \\ &= |f(Q) - f(P)|, \end{aligned}$$

as desired.

Theorem 2. Every nonvertical line has a coordinate system.

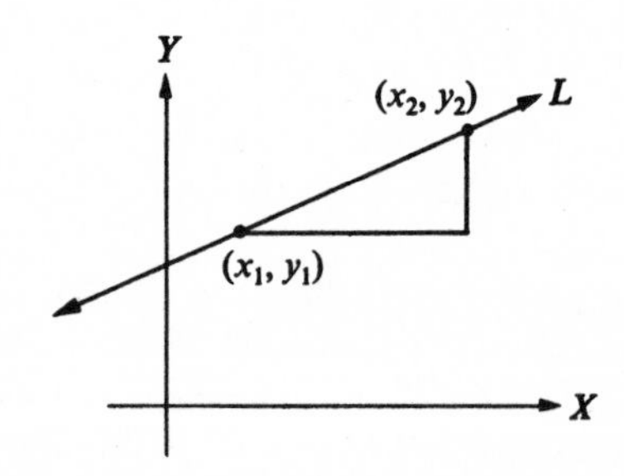

FIGURE 26.1

Proof. Let L be the graph of $y = mx + b$. If (x_1, y_1) and $(x_2, y_2) \in L$, then it is easy to check that

$$\frac{y_2 - y_1}{x_2 - x_1} = m, \qquad y_2 - y_1 = m(x_2 - x_1),$$

and

$$PQ = \sqrt{(x_2 - x_1)^2 + m^2(x_2 - x_1)^2} = \sqrt{(1 + m^2)}\,|x_2 - x_1|.$$

From this we see how to define a coordinate system for L. Let

$$f(x, y) = x\sqrt{1 + m^2}.$$

Then for

$$P = (x_1, y_1), \qquad Q = (x_2, y_2)$$

we have

$$\begin{aligned} PQ &= \sqrt{1 + m^2}\,|x_2 - x_1| \\ &= |x_2\sqrt{1 + m^2} - x_1\sqrt{1 + m^2}\,| \\ &= |f(Q) - f(P)|, \end{aligned}$$

as it should be.

These two theorems give us:

Theorem 3. In the Cartesian model, the ruler postulate holds.

26.3 INCIDENCE AND PARALLELISM

Theorem 1. Every two points of the Cartesian model lie on a line.

Proof. Given $P = (x_1, y_1)$, $Q = (x_2, y_2)$. If $x_1 = x_2$, then P and Q lie on the vertical line $x = a = x_1$. If not, then P and Q lie on the graph of the equation

$$y - y_1 = \frac{y_2 - y_1}{x_2 - x_1}(x - x_1),$$

which is easily seen to be a line.

Theorem 2. Two lines intersect in at most one point.

Proof. Given L_1 and L_2, with $L_1 \neq L_2$. If both are vertical, then they do not intersect at all. If one is vertical and the other is not, then the graphs of

$$x = a, \qquad y = mx + b$$

intersect at the unique point $(a, ma + b)$. Suppose finally, that L_1 and L_2 are the graphs of

$$y = m_1x + b_1, \qquad y = m_2x + b_2.$$

If $m_1 \neq m_2$, very elementary algebra gives us exactly one common solution and hence exactly one intersection point. If $m_1 = m_2$, then $b_1 \neq b_2$, and the graphs do not intersect at all.

We have already observed that if L is the graph of $y = mx + b$, then for every two points (x_1, y_1), (x_2, y_2) of L, we have

$$\frac{y_2 - y_1}{x_2 - x_1} = m.$$

Thus m is determined by the nonvertical line L. As usual, we call m the *slope* of L.

Theorem 3. Every vertical line intersects every nonvertical line. [At the point $(a, ma + b)$.]

By easier algebra we get:

Theorem 4. Two lines are parallel if and only if (1) both are vertical, or (2) neither is vertical, and they have the same slope.

Proof. Given $L_1 \neq L_2$. If both are vertical, then $L_1 \parallel L_2$. If neither is vertical, and they have the same slope, then the equations

$$y = mx + b_1, \qquad y = mx + b_2 \qquad (b_1 \neq b_2)$$

have no common solution, and $L_1 \parallel L_2$.

Suppose, conversely, that $L_1 \parallel L_2$. If both are vertical, then (1) holds. It remains only to show that if neither line is vertical, they have the same slope. Suppose not. Then

$$L_1 \colon y = m_1x + b_1, \qquad L_2 \colon y = m_2x + b_2 \qquad (m_1 \neq m_2).$$

We can now solve for x and y:

$$0 = (m_1 - m_2)x + (b_1 - b_2),$$

$$x = -\frac{b_1 - b_2}{m_1 - m_2},$$

$$y = -m_1\left(\frac{b_1 - b_2}{m_1 - m_2}\right) + b_1.$$

We got this value of y by substituting in the equation of L_1. But our x and y also satisfy the equation of L_2. This contradicts the hypothesis $L_1 \parallel L_2$.

Theorem 5. Given a point $P = (x_1, y_1)$ and a number m, there is exactly one line which passes through P and has slope $=m$.

Proof. The lines L with slope m are the graphs of equations

$$y = mx + b.$$

If L contains (x_1, y_1), then $b = y_1 - mx_1$, and conversely. Therefore our line exists and is unique.

Theorem 6. In the Cartesian model, the Euclidean parallel postulate holds.

Proof. Given a line L and a point $P = (x_1, y_1)$ not on L.

(1) If L is the graph of $x = a$, then the line $L' \colon x = x_1$ is the only vertical line through P, and, by Theorem 3, no nonvertical line is parallel to L. Thus the parallel L through P is unique.

(2) If L is the graph of $y = mx + b$, then the only parallel to L through P is the line through P with slope $=m$. This is unique.

26.4 TRANSLATIONS AND ROTATIONS

By a *translation* of the Cartesian model, we mean a one-to-one correspondence

$$f : E \leftrightarrow E,$$
$$: (x, y) \leftrightarrow (x + a, y + b).$$

Merely by substituting in the distance formula, and observing that a and b cancel out, we have:

Theorem 1. Translations are isometries.

If L is the graph of the equation

$$Ax + By + C = 0,$$

then the points $(x', y') = (x + a, y + b)$ of $f(L)$ satisfy the equation

$$A(x' - a) + B(y' - b) + C = 0,$$

or

$$Ax' + By' + (-aA - bB + C) = 0.$$

This is linear. Thus we have:

Theorem 2. Translations preserve lines.

Since translations preserve lines and distance, they preserve everything defined in terms of lines and distance.

Theorem 3. Translations preserve betweenness, segments, rays, angles, triangles, and angle congruences.

Rotations are harder to describe, because at this stage we have no trigonometry to work with. Let us first try using trigonometry, wishfully, to find out what we ought to be doing, and then find a way to do something equivalent, using only the primitive apparatus that we now have at our disposal in our study of the Cartesian model.

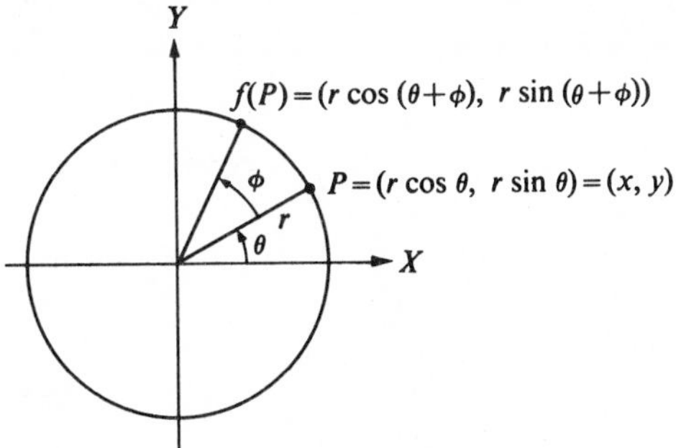

FIGURE 26.2

We want to rotate the Cartesian model through an angle of measure ϕ (Fig. 26.2). Trigonometrically, this can be done by a one-to-one correspondence,

$$f : E \leftrightarrow E,$$

defined as the labels in the figure suggest.

Now

$$\begin{aligned} \cos(\theta + \phi) &= \cos\theta \cos\phi - \sin\theta \sin\phi, \\ \sin(\theta + \phi) &= \sin\theta \cos\phi + \cos\theta \sin\phi. \end{aligned}$$

Let

$$a = \cos\phi, \qquad b = \sin\phi.$$

Now

$$\begin{aligned} r &= \sqrt{x^2 + y^2}, \\ \cos\theta &= \frac{x}{\sqrt{x^2 + y^2}}, \\ \sin\theta &= \frac{y}{\sqrt{x^2 + y^2}}. \end{aligned}$$

We can therefore rewrite our formulas in the form

$$f : (x, y) \leftrightarrow (x', y'),$$

where

$$\begin{aligned} x' &= r\cos(\theta + \phi) \\ &= \sqrt{x^2 + y^2}\left(\frac{x}{\sqrt{x^2 + y^2}}\,a - \frac{y}{\sqrt{x^2 + y^2}}\,b\right) \\ &= ax - by, \end{aligned}$$

and

$$\begin{aligned} y' &= \sqrt{x^2 + y^2}\left(\frac{y}{\sqrt{x^2 + y^2}}\,a + \frac{x}{\sqrt{x^2 + y^2}}\,b\right) \\ &= ay + bx. \end{aligned}$$

Any correspondence of this form, with $a^2 + b^2 = 1$, is called a *rotation* of the Cartesian model.

Theorem 4. Rotations preserve distance.

Proof. We have

$$\begin{aligned} P &= (x_1, y_1), \\ Q &= (x_2, y_2), \\ P' = f(P) &= (ax_1 - by_1, ay_1 + bx_1), \\ Q' = f(Q) &= (ax_2 - by_2, ay_2 + bx_2). \end{aligned}$$

It is merely an exercise in patience to substitute in the distance formula, calculate $P'Q'$, simplify with the aid of the equation $a^2 + b^2 = 1$, and observe that $P'Q' = PQ$. (The reader is warned that $(P'Q')^2$ appears as a sum of twenty terms.)

Solving for x and y in terms of x' and y', we get

$$x = ax' + by', \qquad y = ay' - bx'.$$

Comparing the formulas

$$x' = ax - by, \qquad y' = bx + ay,$$

for f and the corresponding formulas for f^{-1}, we see that these have the same form

$$x = a'x' - b'y', \qquad y = a'y' + b'x',$$

where $a' = a$ and $b' = -b$. Therefore we have the following theorem.

Theorem 5. The inverse of a rotation is a rotation.

Theorem 6. Rotations preserve lines.

Proof. L is the graph of an equation

(1) $x = k$,
(2) $y = k$,

or

(3) $y = mx + k \qquad (m \neq 0)$.

In Case (1), $f(L)$ is the graph of

$$ax' + by' = k,$$

where a and b are not both $=0$, because $a^2 + b^2 = 1$. Therefore L is a line.

In Case (2), $f(L)$ is the graph of

$$ay' - bx' = k,$$

which is a line.

In Case (3), $f(L)$ is the graph of

$$ay' - bx' = max' + mby' + k,$$

or

$$(ma + b)x' + (mb - a)y' + k = 0.$$

If we had both

$$ma + b = 0, \qquad mb - a = 0,$$

then

$$ma^2 + ab = 0, \qquad mb^2 - ab = 0,$$

so that

$$m(a^2 + b^2) = 0,$$

and $m = 0$, contradicting our hypothesis.

As for translations, once we know that rotations preserve lines and distance, it follows that they preserve everything that is defined in terms of lines and distance.

Therefore we have:

Theorem 7. Rotations preserve betweenness, segments, rays, angles, triangles, and angle congruences.

We are going to use rotations in the Cartesian model in much the same way that we used reflections in the Poincaré model, to show that postulates for angle congruence hold. To do this, we shall need to know that every ray starting at the origin $(0, 0)$ can be rotated onto the positive end of the x-axis, and vice versa. By Theorem 5, it will be sufficient to prove the following theorem.

Theorem 8. Let $P = (x_0, 0)$ $(x_0 > 0)$, let $Q = (x_1, y_1)$, and suppose that

$$x_0 = \sqrt{x_1^2 + y_1^2}.$$

Then there is a rotation f such that $f(P) = Q$.

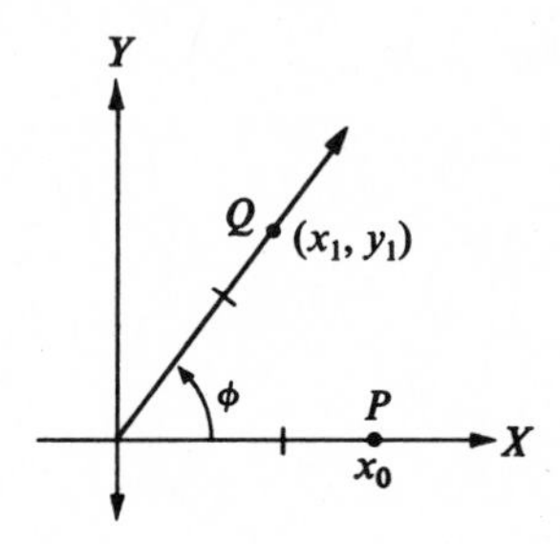

FIGURE 26.3

The equation in the hypothesis says, of course, that P and Q are equidistant from the origin.

As a guide in setting up such a rotation, we note unofficially that we want to rotate E through an angle of measure ϕ, where

$$a = \cos \phi = \frac{x_1}{\sqrt{x_1^2 + y_1^2}},$$

$$b = \sin \theta = \frac{y_1}{\sqrt{x_1^2 + y_1^2}}.$$

Thus the rotation ought to be

$$f : E \leftrightarrow E$$
$$: (x, y) \leftrightarrow (x', y'),$$

where

$$x' = ax - by = \frac{x_1}{\sqrt{x_1^2 + y_1^2}}\, x - \frac{y_1}{\sqrt{x_1^2 + y_1^2}}\, y,$$

$$y' = bx + ay = \frac{y_1}{\sqrt{x_1^2 + y_1^2}}\, x + \frac{x_1}{\sqrt{x_1^2 + y_1^2}}\, y.$$

Obviously $a^2 + b^2 = 1$ in these equations, and so f is a rotation. And

$$f(x_0, 0) = \left(\frac{x_1}{\sqrt{x_1^2 + y_1^2}} x_0, \frac{y_1}{\sqrt{x_1^2 + y_1^2}} x_0\right)$$
$$= (x_1, y_1),$$

which is the result that we wanted.

26.5 PLANE SEPARATION

We shall show first that the plane-separation postulate holds for the case in which the given line is the x-axis. It will then be easy to get the general case.

Let E^+ be the "upper half plane." That is,

$$E^+ = \{(x, y) | y > 0\}.$$

Theorem 1. E^+ is convex.

Proof. Lemma 1, Section 3.4, says that if A, B, and C are points of a line, with coordinates x, y, and z, and $x < y < z$, then A-B-C. (This was proved merely on the basis of the ruler postulate, and we can therefore apply it now.) Since only one of the points A, B, C is between the other two, the lemma has a true converse: if A-B-C, then $x < y < z$ or $z < y < x$.

Consider now two points, $A = (x_1, y_1)$, $C = (x_2, y_2)$ of E^+:

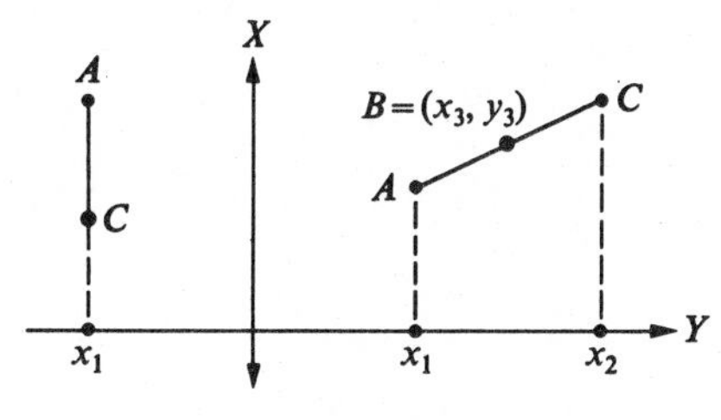

FIGURE 26.4

We need to show that $\overline{AC}$ lies in E^+. That is, if A-B-C, with $B = (x_3, y_3)$, then $y_3 > 0$. Obviously, for the case $x_1 \neq x_2$ we may assume that $x_1 < x_2$, as in the figure; and for the case $x_1 = x_2$, we may assume that $y_1 < y_2$.

In the first case, the line $\overleftrightarrow{AC}$ is the graph of an equation

$$y = mx + b,$$

and has a coordinate system of the form

$$f(x, y) = \sqrt{1 + m^2}\, x.$$

In the second case, the line is the graph of the equation

$$x = x_1$$

and has a coordinate system of the form

$$f(x, y) = y.$$

It is easy to check that in the first case

$$f(A) < f(B) < f(C),$$

so that

$$x_1 < x_3 < x_2.$$

For $m > 0$,

$$mx_1 + b < mx_3 + b < mx_2 + b;$$

for $m < 0$, the inequalities run the other way; but in either case y_2 lies between two positive numbers. In the second case ($x_1 = x_2$), the same result follows even more easily.

Let E^- be the "lower half plane." That is,

$$E^- = \{(x, y) | y < 0\}.$$

Since the function,

$$f : (x, y) \leftrightarrow (x, -y),$$

is obviously an isometry, it preserves segments. Therefore it preserves convexity. Since $f(E^+) = E^-$, we have the following theorem.

Theorem 2. E^- is convex.

It is an easy exercise in algebra to show that if $A = (x_1, y_1) \in E^+$, and $B = (x_2, y_2) \in E^-$, then $\overline{AB}$ contains a point $(x, 0)$ of the x-axis. Thus:

Theorem 3. E and the line $y = 0$ satisfy the conditions for E and L in the plane separation postulate.

Now let L be any line in E, and let $A = (x_1, y_1)$ be any point of L. By a translation f, we can move A to the origin. By a rotation g, we can move the resulting line onto the x-axis. Let

$$H_1 = g^{-1}f^{-1}(E^+), \qquad H_2 = g^{-1}f^{-1}(E^-).$$

Since all of the conditions of the plane separation postulate are preserved under isometries, we have the following theorems.

Theorem 4. E satisfies the conditions of the plane separation postulate.

Theorem 5. Isometries preserve half planes.

Proof. Let H_1 be a half plane with edge L, and let H_2 be the other side of L. If f is an isometry, then $f(L)$ is a line L'. Let

$$H'_1 = f(H_1), \qquad H'_2 = f(H_2).$$

Then H_1' and H_2' are convex, and every segment between two points $f(A)$ of H_1' and $f(B)$ of H_2' must intersect $f(L)$. Therefore H_1' is a half plane with L' as edge.
From Theorem 5 it follows that:

Theorem 6. Isometries preserve interiors of angles.

That is, if I is the interior of $\angle ABC$, then $f(I)$ is the interior of $f(\angle ABC)$.

26.6 ANGLE CONGRUENCES

We want to verify that angle congruence, defined by means of isometries of E onto itself, satisfies the postulates C-6 through C-9, Section 8.1, and also satisfies SAS. Only one of these verifications is trivial.

C-6. For angles, congruence is an equivalence relation.

Proof. (1) $\angle A \cong \angle A$ always, because the identity function $E \leftrightarrow E$ is an isometry. (2) If $\angle A \cong \angle B$, then $\angle B \cong \angle A$, because the inverse of an isometry is an isometry. (3) If $\angle A \cong \angle B$ and $\angle B \cong \angle C$, then $\angle A \cong \angle C$, because the composition of the isometries for which $\angle A \leftrightarrow \angle B$ and $\angle B \leftrightarrow \angle C$ is always an isometry for which $\angle A \leftrightarrow \angle C$.
The other verifications are more difficult. We begin with a lemma.

Lemma 1. Let f be an isometry of E onto itself. If $f(E^+) = E^+$, and $f(P) = P$ for every point P of the x-axis, then f is the identity.

Proof. Let A be the origin $(0, 0)$, and let $B = (1, 0)$. Let $Q = (a, b)$ be any point, and let $f(Q) = (c, d)$. Then

$$AQ = f(A)f(Q), \qquad BQ = f(B)f(Q).$$

Taking the square of each of these distances, we get

$$\begin{aligned} a^2 + b^2 &= c^2 + d^2, \\ (a - 1)^2 + b^2 &= (c - 1)^2 + d^2, \\ a^2 + b^2 - 2a + 1 &= c^2 + d^2 - 2c + 1, \end{aligned}$$

so that $a = c$. Therefore $b^2 = d^2$. Since $f(E^+) = E^+$, b and d are both positive, both zero, or both negative. Therefore $b = d$. Thus $f(Q) = Q$ for every Q, which was to be proved.

Lemma 2. Let A be the origin; let $B = (a, 0)$, $(a > 0)$ be a point of the x-axis; and let $C = (b, c)$ and $D = (d, e)$ be points of E^+ and E^- such that

$$AC = AD, \qquad BC = BD.$$

Then there is an isometry

$$f : E \leftrightarrow E$$

such that $f(A) = A$, $f(B) = B$, $f(C) = D$, and $f(D) = C$.

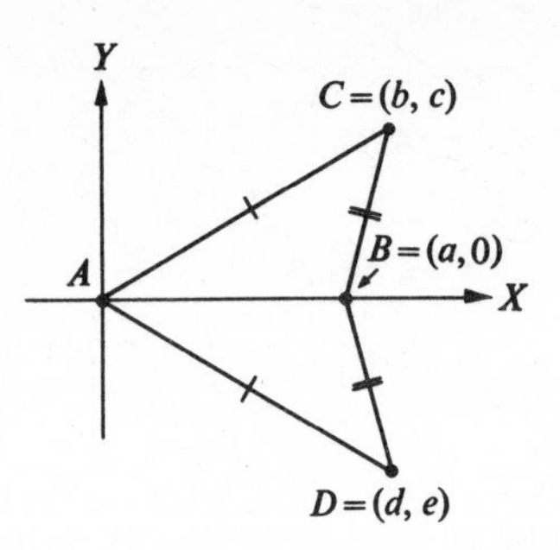

FIGURE 26.5

Proof. We shall show that $d = b$ and $e = -c$. The desired isometry f will then be the function $(x, y) \leftrightarrow (x, -y)$.

Given

$$b^2 + c^2 = d^2 + e^2,$$
$$(b - a)^2 + c^2 = (d - a)^2 + e^2,$$

we have $-2ab = -2ad$. Since $a > 0$, this gives $b = d$. Therefore $c^2 = e^2$. Since $c > 0$ and $e < 0$, we have $e = -c$.

Lemma 3. Given $\angle ABC$, there is an isometry f of E onto itself such that $f(\overrightarrow{BA}) = \overrightarrow{BC}$ and $f(\overrightarrow{BC}) = \overrightarrow{BA}$. That is, the sides of the angle can be interchanged by an isometry.

In the proof, we may suppose that $BA = BC$, since A and C can always be chosen so as to satisfy this condition.

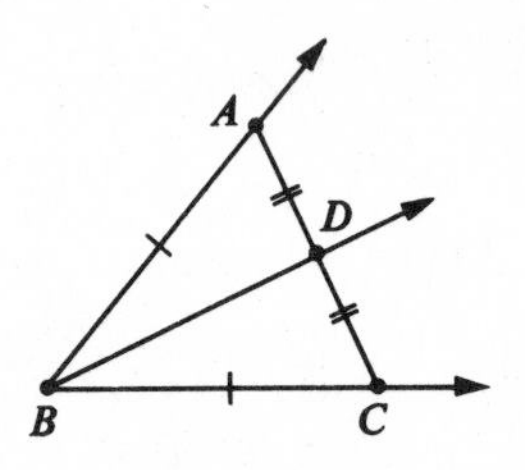

FIGURE 26.6

Let D be the midpoint of $\overline{AC}$. Using a translation followed by a rotation, we get an isometry $g : E \leftrightarrow E$ such that $g(\overrightarrow{BD})$ is the positive end of the x-axis (Fig. 26.7). (First we translate B to the origin, and then we rotate.) By the preceding lemma there is an isometry $h : E \leftrightarrow E$, interchanging A' and C', and leaving B' and D' fixed. Let

$$f = g^{-1}hg.$$

That is, f is the composition of g, h, and g^{-1}. Then f is an isometry; $f(B) = B$, $f(A) = C$ and $f(C) = A$.

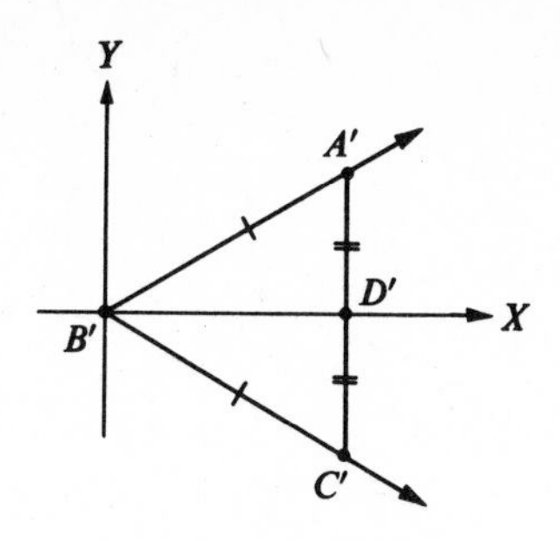

FIGURE 26.7

It is now easy to verify the rest of our congruence postulates. Oddly enough, the easiest is SAS. We put this in the style of a restatement.

SAS. Given $\triangle ABC$, $\triangle A'B'C'$, and a correspondence

$$ABC \leftrightarrow A'B'C'.$$

If (1) $AB = A'B'$, (2) $\angle B \cong \angle B'$, and (3) $BC = B'C'$, then (4) $\angle A \cong \angle A'$, (5) $\angle C \cong \angle C'$, and (6) $AC = A'C'$.

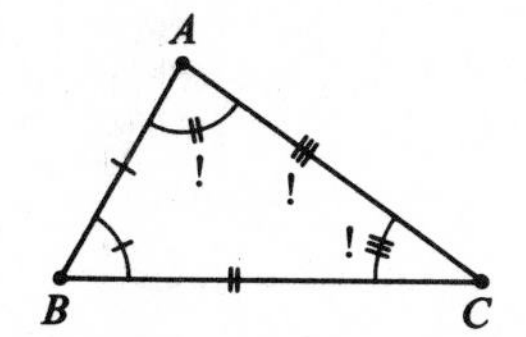

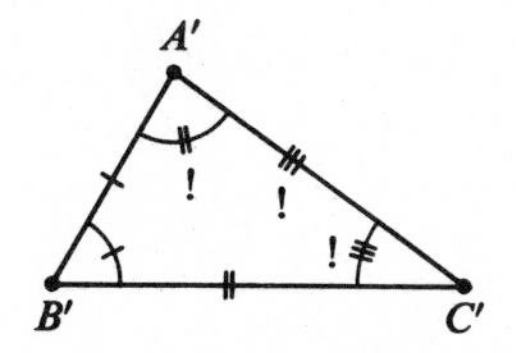

FIGURE 26.8

Proof. By hypothesis (2), there is an isometry $E \leftrightarrow E$, $\angle B \leftrightarrow \angle B'$. By Lemma 3 it follows that there is an isometry

$$\begin{aligned} f : E &\leftrightarrow E \\ : B &\leftrightarrow B' \\ : \overrightarrow{BA} &\leftrightarrow \overrightarrow{B'A'} \\ : \overrightarrow{BC} &\leftrightarrow \overrightarrow{B'C'}. \end{aligned}$$

(If the given isometry moves $\angle B$ onto $\angle B'$ in "the wrong way," then we follow it by an isometry which interchanges the sides of $\angle B'$.) From (1) it follows that $A' = f(A)$ and $C' = f(C)$. Therefore $\angle A' = f(\angle A)$, and $\angle A' \cong \angle A$; $\angle C' = f(\angle C)$, and $\angle C' \cong \angle C$. Also $AC = A'C'$, because f is an isometry.

This proof bears a certain resemblance to Euclid's "proof" of SAS by superposition.

C-7. Let $\angle ABC$ be an angle, let $\overrightarrow{B'C'}$ be a ray, and let H be a half plane whose edge contains $\overrightarrow{B'C'}$. Then there is exactly one ray $\overrightarrow{B'A'}$, with A' in H, such that $\angle ABC = \angle A'B'C'$.

We give the proof merely in outline. It should be understood that all of the functions mentioned are isometries of E onto E, and that the ray R is the positive x-axis.

(1) Take f_1 so that $f_1(\overrightarrow{B'C'}) = R$.

(2) Take f_2 so that $f_2(R) = R$ and $f_2f_1(H) = E^+$. (Of course, if $f_1(H)$ is already $=E^+$, we let f_2 be the identity.)

(3) Take g_1 so that $g_1(\overrightarrow{BC}) = R$.

(4) Take g_2 so that $g_2(R) = R$ and $g_2g_1(A)$ is in E^+.

(5) Let $\angle x = f_1^{-1}f_2^{-1}g_2g_1(\angle ABC)$. Then $\angle x$ is the $\angle A'B'C'$ that we wanted.

(6) Suppose that there are two rays $\overrightarrow{B'A'}$, $\overrightarrow{B'A''}$ satisfying these conditions.

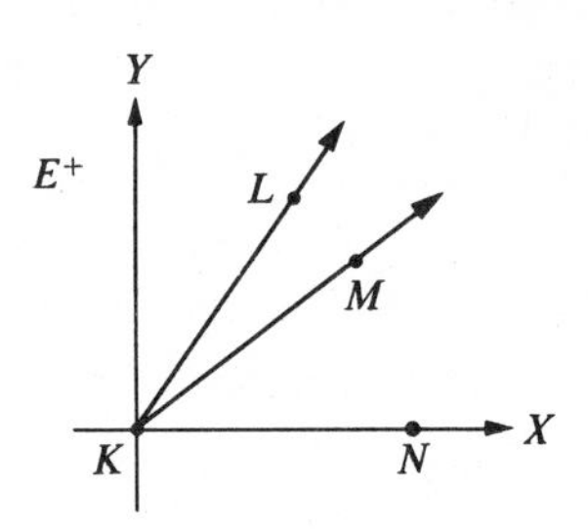

FIGURE 26.9

Then

$$f_2f_1(\overrightarrow{B'A'}) = \overrightarrow{KL}, \qquad f_2f_1(\overrightarrow{B'A''}) = \overrightarrow{KM},$$

where K and M are in E^+ and $\overrightarrow{KL}$ and $\overrightarrow{KM}$ are different rays. Since

$$\angle LKN \cong \angle ABC \cong \angle MKN,$$

we have

$$\angle LKN \cong \angle MKN.$$

Thus there is an isometry f, of E onto itself, such that

$$f(\angle LKN) = \angle MKN.$$

By Lemma 3, f can be chosen so that $f(\overrightarrow{KN}) = \overrightarrow{KN}$ and $f(\overrightarrow{KL}) = \overrightarrow{KM}$. It follows that for each point P of the x-axis, $f(P) = P$. Since isometries preserve half-planes, and $f(L)$ is in E^+, we have $f(E^+) = E^+$. By Lemma 1 it follows that f is the identity. This contradicts the hypothesis $f(\overrightarrow{KL}) = \overrightarrow{KM} \neq \overrightarrow{KL}$.

C-8. If (1) D is in the interior of $\angle BAC$, (2) D' is in the interior of $\angle B'A'C'$, (3) $\angle BAD \cong \angle B'A'D'$, and (4) $\angle DAC \cong \angle D'A'C'$, then (5) $\angle BAC \cong \angle B'A'C'$.

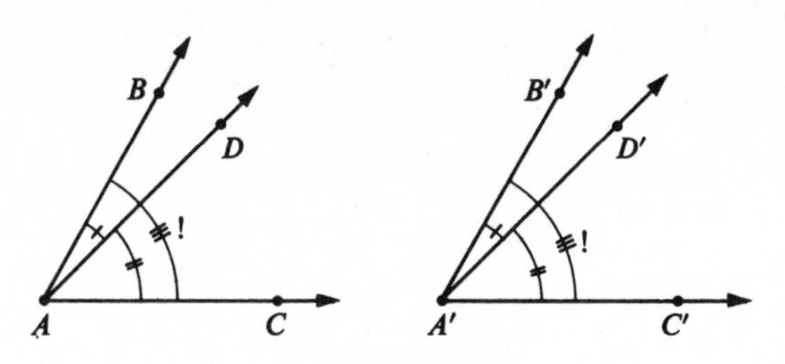

FIGURE 26.10

Proof. (1) By an isometry f, we move $\overrightarrow{AD}$ onto R and B into E^+. (For this we need a translation, followed by a rotation and perhaps a reflection $(x, y) \leftrightarrow (x, -y)$.)

(2) By an isometry g, we move $\overrightarrow{A'D'}$ onto R and B' into E^+.

(3) By the uniqueness condition in the preceding postulate, we know that $f(\overrightarrow{AB}) = g(\overrightarrow{A'B'})$ and $f(\overrightarrow{AC}) = g(\overrightarrow{A'C'})$.

(4) Therefore $f(\angle BAC) = g(\angle B'A'C')$. Therefore $\angle BAC \cong \angle B'A'C'$; the required isometry is $g^{-1}f$.

C-9. If (1) D is in the interior of $\angle BAC$, (2) D' is in the interior of $\angle B'A'C'$, (3) $\angle BAD \cong \angle B'A'D'$ and (4) $\angle BAC \cong \angle B'A'C'$, then (5) $\angle DAC \cong \angle D'A'C'$.

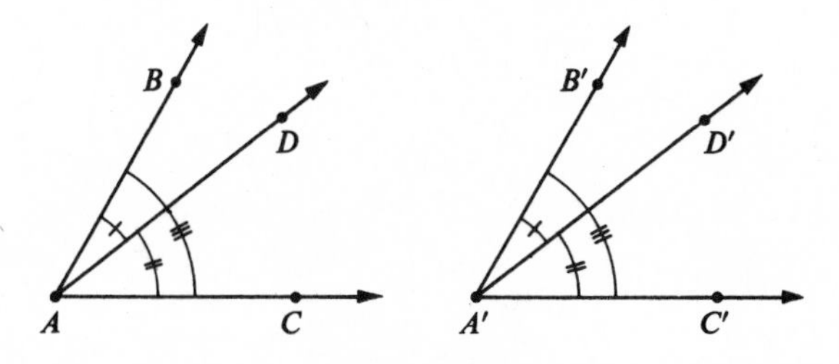

FIGURE 26.11

Proof. Let f be the isometry given by (4), so that $f(\angle BAC) = \angle B'A'C'$. By Lemma 3 we may suppose that $f(\overrightarrow{AB}) = \overrightarrow{A'B'}$ and $f(\overrightarrow{AC}) = \overrightarrow{A'C'}$. Then surely $f(\angle BAD) \cong \angle BAD$. The uniqueness condition in C-7 therefore tells us that $f(\overrightarrow{AD}) = \overrightarrow{A'D'}$. Therefore $f(\angle DAC) = \angle D'A'C'$, and $\angle DAC \cong \angle D'A'C'$, which was to be proved.

Chapter 27
The Postulational Method

27.1 INTRODUCTION

In this chapter we give a general discussion of the ways in which sets of postulates are used in mathematics, with illustrations from preceding chapters. The general discussion has been postponed until now precisely so that the illustrations could be cited. The postulational method is used in a number of quite different ways, and the distinctions among these are rather difficult to explain in the abstract.

27.2 POSTULATES CONSIDERED AS SELF-EVIDENT TRUTHS

In the time of Euclid, and for over two thousand years thereafter, the postulates of geometry were thought of as self-evident truths about physical space; and geometry was thought of as a kind of purely deductive physics. Starting with the truths that were self-evident, geometers considered that they were deducing other and more obscure truths without the possibility of error. (Here, of course, we are not counting the casual errors of individuals, which in mathematics are nearly always corrected rather promptly.) This conception of the enterprise in which geometers were engaged appeared to rest on firmer and firmer ground as the centuries wore on. As the other sciences developed, it became plain that in their earlier stages they had fallen into fundamental errors. Meanwhile the "self-evident truths" of geometry continued to look like truths, and continued to seem self-evident.

With the development of hyperbolic geometry, however, this view became untenable. We then had two different, and mutually incompatible, systems of geometry. Each of them was mathematically self-consistent, and each of them was compatible with our observations of the physical world. From this point on, the whole discussion of the relation between geometry and physical space was carried on in quite different terms. We now think not of a unique, physically "true" *geometry*, but of a number of mathematical *geometries*, each of which may be a good approximation of physical space, and each of which may be useful in various physical investigations. Thus we have lost our faith not only in the idea that simple and fundamental truths can be relied upon to be self-evident, but also in the idea that geometry is an aspect of physics.

This philosophical revolution is reflected, oddly enough, in the differences between the early passages of the Declaration of Independence and the Gettysburg

Address. Thomas Jefferson wrote:

"... We hold these truths to be self-evident, that all men are created equal, that they are endowed by their creator with certain unalienable rights, that among these are Life, Liberty and the pursuit of Happiness ...."

The spirit of these remarks is Euclidean. From his postulates, Jefferson went on to deduce a nontrivial theorem, to the effect that the American colonies had the right to establish their independence by force of arms.

Lincoln spoke in a different style:

"Fourscore and seven years ago our fathers brought forth on this continent a new nation, conceived in liberty and dedicated to the proposition that all men are created equal."

Here Lincoln is referring to one of the propositions mentioned by Jefferson, but he is not claiming, as Jefferson did, that this proposition is self-evidently true, or even that it is true at all. He refers to it merely as a proposition to which a certain nation was dedicated. Thus, to Lincoln, this proposition is a *description* of a certain aspect of the United States (and, of course, an aspect of himself). (I am indebted for this observation to Lipman Bers.)

This is not to say that Lincoln was a reader of Lobachevsky, Bolyai or Gauss, or that he was influenced, even at several removes, by people who were. It seems more likely that a shift in philosophy had been developing independently of the mathematicians, and that this helped to give mathematicians the courage to undertake non-Euclidean investigations and publish the results.

At any rate, modern mathematicians use postulates in the spirit of Lincoln. The question whether the postulates are "true" does not even arise. Sets of postulates are regarded merely as *descriptions* of mathematical structures. Their value consists in the fact that they are practical aids in the study of the mathematical structures that they describe.

27.3 CATEGORIC POSTULATE SETS

It sometimes happens that a postulate set gives a complete description of a mathematical structure, in the sense that any two structures that satisfy all of the postulates are essentially the same. Rather than attempting to give a definition of the phrase "essentially the same," let us look at an example.

Let

$$[F, +, \cdot]$$

and

$$[F', +', \cdot']$$

be two algebraic structures satisfying the field postulates of Section 1.3, and satisfying the further postulate that each of the sets F and F' have exactly two elements. Let 0 be the element of F given by A-3, and let $0'$ be the element of F' given by A-3. Similarly, let 1 and $1'$ be the elements of F and F' given by M-3. We can then set up a one-to-one correspondence

$$\begin{aligned} f &: 0 \leftrightarrow 0', \\ &: 1 \leftrightarrow 1'; \end{aligned}$$

it is easy to check that f preserves both sums and products, both ways. This is what we mean when we say that $[F, +, \cdot]$ and $[F', +', \cdot']$ are *essentially the same*. For algebraic structures, this relation is called *isomorphism*. Thus a set of algebraic postulates is *categoric* if any two algebraic structures that satisfy the postulates are isomorphic.

Obviously the postulates for a field are not categoric, and neither are the postulates for an ordered field; they are satisfied by the rationals, by the surds, and by the real numbers, as well as the non-Archimedean ordered field described in Chapter 28.

Similarly, the postulates for synthetic plane geometry are *not* categoric. They are satisfied by a metric plane, in which the ruler postulate tells us that we have one-to-one coordinate functions

$$f : L \leftrightarrow \mathbf{R}$$

for every line. The same postulates are also satisfied in the surd plane, in which every line is everywhere full of holes. We can, however, get a categoric postulate set by adding two more postulates.

We recall the metrization theorem of Section 20.6. Given a plane

$$[E, \mathcal{L}, \cong, \mathcal{B}]$$

satisfying the synthetic postulates, and also satisfying the geometric form of the Archimedean postulate, we can always introduce a distance function

$$d : E \times E \to \mathbf{R},$$

which gives back to us the congruence and betweenness relations that we started with. (It would be worthwhile to review Section 20.6, at this point, because we are going to make heavy use of it.) Under the distance function given by the metrization theorem, the ruler postulate is almost satisfied; we have "coordinate functions"

$$f : L \to \mathbf{R}$$

such that

$$d(P, Q) = |f(P) - f(Q)|,$$

but we have no guarantee that all real numbers get used as coordinates. That is, the coordinate functions may be functions *into* $\mathbf{R}$, and may fail to be one-to-one correspondences. If it happens that all coordinate systems on all lines are one-to-one correspondences, then we say that the plane that we started with is *complete in the sense of Dedekind*.

This suggests a way to get a categoric set of postulates for synthetic plane geometry: we should add, to the usual postulates, the conditions:

(I) E is Archimedean.
(II) E is complete in the sense of Dedekind.

Any two structures

$$[E, \mathcal{L}, \cong, \mathcal{B}], \qquad [E', \mathcal{L}', \cong', \mathcal{B}']$$

which satisfy all of these conditions are essentially the same. Oddly enough, this is not hard to prove, once we have come this far in this book.

By the methods of Chapter 17, we set up coordinate systems in E and E'. Thus we have correspondences

$$E \leftrightarrow \mathbf{R} \times \mathbf{R}, \qquad P \leftrightarrow (x, y)$$

and

$$E' \leftrightarrow \mathbf{R} \times \mathbf{R}, \qquad P' \leftrightarrow (x, y).$$

These really are one to one, because both of our planes are complete in the sense of Dedekind. For each point P of E, we let $P' = f(P)$ be the point of E' which has the same coordinates (x, y) as P. Then

(1) *f preserves distance.*

That is, $d(P, Q) = d'(P', Q')$, for all points P, Q of E.

Proof. Let P and Q have coordinates (x_1, y_1) and (x_2, y_2). Then P' and Q' have the same coordinates (x_1, y_1) and (x_2, y_2). Therefore

$$d(P, Q) = \sqrt{(x_2 - x_1)^2 + (y_2 - y_1)^2} = d'(P', Q').$$

(2) *f preserves lines.*

That is, L is a line in E if and only if $f(L)$ is a line in E'.

Proof. If L is a line in E, then L is the graph of a linear equation

$$Ax + By + C = 0.$$

It then follows that $f(L)$ is the graph of the same linear equation. Therefore $f(L)$ is a line in E'. The same proof works in reverse.

(3) *f preserves betweenness.*

That is, A-B-C in E if and only if A'-B'-C' in E'.

Proof. Each of the following statements is equivalent to the next:
(a) A-B-C in E.
(b) A, B, and C are collinear, and $d(A, B) + d(B, C) = d(A, C)$.
(c) A', B', and C' are collinear, and $d'(A', B') + d'(B', C') = d'(A', C')$.
(d) A'-B'-C' in E'.

(4) *f preserves segments, rays, and angles,* because these are defined in terms of betweenness.

(5) *f preserves congruence between segments.*

Proof. Each of the following statements is equivalent to the next:
(a) $\overline{AB} \cong \overline{CD}$.
(b) $d(A, B) = d(C, D)$.
(c) $d'(A', B') = d'(C', D')$.
(d) $\overline{A'B'} \cong \overline{C'D'}$.

(6) *f preserves congruence between angles.*

That is, $\angle ABC \cong \angle DEF$ if and only if $f(\angle ABC) \cong' f(\angle DEF)$.

Proof. First we observe that

$$f(\angle ABC) = \angle A'B'C'$$

and

$$f(\angle DEF) = \angle D'E'F',$$

by (4). We are free to choose D and F on the sides of $\angle DEF$, so that $DE = AB$ and $EF = BC$. By SAS it follows that $AC = DF$. Therefore

$$D'E' = A'B', \qquad E'F' = B'C', \qquad A'C' = D'F'$$

because f preserves distance. (Here we are using the short notation for distance in both of our planes.) By SSS it follows that

$$\triangle A'B'C' \cong \triangle D'E'F',$$

and

$$\angle A'B'C' \cong \angle D'E'F',$$

which was to be proved.

Thus f preserves all the structure mentioned in our postulates. This means that our postulates for synthetic plane geometry became categoric once we had added (1) the Archimedean postulate and (2) the Dedekind postulate.

A categoric postulate set is a sort of arch of triumph. When we are able to write such a postulate set for a particular mathematical structure, this means that we have a complete understanding of its essential properties. Note, for example, in the case of synthetic plane geometry, that we did not know what conditions to add to make our postulates categoric, until we had gone through the deep and difficult discussion in Chapter 20.

27.4 THE USE OF POSTULATE SETS AS CODIFICATIONS

In fact, categoric postulate sets are rare. Most of the time, when we write down a set of postulates, we do so not to get a complete description of a particular mathematical system, but for precisely the opposite purpose. Most of the time, the value of the postulates lies in their generality: they describe a common aspect of various mathematical systems which may have little else in common.

One striking example of this is the idea of a group. (A *group*, of course, is a pair $[F, \cdot]$ satisfying M-1 through M-4 of Section 1.3.) Once we have proved a theorem about groups in general, on the basis of these four postulates, we are free to apply the theorem in an immense variety of contexts. This process gives an efficient codification of mathematics; it spares us the job of repeating essentially the same proof over and over. (Often it leads us to simpler proofs, because it tends to protect us from being distracted by irrelevancies.)

To some extent, postulates have been used in this way in this book. For example, the postulates of metric absolute plane geometry are not categoric; they allow the possibility that the plane is either Euclidean or hyperbolic. In Chapters 1 through 7, and in Chapter 10, we used neither of the two possible parallel postulates. We were then free to use the resulting theorems both in Euclidean geometry (Chapters 11 through 23, and 25) and in hyperbolic geometry (Chapter 24). The resulting economy was considerable. If we had introduced the Euclidean parallel postulate earlier, it would have been necessary to do some of our work over again in Chapter 24. Suppose, for example, that we had postponed the study of geometric inequalities until after Chapter 11, and had proved the exterior angle theorem in the way suggested by the following figure:

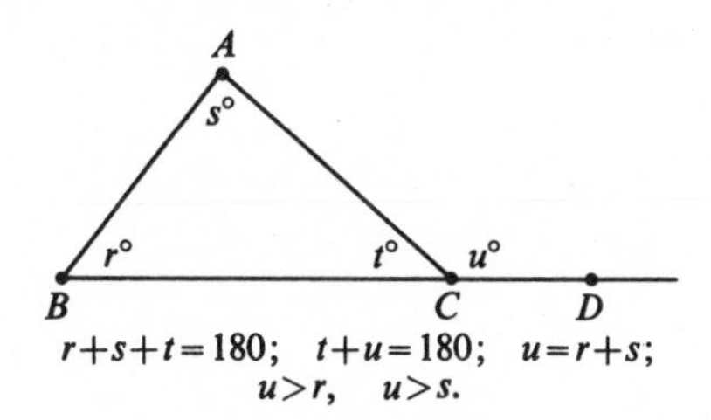

FIGURE 27.1

It would then have been necessary to develop the theory of geometric inequalities all over again in Chapter 24.

Similarly, we used the postulates for a Euclidean ordered field in Chapters 1, 3, 6, and Chapters 10 through 18. These postulates allow both the complete real number system and the surd field, and therefore our results could be used in both Chapters 19 and 20. In Chapter 20 we introduced the Dedekind postulate, at the point where we really needed it. Thus, at most points in this book, the usefulness of our postulate sets has been due to the fact that they were *not* categoric.

27.5 THE USE OF POSTULATES TO KEEP THE RECORD STRAIGHT

Often we use postulates when they aren't logically necessary at all. For example, we showed in Chapter 14 that all the area postulates of Chapter 13 were superfluous, because we could prove that in any metric geometry there has to be an area function satisfying these area postulates. The postulates for volume, in Chapter 23, are also superfluous, for the same sort of reason, although we have not proved the fact in this book.

In each of these cases, we introduced new postulates simply in order to make it clear exactly what was being assumed at a particular stage of our investigations.

Chapter 28
An Example of an Ordered Field which is not Archimedean

In Chapter 1 we postulated that the real number system had the Archimedean property. In fact, this postulate was necessary; it is not true that every ordered field satisfies the Archimedean postulate. Here we give an example of a field which does not.

First let **P** be the set of all polynomials with real coefficients. Thus **P** includes the "zero polynomial," which is $=0$ for every x, the constant polynomials $f(x) = a_0$, and the polynomials of degree $n > 0$, of the form

$$f(x) = a_n x^n + a_{n-1}x^{n-1} + \cdots + a_1 x + a_0,$$

where $a_n \neq 0$. In any case, a_n is called the leading coefficient. Thus the leading coefficient of $f(x) = 2x^2 - 3x + 4$ is 2, and the leading coefficient of a constant polynomial $f(x) = a_0$ is the same constant a_0.

For the purposes of this chapter, a polynomial will be called *positive* if it takes on only positive values, when x is sufficiently large. To be more precise, f is *positive* if there is a number k such that

$$f(x) > 0 \qquad \text{for every} \quad x > k.$$

For example, $f(x) = x^2 - 2$ is positive, because

$$x^2 - 2 > 0 \qquad \text{for every} \quad x > \sqrt{2}.$$

We are using $k = \sqrt{2}$. Similarly, $f(x) = x^4 + x^2 + 1$ is positive; here any number at all can be used as k, because $x^4 + x^2 + 1 > 0$ for every x.

Theorem 1. $f > 0$ if and only if the leading coefficient in f is >0.

Proof. For constant polynomials, this is obvious. Suppose, then, that

$$f(x) = a_n x^n + a_{n-1}x^{n-1} + \cdots + a_1 x + a_0,$$

with $a_n > 0$. Then

$$f(x) = a_n x^n \left[1 + \frac{a_{n-1}}{a_n x} + \frac{a_{n-2}}{a_n x^2} + \cdots + \frac{a_0}{a_n x^n}\right].$$

Here the first factor is always >0 when $x > 0$. And the expression in the brackets is >0 whenever

$$\left|\frac{a_{n-1}}{a_n x} + \frac{a_{n-2}}{a_n x^2} + \cdots + \frac{a_0}{a_n x_n}\right| < 1.$$

Surely this holds when x is greater than a certain k, because as $x \to \infty$, each of the terms on the left $\to 0$. (For a complete proof of this, see a calculus book.)

On the other hand, if $a_n < 0$, then $a_n x^n < 0$ when $x > 0$, and the bracket is still >0 when x is greater than a certain k.

Thus the positive polynomials are simply the ones that have positive numbers as their leading coefficients.

A polynomial f is called *negative* if there is a number k such that

$$f(x) < 0 \qquad \text{for every} \quad x > k.$$

From Theorem 1 it follows immediately that:

Theorem 2. Every polynomial (other than the zero polynomial) is either positive or negative. If $f > 0$, then $-f < 0$, and conversely.

The algebraic system formed by our set **P** of polynomials has nearly all the properties that we are looking for. It has a 0, namely, the polynomial $f \equiv 0$ which is $=0$ for every x. It has a 1, namely, the polynomial $g \equiv 1$. Addition and multiplication are associative, commutative, and distributive, because the real numbers have these properties. We easily define an order relation, by defining

$$f < g$$

to mean that

$$g - f > 0.$$

Thus $f < g$ if $f(x) < g(x)$ for every x greater than a certain k. It can also be checked that Conditions O-1, O-2, AO-1, and MO-1 hold. Thus **P** forms an *ordered commutative ring with unity:* it satisfies all of the postulates for an ordered field, with the sole exception of the postulate which says that every $f \neq 0$ has a reciprocal.

(We omit the details of these verifications, but will give the analogous details presently, for the ordered *field* which we shall finally be interested in.)

On the other hand, **P** is surely not Archimedean. Take, for example

$$e = f, \qquad f(x) = 1 \qquad \text{for every} \quad x,$$

and

$$M = g, \qquad g(x) = x \qquad \text{for every} \quad x.$$

No matter what the integer n may be, we have

$$n < x \qquad \text{when} \qquad x > n.$$

Therefore

$$ne < M$$

for every n, and our e and M do not satisfy the Archimedean postulate. We may say that $f(x) = x$ is "infinitely large compared with" $g(x) = 1$.

In the same way, $x^2 > nx$ when x is large enough. Therefore, if $e(x) = x$, $M(x) = x^2$, then

$$ne < M \qquad \text{for every} \quad n.$$

Thus there is an ordered ring which is not Archimedean. To get a non-Archimedean ordered *field*, we use the simple device of forming quotients of polynomials.

By a *rational function* we mean a function r of the form

$$r(x) = \frac{f(x)}{g(x)},$$

where f and g are polynomials and $g \neq 0$. A rational function r is called *positive* if there is a number k such that

$$r(x) > 0 \qquad \text{for every} \quad x > k.$$

In this case we write $r > 0$. The function r is called *negative* if there is a number k such that

$$r(x) < 0 \qquad \text{for every} \quad x > k.$$

Theorem 3. Every rational function (other than 0) is either >0 or <0.

Proof. Given

$$r(x) = \frac{f(x)}{g(x)},$$

where f is not the zero polynomial. We can surely choose f and g so that $g > 0$. (If this does not already hold, we multiply in numerator and denominator by -1.) Thus we have a k_1 such that

$$g(x) > 0 \qquad \text{for every} \quad x > k_1.$$

Suppose now that $f > 0$. Then there is a k_2 such that

$$f(x) > 0 \qquad \text{for every} \quad x > k_2.$$

Let k be the larger of the numbers k_1 and k_2. If $x > k$, then we have $x > k_1$ and $x > k_2$. Therefore $f(x) > 0$ and $g(x) > 0$. Hence

$$r(x) = \frac{f(x)}{g(x)} > 0 \qquad \text{for every} \quad x > k.$$

Similarly, if $f < 0$ and $g > 0$ it turns out that $r < 0$.

Theorem 4. If $r > 0$ and $s > 0$, then $r + s > 0$ and $rs > 0$.

Proof. Given

$$r(x) > 0 \qquad \text{for every} \quad x > k_1$$
$$s(x) > 0 \qquad \text{for every} \quad x > k_2.$$

Let k be the larger of the numbers k_1 and k_2. For every $x > k$, we then have

$$r(x) + s(x) > 0,$$

and

$$r(x)s(x) > 0.$$

Therefore $r + s > 0$ and $rs > 0$, which was to be proved.

It is easy, of course, for two rational functions to be essentially the same. For example, consider

$$r_1(x) = \frac{x(x-2)}{(x-1)(x-2)},$$

$$r_2(x) = \frac{x(x-3)}{(x-1)(x-3)}.$$

The first of these is defined except at $x = 1$ and $x = 2$; and the second is defined except at $x = 1$ and $x = 3$. Wherever both functions are defined, they have the same value. Two rational functions which are related in this way are called *equivalent*. More precisely,

$$r_1 \sim r_2$$

if

$$r_1(x) = r_2(x),$$

except perhaps at a finite number of points. The following theorem is easy to see.

Theorem 5. If $r_1 \sim r_2$ and $s_1 \sim s_2$, then $r_1r_2 \sim s_1s_2$ and $r_1 + r_2 \sim s_1 + s_2$. If $r_1 \sim r_2$ and $r_1 > 0$, then $r_2 > 0$.

The set of all rational functions equivalent to r is denoted by $\bar{r}$. Thus

$$\bar{r} = \{s | s \sim r\}.$$

By the preceding theorem, we can state the following definitions.

(1) $\bar{r} > 0$ if $r > 0$.
(2) $\bar{r} + \bar{s} = \overline{r + s}$.
(3) $\bar{r} \cdot \bar{s} = \overline{rs}$.

The point is that positiveness, sums, and products depend only on the equivalence classes $\bar{r}$, $\bar{s}$, and do *not* depend on their representatives r and s. From Theorem 4 we now get:

Theorem 6. If $\bar{r} > 0$ and $\bar{s} > 0$, then $\bar{r} + \bar{s} > 0$ and $\overline{rs} > 0$.

Now let $\mathbf{F}$ be the set of all equivalence classes $\bar{r}$. We assert that $\mathbf{F}$ forms a field.

The associative, commutative, and distributive laws hold in $\mathbf{F}$, because they hold for real numbers. For example,

$$(\overline{rs})\bar{t} = \bar{r}(\overline{st}),$$

because for all but a finite number of real numbers x, we have

$$[r(x)s(x)]t(x) = r(x)[s(x)t(x)].$$

Therefore

$$(rs)t \sim r(st).$$

Hence

$$(\overline{rs})\bar{t} = \bar{r}(\overline{st}),$$

by definition of multiplication for $\bar{r}$, $\bar{s}$, $\bar{t}$. Similarly for the other postulates which merely state algebraic identities.

The set $\mathbf{F}$ contains a 0 and a 1, namely $\bar{0}$ and $\bar{1}$. Also $-\bar{r} = \overline{-r}$ and

$$\frac{1}{\bar{r}} = \overline{\left(\frac{1}{r}\right)}.$$

Obviously $\mathbf{F}$ is closed under addition and multiplication, because sums and products of rational functions are always rational functions.

We must now define an order relation in $\mathbf{F}$, and show that $\mathbf{F}$ forms an ordered field.

Given $\bar{r}$, $\bar{s}$, $\mathbf{F}$. If

$$\bar{s} - \bar{r} > 0,$$

then (by definition),

$$\bar{r} < \bar{s}.$$

For every $\bar{r}$ and $\bar{s}$, we have exactly one of the conditions

$$\bar{s} - \bar{r} > 0, \qquad \bar{s} - \bar{r} = 0, \qquad \bar{s} - \bar{r} < 0.$$

Therefore we have exactly one of the conditions

$$\bar{r} < \bar{s}, \qquad \bar{r} = \bar{s}, \qquad \bar{s} < \bar{r}.$$

Thus our relation $<$ satisfies O-1. If

$$\bar{r} < \bar{s} \quad \text{and} \quad \bar{s} < \bar{t},$$

then

$$\bar{s} - \bar{r} > 0 \quad \text{and} \quad \bar{t} - \bar{s} > 0.$$

By Theorem 6, we have

$$(\bar{s} - \bar{r}) + (\bar{t} - \bar{s}) > 0,$$

so that

$$\bar{t} - \bar{r} > 0,$$

and

$$\bar{r} < \bar{t}.$$

Therefore the relation $<$ satisfies O-2.

It remains to verify AO-1 and MO-1.

Given

$$\bar{r} < \bar{s},$$

we want to conclude that

$$\bar{r} + \bar{t} < \bar{s} + \bar{t}.$$

The first condition says that

$$\bar{s} - \bar{r} > 0.$$

The second condition says that

$$(\bar{s} + \bar{t}) - (\bar{r} + \bar{t}) > 0.$$

Obviously these are equivalent. Therefore AO-1 holds; we already know from Theorem 6 that MO-1 holds. Therefore **F** forms an ordered field.

But **F** is not Archimedean. To show this, we proceed in exactly the same way that we did for **P**. Let e be the function which is $=1$ for every x, and let M be the function which is $=x$ for every x. For every positive integer n, $ne(x) = n$ for every x. Therefore, for every n we have

$$M(x) - ne(x) = r_n(x) = x - n.$$

Now $x - n > 0$ when $x > n$. Therefore r_n is a *positive* function for every n. Therefore, for every n, we have

$$\overline{M} - n\bar{e} > 0.$$

Therefore

$$n\bar{e} < \overline{M}$$

for every n. Thus **F** is not an Archimedean ordered field; $\bar{e}$ is so exceedingly small, compared with $\overline{M}$, that no integral multiple of $\bar{e}$ is $>\overline{M}$.

In the same way, if we let

$$r_m(x) = x^m,$$

it is rather easy to see that the equivalence classes $\bar{r}_1, \bar{r}_2, \ldots$ are not related in an Archimedean fashion in **F**. In fact, *every* integral multiple $n\bar{r}_1$ of $\bar{r}_1$ is $<\bar{r}_2$; we always have $n\bar{r}_2 < \bar{r}_3$, and so on.

Chapter 29
The Theory of Numbers

We recall, from Chapter 1, that the set **N** of positive integers is defined by the following three conditions.

(1) **N** contains 1.
(2) **N** is closed under the operation of adding 1.
(3) **N** is the intersection of all sets of numbers satisfying (1) and (2).

On the basis of this definition of **N**, we immediately got the following:

Induction Principle. Let S be a set of numbers. If (1) S contains 1, and (2) S is closed under the operation of adding 1, then (3) S contains all of the positive integers.

We then defined the set **Z** of integers as the set whose elements are the positive integers, their negatives, and 0, and we showed that **Z** formed a commutative ring with unity.

We shall now investigate the divisibility and factorization properties of the integers. Our first step is to observe that we can always divide one positive integer by another, getting a quotient q and a remainder r. More precisely, without mentioning division, we can state the following theorem.

Theorem 1. Let n and a be positive integers. Then n can be expressed in the form

$$n = aq + r,$$

where

$$0 \leqq r < a.$$

The easiest formal proof is by induction. Take a as fixed, and let S be the set of all positive integers n which can be expressed in the desired form. Then S contains 1, because

$$1 = a \cdot 0 + 1 \qquad (a > 1)$$

or

$$1 = 1 \cdot 1 + 0 \qquad (a = 1).$$

And S is closed under the operation of adding 1. Given

$$n = aq + r, \qquad 0 \leqq r < a,$$

we have

$$n + 1 = aq + (r + 1),$$

which is what we wanted, unless $r = a - 1$. If $r = a - 1$, then

$$\begin{aligned} n + 1 &= aq + a \\ &= a(q + 1), \end{aligned}$$

which has the desired form.

An integer $d \neq 0$ is a *divisor* of an integer a if $a = dq$ for some integer q; that is, d divides a if a/d is an integer. If d divides both a and b, then d is a *common divisor* of a and b. If it is also true that every common divisor of a and b is also a divisor of d, then d is a *greatest common divisor* of a and b, and we write

$$d = \gcd(a, b).$$

Note that this definition does not say merely that (1) d divides both a and b, and (2) d is the largest number that divides both a and b. When we say that $d = \gcd(a, b)$, we mean that d is "largest" in the sense of divisibility; that is, every common divisor of a and b must not only be $\leqq d$ but must also be a *divisor* of d. Thus, while it is plain that no pair of numbers a, b can have more than one gcd, it is not plain that a and b have any gcd's at all; and so the following theorem is not trivial.

Theorem 2. Any two positive integers have a greatest common divisor.

Proof. Let a and b be the two numbers. Consider the set D of all positive integers that can be written in the form

$$Ma + Nb,$$

where M and N are any integers, positive, negative, or zero. Obviously D is not empty, because the positive integer a can be written as

$$1 \cdot a + 0 \cdot b.$$

By the well-ordering principle, D has a least element d. We shall prove that

$$d = \gcd(a, b).$$

(1) *d divides b.* Suppose that d does not divide b. Then

$$b = dq + r, \qquad 0 \leqq r < d.$$

We know that

$$Ma + Nb = d,$$

for some integers M, N, and

$$r = b - dq.$$

Therefore

$$\begin{aligned} r &= b - (Ma + Nb)q \\ &= (-Mq)a + (1 - Nq)b. \end{aligned}$$

Therefore $r \in D$. This is impossible, because $0 \leqq r < d$, and d was supposed to be the least positive element of D. Therefore d divides b. In exactly the same way we get

(2) *d divides a.*

Therefore d is a common divisor of a and b. If e is any other common divisor of a and b, then

$$d = Mpe + Nqe = (Mp + Nq)e,$$

so that e divides d.

Note that in the proof of Theorem 2, we had $d \in D$. This gives us the following theorem.

Theorem 3. Let a and b be any positive integers. Then there are integers M, N such that

$$Ma + Nb = \gcd(a, b).$$

We can now verify a familiar theorem.

Theorem 4. Every rational number can be expressed as a fraction in lowest terms. That is, given a rational number p/q, there are always integers r and s such that

$$\frac{p}{q} = \frac{r}{s}$$

and

$$\gcd(r, s) = 1.$$

To prove this, we let

$$d = \gcd(p, q),$$

so that

$$p = rd, \qquad q = sd$$

for some positive integers r and s. Now r and s are the integers that we wanted. Obviously

$$\frac{p}{q} = \frac{rd}{sd} = \frac{r}{s}.$$

And since

$$Mp + Nq = d$$

for some M, N, we have

$$Mrd + Nsd = d,$$

and

$$Mr + Ns = 1,$$

so that

$$\gcd(r, s) = 1.$$

A *prime* number is a positive integer $p > 1$ whose only positive divisors are itself and 1.

Theorem 5. If n divides ab, and $\gcd(n, a) = 1$, then n divides b.

(Here n, a, and b are positive integers.)

Proof. There are integers M and N such that

$$Mn + Na = 1.$$

Therefore

$$Mnb + Nab = b.$$

Since n divides both Mnb and Nab, it divides their sum.

Theorem 6. Let a and b be positive integers, and let p be a prime. If p divides ab, then either p divides a or p divides b.

Proof. We need to show that if p does not divide a, then p divides b. If p does not divide a, then $\gcd(p, a) = 1$, because p is a prime. By the preceding theorem, p divides b.

The above is all of the number theory that we shall really need for the purposes of this book. Once we have gotten this far, however, we may as well prove the unique factorization theorem.

Theorem 7. Every natural number greater than one can be expressed as a product of primes.

Here repeated factors are allowed. For example,

$$12 = 2 \cdot 3 \cdot 2$$

is a product of primes.

Proof. Suppose that the theorem is false. Then some number is not a product of primes. Let n be the smallest such number. Then n is not a prime. Therefore n has some divisor a, different from n and from 1. Thus

$$n = ab, \qquad 1 < a < n, \qquad 1 < b < n.$$

Since n was the smallest number for which the theorem fails, a and b are products of primes. Therefore n is also a product of primes, and this contradicts our hypothesis.

It follows that every natural number n can be expressed in the form

$$n = p_1^{a_1} p_2^{a_2} \cdots p_k^{a_k},$$

where the p_i's are different primes, and the a_i's are natural numbers. Such a factorization is in *standard form* if (1) $a_i > 0$ for each i, and (2) the p_i's are written in order of magnitude, so that

$$p_1 < p_2 < \cdots < p_k.$$

Theorem 8. The factorization of a natural number into primes is unique, except for the order of the factors.

Proof. Suppose that

$$n = p_1^{a_1}p_2^{a_2}\cdots p_k^{a_k} = q_1^{b_1}q_2^{b_2}\cdots q_j^{b_j},$$

where both of the indicated factorizations are in standard form. Suppose, as an induction hypothesis, that every number less than n has only one prime factorization in standard form. We shall show, on this basis, that $k = j$, $p_i = q_i$ for each i, and $a_i = b_i$ for each i.

Every prime factor p of n divides one of the factors $p_i^{a_i}$. Therefore p divides some p_i. Therefore p *is* some p_i. Therefore p_1 is the smallest prime factor of n. For the same reason, q_1 is the smallest prime factor of n. Therefore $p_1 = q_1$. Therefore

$$\frac{n}{p_1} = p_1^{a_1-1}p_2^{a_2}\cdots p_k^{a_k} = q_1^{b_1-1}q_2^{b_2}\cdots q_j^{b_j}.$$

But $n/p_1 < n$. Therefore n/p_1 has only one factorization in standard form. Therefore $k = j$, $p_i = q_i$ for each i, and $a_i = b_i$ for each i. Therefore our "two" factorizations of n were the same all along, which was to be proved.

Problem Set

*1. Given two natural numbers r_1, r_2. Let q_1 and r_3 be such that

$$r_1 = r_2q_1 + r_3, \qquad 0 \leqq r_3 < r_2.$$

(That is, divide r_2 into r_1, to get a quotient and a remainder.) Let q_2 and r_4 be such that

$$r_2 = r_3q_2 + r_4 \qquad (0 \leqq r_4 < r_3).$$

Proceed in this way until you get a last positive remainder r_n:

$$\begin{aligned} r_1 &= r_2q_1 + r_3 \\ r_2 &= r_3q_2 + r_4 \\ &\vdots \\ r_i &= r_{i+1}q_i + r_{i+2} \\ &\vdots \\ r_{n-2} &= r_{n-1}q_{n-2} + r_n \\ r_{n-1} &= r_nq_{n-1}. \end{aligned}$$

Of course, the process must terminate, because the r_i's are all positive, and they form a decreasing sequence.

Show that

$$r_n = \gcd(r_1, r_2).$$

*2. Show that if n is a natural number, then $\sqrt{n}$ is either a natural number or an irrational number.

*3. Theorem 2 tells us that certain integers M and N must exist, but it gives us no help in finding such a pair of numbers. Describe a scheme for finding them.

4. Find integers M and N, so that

$$41M + 31N = 1.$$

5. In the proof of Theorem 8, we appealed tacitly to the following theorem.

Theorem. If a and b are positive integers, and $a > 1$, then $b < ab$. Prove this.

6. Justify the following statement from Problem 1: "Of course, the process must terminate, because the r_i's are all positive, and they form a decreasing sequence." (Remember the well-ordering principle.)

Chapter 30
The Theory of Equations

For the minimum purposes of this book, it would suffice to discuss the theory of equations only for cases where all of the roots of our equations are real. But this would be hopelessly artificial and somewhat misleading. Throughout this chapter, therefore, when we speak of *numbers*, we allow the possibility that the numbers are complex, unless the contrary is explicitly stated. The set of all complex numbers is denoted by **C**. We assume that **C** forms a field.

As usual, a *polynomial of degree n* is a function

$$f : \mathbf{C} \to \mathbf{C}$$

of the form

$$f(z) = a_n z^n + a_{n-1} z^{n-1} + \cdots + a_1 z + a_0.$$

For $n \geqq 1$, we require that $a_n \neq 0$. Thus the zero polynomial is allowed, as a polynomial of degree 0, but we don't call $0 \cdot z^2 + z + 1$ a polynomial of degree 2.

Note that we are allowing the coefficients a_i to be complex.

Theorem 1. Let f be a polynomial of degree $n \geqq 1$, and let z_0 be any number. Then f can be expressed in the form

$$f(z) = q(z)(z - z_0) + r,$$

where q is a polynomial of degree $n - 1$ and $r \in \mathbf{C}$.

For reasons which will soon be plain, it is imperative to avoid mentioning division in stating this theorem. The proof is by induction. Let S be the set of all positive integers n for which it is true that every polynomial of degree $\leqq n$ has the desired property. Then S contains 1, because

$$a_1 z + a_0 = a_1(z - z_0) + (a_1 z_0 + a_0).$$

Here $q(z) = a_1$ for every z, and $r = a_1 z_0 + a_0$.

If $n \in S$, then $n + 1 \in S$.

Proof. Given

$$f_{n+1}(z) = a_{n+1} z^{n+1} + a_n z^n + \cdots + a_1 z + a_0.$$

Then

$$\begin{aligned} f_{n+1}(z) &= a_{n+1} z^n (z - z_0) + (a_{n+1} z_0 + a_n) z^n + \cdots + a_1 z + a_0 \\ &= a_{n+1} z^n (z - z_0) + f_n(z). \end{aligned}$$

Here f_n is a polynomial of degree $\leqq n$. Therefore

$$f_n(z) = q(z)(z - z_0) + r,$$

and

$$f_{n+1}(z) = [a_{n+1}z^n + q(z)](z - z_0) + r,$$

which has the desired form. Therefore $n + 1 \in S$, and S contains all the positive integers.

In this theorem, when $r = 0$, we have

$$f(z) = q(z)(z - z_0).$$

Here we say that $z - z_0$ *divides* $f(z)$. In the equation

$$f(z) = q(z)(z - z_0) + r,$$

we set $z = z_0$. This gives

$$f(z_0) = r.$$

Thus we have:

Theorem 2. *The Remainder Theorem.* If

$$f(z) = q(z)(z - z_0) + r,$$

then

$$f(z_0) = r.$$

Theorem 3. *The Factor Theorem.* If z_0 is a root of the equation,

$$f(z) = a_nz^n + a_{n-1}z^{n-1} + \cdots + a_1z + a_0 = 0,$$

then $z - z_0$ divides $f(x)$, and conversely.

Note that we could not have proved Theorems 2 and 3 by first writing

$$\frac{f(z)}{z - z_0} = q(z) + \frac{r}{z - z_0},$$

and then multiplying by $z - z_0$ to get the equation that we really want. The division is valid only if $z - z_0 \neq 0$, and $z = z_0$ happens to be the very value that we are interested in.

An algebraic equation of degree n ($n \geqq 1$) is an equation of the form

$$a_nx^n + a_{n-1}x^{n-1} + \cdots + a_1x + a_0 = 0,$$

where all of the coefficients a_i are *integers*, and where $a_n \neq 0$.

The fundamental theorem of algebra, due to C. F. Gauss, asserts that every algebraic equation has at least one root in the field of complex numbers. Of course there may not be any roots in the field of real numbers. Moreover, the only roots that are easy to find, for $n > 2$, are the rational ones. The method of finding these is based on the following theorem.

Theorem 4. Let $x = p/q$, in lowest terms, be a root of the equation

$$a_n x^n + a_{n-1}x^{n-1} + \cdots + a_1 x + a_0 = 0.$$

Then p divides a_0 and q divides a_n.

Proof. We have

$$\frac{a_n p^n}{q^n} + \frac{a_{n-1}p^{n-1}}{q^{n-1}} + \cdots + \frac{a_1 p}{q} + a_0 = 0.$$

Therefore

$$a_n p^n + a_{n-1}p^{n-1}q + \cdots + a_1 p q^{n-1} + a_0 q^n = 0.$$

We know that gcd $(p, q) = 1$. Hence p and q have no prime factor in common. Thus q and p^n have no prime factor in common. Therefore gcd $(p^n, q) = 1$.

Since q divides every term in the equation after the first term $a_n p^n$, it follows that q divides $a_n p^n$. Since gcd $(p^n, q) = 1$, it follows by Theorem 5 of Chapter 29 that q divides a_n.

In exactly the same way, we see that p divides a_0.

The applications of this theorem can be tedious; we have only a finite number of things to try, but finite numbers can be large. For example, given the equation

$$8x^3 - x^2 - 27 = 0,$$

the only possible rational roots are the numbers $\pm p/q$, where $q = 1, 2, 4, 8$ and $p = 1, 3, 9, 27$. This gives 32 possibilities. On the other hand the theorem sometimes enables us to conclude very quickly that an equation has no rational roots at all. Consider

$$x^3 - 2x + 2 = 0.$$

Here the only possible rational roots are 1, -1, 2, and -2. None of them works. Therefore there are no rational roots.

Finally, let us recall one of the consequences of the factor theorem. Given a polynomial

$$f_n(x) = x^n + a_{n-1}x^{n-1} + \cdots + a_1 x + a_0.$$

If x_1 is a root of $f_n(x) = 0$, then we have

$$f_n(x) = (x - x_1)f_{n-1}(x).$$

If x_2 is a root of $f_{n-1}(x) = 0$, then $(x - x_2)$ is a divisor of $f_{n-1}(x)$. Thus

$$f_n(x) = (x - x_1)(x - x_2)f_{n-2}(x).$$

In a finite number of such steps, we get a factorization,

$$f_n(x) = (x - x_1)(x - x_2)\cdots(x - x_n).$$

Each of the numbers x_i is a root of the equation $f_n(x) = 0$. And no other number is a root, because a product is $=0$ only if one of the factors is $=0$. (This principle applies, of course, to complex roots as well as real roots.)

It is important to remember, in all of the above discussions, that the numbers x_i are not necessarily different. If we collect the repeated factors of $f_n(x)$, we get a factorization of the form

$$f_n(x) = (x - b_1)^{k_1}(x - b_2)^{k_2} \cdots (x - b_m)^{k_m}.$$

We then say that each b_i is a root of *multiplicity* k_i; and we observe that the *sum of the multiplicities* of the roots is the degree of the equation.

People often refer to this fact by saying that "every algebraic equation of degree n has n roots." But the latter statement, taken at face value, is silly. There are simple examples to show that the number of roots may be any integer from 1 to n. For example, the equation $x^{10} = 0$ is of degree ten, but it has only one root, namely, zero.

The upper bound n, for the number of roots, tells us that two polynomials can never be alike unless they look alike.

Theorem 5. If

$$a_nx^n + a_{n-1}x^{n-1} + \cdots + a_1x + a_0 = b_nx^n + b_{n-1}x^{n-1} + \cdots + b_1x + b_0,$$

for every x, then $a_i = b_i$ for every i from 0 to n.

Proof. Every number is a root of the equation

$$(a_n - b_n)x^n + (a_{n-1} - b_{n-1})x^{n-1} + \cdots + (a_1 - b_1)x + (a_0 - b_0) = 0.$$

Therefore this equation cannot have a positive degree, so that $a_i = b_i$ for $i > 0$. The equation must therefore take the form $a_0 = b_0$. Therefore $a_i = b_i$ for every i from 0 to n.

Finally, a remark on the fundamental theorem of algebra. It may seem that we have used this theorem, to prove Theorem 5, and in fact we did use it to show that every polynomial

$$f_n(x) = x^n + a_{n-1}x^{n-1} + \cdots + a_1x + a_0$$

has a factorization

$$f_n(x) = (x - x_1)(x - x_2) \ldots (x - x_n).$$

But in the proof of Theorem 5, all that we really need is the following.

Lemma. Every equation of the form

$$f_n(x) = x^n + a_{n-1}x^{n-1} + \cdots + a_1x + a_0 = 0$$

has at most n roots.

This lemma has an elementary proof, as follows. If the equation has no roots at all, then we are done. If it has a root x_1, then

$$f_n(x) = (x - x_1)f_{n-1}(x),$$

for some polynomial f_{n-1}, of degree $n - 1$ and with leading coefficient 1. If the equation $f_{n-1}(x) = 0$ has no roots, then x_1 is the only root of $f_n(x) = 0$. If $f_{n-1}(x_2) = 0$ for some x_2, then we get

$$f_n(x) = (x - x_1)(x - x_2)f_{n-2}(x).$$

We proceed in this way as far as we can. If at every stage

$$f_n(x) = (x - x_1)(x - x_2) \cdots (x - x_i)f_{n-i}(x) \qquad (i < n)$$

we find that $f_{n-i}(x) = 0$ has a root, then we get a complete factorization of $f_n(x)$ into linear factors. If at some stage the equation $f_{n-i}(x) = 0$ has no roots, then the original equation has at most i roots, with $i \leqq n$. The lemma follows.

Similarly, the following is elementary:

Lemma. Given a cubic polynomial $f_3(x) = x^3 + a_2x^2 + a_1x + a_0$, with real coefficients. If the equation $f_3(x) = 0$ has a root x_1, then f_3 has a complete factorization.

Proof. We know that $f_3(x)$ has a factorization of the form $(x - x_1)f_2(x)$, where $f_2(x)$ is quadratic. By the quadratic formula, $f_2(x)$ has a root. Now complete the factorization.

Chapter 31
Limits of Sequences

THE DEFINITION OF A LIMIT FOR SEQUENCES

Given a sequence,

$$a_1, a_2, \ldots,$$

of real numbers. When we write

$$\lim_{n\to\infty} a_n = a,$$

this means, roughly speaking, that when n is very large, then a_n is very close to a. For example

$$\lim_{n\to\infty} \frac{1}{n} = 0,$$

$$\lim_{n\to\infty} \left(\frac{n+1}{n}\right) = \lim_{n\to\infty} \left(1 + \frac{1}{n}\right) = 1,$$

and so on.

Let us now try to frame a definition of this idea, in a sufficiently exact form to enable us to prove things about it.

In the first place, when we say that a_n is close to a, this means that $|a_n - a|$ is small. The idea, then, is that we can make $|a_n - a|$ as small as we please, merely by making n large enough.

To say how small we want $|a_n - a|$ to be, we should name a positive number, say e, and then demand that

$$|a_n - a| < e.$$

To explain what integers n are large enough, we should name a positive integer N, and require that $n > N$. In these terms, when we say that

$$\lim_{n\to\infty} a_n = a,$$

we are saying that no matter what number $e > 0$ is given, there is always an integer N with the property that $|a_n - a| < e$ for every $n > N$. This suggests the following definition.

DEFINITION. $\lim_{n\to\infty} a_n = a$ means that for every $e > 0$ there is an integer N such that if

$$n > N,$$

then

$$|a_n - a| < e.$$

Let us try this out on some simple examples. First we shall show, *using the above definition*, that

(1) $\lim_{n\to\infty} 1/n = 0$.

By definition, this says that

(2) for every $e > 0$, there is an integer N such that if

$$n > N,$$

then

$$\left|\frac{1}{n} - 0\right| < e.$$

The desired N is easy to find: $1/n < e$ means merely that $n > 1/e$. By the Archimedean postulate, some integer N is greater than $1/e$. For $n > N$, we have $1/n < e$, as desired.

(The use of the Archimedean postulate was essential in this proof. In fact, the statement that $\lim_{n\to\infty} 1/n = 0$ is precisely *equivalent* to the Archimedean postulate.

Let us try the same thing for

$$\lim_{n\to\infty} \frac{n+1}{n} = 1.$$

Given $e > 0$, we want an integer N such that if

$$n > N,$$

then

$$\left|\frac{n+1}{n} - 1\right| < e.$$

The second inequality is equivalent to

$$\frac{1}{n} < e.$$

We can therefore let N be any integer greater than $1/e$, as before.

An *upper bound* of a sequence $a_1, a_2, \ldots$ is simply an upper bound of the set $\{a_n\}$. A sequence which has an upper bound is said to be *bounded above.* A *lower bound* of a sequence $a_1, a_2, \ldots$ is a lower bound of the set $\{a_n\}$. A sequence which has a lower bound is said to be *bounded below.* A sequence is *bounded* if it is both bounded above and bounded below. This is equivalent to the statement that there is a number b such that $|a_n| \leqq b$ for every n. If

$$a_1 \leqq a_2 \leqq \cdots \leqq a_n \leqq a_{n+1} \leqq \cdots,$$

then the sequence is called *increasing.* (If the strict inequality $a_n < a_{n+1}$ always holds, then the sequence is called *strictly increasing.* But this idea does not come up very often.)

The following theorem is a consequence of the Dedekind postulate.

Theorem 1. If a sequence is increasing, and has an upper bound, then it has a limit.

Proof. Let the sequence be

$$a_1, a_2, \ldots .$$

Since $\{a_n\}$ has an upper bound, it follows by the Dedekind postulate that $\{a_n\}$ has a least upper bound. Let

$$a = \sup \{a_n\}.$$

We shall show that

$$\lim_{n\to\infty} a_n = a.$$

Let e be any positive number. Then

$$a_N > a - e$$

for some integer N, because $a - e$ is not an upper bound of $\{a_n\}$. If $n > N$, then $a_n \geqq a_N$, because the sequence is increasing. Thus, if

$$n > N,$$

then

$$a_n > a - e.$$

But $a_n < a + e$ for every n, because a is an upper bound of $\{a_n\}$. If

$$a - e < a_n < a + e,$$

then

$$-e < a_n - a < e,$$

and

$$|a_n - a| < e.$$

Thus if

$$n > N$$

then

$$|a_n - a| < e.$$

That is, the number N that we found is of the sort we were looking for.

The other fundamental theorems on limits of sequences are as follows.

Theorem 2. If $\lim_{n\to\infty} a_n = a$, and $\lim_{n\to\infty} b_n = b$, then $\lim_{n\to\infty} (a_n + b_n) = a + b$.

Theorem 3. If $\lim_{n\to\infty} a_n = a$ and $\lim_{n\to\infty} b_n = b$, then $\lim_{n\to\infty} (a_n \cdot b_n) = ab$.

Theorem 4. If $\lim_{n\to\infty} a_n = a$, and $a_n \leqq k$ for every n, then $a \leqq k$.

Theorem 5. If $\lim_{n\to\infty} a_n = a$, and $\lim_{n\to\infty} b_n = b$, and $a_n \leqq b_n$ for every n, then $a \leqq b$.

Theorem 6. *The Squeeze Principle.* If

$$a \leqq b_n \leqq a_n$$

for every n, and

$$\lim_{n\to\infty} a_n = a,$$

then

$$\lim_{n\to\infty} b_n = a.$$

Of course the same conclusion follows if $a_n \leqq b_n \leqq a$ for every n. We shall refer to both of these theorems as the squeeze principle.

We shall not prove these theorems. Instead, we give in the following problem set a sequence of theorems that should lead you to the proofs by fairly easy stages.

Problem Set

1. Show that

$$\lim_{n\to\infty} a_n = a$$

if and only if

$$\lim_{n\to\infty} (a_n - a) = 0.$$

2. Show that if $\lim_{n\to\infty} a_n = 0$ and $\lim_{n\to\infty} b_n = 0$, then $\lim_{n\to\infty} (a_n + b_n) = 0$. [*Hint:* You want $|a_n + b_n| < e$. This will hold if the inequalities $|a_n| < e/2$ and $|b_n| < e/2$ *both* hold.]

3. Prove Theorem 2.

4. Show that if $\lim_{n\to\infty} a_n = a$, then the sequence is bounded.

5. Show that if $\lim_{n\to\infty} a_n = 0$ and $\{b_n\}$ is bounded, then $\lim_{n\to\infty} a_n b_n = 0$. [*Hint:* If $|b_n| < b$, and $|a_n| < e/b$, then $|a_n b_n| < e$.]

6. Show that if $\lim_{n\to\infty} a_n = a$ and $\lim_{n\to\infty} b_n = b$, then

$$\lim_{n\to\infty} [a_n(b_n - b)] = 0.$$

7. Show that if $\lim_{n\to\infty} a_n = a$ and $\lim_{n\to\infty} b_n = b$, then

$$\lim_{n\to\infty} [b_n(a_n - a) + a(b_n - b)] = 0.$$

8. Prove Theorem 3.

9. Prove Theorem 4. [*Hint:* Suppose that $a > k$, let $a - k = e > 0$, and use e in the definition of a limit.]

10. Prove Theorem 5.

11. Prove Theorem 6.

Chapter 32
Countable and Uncountable Sets

32.1 FINITE AND COUNTABLE SETS

By a *segment of the integers* we mean a set of the form

$$I_n = \{1, 2, 3, \ldots, n\}.$$

A *finite sequence* is a function whose domain is a set I_n. Usually we write sequences in the subscript notation

$$a_1, a_2, \ldots, a_n;$$

here for each i, a_i is the object which corresponds to i under the action of the function.

Let A be a set. If there is a one-to-one correspondence

$$f: I_n \leftrightarrow A$$

between A and a set I_n, then we say that A is *finite* and *has n elements,* and we write

$$A \sim I_n.$$

In general, when we write

$$A \sim B,$$

this means that there is a one-to-one correspondence between the set A and the set B. In this case, we say that A and B are *equivalent sets.*

As usual, by an *infinite sequence* (or simply a *sequence*), we mean a function whose domain is the entire set $\mathbf{N}$ of natural numbers. In the subscript notation, we write sequences as

$$a_1, a_2, \ldots;$$

here a_i is the object corresponding to the natural number i.

A set A is called *countable* if there is a sequence in which every element of A appears at least once. Note that since repetitions are allowed, every finite set is countable; we merely repeat one of its elements over and over. If

$$A \sim \mathbf{N},$$

then we say that the set A is *countably infinite.* Thus the set $\mathbf{N}$ itself is countably infinite, *ex officio*; we let each natural number n correspond to itself. And the set **Z** of integers is countably infinite; we can arrange it in the sequence

$$0, 1, -1, 2, -2, \ldots, n, -n, \ldots,$$

in which every integer appears exactly once.

Theorem 1. Every countable set is either finite or countably infinite.

That is, if the elements of the set A can be arranged in a sequence

$$a_1, a_2, \ldots, a_n, \ldots,$$

with repetitions allowed, then either $A \sim I_n$ for some n or $A \sim \mathbf{N}$. To prove this, we merely need to eliminate the repetitions from the sequence which is given by hypothesis, thus getting a sequence

$$b_1, b_2, \ldots, b_n$$

or

$$b_1, b_2, \ldots, b_n, \ldots,$$

in which every a_i appears exactly once. The method of doing this is fairly obvious. Let

$$b_1 = a_1.$$

Given

$$b_1, b_2, \ldots, b_i,$$

we look to see whether every one of the a_n's has already been listed. If so, we have finished; $n = i$, and the set A is finite. If not, we let b_{i+1} be the first term of the a-sequence which has not been listed so far.

If this process terminates, then A is finite. If the process does not terminate, we get an infinite sequence

$$b_1, b_2, \ldots,$$

with no repetitions. The new sequence includes all the elements of A, because for each n, a_n is one of the objects $b_1, b_2, \ldots, b_n$. [*Query:* Under what conditions will $a_n = b_n$ for a particular n? Under what conditions will it be true that $a_n = b_n$ for every n?]

Theorem 2. The union of a countable collection of countable sets is countable.

Proof. Given

$$A = A_1 \cup A_2 \cup \cdots \cup A_n \cup \cdots,$$

where each A_i can be arranged in a sequence

$$A_i\colon \quad a_{i1}, a_{i2}, a_{i3}, \ldots.$$

Let us regard the objects a_{ij} as forming a doubly infinite array, as follows:

$$\begin{matrix} a_{11} & a_{12} & a_{13} \cdots \\ a_{21} & a_{22} & a_{23} \cdots \\ a_{31} & a_{32} & a_{33} \cdots \\ \vdots & & \end{matrix}$$

In this array, consider the diagonal sequences

$$\begin{array}{l} a_{11}, \\ a_{12}, a_{21}, \\ a_{13}, a_{22}, a_{31}, \\ \vdots \\ a_{1n}, a_{2,n-1}, a_{3,n-2}, \ldots, a_{n1} \\ \vdots \end{array}$$

Laying these finite sequences end to end, we get a single sequence which includes all of the a_{ij}'s.

Note that this theorem and its proof include the cases of (1) a finite collection of countably infinite sets, (2) a countably infinite collection of finite sets, and also the two other possibilities. The definition of a countable set was stated in such a way as to permit us to take care of all these cases at once. It is better to apply Theorem 1 at the times when we really need it than to worry about eliminating repetitions at most stages of most proofs.

PROBLEM SET 32.1

1. Show that the rational numbers between 0 and 1 form a countable set.
2. Show that the set **Q** of all rational numbers is a countable set.
3. Let E_Q be the set of all points (x, y) in a coordinate plane for which x and y are both rational. Show that E_Q is countable.

32.2 THE COUNTABILITY OF THE SURD FIELD

We recall, from p. 234, the definition of a surd. A real number x is a *surd* if we can calculate x by a finite number of additions, subtractions, multiplications, divisions, and extractions of square roots, starting with 0 and 1. S is the set of all surds.

We shall show, using Theorem 2 of the preceding section, that S is countable.

Let S_n be the set of all surds that can be calculated by n operations of the sort that are allowed. Then S_n is finite for every n. The proof is by induction.

(1) S_o is finite. (Because its only elements are 0 and 1.)

(2) If S_n is finite, then so also is S_{n+1}.

Proof. Let k_n be the number of elements in S_n. The elements of S_{n+1} are the numbers of the form $x + y$, $x - y$, xy, x/y, $\sqrt{x}$, where x and y belong to S_n. For each of the first four of these forms, there are surely no more than k_n^2 possibilities, because there are at most k_n choices for x and k_n choices for y. And there are at most k_n possibilities for $\sqrt{x}$. Therefore the number of elements in S_{n+1} is surely no greater than $4k_n^2 + k_n$ (and, in fact, it is easy to see that this is a gross overestimate). Therefore S_{n+1} is finite.

Since

$$S = S_1 \cup S_2 \cup \cdots \cup S_n \cup \cdots,$$

it follows by Theorem 2, Section 32.1, that S is countable.

32.3 PROOF THAT THE REAL NUMBERS ARE UNCOUNTABLE

To show that the real numbers cannot be arranged in a sequence, we need a preliminary result.

By a *closed interval* (of the real numbers), we mean a set

$$[a, b] = \{x | a \leqq x \leqq b\}.$$

By a *nested sequence* of closed intervals we mean a sequence

$$[a_1, b_1], [a_2, b_2], \ldots$$

in which

$$[a_{i+1}, b_{i+1}] \subset [a_i, b_i]$$

for every i.

Theorem 1. The intersection of a nested sequence of closed intervals is not empty.

That is, some number $\bar{x}$ lies in every interval in the sequence. The proof is as follows:

(1) Let $A = \{a_i\}$, and let $B = \{b_i\}$.

Then

$$a_i < b_i$$

for every i, by definition of an interval. And

$$a_i \leqq a_{i+1} < b_{i+1} \leqq b_i,$$

because the sequence is nested. If $i \leqq j$, then

$$a_i < b_j,$$

because

$$a_i \leqq a_{i+1} \leqq \cdots \leqq a_j < b_j.$$

Similarly, $a_i < b_j$ if $i \geqq j$. Therefore *every b_i is an upper bound of A*.

(2) Let $\bar{x}$ be the supremum of A, that is, the least upper bound of A. Then

$$a_i \leqq \bar{x} \qquad \text{for every } i,$$

because $\bar{x}$ is *an* upper bound of A. And

$$\bar{x} \leqq b_i \qquad \text{for every } i,$$

because $\bar{x}$ is the *least* upper bound of A. Therefore

$$x \in [a_i, b_i] \qquad \text{for every } i,$$

which was to be proved.

Theorem 2. The set **R** of all real numbers is uncountable.

Proof. Suppose that **R** can be arranged in a sequence

$$x_1, x_2, \cdots, x_n, \cdots.$$

We set up a nested sequence of closed intervals, in the following way.

(1) Let $[a_1, b_1]$ be any interval not containing x_1.

(2) Given a finite sequence

$$[a_1, b_1], [a_2, b_2], \ldots, [a_n, b_n],$$

nested as far as it goes, such that $[a_n, b_n]$ contains none of the numbers $x_1, x_2, \ldots, x_n$. Let $[a_{n+1}, b_{n+1}]$ be any interval which lies in $[a_n, b_n]$ and does not contain x_{n+1}.

Now let I be the intersection of all the intervals in the sequence. Thus

$$I = [a_1, b_1] \cap [a_2, b_2] \cap \cdots \cap [a_n, b_n] \cap \cdots.$$

Then I is empty. The reason is that every real number x is supposed to be equal to x_n for some n; and x_n cannot belong to I, because x_n does not belong to $[a_n, b_n]$.

But I cannot be empty, by Theorem 1. Thus we have a contradiction, and Theorem 2 must be true.

We found in Chapter 19 that some real numbers (for example, cos 20° and $\sqrt[3]{2}$) are not surds. The results of this section and the preceding one give an independent proof of the same fact: it is impossible for S and **R** to be the same set, because S is countable and **R** is not. In fact, since the nonsurds are more numerous, it is, in a way, a remarkable accident for a real number chosen at random to be a surd. The surds are more familiar, but this is not because they are more common; it is merely because they are easier to describe.

Problem Set 32.3

1. In the preceding paragraph, it is stated that the nonsurds are more numerous than the surds. Justify this statement, by showing that $\mathbf{R} - S$ is uncountable.

Index